Carnegie Learning™

THE COGNITIVE TUTOR® COMPANY

SAMS COPY
jmdeancmu@gmail.

# INTEGRATED MATH III

Cognitive Tutor® Student

## FOREWORD TO THE STUDENT

Congratulations! You are about to begin the exciting journey through the world of mathematics: the language of science and technology. As you sit in front of the television, computer screen or video game, ride in an automobile, fly in a plane, talk on a cellular phone or use practically any other tools of modern society, realize that mathematics was critical in its invention, design and production.

The modern workplace today demands employees who are technologically literate, work well in teams and are self-starters. So, we have designed a course that combines state-of-the-art computer software with collaborative classroom activities. While you use the Cognitive Tutor software, it actually learns about you as you learn the mathematics. You'll receive "just-in-time" instruction from the Tutor and it will always make sure you're ready for the next problem. In the classroom you will collaborate with your peers to explore and solve real-world problem scenarios. Working in groups, you'll be answering questions by using multiple representations to analyze and solve the problem. Then, as a group, you will present your solutions to the class. Throughout the whole process, your teacher will serve as a facilitator and guide in support of your learning. As a result, you will become a self-sufficient learner; moving through the software and text at your own rate, using the Tutor's extensive help system and discovering solutions to problems you never thought possible.

Have fun Learning by Doing!

William S. Hadley,
Author, Vice President of Educational Services and Chief Academic Officer
Carnegie Learning, Inc.

*P.S. Have questions about the curriculum? Ask me directly by sending an email to billhelp@carnegielearning.com. I will be happy to send you a response within 5 to 7 days. I look forward to hearing from you!*

### CONTRIBUTING AUTHORS

William S. Hadley
David Dengler
Matthew R. Freedman
Corinne M. Murawski
Lora J. Shapiro

Jaclyn Snyder
Pittsburgh Public Schools

### CONTRIBUTING EDITORS

Corinne Murawski
Kelly Clay-Slack

### REVIEWERS

Theodore R. Hodgson
Dept. of Mathematical Sciences
Montana State University
Bozeman, MT

Signe E. Kastberg
IUPUI School of Education
Indianapolis, IN

ISBN-13 978-1-932409-19-2
ISBN-10 1-932409-19-X

Student Text

Printed in the United States of America.
2-2007-BRR

Carnegie Learning, Inc. | Pittsburgh, PA
Phone 888.851.7094 | Fax 412.690.2444
www.carnegielearning.com

# Table of Contents

## Unit 1: Behaviors of Functions



## Unit 2: Continuous and Discontinuous Functions



## Unit 3: Functions, Relations, and Inverses

## Unit 4: Polynomial and Rational Functions



## Unit 5: Statistics and Probability

## Unit 6: Circles



## Unit 7: Periodic Functions

## Unit 8: Conic Sections



## Unit 9: Three-Dimensional Geometry

## Unit 10: Compositions of Transformations

# Unit 1 : Behavior of Functions

## Name that Function

input

In the 1950's a TV game show called *Name that Tune* debuted. The object of the game was for the contestant to name a tune after hearing as few notes as possible.

You will be playing a mathematical version of this game called *Name that Function.* Instead of naming a tune, you will have to name the type of function or family of functions. For example, given the following graph,

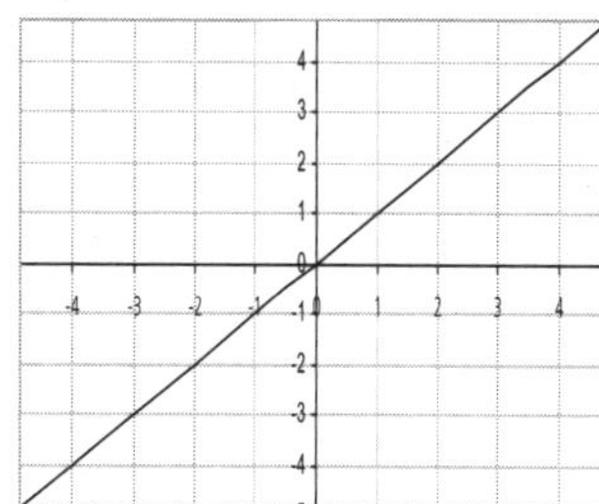

you would identify it as the graph of a linear function or say the graph belongs to the family of linear functions.

In the first round of the game, you are asked to identify the function below.

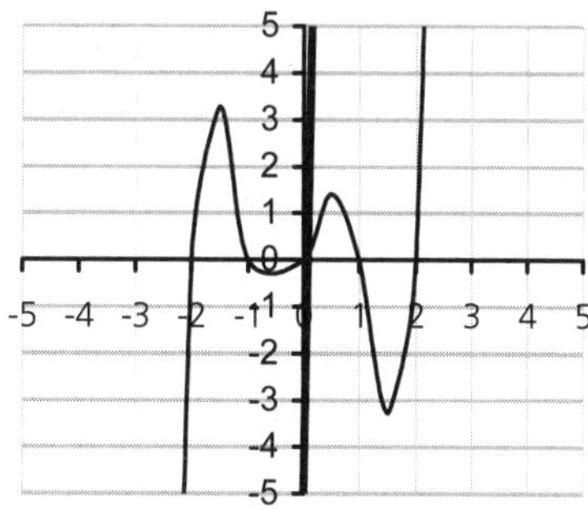

It is likely that you may not be able to identify the specific function. However, you can use information you do know about functions to help you connect the graph to its family of functions? List some of your observations.

## It's A Function Of....

process: model

1. Give an example of a non-constant linear function written in algebraic form.

   a. What is the value of the slope for the function you have written?

   b. What is the y-intercept for the function you have written?

   c. Draw the graph of a function you generated.

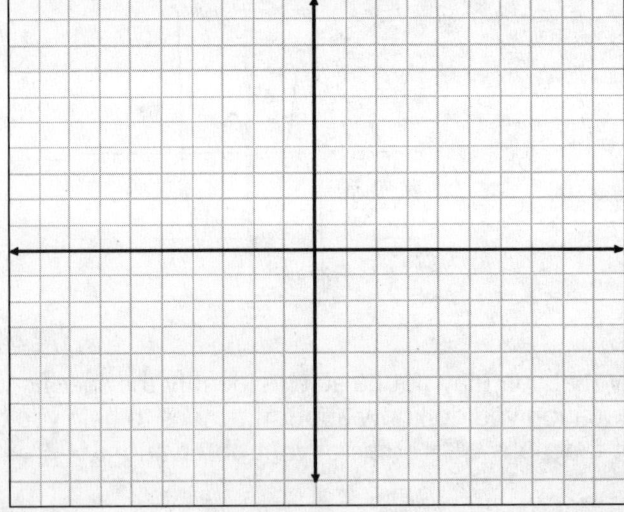

## It's A Function Of....

process: model

2. Give an example of a quadratic function written in standard form.

   a. Draw the graph of a function you generated.

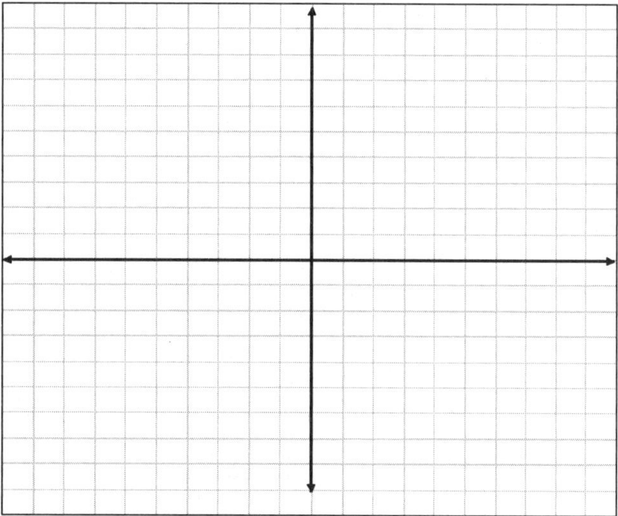

   b. Are there x and y-intercepts? If so, indicate what they are?

   c. Does the function reach a maximum or minimum value? If so, indicate which it is.

## It's A Function Of....

input

Functions such as linear and quadratic functions belong to a broader class of functions known as polynomial functions. A **polynomial function** is defined as

$$y = f(x) = a_n x^n + a_{n-1} x^{n-1} + \ldots + a_1 x^1 + a_0$$

where $a_n$ represents the real number constant coefficient associated to the variable of degree n. The **degree** of a polynomial function is defined by the term of highest power. For example, $f(x) = 2x^4 - 3x^3 + x^2 - 5x + 1$ is a fourth degree polynomial where $a_4 = 2$, $a_3 = -3$, $a_2 = 1$, $a_a = -5$ and $a_0 = 1$.

Give an example for each type of function listed below. In each example, indicate the value of $a_n$, $a_{n-1}$ etc.

a. First degree polynomial function.

b. Second degree polynomial function.

c. Third degree polynomial function

d. Fourth degree polynomial function

e. Fifth degree polynomial function

In earlier courses, you modeled situations using first degree and second degree polynomial functions. In this unit, you will become familiar with the family of polynomial functions by identifying key characteristics and behaviors as read from the graph of the function.

input

## It's A Function Of....

To associate the graph of a function to its family, it is useful to study the graph in terms of the following key behaviors and characteristics. These include:

- Domain and Range
- Intercepts
- Rate of Change
- Extrema (Maxima and Minima)
- Intervals of Increase and Decrease
- End Behavior

In your previous studies, you examined the domain and range, intercepts, rate of change and extrema. Based on what you know, define each of these concepts in your own words.

Domain and Range:

Intercepts:

Rate of Change:

Extrema:

The two behaviors to which you had not already been introduced are

Intervals of Increase and Decrease

The **intervals of increase and decrease**, are, reading from left to right, defined as the set of values for which the function will either increase or decrease. The point at which the function changes from increasing to decreasing or vice versa is called the **Point of Inflection.**

End Behavior

The **end behavior** of a function represents the behavior of the function for large positive and large negative values.

While this list of key behaviors is not exhaustive, it is satisfactory for the study of most functions.

## It's A Function Of....

analysis

3. Describe the graphs of the linear functions based on their key behaviors and characteristics.

| Behaviors | Graph I 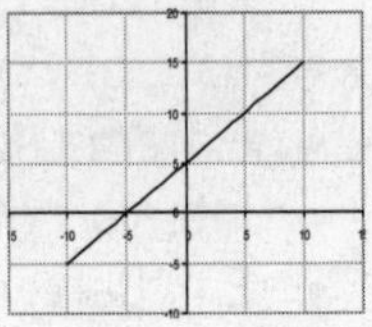 | Graph II 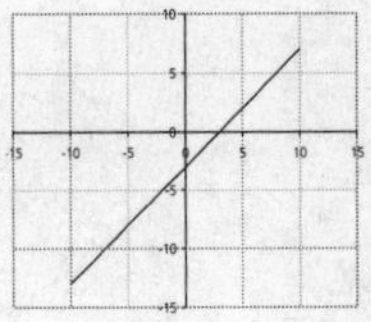 |
|---|---|---|
| **Domain and Range** For what values of x is the function defined? For what values of y is the function defined? | | |
| **Intercepts** How many times does the graph intersect the y-axis? The x-axis, i.e. how many real zeros are possible? | | |
| **Rate of Change** Describe the change in the dependent variable (y) with respect to the change in the independent variable (x). Is the rate of change constant or variable? Is the rate of change positive or negative? | | |
| **Extrema** At what point does the function reach a relative maximum or minimum value? | | |
| **Intervals of Increase or Decrease** Reading from left to right, define the interval(s) for which the function is increasing or decreasing. | | |
| **End Behavior** What happens to the graph of the function as x gets very large in the positive direction or very large in the negative direction? | | |

analysis

## It's A Function Of....

4. Describe the graphs of the quadratic functions based on their key behaviors and characteristics.

| Behaviors | Graph I 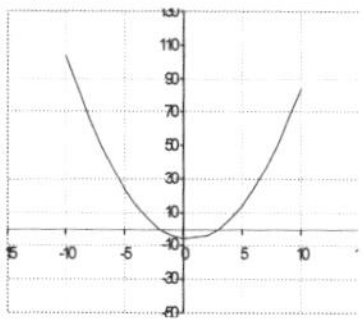 | Graph II 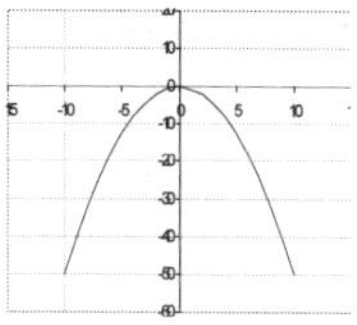 | Graph III 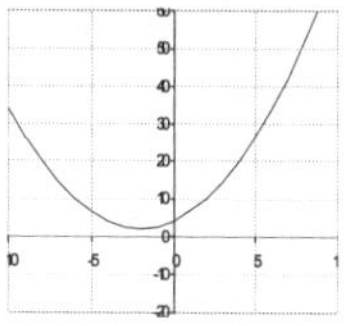 |
|---|---|---|---|
| Domain and Range | | | |
| Intercepts | | | |
| Rate of Change | | | |
| Extrema | | | |
| Intervals of Increase and Decrease | | | |
| End Behavior | | | |

analysis

## It's A Function Of....

5. Generate the graphs of three additional quadratic functions. Write the functions in symbolic form and describe each of the graphs in terms of the key behaviors and characteristics.

| **Behaviors** | **Graph I** | **Graph II** | **Graph III** |
|---|---|---|---|
| **Domain and Range** | | | |
| **Intercepts** | | | |
| **Rate of Change** | | | |
| **Extrema** | | | |
| **Intervals of Increase and Decrease** | | | |
| **End Behavior** | | | |

analysis

## It's A Function Of....

6. Based on your analysis of the all the graphs, what generalizations can you make about non-constant linear functions with respect to
   a. Domain and Range
   b. Intercepts
   c. Rate of Change
   d. Extrema
   e. Intervals of Increase and Decrease
   f. End Behavior

analysis

## It's A Function Of....

7. Based on your analysis of the all the graphs, what generalizations can you make about quadratic functions with respect to
   a. Domain and Range

   b. Intercepts

   c. Rate of Change

   d. Extrema

   e. Intervals of Increase and Decrease

   f. End Behavior

## It's A Function Of....

analysis

8. How many real zeros are possible for a
   a. Non-constant linear function? Explain your reasoning.

   b. Quadratic function? Explain your reasoning.

9. How many relative maxima or minima are possible for
   a. Non-constant linear function?

   b. Quadratic function?

10. Why do quadratic functions have relative maxima or minima, but linear functions do not?

11. Based on what you have observed with linear and quadratic functions, how does the degree of the function relate to the
    a. Possible number of zeros?

    b. Possible number of relative maxima and minima?

12. How does the end behavior differ for a first degree (odd function) and a second degree (even) function? Explain.

## Getting the Third Degree

analysis

1. Graphs of cubic functions are shown below. Read the graphs of the functions based on their key behaviors and characteristics.

| **Behaviors** | **Graph I** | **Graph II** | **Graph III** |
|---|---|---|---|
| **Domain and Range** | | | |
| **Intercepts** | | | |
| **Rate of Change** | | | |
| **Extrema** | | | |
| **Intervals of Increase and Decrease** | | | |
| **End Behavior** | | | |

## Getting the Third Degree

analysis

2. Generate the graphs of three additional cubic functions. Write the function in symbolic form and describe each of the graphs in terms of the key behaviors and characteristics.

| Behaviors | Graph I | Graph II | Graph III |
|---|---|---|---|
| Domain and Range | | | |
| Intercepts | | | |
| Rate of Change | | | |
| Extrema | | | |
| Intervals of Inc. and Dec. | | | |
| End Behavior | | | |

## Getting the Third Degree

analysis

3. Based on your analysis of all the graphs, what generalizations can you make about cubic functions with respect to the following characteristics:

   a. Domain and Range

   b. Intercepts

   c. Rate of Change

   d. Relative Maxima and Minima

   e. Intervals of Increase and Decrease

   f. End Behavior

## Getting the Third Degree

analysis

4. Compare the domain and range of a cubic function to a:
   a. Non-constant linear function

   b. Quadratic function

5. Compare the number of x and y intercepts of a cubic function to a:
   a. Non-constant linear function

   b. Quadratic function

6. Compare the number of relative maxima and minima of a cubic function to a:
   a. Non-constant linear function

   b. Quadratic function

analysis

## Getting the Third Degree

7. Compare the end behavior of a cubic function to a:

   a. Non-constant linear function

   b. Quadratic function

8. How does the degree of the function relate to the maximum number of zeros?

9. How does the degree of the function relate to the number of relative maxima and minima?

   a. If there is a relative maximum and a relative minimum, describe the intervals of increase and decrease.

10. How does the degree of the function affect its end behavior?

## Fourth Degree and Beyond

process: model

1. Using your graphing calculator, graph the following fourth degree polynomial functions, known as **quartic** functions. Sketch their graphs on the grid provided and describe the domain and range, relative maxima and minima, rate of change, intervals of increase and decrease and end behavior.

| Functions | Behaviors |
|---|---|
| a. $f(x) = x^4 - 8$ | |
| b. $f(x) = x^4 - 4x^3 - 4x^2 + 16x$ | |

## Fourth Degree and Beyond

process: model

| Functions | Behaviors |
|---|---|
| c. $f(x) = -x^4 + 4x^3 - 3x^2 - 14x - 8$ 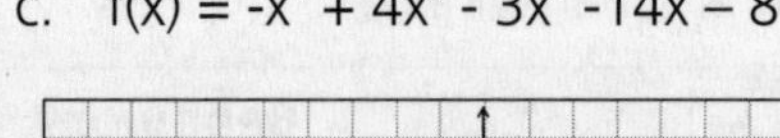 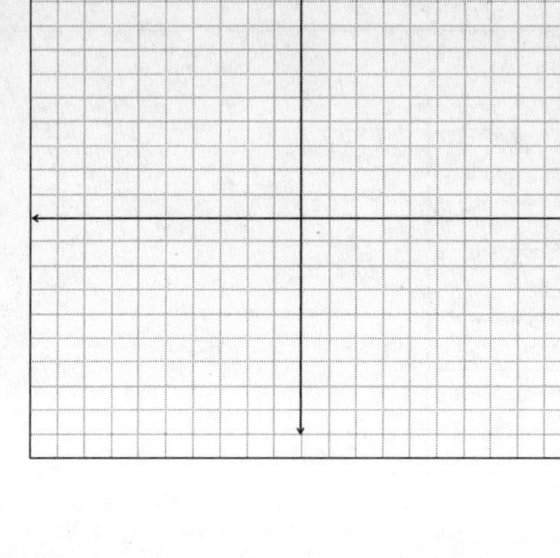 | |
| d. $f(x) = x^4 - 5x^3 + 6$ 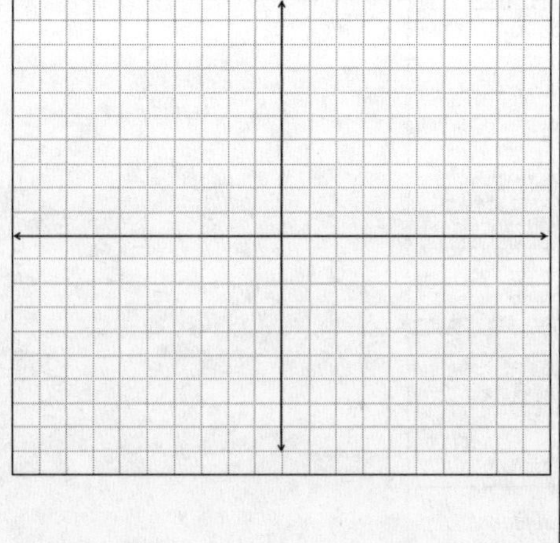 | |

## Fourth Degree and Beyond

analysis

2. Using the graphs, provide a detailed description of the behavior of quartic functions.

3. What is the maximum and minimum number of real zeros that are possible for a quartic function? Explain your reasoning.

4. What is the maximum and minimum number relative maxima or minima possible for a quartic function?

5. How does the degree of the quartic function affect its end behavior?

## Fourth Degree and Beyond

analysis

6. Compare the domain and range of a quartic function to a:
   a. Non-constant linear function

   b. Quadratic function

   c. Cubic function

7. Compare the number of x and y intercepts of a quartic function to a:
   a. Non-constant linear function

   b. Quadratic function

   c. Cubic function

8. Compare the number of relative maxima and minima of a quartic function to a:
   a. Non-constant linear function

   b. Quadratic function

   c. Cubic function

## Fourth Degree and Beyond

analysis

9. Compare the end behavior of a quartic function to a:
   a. Non-constant inear function

   b. Quadratic function

   c. Cubic function

10. Given a fifth degree polynomial or **quintic** function, how many real zeros are possible?

11. How many relative maxima or minima are possible for a fifth degree polynomial function?

## Fourth Degree and Beyond

analysis

12. In general what is the relationship between the degree of a polynomial function and the maximum number of zeros?

13. In general, what is the relationship between the degree of a polynomial function and the number of relative maxima and minima.

14. In general, what is the difference in the end behavior of function of even degree and function of odd degree?

15. Using your graphing calculator, determine if it is possible to generate a fifth degree function with one real zero. Draw its graph.

process: model

## Fourth Degree and Beyond

16. Using your graphing calculator, give an example of a fifth degree polynomial function that has three real zeros. Draw its graph.

extension

## Fourth Degree and Beyond

17. Why does every odd degree polynomial function have at least one x-intercept?

process: model

## Fourth Degree and Beyond

18. Identify a polynomial function that has five zeros.

19. Identify a function of degree five that has
    a. One zero
    b. Two zeros
    c. Three zeros
    d. Four zeros
    e. Five zeros

process: summary

## Fourth Degree and Beyond

20. Discuss the number of zeros, the number of maxima and minima and the end behavior of an eighth degree polynomial function? A twelfth degree polynomial function? A twentieth degree polynomial function? Any polynomial of even degree?

21. Discuss the number of zeros, the number of maxima and minima and the end behavior of seventh degree polynomial function? An eleventh degree polynomial function? A seventeenth degree polynomial function? Any polynomial of odd degree?

22. Is this statement accurate: *It is sufficient to know the term of highest degree of the polynomial function to know about its domain and range, the maximum possible number of extrema and intercepts.* Explain your reasoning.

process: model

## Let's Play Name That Function

6. Write a scenario to fit each of the functions.

   a. $f(x) = 15x + 150$ for $x > 0$.

   b. $f(x) = -4x^2 - 40x$ for $x > 0$ and $y > 0$.

   c. $f(x) = x^3 - 4x^2$ for $x \geq 0$.

process: model

## Let's Play Name That Function

7. Points of a specific function are displayed in the table below. Plot the points on the graph. (Set the intervals along the y axis at 10 units).

| x | y |
|---|---|
| 1 | 6.5 |
| 2 | 16 |
| 3 | 19.5 |
| 4 | 20 |
| 5 | 20.5 |
| | |
| | |
| | |

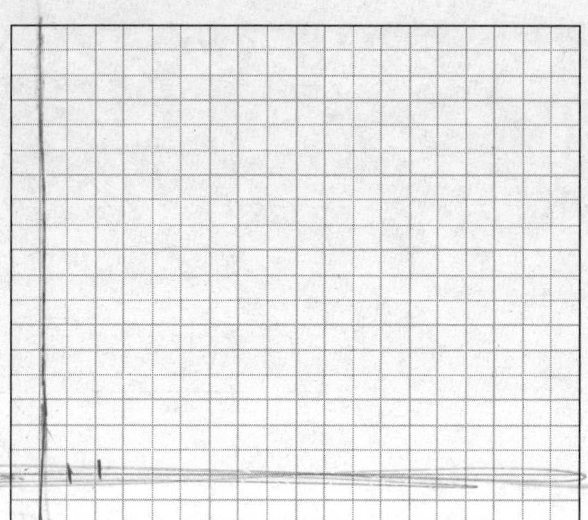

a. Based on the points displayed, predict and describe the end behavior of the function.

b. What equation type do you think will best model the data?

c. Add the following points to the graph: (7, 33.5), (8, 52), (9, 82.5)

d. Based on the points now displayed, predict and describe the end behavior of the function.

e. What equation type do you think best models the data now?

extension

## Let's Play Name That Function

10. Even though every data set can be modeled with a polynomial function, do you believe that a polynomial function is always the best fit? Explain.

11. The product of three linear factors $(x + 3)(x - 5)(x + 1) = 0$ will generate a cubic function with zeros at $x = -3$, $x = 5$ and $x = 1$. Verify whether this statement is true or false. If you had to make a generalization about the factors of a polynomial and the zeros of the function, what would it be?

# Unit 2: Continuous and Discontinuous Functions

## Contents

## Down on Main Street

input

Below is a map of part of Oxford, Ohio. Like many towns, the streets in Oxford are designed in a grid-like pattern.

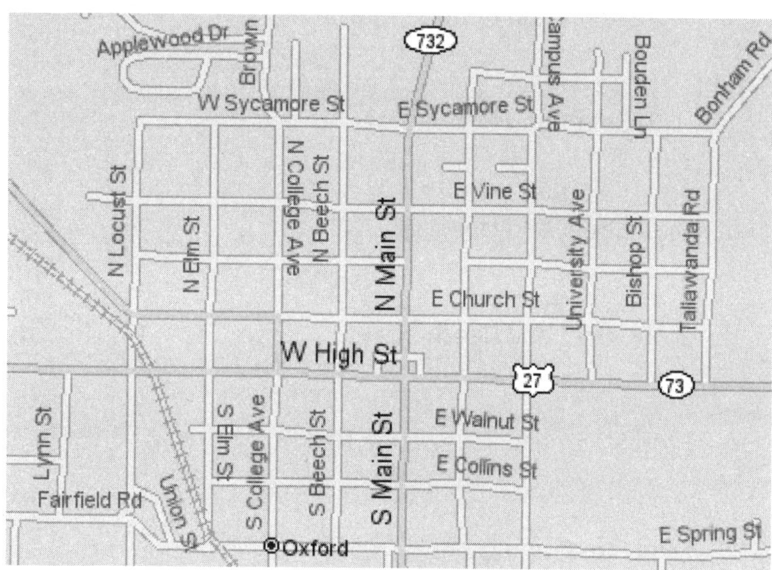


If you consider High Street and Main Street the axes of the grid, you can superimpose a coordinate plane on the map as follows:

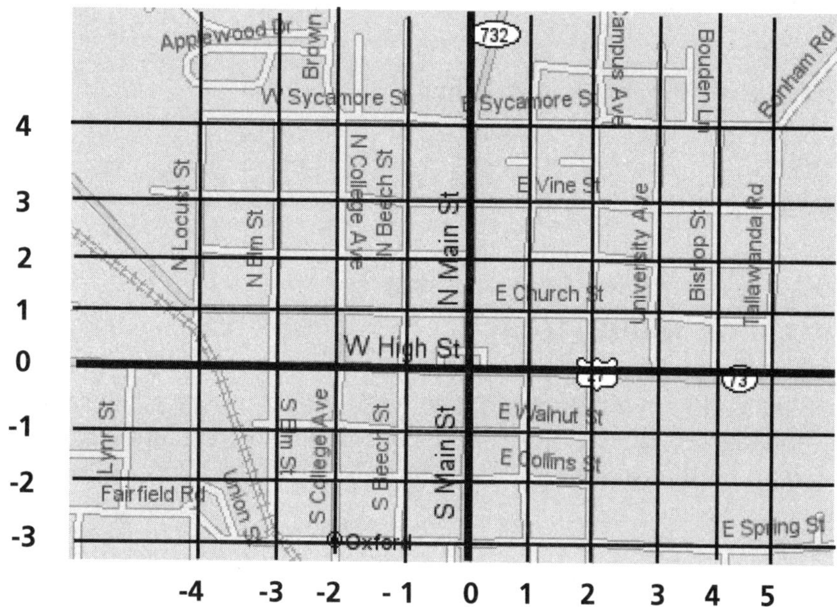

## Down on Main Street

process: evaluate

1. You are walking on High Street. How many blocks are you from Main Street if you are located at the intersection of:
   a. High Street and Bishop Street?
   b. High Street and Elm Street?
   c. High Street and University Avenue?
   d. High Street and Beech Street?
   e. High Street and Tallawanda Road?
   f. High Street and Locust Street?
   g. High Street and Campus Avenue?
   h. High Street and Main Street?

## Down on Main Street

analysis

2. Carla and Jim disagree about some of their answers. Carla says that if you are at the intersection of High Street and College Avenue, you are 2 blocks from Main Street. Jim says that you are -2 blocks from Main Street. Who is correct? Why?

process: model

## Down on Main Street

3. Complete the table below based on your answers to question 1. Use the values on the x-axis of the coordinate grid superimposed on the map to represent the location of a cross street.

| Cross Street | Location of Cross Street | Number of Blocks from Main Street |
|---|---|---|
| | | Blocks |
| Bishop | 4 | 4 |
| Elm | -3 | 3 |
| University | | |
| Beech | | |
| Tallawanda | | |
| Locust | | |
| Campus | | |
| Main | | |

process: model

## Down on Main Street

4. Plot the points from your table on the coordinate plane and connect them to sketch a graph. Remember to label your axes.

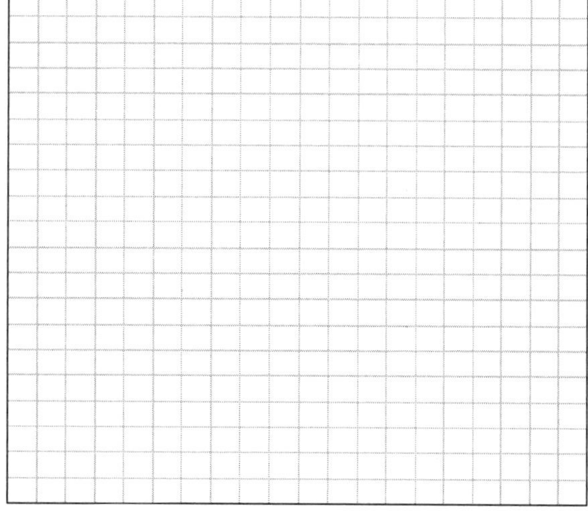

process: evaluate

## Down on Main Street

5. Describe the relation that you have just graphed by answering the following questions.

   a. Is the relation a function? Explain.

   b. What is the domain and range?

   c. What are the x- and y-intercepts?

   d. In which intervals does the function increase? In which intervals does it decrease?

   e. How can you describe the rate of change?

   f. Are there any maximum or minimum values?

   g. How can you describe the end behavior?

input

## Down on Main Street

The function used to compute the distance to Main Street is the absolute value function. The **absolute value** of a number is its distance from zero on a number line regardless of direction.

The absolute value of 4 is 4 because it is four units from zero on the number line. The absolute value of -3 is 3 because it is three units from zero on the number line. The absolute value of 0 is 0 because it is zero units from zero on the number line.

The absolute value of $x$ is written as $|x|$. For example, $|4| = 4$, $|-3| = 3$, and $|0| = 0$.

The **parent function** for **absolute value** is $\boldsymbol{f(x) = |x|}$. You graphed this function in question 4.

process: practice

## Down on Main Street

6. Evaluate each of the following absolute values.

a. $|-2|$

b. $|0|$

c. $|25|$

d. $|-1245|$

e. $|-4.25|$

f. $\left|\frac{1}{-2}\right|$

g. $|9 - 11|$

h. $|9| - |11|$

i. $|2 \cdot 7 - 5(5)|$

j. $|3^2 - 11|$

k. $|-7 -8|$

l. $|10^2 - 11^2|$

analysis

## Down on Main Street

7. What numbers have an absolute value of 15?

8. What numbers have an absolute value of -5?

9. What numbers have an absolute value of 0?

summary

## Down on Main Street

10. Describe the relationship between a number and its absolute value.

11. Describe the absolute value graph. Why does the absolute value graph look like it does?

## Absolutely Functional

process: model

The graphs below represent the linear, absolute value and quadratic parent functions.

$f(x) = x$

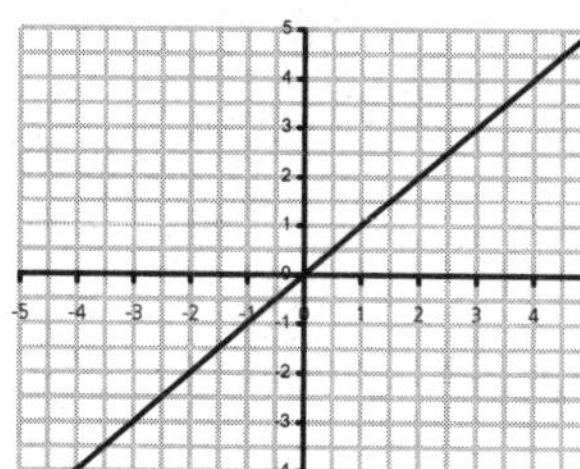

$f(x) = |x|$

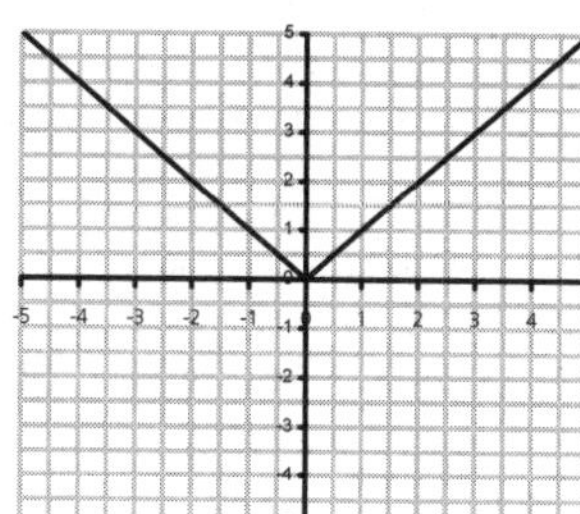

$f(x) = x^2$

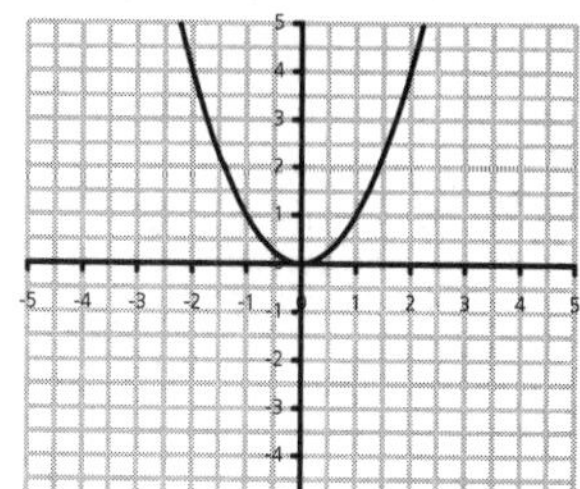

1. In your own words, describe how each of the above graphs is similar to each other and how they are different from each other.

## Absolutely Functional

process: model

2. Graph the following functions using your calculator. Then sketch the graphs on the grids below.

a. $y = x + 3$

d. $y = x - 2$

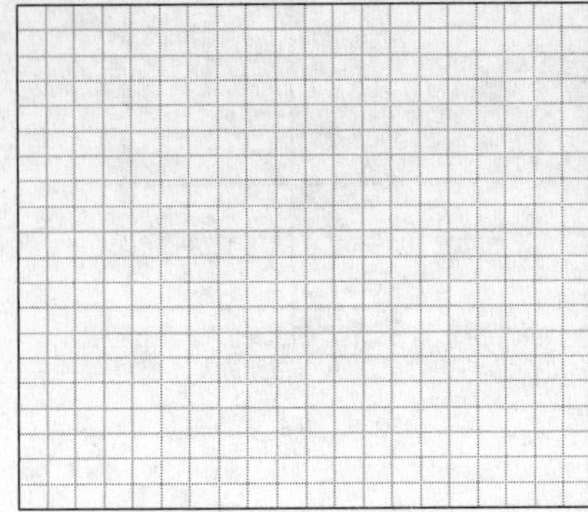

b. $y = |x| + 3$

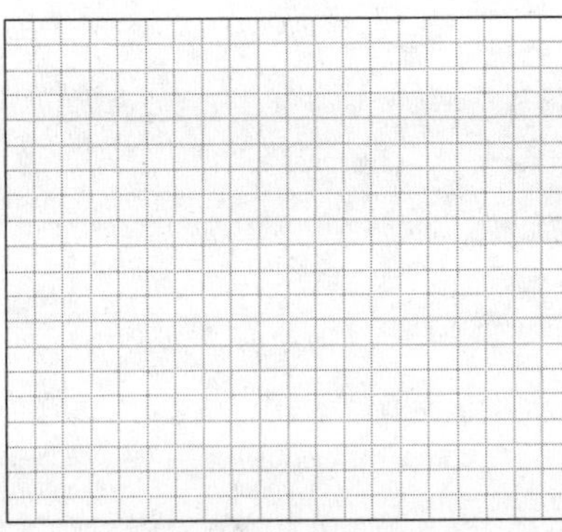

e. $y = |x| - 2$

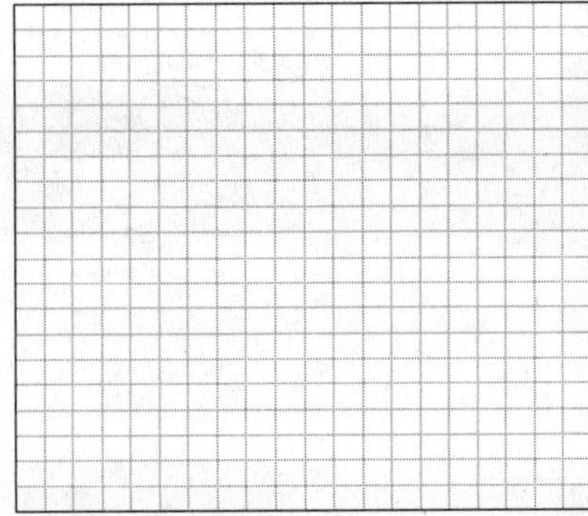

c. $y = x^2 + 3$

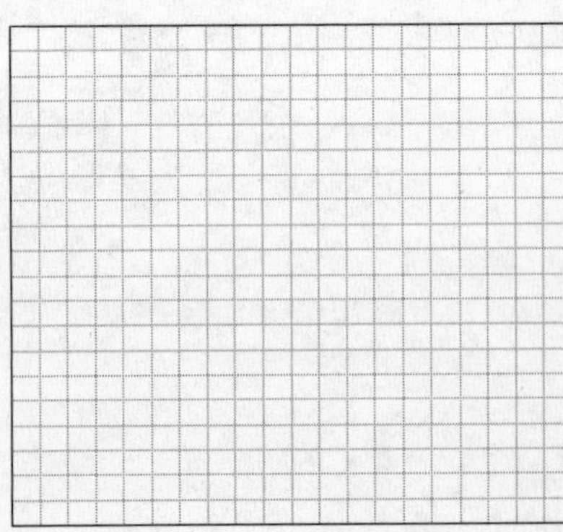

f. $y = x^2 - 2$

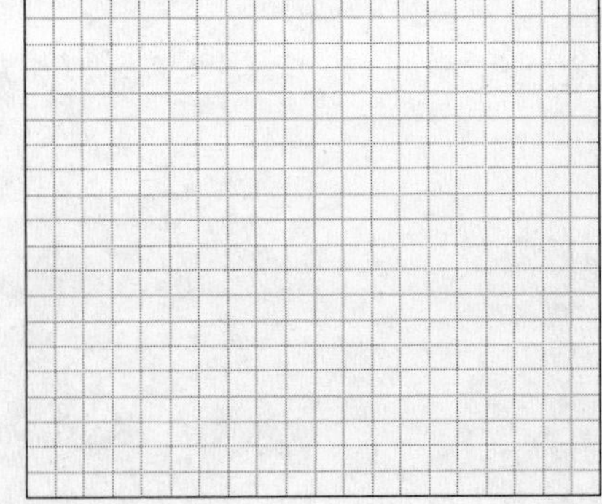

## Absolutely Functional

analysis

3. Describe each of the equations on the previous page in relation to their parent functions. How are they similar? How are they different?

4. When the equation of a line is written in the form of $y = x + k$, how does the graph compare to the graph of $y = x$?

5. When the equation of an absolute value function is written in the form $y = |x| + k$, how does the graph compare to the graph of $y = |x|$?

6. When the equation of a parabola is written in the form of $y = x^2 + k$, how does the graph compare to the graph of $y = x^2$?

## Absolutely Functional

process: model

7. Graph the following functions using your calculator. Then sketch the graphs on the grids below.

a. $y = x^2$

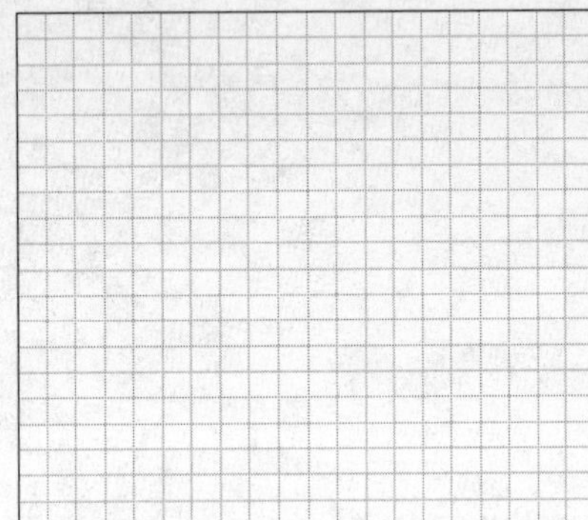

d. $y = |x|$

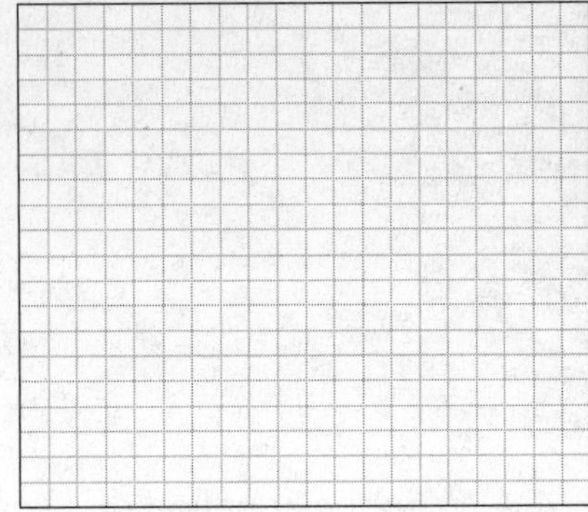

b. $y = (x + 3)^2$

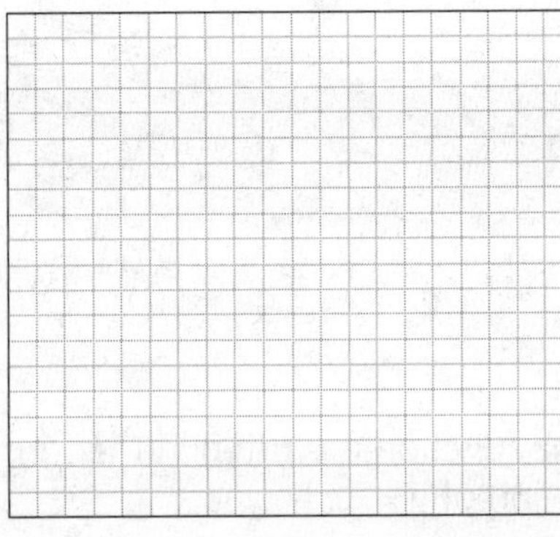

e. $y = |x + 3|$

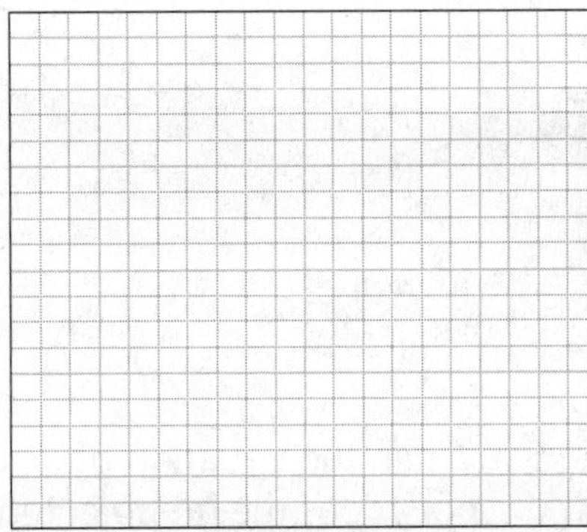

c. $y = (x - 2)^2$

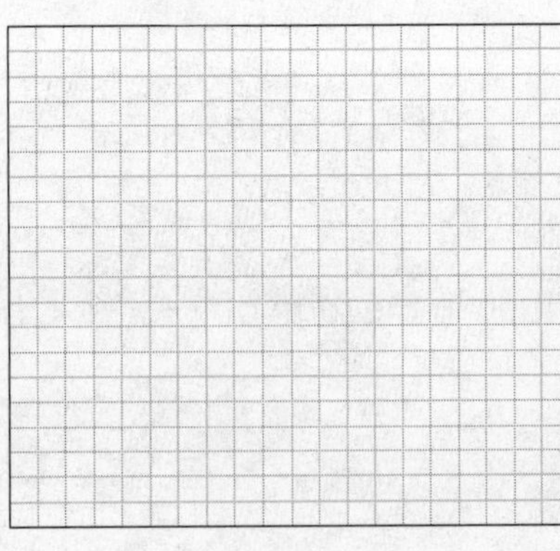

f. $y = |x - 2|$

## Absolutely Functional

analysis

8. Give at least one similarity and at least one difference among the graphs of the three quadratics on the previous page.

9. How do the graphs of the three absolute value equations compare to one another?

10. When the equation of a parabola is written in the form of $y = (x + \text{h})^2$, how does the graph compare to the graph of $y = x^2$?

11. When the equation of an absolute value function is written in the form of $y = | x + \text{h} |$, how does the graph compare to the graph of $y = | x |$?

process: model

## Absolutely Functional

12. Graph each of the absolute value functions below. The parent function is already graphed on each grid. Use your graphing calculator for help if you need it.

a. $y = 2|x|$

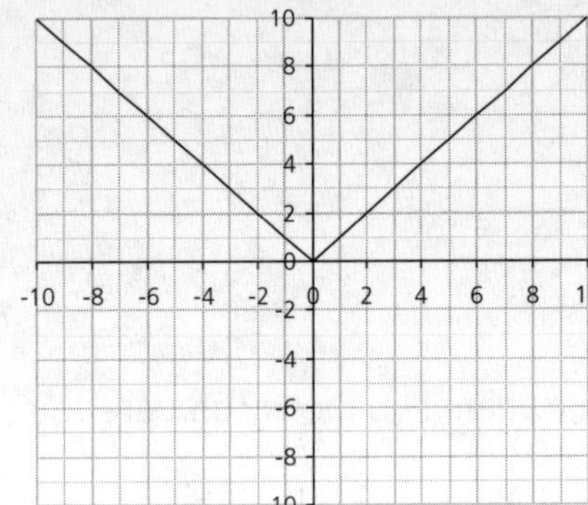

b. $y = \frac{1}{2}|x|$

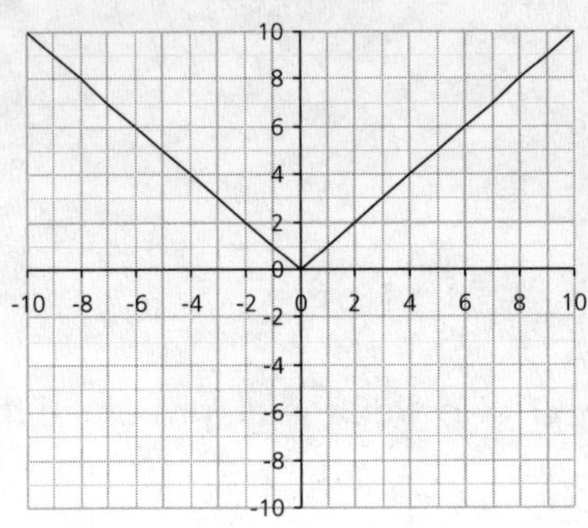

c. $y = -|x|$

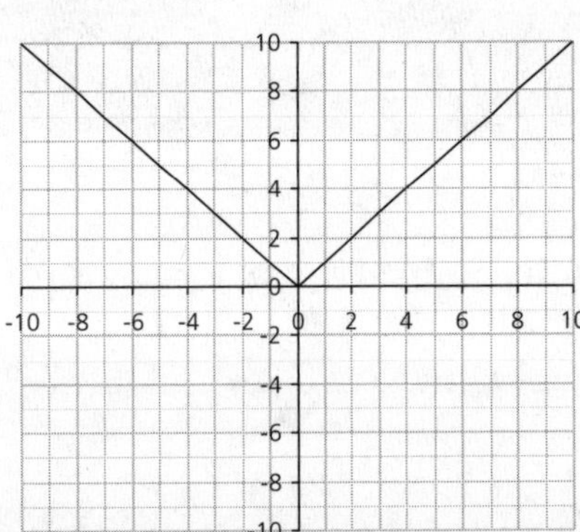

d. $y = -2|x|$

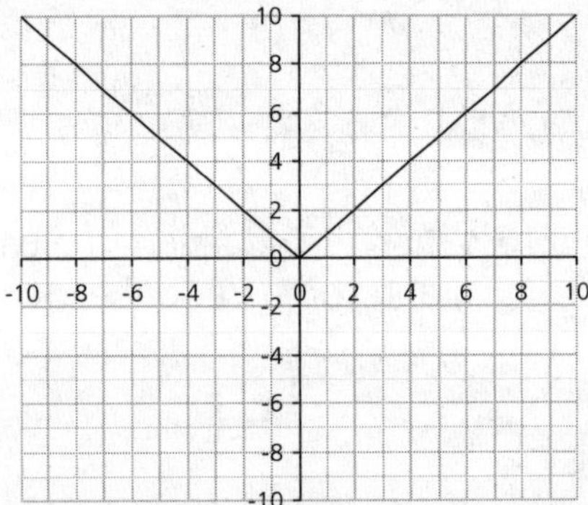

13. Describe how the graphs of the transformed functions compare to the graph of the parent function. A comparison of slopes will be helpful.

input

## Absolutely Functional

Transformations on the absolute value function are applied as with other functions. A **general absolute value function** can be written as $\mathbf{f(x) = a|x - h| + k}$. As with other functions, *a* represents the dilation factor, *h* represents the horizontal translation, and *k* represents the vertical translation. Also like other functions, if the *a* value is negative, the function is reflected over the x-axis.

process: model

## Absolutely Functional

14. How is the graph of $f(x) = |x|$ transformed for the function $f(x) = -2|x - 3| + 1$?
    a. Is the *a* value positive or negative? Is the function reflected over the x-axis?

    b. Does the *a* value indicate a dilation? By what factor?

    c. What is the value of *h*? How is the function translated horizontally?

    d. What is the value of *k*? How is the function translated vertically?

    e. Sketch the graph of the transformed function based on your answers.

process: practice

## Absolutely Functional

Sketch the graph of each absolute value function below by transforming the parent absolute value function. Write a description of the transformations from the parent function, $f(x) = x$.

15. $f(x) = -|x|$

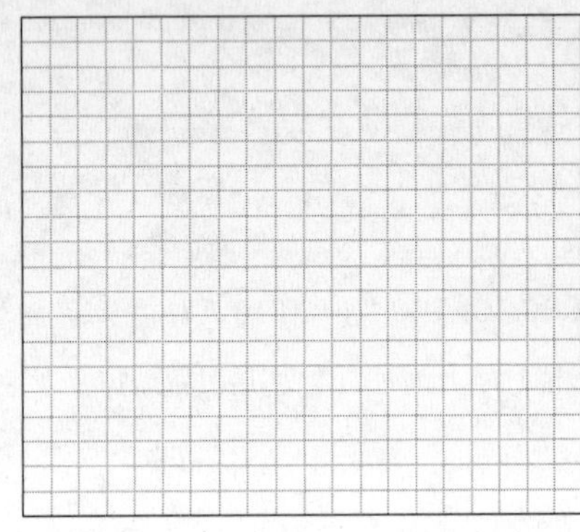

16. $f(x) = 2\,|x|$

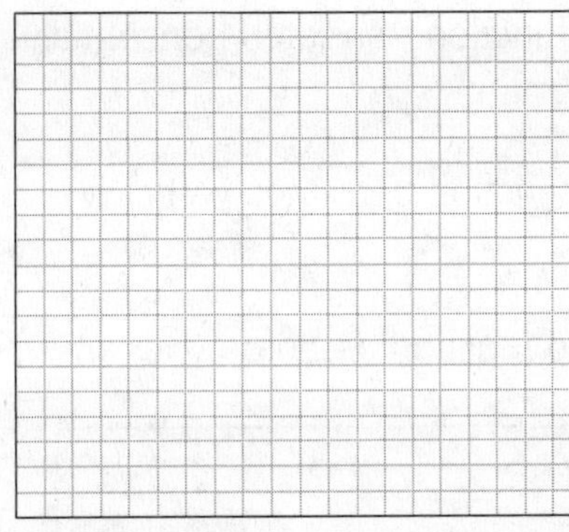

17. $f(x) = |x| - 2$

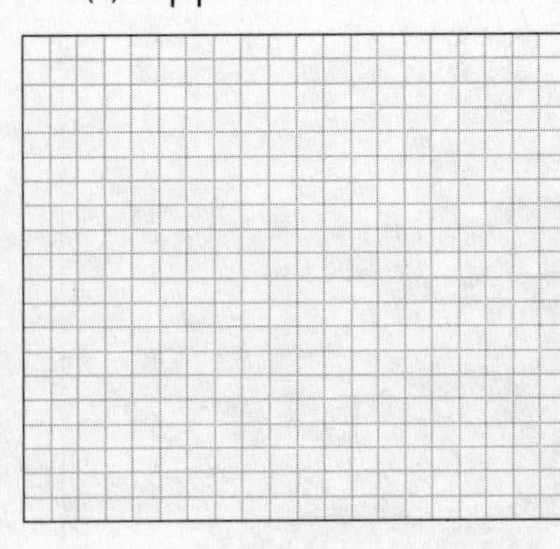

process: practice

## Absolutely Functional

18. $f(x) = |x + 3|$

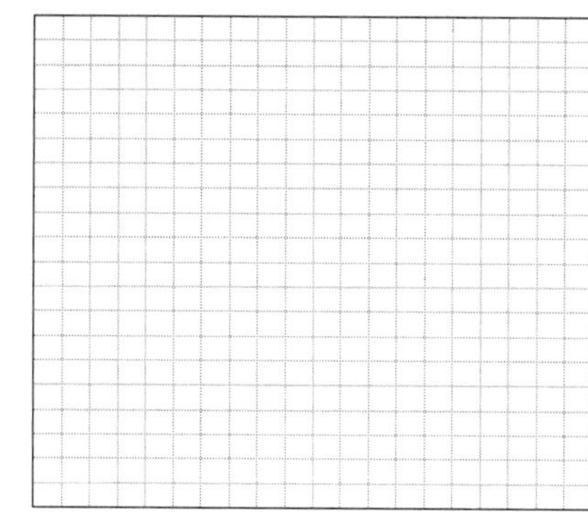

19. $f(x) = -\frac{1}{2}|x - 2|$

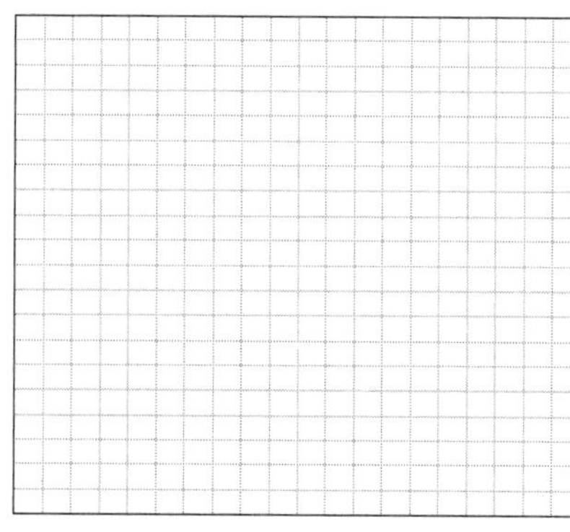

20. $f(x) = |x + 1| + 3$

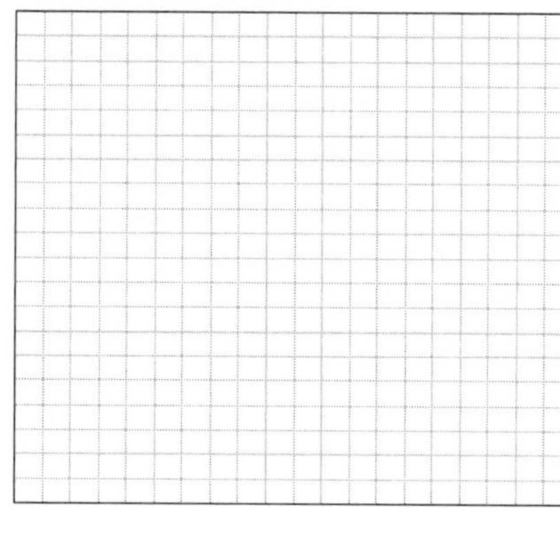

process: model

## Absolutely Functional

21. Answer the questions below based on the following graph.

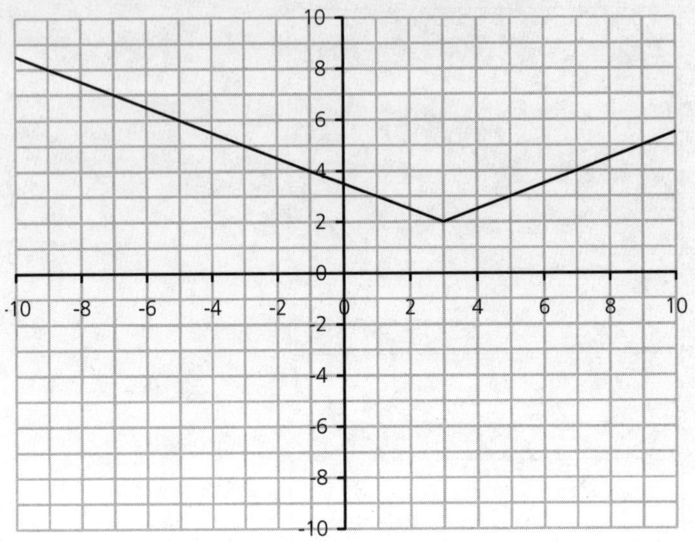


a. Has the function been translated vertically from the parent function, $y = |x|$? If so, how and what is the value of $k$ in the general absolute value function $y = a|x - h| + k$?

b. Has the function been translated horizontally from the parent function, $y = |x|$? If so, how and what is the value of $h$ in the general absolute value function $y = a|x - h| + k$?

c. Has the function been dilated horizontally or reflected over the x-axis from the parent function, $y = |x|$? If so, what is the dilation factor? What is the value of $a$ in the general absolute value function $y = a|x - h| + k$?

d. What is the algebraic function for the graph?

process: practice

## Absolutely Functional

Write the algebraic function for each graph shown based on the transformations performed on the parent absolute value function. Also, write a description of the transformations from the parent function, $f(x) = x$.

22.

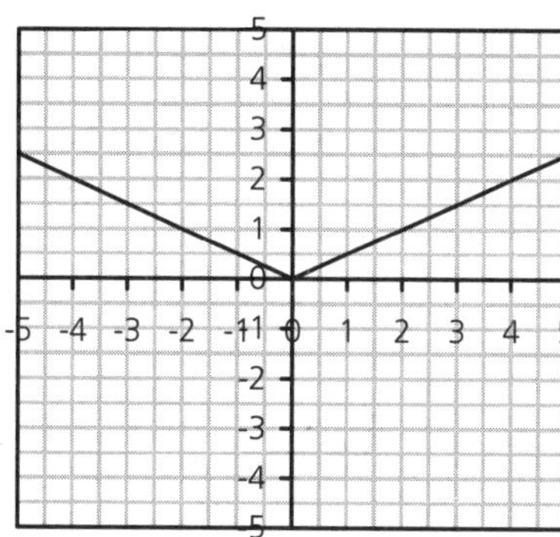

23.

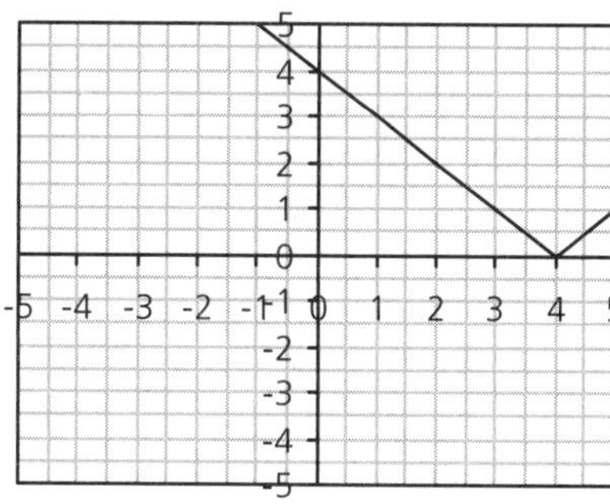

24.

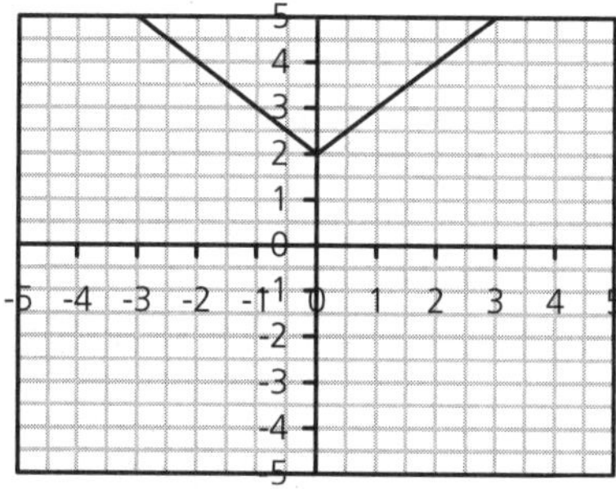

process: practice

## Absolutely Functional

25.

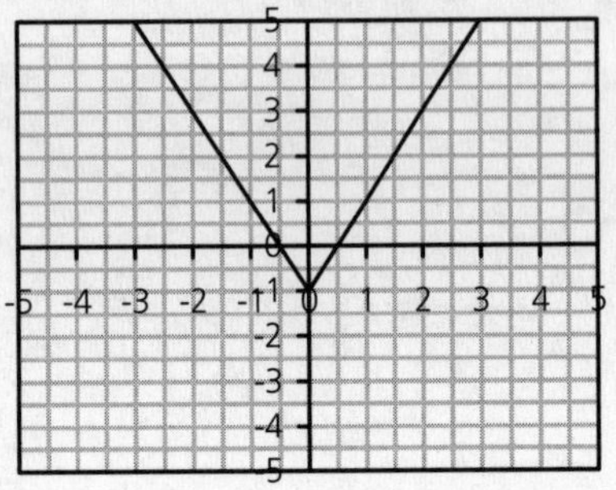

26.

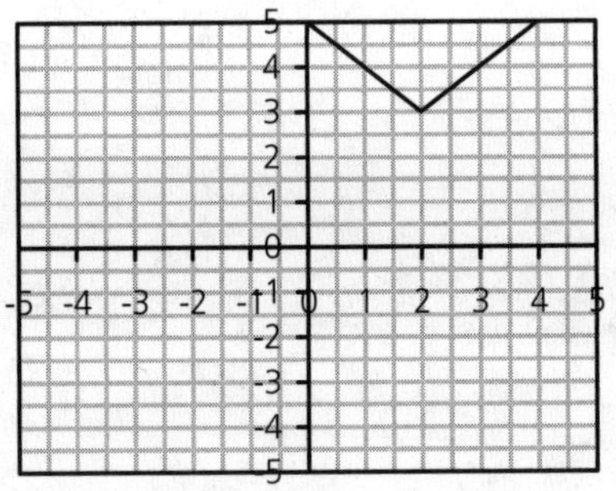

27.

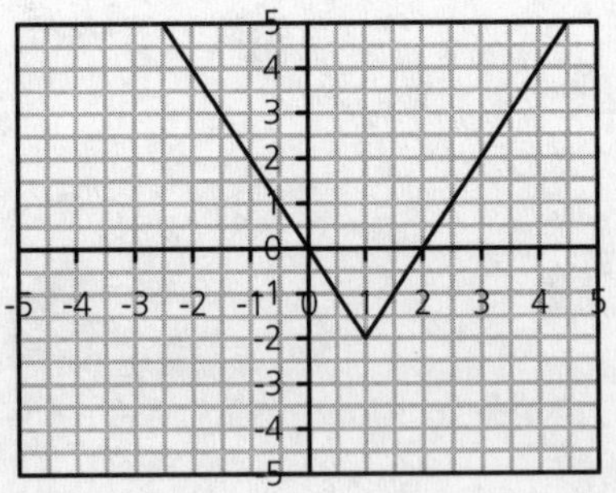

summary

## Absolutely Functional

28. The general absolute value function is written as $f(x) = a|x - h| + k$.

    a. How does changing the value of *a* affect the graph of the function?

    b. How does changing the value of *h* affect the graph of the function?

    c. How does changing the value of *k* affect the graph of the function?

29. The absolute value function can be written as two functions each with a smaller domain. Consider the graph of the parent absolute value function, $f(x) = |x|$.

    a. Where is a natural point to break the function into two pieces?

    b. What is the domain of the piece farther to the left on the graph?

    c. What is the algebraic formula for the piece farther to the left on the graph?

    d. What is the domain for the piece of the function farther to the right on the graph?

    e. What is the algebraic formula for the piece farther to the right on the graph?

    f. When a function is defined in pieces, we usually write the function by placing the pieces and corresponding domains in brackets. Enter your answers to parts b through e below to create the piecewise function for $f(x) = |x|$.

$f(x) = \begin{cases} \underline{\qquad\qquad} & \underline{\qquad\qquad} \\ \text{answer to part c} & \text{answer to part b} \\ \underline{\qquad\qquad} & \underline{\qquad\qquad} \\ \text{answer to part e} & \text{answer to part d} \end{cases}$

## A Moving Experience

input

A moving truck starts traveling along a straight, local road at an average speed of 50 miles per hour. Two hours later, the driver of a passing car gets the truck driver's attention and tells him that some items have fallen from the moving truck several miles back. The truck driver backtracks at an average speed of 30 miles per hour finding the lost items after 30 minutes. He parks the truck and spends another 30 minutes gathering the items and securing them on the truck. He then rushes to his destination at an average speed of 60 miles per hour. An hour later, he reaches his destination.

## A Moving Experience

process: evaluate

1. How far is the truck from its original location after:
   a. 1 hour?

   b. 90 minutes?

   c. 2 hours?

   d. 2.5 hours?

   e. 3.5 hours?

process: evaluate

## A Moving Experience

2. When will the truck be:
   a. 50 miles from its original location?

   b. At its final destination?

   c. 100 miles from its original location?

process: model

## A Moving Experience

3. Using the information you know about the moving truck situation, complete the following table.

| Labels | Time | Distance from Starting Location |
|---|---|---|
| Units | Hours | Miles |
| | 0 | |
| | 0.5 | |
| | 1 | |
| | 1.5 | |
| | 2 | |
| | 2.5 | |
| | 3 | |
| | 3.5 | |
| | 4 | |

process: model

## A Moving Experience

4. Plot the points from your table and connect them in order to graph the situation on the coordinate plane below. Remember to label your axes.

analysis

## A Moving Experience

5. Based on your table and graph, which answers, if any, to questions 1 or 2 would you like to change or modify? If you are adjusting answers, why did you decide to do so?

process: model

## A Moving Experience

6. There are four distinct parts or pieces to the graph of this one function.

   a. What algebraic equation models the function for the first 2 hours (when $0 \le x < 2$)?

   b. What algebraic equation models the function for the next half hour (when $2 \le x < 2.5$)?

   c. What algebraic equation models the function for the next half hour (when $2.5 \le x < 3$)?

   d. What algebraic equation models the function for the final hour (when $3 \le x \le 4$)?

input

## A Moving Experience

A function like the one describing this problem situation is called a **piecewise linear function** because it can be divided into pieces that are represented by different linear functions. Typically, the function is written with the pieces shown within a brace followed by the domain for each piece. Match your answers from question 6 to each domain below to create the piecewise function for this problem.

$$f(x) = \begin{cases} \underline{\hspace{4cm}} \text{ (answer to part a)} & 0 \le x < 2 \\ \underline{\hspace{4cm}} \text{ (answer to part b)} & 2 \le x < 2.5 \\ \underline{\hspace{4cm}} \text{ (answer to part c)} & 2.5 \le x < 3 \\ \underline{\hspace{4cm}} \text{ (answer to part d)} & 3 \le x \le 4 \end{cases}$$

## Comparing Long Distance Rates

input

You have been told that if you want to call your friends who have moved out of town, you will have to pay for the calls. Of course, since you are paying, you want to make your calls using the least expensive long distance plan. You find three plans that are available with no sign up or long-term commitment by simply dialing a long distance code starting with 10-10 before dialing the telephone number. The three plans are as follows:

**AmeriCall:** Each call costs $1.00 for up to 30 minutes and 5¢ for each additional minute.

**TeleTalk:** Each call costs 50¢ for up to 10 minutes and 5¢ for each additional minute.

**U-Dial:** Each call costs 6¢ per minute.

## Comparing Long Distance Rates

process: model

1. Complete the following table of values to compare the cost for phone calls of various lengths.

| Duration of Phone Call | Cost using AmeriCall | Cost using TeleTalk | Cost using U-Dial |
|---|---|---|---|
| Minutes | | | |
| 5 | | | |
| 10 | | | |
| 15 | | | |
| 20 | | | |
| 25 | | | |
| 30 | | | |
| 35 | | | |
| 40 | | | |

## Comparing Long Distance Rates

process: model

2. Using one coordinate plane, create a graph for the cost of each of the phone service plans.

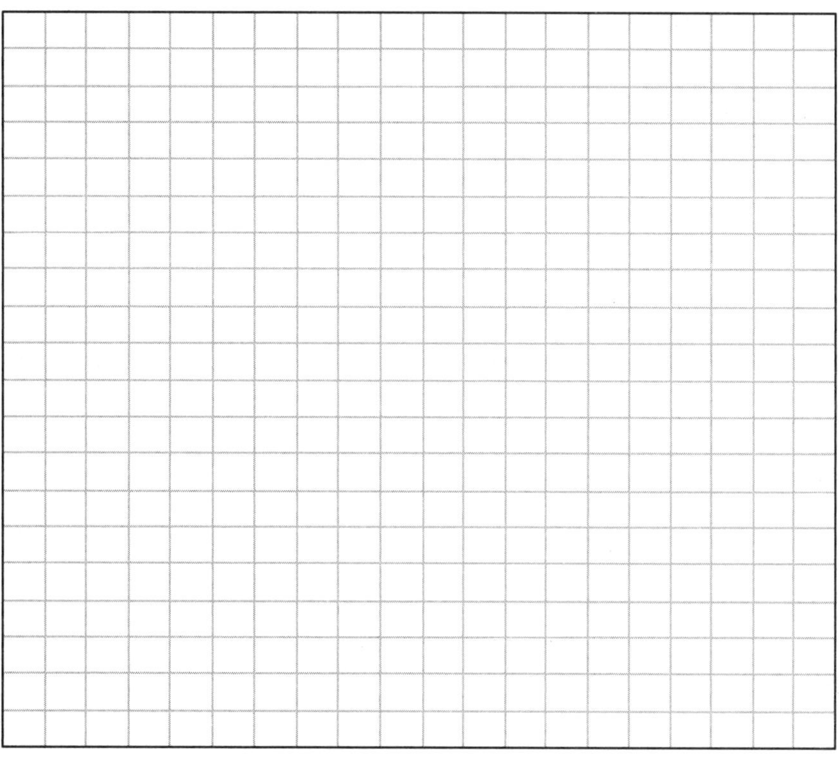

## Comparing Long Distance Rates

analysis

3. If you consider only the cost per minute for a phone call, U-Dial seems to charge the most at 6 cents per minute. Why might someone choose the U-Dial plan knowing this?

4. TeleTalk seems to have one of the lowest costs per minute at 5 cents per minute. Why might someone choose not to use the TeleTalk plan?

5. Based on your work so far, which plan would you choose for your long distance calls? Why?

## Comparing Long Distance Rates

process: model

6. Define a function to represent the cost of a phone call for each of the plans. Use a piecewise function when appropriate.

## Comparing Long Distance Rates

process: evaluate

7. Determine the points of intersection of the functions you defined.

8. For what range of call lengths is each of the plans the least expensive?

## Comparing Long Distance Rates

analysis

9. Is it better to use one phone plan for all of your long distance calls or to use different plans at different times? Why?

10. If you were to use different plans at different times, how would you determine which plan to use for each call?

11. If you had to choose only one of the three plans to use for all of your calls, which would it be and why? Think about how long your phone calls last.

## Media Mail

input

The United States Postal Service has different postal rates for different types of mailings. The Media Mail (or Book Rate) description and rates effective starting June 30, 2002 are as follows.

> Media Mail is used for books, film, manuscripts, printed music, printed test materials, sound recordings, play scripts, printed educational charts, loose-leaf pages and binders consisting of medical information, videotapes, and computer recorded media such as CD-ROMs and diskettes. Media Mail cannot contain advertising. The maximum size is 108 inches in combined length and distance around the thickest part.

| Weight Not Over | Media Mail Rates |
|---|---|
| Pounds | $ |
| 1 | 1.42 |
| 2 | 1.84 |
| 3 | 2.26 |
| 4 | 2.68 |
| 5 | 3.10 |
| 6 | 3.52 |
| 7 | 3.94 |
| 8 | 4.24 |
| 9 | 4.54 |
| 10 | 4.84 |
| 11 | 5.14 |
| 12 | 5.44 |

## Media Mail

analysis

1. What patterns do you notice in the data for the Media Mail rates?

process: model

## Media Mail

2. Create a scatter plot of the data for this problem.

process: model

## Media Mail

3. Write a piecewise linear function to describe the Media Mail rates data.

## Media Mail

process: evaluate

4. The table provided for the Media Mail rates indicates "Weight Not Over." What does this mean?

5. Complete the table of values below. Provide the actual cost for a mailing based on the given weights.

| Weight Not Over | Media Mail Cost: |
|---|---|
| **Pounds** | **$** |
| 0.25 | |
| 0.5 | |
| 0.75 | |
| 1 | |
| 1.25 | |
| 1.5 | |
| 1.75 | |
| 2 | |
| 2.25 | |
| 2.5 | |
| 2.75 | |

6. Plot the values from this table on the scatter plot you generated in question 2.

## Media Mail

process: evaluate

7. Refer back to the piecewise function that you wrote in question 3. Does the function make sense for values of x such as 1.5 or 2.25? Why or why not?

8. Write a paragraph describing the function that models the media mail rates data. Consider domain and range, intercepts, maximum or minimum values, end behavior, intervals of increase or decrease and rate of change.

## Media Mail

input

The model which describes the Media Mail situation belongs to the family of functions called step functions. **Step functions** are functions with graphs that look like steps.

The Media Mail situation is described by a step function called the **rounding-up function** or **ceiling function**. In a ceiling function, the *x* value is paired with the smallest integer greater than or equal to *x*.

The rounding up of *x* is denoted as $\lceil x \rceil$. For example, $\lceil -3.9 \rceil = -3$, $\lceil 2 \rceil = 2$, and $\lceil 5.4 \rceil = 6$. In the problem situation, this would mean that the charge for mailing a package weighing 5.4 pounds would be calculated as a package weighing 6 pounds.

The Media Mail situation is defined by the function,

$$f(x) = \begin{cases} 1 + 0.42\lceil x \rceil & 0 < x \le 7 \\ 1.84 + 0.3\lceil x \rceil & x > 7 \end{cases}$$

## Media Mail

input

The **parent ceiling function** is $f(x) = \lceil x \rceil$ and has the following graph:

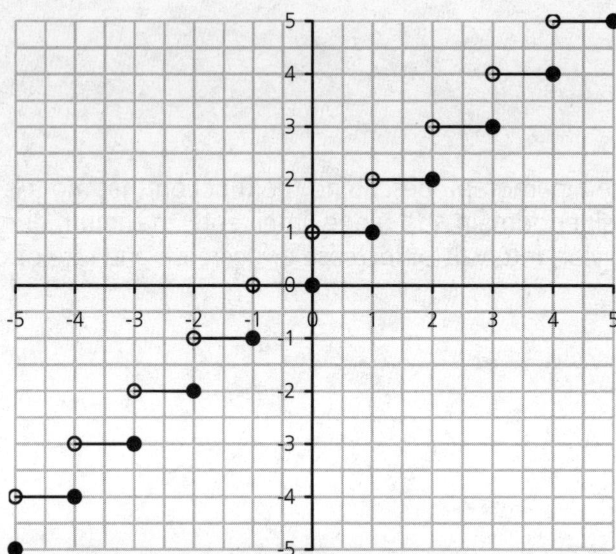


Notice that the segments on the graph have endpoints with open and closed circles. The closed circles indicate points on the graph that are included in the function. The open circles indicate points on the graph that are not included in the function.

The ceiling function is **discontinuous** because its graph cannot be drawn without lifting your pencil off the paper. The points at which you would lift your pencil off the paper are **points of discontinuity.** A function is **continuous** if it has no points of discontinuity.

## The Bike Rental Shop

input

While on vacation, you decide to rent a bicycle to tour the area. The local bike rental shop offers a rental rate of $25 per day. You pay the daily rental fee for each day or part of a day that you have the bicycle.

## The Bike Rental Shop

process: model

1. Complete the following table of values for the bike rental situation.

| Time | Rental Charge |
|---|---|
| Days | $ |
| 0.5 | |
| 1 | |
| 1.25 | |
| 1.5 | |
| 1.75 | |
| 2 | |
| 2.5 | |
| 3 | |
| 3.5 | |
| 4 | |
| | |
| | |

2. Write a formula for the problem situation using the ceiling function notation as appropriate.

process: model

## The Bike Rental Shop

3. Create a graph of the situation.

analysis

## The Bike Rental Shop

4. How would the formula and graph change if the bike rental company charged a onetime $5 insurance fee?

## The Bike Rental Shop

analysis

5. How would the formula and graph change if the bike rental company ran an off-season special and charged only half their normal fee?

6. Is the function, $f(x) = 25\lceil x \rceil$ the same as the function, $g(x) = \lceil 25x \rceil$? Explain your answer.

## The Bike Rental Shop

summary

7. Answer the following questions based on the function you generated for the Bike Rental problem.

   a. What are the domain and range of the function?

## The Bike Rental Shop

summary

b. What are the x-intercepts and y-intercepts of the function and what do they represent in the problem situation?

c. How can you describe the rate of change in the function?

d. What are the maxima and minima of the function?

e. What are the intervals of increase or decrease for the function?

f. Describe the end behavior of the function.

g. Is this function continuous or discontinuous? If it is discontinuous, what are the points of discontinuity?

## The Bike Rental Customer

input

In the last scenario, we examined the bike rental problem. We calculated the fee based on the duration of the rental. However, as a customer renting a bicycle you really want to know the maximum amount of time you can rent a bicycle based on how much money you have.

## The Bike Rental Customer

process: evaluate

1. What is the maximum amount of time that a customer could rent a bicycle with:
   a. \$125?
   b. \$92?
   c. \$21.50?
   d. \$212.25?
2. How did you determine your answers in question 1?

analysis

## The Bike Rental Customer

3. How is this situation mathematically different from The Bike Rental Shop situation?

4. Would the customer's perspective of determining how long a bike can be rented based on the amount of money the customer has be represented by a ceiling function? If so, write the function. If not, explain why not.

process: model

## The Bike Rental Customer

5. Complete the following table of values for the bike rental customer situation.

| Money to Spend | Rental Time |
|---|---|
| $ | Days |
| 0 | |
| 10 | |
| 25 | |
| 37.50 | |
| 40 | |
| 50 | |
| 60 | |
| 68.50 | |
| 75 | |
| 90 | |
| 99 | |
| 100 | |

process: model

## The Bike Rental Customer

6. Create a graph of the situation.

summary

## The Bike Rental Customer

7. Describe the key behaviors or characteristics of this situation.
   a. Is this relationship a function? Why or why not?

   b. What is the domain and range?

summary

## The Bike Rental Customer

c. What are the intercepts and what do they represent in the problem situation?

d. How can you describe the rate of change?

e. What are the maxima and minima?

f. What are the intervals of increase or decrease?

g. Describe the end behavior.

h. Is this situation continuous or discontinuous? If it is discontinuous, what are the points of discontinuity?

input

## The Bike Rental Customer

This situation is described by a step function called the **greatest integer function.** The greatest integer function is also known as the **rounding down function,** or **floor function**. In a floor function, the *x* value is paired with the greatest integer less than or equal to *x*. The rounding down of *x* is denoted $\lfloor x \rfloor$. For example, $\lfloor 5.4 \rfloor = 5$, $\lfloor -3.9 \rfloor = -4$, and $\lfloor 2 \rfloor = 2$.

The parent greatest integer function is $f(x) = \lfloor x \rfloor$ and has the following graph:

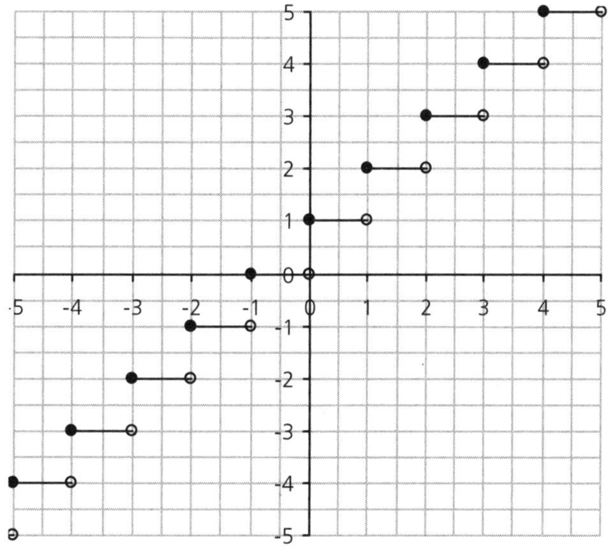

process: model

## The Bike Rental Customer

8. Explain in words how to calculate the maximum number of days of rental for any given amount of money.

9. Write an algebraic equation representing this situation using the greatest integer function.

## The Bike Rental Customer

analysis

10. Someone working at the bike rental shop calculates the cost for the number of days a bicycle is rented. The customer takes that cost and computes the number of days that a bike can be rented. Will the customer get the same number of days the bike rental shop worker used in the calculation? Why or why not?

11. In the Bike Rental Customer problem, the function is modeled with the equation, $y = \left\lfloor \frac{x}{25} \right\rfloor$. Is that the same as $y = \frac{\lfloor x \rfloor}{25}$? Explain.

12. Name some situations in which you would use the greatest integer or floor function. That is, in what situations in life do you round down?

13. Name some situations in which you would use the ceiling function?

## Working at the Mall

input

You landed a part-time job working at the food court at the mall. Each week, your take-home pay is about $80.

## Working at the Mall

process: evaluation

1. If you save all of your money, how much will you have saved after:
   a. 2 weeks?
   b. 1 month?
   c. 1 year?
2. How long will it take you to save:
   a. $1000?
   b. $2500?
   c. $10,000?

## Working at the Mall

process: model

3. Write an algebraic equation to model the situation.

4. Create a table and graph for the situation.

| Quantity Name | | |
|---|---|---|
| Units | | |
| | | |
| | | |
| | | |
| | | |
| | | |
| | | |
| | | |

process: analysis

## Working at the Mall

5. Is this situation modeled by a function? If so, in which family of functions is the situation classified? Explain your answer.

6. If you already saved $200 before starting the job, how would the equation and graph change?

process: evaluate

## Working at the Mall

7. What is the domain and range of the algebraic equation? What is the domain and range of the problem situation?

8. What are the x- and y-intercepts? What do they represent in the problem situation?

9. What are the intervals of increase or decrease?

10. What is the end behavior?

11. Is the function continuous or discontinuous? If it is discontinuous, list any points of discontinuity.

input

## Working at the Mall

The Working at the Mall problem is an example of direct variation. A **direct variation** is a linear function that can be written in the form $y = kx, where\ x \neq 0$. With direct variation, if one variable increases or decreases, the other will also. The **constant of variation**, *k*, represents the rate of change for the function. In the Working at the Mall problem, the constant of variation is the savings per week of $80.

analysis

## Working at the Mall

12. Your club is selling candy bars to make money. For each candy bar sold, your club earns fifty cents.

    a. Write an algebraic equation for the situation.

    b. Is this an example of direct variation? Explain.

13. You are on a game show. For every question you answer correctly, you earn 100 points.

    a. Write an algebraic equation for the situation.

    b. Is this an example of direct variation? Explain.

14. When you order pizza, the restaurant charges you $6.00 plus $1.25 per topping.

    a. Write an algebraic equation for the situation.

    b. Is this an example of direct variation? Explain.

## Saving for a Car

input

You want to purchase a nice, used car, but it costs $10,000. With your current job, you will need to work for more than 2 years to save that much money. So you start applying for other jobs hoping to make more money.

## Saving for a Car

process: evaluate

1. How long would it take to earn $10,000 if you could save:
   a. $50 per week?

   b. $100 per week?

   c. $200 per week?

   d. $250 per week?

   e. $300 per week?

process: model

## Saving for a Car

2. Complete the table and construct a graph for the situation.

| Weekly Savings | Time to Save $10,000 |
|---|---|
| $ | weeks |
| 50 | |
| 100 | |
| 150 | |
| 200 | |
| 250 | |
| 300 | |
| 350 | |
| 400 | |

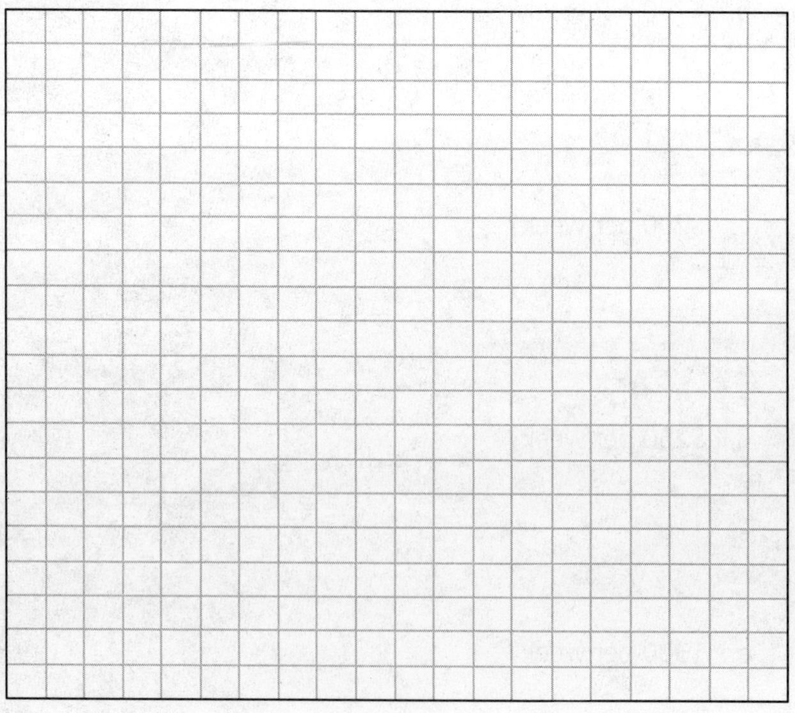

## Saving for a Car

analysis

3. Can this situation be modeled by a function? Explain your answer.

4. If this situation can be modeled by a function, what type of function will model this situation? How do you know?

## Saving for a Car

process: model

5. Write an algebraic equation to model the situation.

## Saving for a Car

analysis

6. What happens to the graph as *x* gets closer to zero? What happens to the graph as *x* gets very large?

analysis

## Saving for a Car

7. Will the graph have any x- or y-intercepts? Explain.

input

## Saving for a Car

The situation in this problem is modeled with a rational function. A function that can be written as the quotient of two polynomials is a **rational function**. You have already seen some specific types of rational functions. The linear, quadratic, cubic, and higher-order polynomial functions are types of rational functions. Rational functions can be either continuous or discontinuous and their graphs can have many different shapes.

Even though the Saving for a Car situation makes sense only for positive values (in the first quadrant), looking at the graph of the function for all real values of *x* will help you understand this rational function.

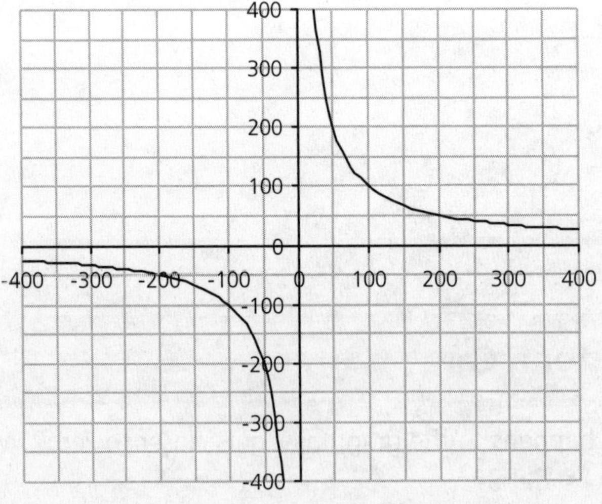

summary

## Saving for a Car

8. Describe the rational function, $y = \frac{10000}{x}$ that is graphed on the previous page.

   a. What are the domain and range of the function?

   b. What are the x- and y-intercepts of the function?

   c. What are the intervals of increase and decrease?

   d. Describe the end behavior of the function.

   e. Does the function have asymptotes? If so, what are they?

   f. In the function continuous or discontinuous? If it is discontinuous, name any locations of discontinuity.

input

## Saving for a Car

The Saving for a Car problem is an example of inverse variation. An **inverse variation** is a function that can be written in the form $y = \frac{k}{x}$. With inverse variation, if one variable increases, the other variable decreases. The **constant of variation**, $k$, represents the constant product of the two variables. In the Saving for a Car problem, the constant of variation represented the product of the number of weeks worked and the amount saved each week, or $10,000.

## An Age Old Problem

input

One day, Dan Petersen was having a disagreement with his father. During their discussion, Mr. Petersen said, "You should listen to me. I know more. I've lived longer. You're not even half my age." That got Dan thinking about when he would be half of his father's age. Right now, Dan is 16 years old and his father is 36 years old.

## An Age Old Problem

process: evaluation

1. Calculate Dan's age as a percentage of his father's age.

2. How old was Dan's father when Dan was born?

3. When will Dan be half of his father's age?

4. When will Dan be three-quarters of his father's age?

5. Will Dan ever be as old as his father?

process: model

## An Age Old Problem

6. If $x$ represents Dan's age, what expression represents his father's age?

7. What expression represents the ratio of Dan's age to his father's age?

analysis

## An Age Old Problem

8. What is the domain and range for this problem situation?

9. What is the domain and range of the mathematical function, $y = \frac{x}{x+20}$?

10. Are your answers for questions 8 and 9 the same? Explain.

## An Age Old Problem

process: model

11. Complete the following table of values.

| x | $\frac{x}{x+20}$ | x | $\frac{x}{x+20}$ |
|---|---|---|---|
| -70 | | 10 | |
| -60 | | 15 | |
| -50 | | 20 | |
| -40 | | 30 | |
| -30 | | 40 | |
| -25 | | 50 | |
| -15 | | 60 | |
| -10 | | 70 | |

## An Age Old Problem

process: model

12. Create a graph using the values from the table.

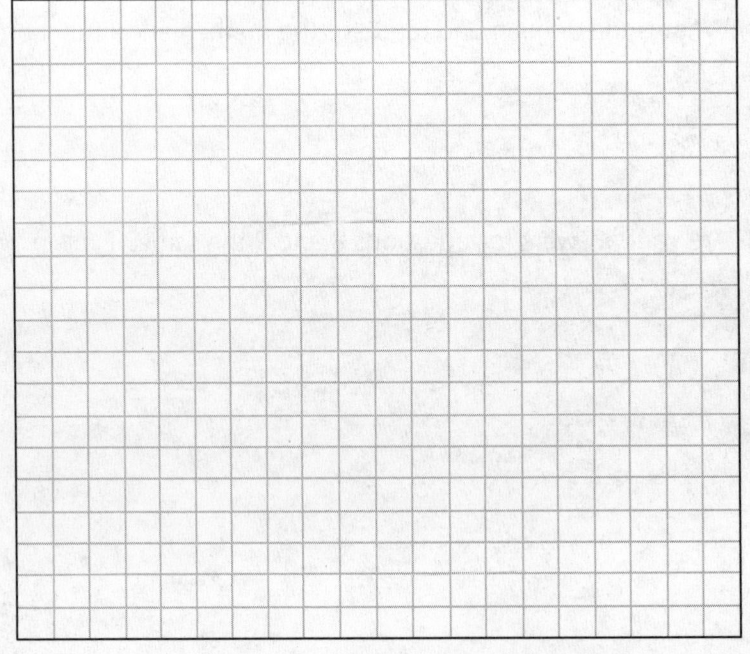

process: summary

## An Age Old Problem

13. What are the x- and y-intercepts?

14. What are the intervals of increase and decrease?

15. What are the asymptotes?

16. Describe the end behavior.

17. Describe the behavior near the vertical asymptote.

18. Explain why the function has the asymptotes it does in terms of the mathematical function and the problem situation.

19. Is the function continuous or discontinuous? If it is discontinuous, name any locations of discontinuity.

## An Age Old Problem

analysis

20. Is the situation in this problem an example of direct variation, inverse variation or neither? Explain your answer.

21 What type of function models this situation?

22. How would the equation and graph change if Dan's father was 30 years older than Dan?

process; evaluate

## Rational Functions: Discontinuity

The graphs of rational functions do not all look similar. The ability to identify the domain and discontinuities will help you understand rational functions. For each of the following rational functions and their graphs, state whether the function is continuous or discontinuous, determine the domain, and list any locations of discontinuity.

1. $f(x) = \dfrac{1}{(X+2)(X-2)}$

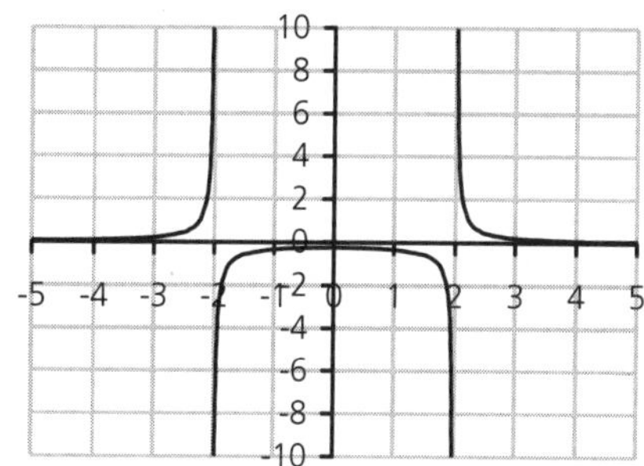

2. $g(x) = \dfrac{(x+5)(x-2)}{x-2}$

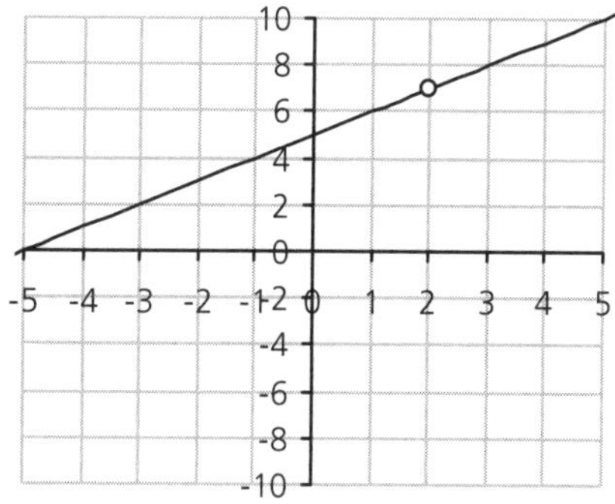

3. $h(x) = \dfrac{x+3}{x-1}$

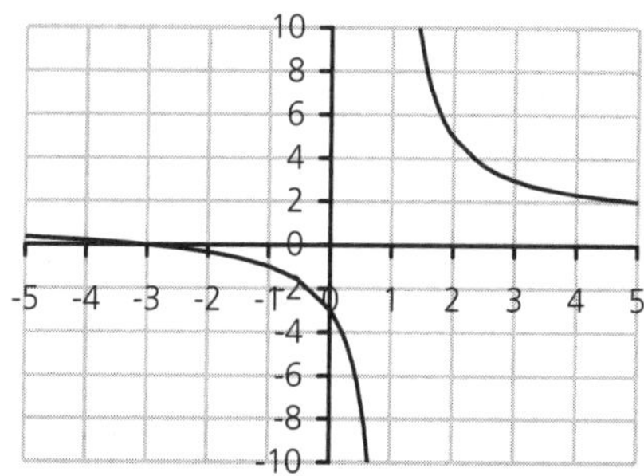

process; evaluate

## Rational Functions: Discontinuity

4. $j(x) = \dfrac{8}{x^2 + 1}$

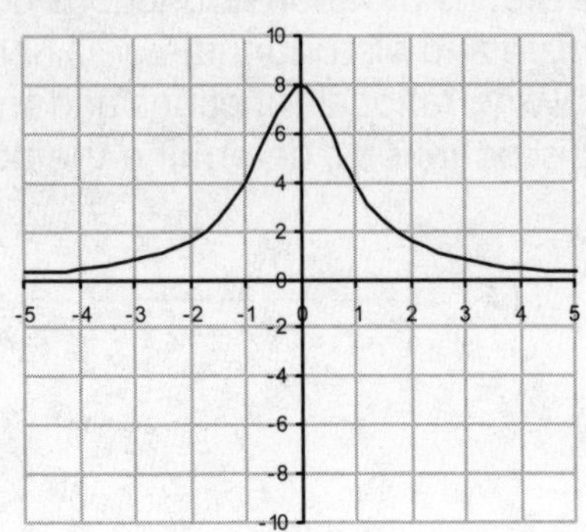

5. $k(x) = \dfrac{x^2 - x - 6}{x - 3}$

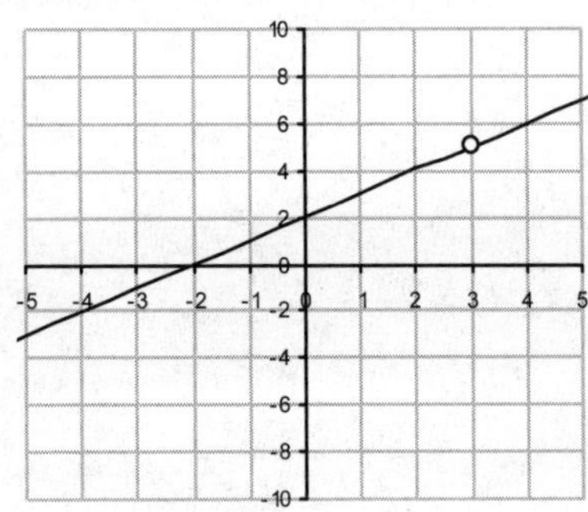

6. $m(x) = \dfrac{2}{x^2 - 9}$

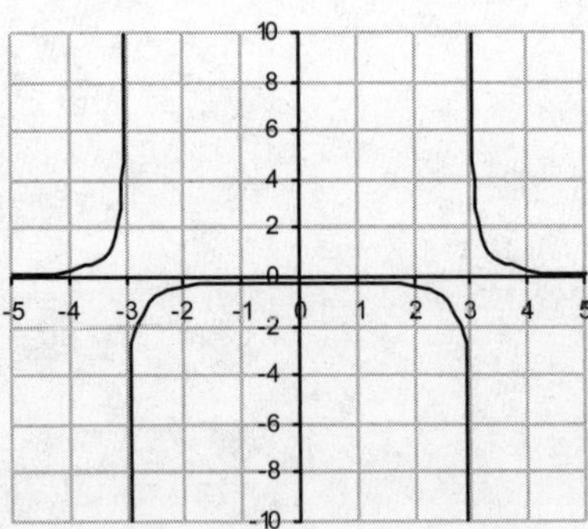

analysis

## Rational Functions: Discontinuity

7. Based on your work so far, how can you determine the domain of a rational function?

8. Based on your work so far, how can you determine whether a rational function is continuous or discontinuous? How can you determine any locations of discontinuity?

input

## Rational Functions: Discontinuity

Discontinuous rational functions can have two different types of discontinuity.

**Removable discontinuity** or **point discontinuity** occurs when a function has a hole at *x*, but is otherwise continuous like in the function $g(x) = \frac{x^2 - x - 6}{x - 3}$. The discontinuity is called removable because the function can be made continuous by defining its value at that one point.

The function would be continuous if written in a simpler form by factoring and canceling factors as shown below.

$$g(x) = \frac{x^2 - x - 6}{x - 3} = \frac{(x-3)(x+2)}{x - 3} = x + 2$$

**Essential discontinuity** occurs when a function has a discontinuity that cannot be removed with the insertion of a single point. An example is the discontinuity around the vertical asymptotes in the function, $m(x) = \frac{2}{x^2 - 9}$ from question 6.

process: practice

## Rational Functions: Discontinuity

For each rational function below, determine the domain of the function, state whether the function is continuous or discontinuous, identify any locations of discontinuity, and classify them as essential or removable.

9. $f(x) = \dfrac{1}{x^2}$

10. $g(x) = \dfrac{3}{(x+4)(x-4)}$

11. $h(x) = \dfrac{x+1}{(x+1)(x-3)}$

12. $j(x) = \dfrac{x^2+3x}{x^2+5x+6}$

13. $k(x) = \dfrac{2x^2}{3x+5}$

process: practice

## Rational Functions: Discontinuity

14. $m(x)=\dfrac{2x+1}{3x-2}$

15. $n(x)=2x^2+3$

16. $p(x)=\dfrac{x}{x^2-1}$

17. $q(x)=\dfrac{2x^2-5x-3}{x-3}$

18. $r(x)=\dfrac{x+3}{x^2+9}$

19. $t(x)=\dfrac{x+4}{x^2-3x}$

analysis

## Rational Functions: Discontinuity

20. Consider the function $f(x) = \begin{cases} \dfrac{x^2 + 3x - 4}{x - 1} & \text{if } x \neq 1 \\ 5 & \text{if } x = 1 \end{cases}$.

    Is *f* a continuous function? Explain.

21. Consider the function $g(x) = \begin{cases} \dfrac{x^2 - 7x + 10}{x - 5} & \text{if } x \neq 5 \\ 5 & \text{if } x = 5 \end{cases}$.

    Is *g* a continuous function? Explain.

22. Write an equation for the function below that has a point of removable discontinuity at (1, -2). *Hint: First think about the denominator, and then determine an appropriate numerator.*

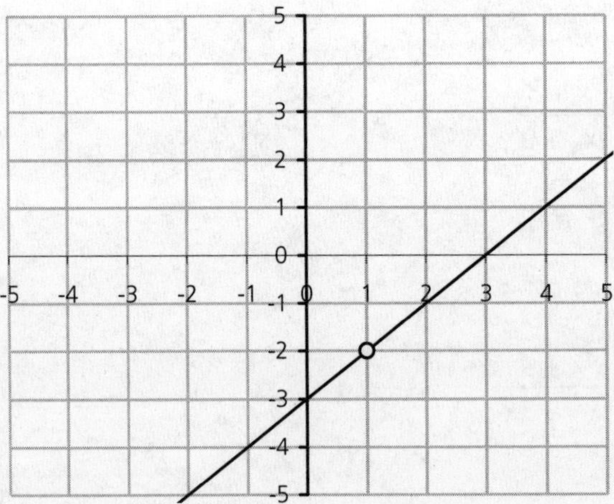

process: evaluate

## Rational Functions: Asymptotes

In addition to understanding continuity, the ability to identify horizontal and vertical asymptotes will help you understand rational functions. For each of the following rational functions and their graphs, determine any values of *x* for which the denominator is zero and identify any vertical asymptotes of the graph.

1. $f(x) = \dfrac{2}{(x-1)(x+2)}$

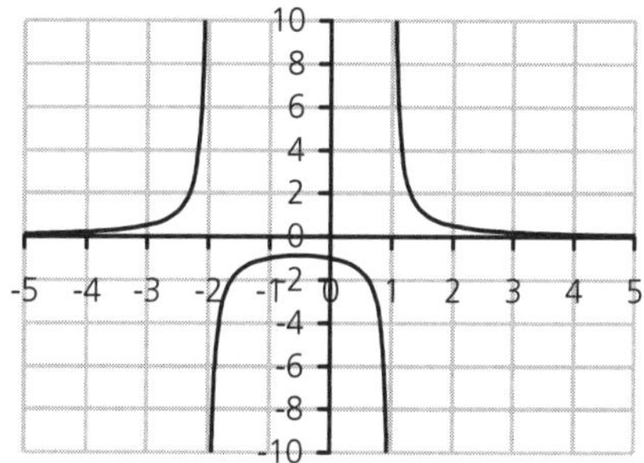

2. $g(x) = \dfrac{x-1}{x-2}$

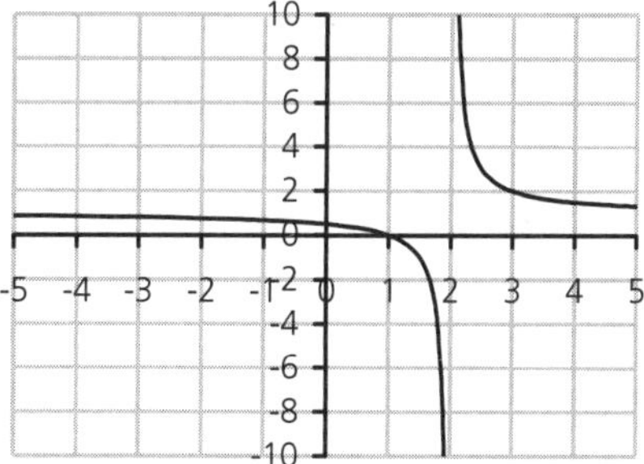

3. $h(x) = \dfrac{2x+3}{x+1}$

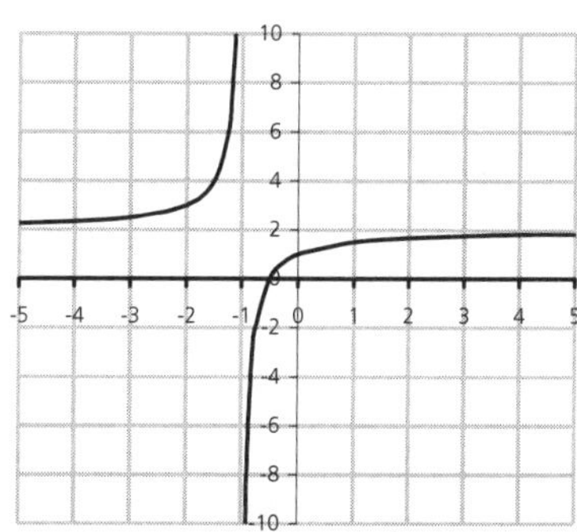

process: evaluate

## Rational Functions: Asymptotes

4. $j(x) = \dfrac{6}{x^2+1}$

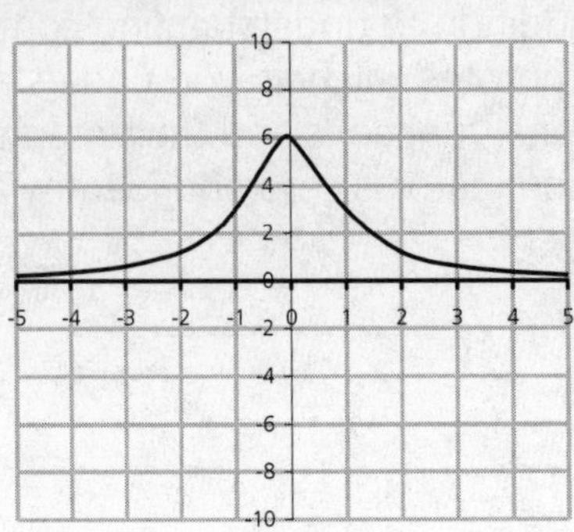

5. $k(x) = \dfrac{2}{x^2-1}$

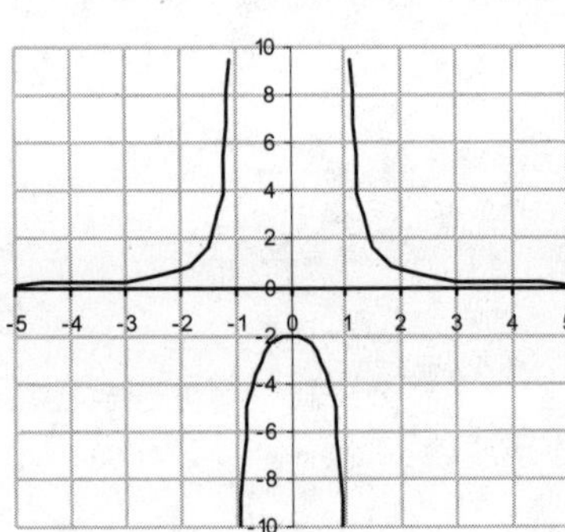

process: summary

## Rational Functions: Asymptotes

6. How can you determine the vertical asymptotes of a rational function given its algebraic equation?

7. When looking at an algebraic equation of a rational expression, how can you tell the difference between a point of removable discontinuity (with a hole in the graph) and essential discontinuity (with a vertical asymptote)?

## Rational Functions: Asymptotes

input

8. An important characteristic of a function is its end behavior: what happens to a function as $x$ gets very small or very large. Complete the following table and consider each function's end behavior.

| x | $f(x) = \frac{3x}{2x^2 + 1}$ | $g(x) = \frac{3x^2}{2x^2 + 1}$ | $j(x) = \frac{3x^3}{2x^2 + 1}$ |
|---|---|---|---|
| -300.00 | | | |
| -200.00 | | | |
| -150.00 | | | |
| -100.00 | | | |
| -50.00 | | | |
| -30.00 | | | |
| -20.00 | | | |
| -10.00 | | | |
| -8.00 | | | |
| -6.00 | | | |
| -4.00 | | | |
| -2.00 | | | |
| 0.00 | | | |
| 2.00 | | | |
| 4.00 | | | |
| 6.00 | | | |
| 8.00 | | | |
| 10.00 | | | |
| 20.00 | | | |
| 30.00 | | | |
| 50.00 | | | |
| 100.00 | | | |
| 150.00 | | | |
| 200.00 | | | |
| 300.00 | | | |

## Rational Functions: Asymptotes

input

Below are the graphs of the functions from the table on the previous page.

$$f(x) = \frac{3x}{2x^2 + 1}$$

Degree of numerator is less than the degree of the denominator.

Horizontal asymptote: $y = 0$

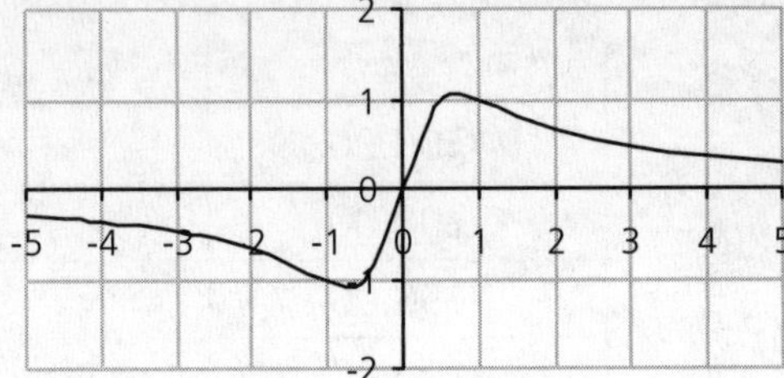

$$g(x) = \frac{3x^2}{2x^2 + 1}$$

Degree of numerator equals the degree of the denominator.

Horizontal asymptote: $y = \frac{3}{2}$

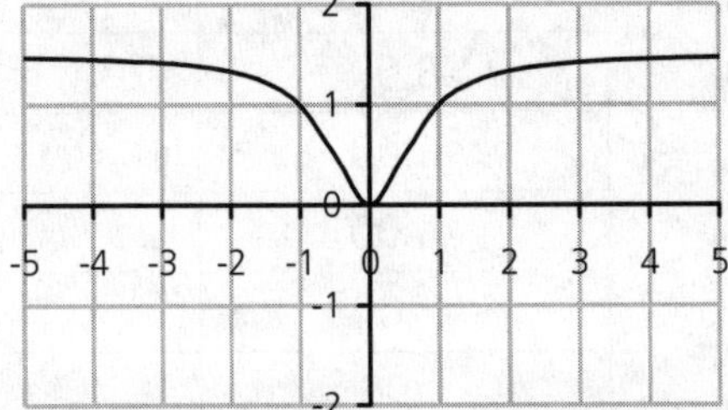

$$j(x) = \frac{3x^3}{2x^2 + 1}$$

Degree of numerator is greater than the degree of the denominator.

No horizontal asymptote

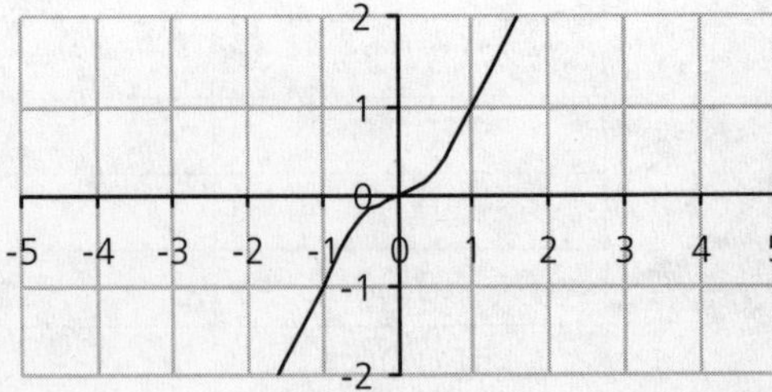

## Rational Functions: Asymptotes

analysis

9. For the function, $f(x)=\dfrac{3x}{2x^2+1}$:

   a. What happens to values of $f(x)$ in the table as $x$ gets very small or very large?

   b. What happens on the graph as $x$ gets very small or very large?

10. For the function, $g(x)=\dfrac{3x^2}{2x^2+1}$:

    a. What happens to values of $g(x)$ in the table as $x$ gets very small or very large?

    b. What happens on the graph as $x$ gets very small or very large?

11. For the function, $j(x)=\dfrac{3x^3}{2x^2+1}$:

    a. What happens to values of $j(x)$ in the table as $x$ gets very small or very large?

    b. What happens on the graph as $x$ gets very small or very large?

process: summary

## Rational Functions: Asymptotes

12. The horizontal asymptotes of a rational function depend on the degree of the numerator and denominator. How can you determine the horizontal asymptotes if:

    a. The degree of the numerator is less than the degree of the denominator.

    b. The degree of the numerator is equal to the degree of the denominator.

    c. The degree of the numerator is greater than the degree of the denominator.

analysis

## Rational Functions: Asymptotes

13. Why do the answers you gave in question 8 make sense? Hint: Consider what happens to the numerator and denominator as x gets very large.

## Rational Functions: Graphing

input

Knowing how to determine intercepts, discontinuity, and the location of asymptotes for a rational function will help you graph the rational function. Below is a summary of what you need to know about discontinuity and asymptotes of rational functions.

**Y-Intercept:**

- The y-intercept (if it exists) can be determined by evaluating f(0).

**X-Intercepts:**

- X -intercepts are values in the domain of the function for which $f(x) = 0$. Since a rational function has a value of zero if the numerator is zero, the x-intercepts (if any exist) can be determined by setting the numerator equal to zero and solving the equation.

**Discontinuity:**

- A rational function is continuous if the denominator has no zeros.
- A rational function has removable discontinuity if the denominator and numerator have the same zero.
- A rational function has essential discontinuity if the denominator has a zero that is not shared by the numerator.

**Vertical Asymptotes:**

- A rational function has a vertical asymptote at any location of essential discontinuity.

**Horizontal Asymptotes:**

- A rational function has a horizontal asymptote at $y = 0$ if the degree of the numerator is less than the degree of the denominator.
- A rational function has a horizontal asymptote at $y = \frac{\text{coefficient of leading term of numerator}}{\text{coefficient of leading term of denominator}}$ if the degree of the numerator is equal to the degree of the denominator.
- A rational function no horizontal asymptotes if the degree of the numerator is greater than the degree of the denominator.

## Rational Functions: Graphing

process: model

1. Sketch the graph for the rational function, $f(x) = \dfrac{x+3}{2x-1}$ by hand as explained in the following steps:

   a. Find the x- and y-intercept and plot them on the coordinate grid below.

   b. Determine any locations of discontinuity. If there is removable discontinuity, plot the point with an open circle. If there is essential discontinuity, sketch the vertical asymptote with a dashed vertical line.

   c. Determine the horizontal asymptote and sketch it with a dashed horizontal line.

   d. Find and plot at least one point between and one point beyond each x-intercept and vertical asymptote.

   e. Complete the graph by drawing smooth curves.

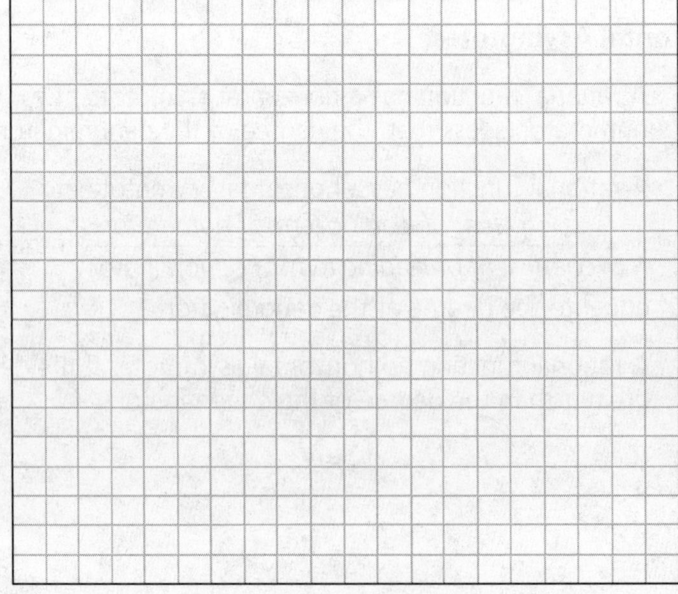

process: model

## Rational Functions: Graphing

Use what you know about intercepts, continuity, asymptotes, and locating points to sketch a graph for each of the following rational functions. Check your work using a graphing calculator.

2. $g(x) = \dfrac{-3}{x+2}$

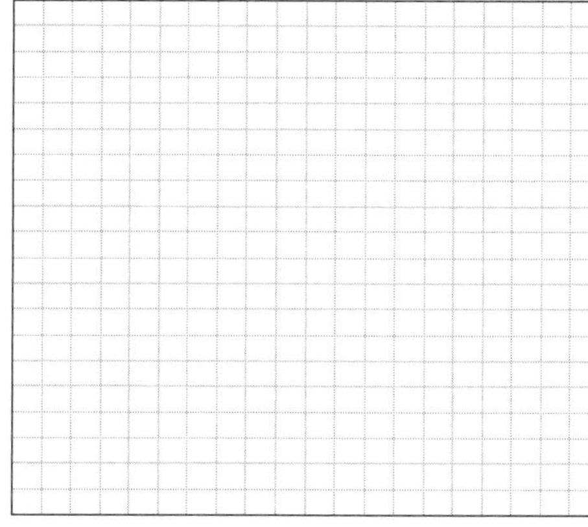

3. $h(x) = \dfrac{2x-1}{x+1}$

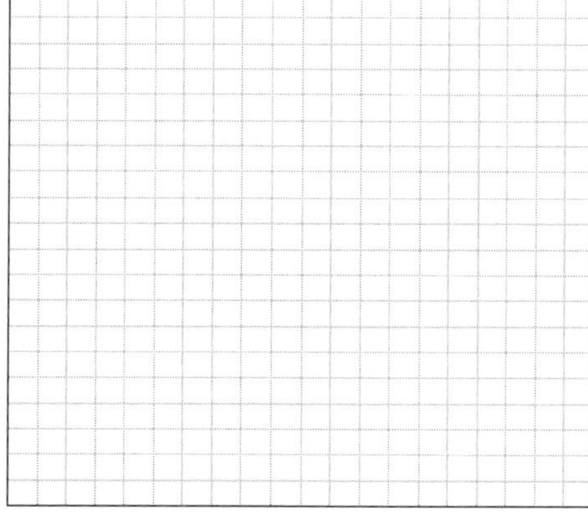

## Rational Functions: Graphing

process: model

4. $j(x) = \dfrac{1}{x+4}$

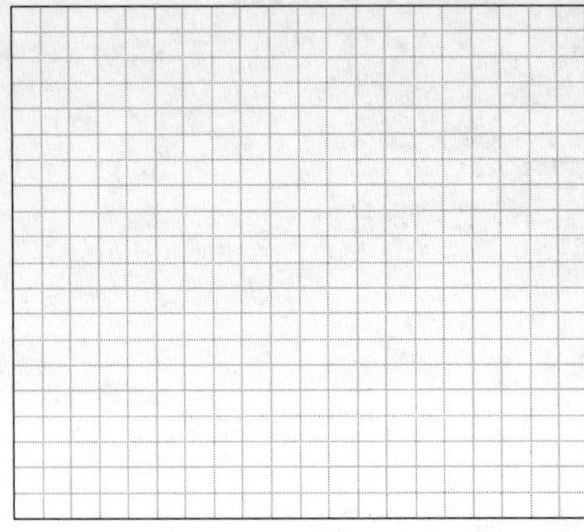

5. $k(x) = \dfrac{1}{x+4} + 2$

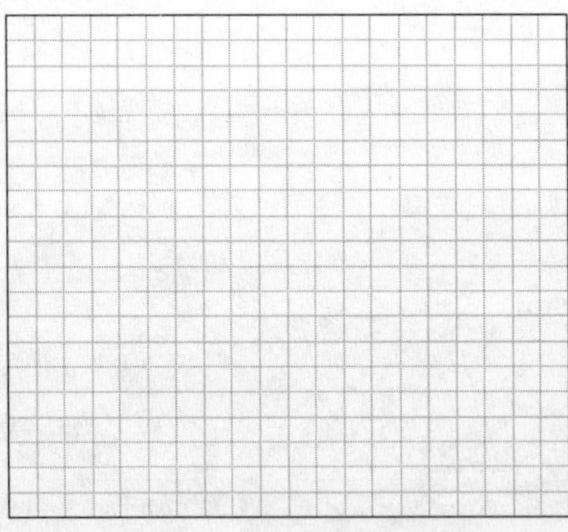

## Light Intensity

input

Light from a car headlight or a flashlight diminishes with distance. When driving at night, you can only see a certain distance down the road. Beyond that, everything appears dark.

For this problem, you will conduct an experiment to determine how light diminishes with distance. You will need a small flashlight with a beam you can focus and a metric ruler or meter stick to measure distances.

A flashlight shone on the wall will create a circle of light. As you move the flashlight farther from the wall, the circle of light becomes bigger and less intense.

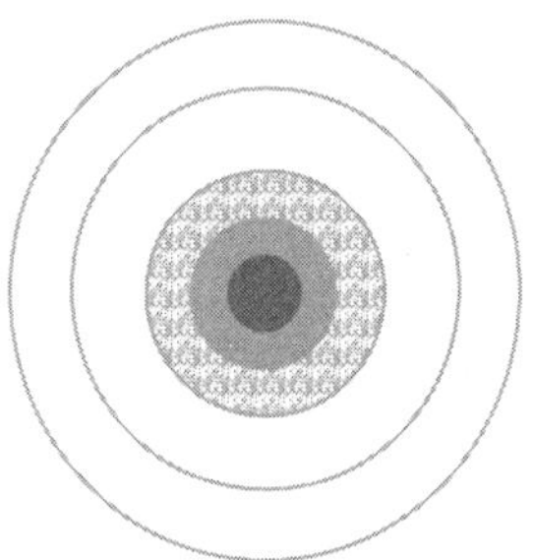

To perform the experiment, you will need to measure the circle of light created on the wall as the flashlight is placed at varying distances from the wall. It will help to place the flashlight on a flat desk or table near the wall.

Place the flashlight 5 cm from the wall and measure the diameter of the circle of light on the wall. Record the data. Now move the flashlight to a distance of 10 cm from the wall. Measure the diameter of the circle and record the data. Repeat the experiment for distances of 15, 20, and 25 cm.

## Light Intensity

process: model

1. Complete the first two columns of the table below using data from the experiment.

| **Distance** | **Diameter** | **Area** | **Intensity** |
|---|---|---|---|
| **cm** | **cm** | **square cm** | **lumen/cm$^2$** |
| | | | |
| | | | |
| | | | |
| | | | |
| | | | |

## Light Intensity

process: model

2. To determine the amount of light cast, calculate the area of the circles of light and enter those values in your table.

3. A lumen is a way of measuring how much light is cast. Since you do not know the exact output of the flashlight, assume it is a convenient number like 100 lumens. Calculate the intensity at each distance by dividing 100 lumens by the area of the circle of light. Record these values in your table.

4. Create a scatterplot of your data, using distance on the x-axis and intensity on the y-axis.

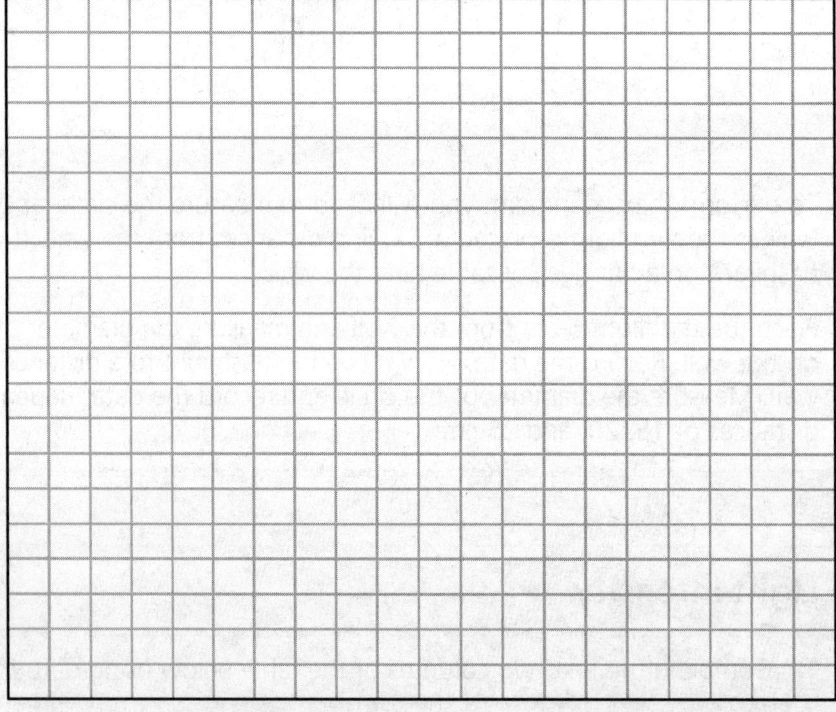

input

## Light Intensity

When you moved the flashlight farther from the wall, the circle of the beam of light became larger. Because the light output of the beam was constant, the intensity of the light became much less. You can see from your table and graph that as you double the distance of the light (or increase it by a factor of 2), the intensity decreases by about a factor of 4.

When doubling the independent value results in a decrease by a factor of 4 in the dependent value, the relationship is a type of rational function called an **inverse square function**.

Why do you suppose this relationship is called an inverse square relationship?

process: model

## Light Intensity

We can model the light intensity situation with a rational function. Recall that a rational function is a function that can be written as the quotient of polynomials. The function to model the situation is: $\text{light intensity} = \dfrac{\text{number of lumens}}{\text{area of circle}}$.

5. What is the number of lumens?

6. To determine a model for the area of the circle of light based on the distance, we can use our data and perform a quadratic regression using a calculator. Why is a quadratic regression appropriate?

process: model

## Light Intensity

7. What is the result of a quadratic regression on the data for the distance and the area of the circle of light?

8. Write a function, $f(x)$, for determining light intensity based on the distance.

9. Calculate each of the following using your function from question 8.
   a. $f(5)$
   b. $f(10)$
   c. $f(15)$
   d. $f(20)$

10. Is the function a good model for the data? Why or why not?

## The Hum Bug

input

Over the summer you noticed strange, noisy bugs in the field near your house. Each day, there seem to be more and more bugs in the field. You do a little research and find out that this insect is called a hum bug because of the humming sound it makes. You also find someone who has studied hum bug population growth in your area. The data they provided is summarized in the table below.

| Time | Population |
|---|---|
| Days | Hum Bugs |
| 0 | 50 |
| 1 | 100 |
| 2 | 250 |
| 3 | 500 |
| 4 | 1200 |
| 5 | 2500 |
| 6 | 5250 |
| 7 | 10000 |
| 8 | 18000 |

| Time | Population |
|---|---|
| Days | Hum Bugs |
| 9 | 28000 |
| 10 | 37000 |
| 11 | 43000 |
| 12 | 46500 |
| 13 | 48500 |
| 14 | 49250 |
| 15 | 49750 |
| 16 | 49850 |
| 17 | 49950 |

## The Hum Bug

process: model

1. Create a scatter plot for the data and sketch a curve through the points.

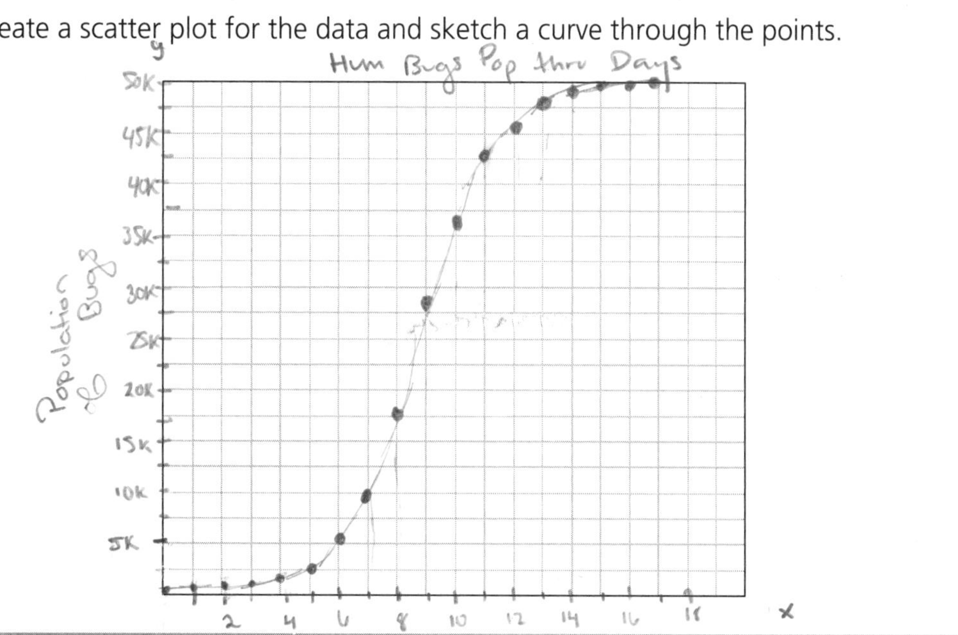


time (in days)

## The Hum Bug

**analysis**

2. Describe the overall shape of your graph.

   the shape of the graph looks like a very stretched out 'S.' It starts out slow, grows exponentially and then flattens out. It's a logistic graph.

3. Where on the graph is the hum bug population beginning to grow?

   The initial growth stage is from 0-6 days.

4. Where on the graph is the greatest increase in the hum bug population?

   The greatest exponential growth stage is from 6-10 days.

5. Where on the graph does the hum bug population growth seem to slow down?

   It dampens from the 11th to the 14 days

6. Where on the graph does the hum bug population seem to level off?

   It seems to level off around the 14th day to the 17th day

7. What factors in the environment might have caused the hum bug population to grow as it did?

   It grows because when they first come to a new area and reproduce a great amount. But then later it levels off b/c of the environment's carrying capacity

analysis

## The Hum Bug

8. If insect populations can grow very quickly, why have they not just taken over the world?

They have not taken over the world because insects have short life + although they may rapidly reproduce, they cannot live forever. Also insects are easily others' prey before predators. There's a limit to the bugs' growth. Eventually it reaches "its carrying capacity"

input

## The Hum Bug

The function graphed in this situation is closely modeled by the equation $f(x)=\dfrac{50000}{1+1000e^{-0.8x}}$ which is part of the family of logistic functions.

**Logistic functions** are functions which can be written in the form $y=\dfrac{C}{1+Ae^{-Bx}}$.

The S-shaped graph of a logistic function has four distinct intervals: the initial growth stage, the exponential growth stage, the dampened growth stage, and the equilibrium stage. The value the graph approaches in the equilibrium stage is called the **carrying capacity**.

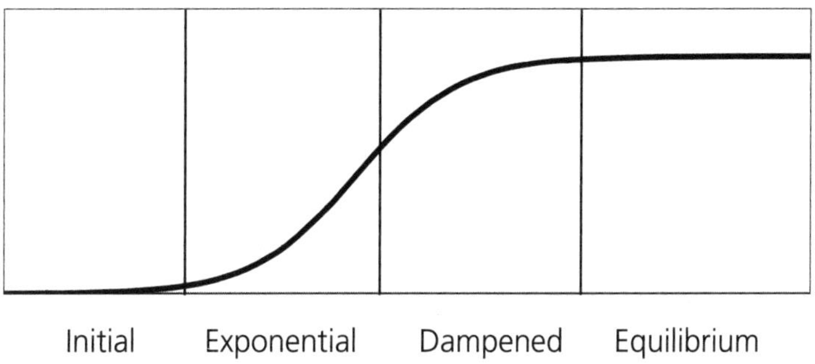


Logistic functions are good models of biological population growth, the spread of diseases, the spread of information, and sales of new products over time.

## The Hum Bug

analysis

Type the function $f(x)=\frac{50000}{1+1000e^{-0.8x}}$ into your graphing calculator and view the graph.

9. Change the numerator of the function so that it becomes $f(x)=\frac{40000}{1+1000e^{-0.8x}}$. How did the graph change?

It changes the limit. It flattens out sooner - before it slowed down at 50K, now it slows down at 40K.
* Same basic shape

original
modified

10. Change the numerator again so the function becomes $f(x)=\frac{30000}{1+1000e^{-0.8x}}$. How did the graph change?

It changes the limit again. It flattens out even sooner at 30K from the original 50K.
* same basic shape

org
v2
v3

11. What does the value of *a* (the numerator) in the logistic function $y=\frac{a}{1+be^{-cx}}$ represent on the graph?

(carrying capacity)

upper
a= the limit. It represents where the function slows down & flattens out after the exponential growth. The more a decreases/increases, the limit follows the decrease/increase with it.

## Sick of School

input

Logistic functions can be used to model the spread of disease. The data you collect in this experiment will model the spread of a disease. After collecting the data, you will be able to analyze it by using a logistic function.

In a small school, there are 100 students. One day, Zelda Zero, a student, comes to the school infected with a virus that causes mild cold like symptoms. On the first day that he has the virus, he has the potential to transmit the virus to one other person. On subsequent days, each infected person has the potential to transmit the virus to one other person. If the virus is transmitted to someone who has already had the virus, the person will not become ill again.

$2^x$

1, 2, 4, 8, 16, 32 . . .

## Sick of School

analysis

Before starting the experiment, take a few minutes to think about what will happen in the situation and answer the following questions.

1. How long will it take before most of the school is infected with the virus? Why?

   It will take about 7 days for most of the school to be infected, because $2^7 = 128$, and since it's over 100 that means the entire population would be sick then.

   $2^x = 100$

   $\frac{\log 100}{\log 2} = \frac{x \log 2}{\log 2}$

   $x = 6.64$

2. Is there a limit to the number of people who can be infected? Explain your answer.

   Yes because there is only 100 people at the school, and once someone is sick they can't catch the illness again.

## Sick of School

input

To keep things simple for the experiment, we will represent the 100 students in the school with numbers from 0 to 99. Using the list of numbers on the following page, you can keep track of students infected with the virus by crossing off their number. You'll notice that zero is already crossed off because Zelda Zero is already infected.

To simulate students getting infected with the virus, you can use the random number generator on a graphing calculator. Generate a random number between 0 and 99 on a TI-83 calculator, by typing int(rand*100) and pressing ENTER. After getting the first random number, you can get more random numbers simply by pressing ENTER.

On the first day, you will potentially infect one more person by choosing one random number. On subsequent days, you will potentially infect a number of people equal to the total number infected to that point by generating that many more random numbers.

Conduct the experiment following these directions.

- Use the graphing calculator to generate random numbers indicating who gets infected.
- Keep track of infected students by crossing out numbers as they are randomly selected. If a number is already crossed out, do not generate an alternate number.
- For each day, generate as many random numbers as there are currently infected students.
- At the end of each round of random numbers representing one day, record the total number infected (crossed out numbers) in the table on the next page.
- Repeat the process. Work until most students are infected (you have most of your numbers crossed off).

analysis

## Sick of School

6. What is the carrying capacity for this function?

7. Why does your answer to question 6 make sense as the carrying capacity?

process: model

## Sick of School

8. Use your calculator to find the equation of a logistic function that will model the data you collected for this problem.
   a. Enter your data.
   b. Find the equation by performing a logistic regression.
   c. Fill in the approximate values of a, b, and c in the general logistic function, $y = \frac{C}{1 + Ae^{-Bx}}$, to model the spread of the virus.

$$y = \frac{\square}{1 + \square e^{-\square x}}$$

## Sick of School

analysis

9. How do you think the experiment and results would be different if there were 200 people in the school?

10. Could you model the spread of the virus with another type of function (e.g. linear, quadratic, or exponential)? Why or why not?

## I Heard a Rumor

input

Last spring, 200 students attended the prom and after-prom. At the beginning of the evening, one student started spreading a rumor that the student who had been voted most likely to succeed was about to fail math class and might not graduate. During that evening, the number of students who heard the rumor continued to grow. The number of students who heard the rumor is modeled with the function $f(x) = \frac{200}{1 + 200e^{-0.75x}}$ where x represent the number of hours since the start of the prom.

## I Heard a Rumor

process: evaluate

1. About how many students know about the rumor at the start of the prom?

2. About how many students know about the rumor after 2 hours?

3. About how many students know about the rumor after 6 hours?

4. When will half of the students at the prom know about the rumor?

5. When will most of the students at the prom know about the rumor?

## I Heard a Rumor

Process: model

6. Create a table of values and a graph for the function.

| Time | Students That Know About the Rumor |
|---|---|
| Hours | Students |
| 0 | |
| 1 | |
| 2 | |
| 3 | |
| 4 | |
| 5 | |
| 6 | |
| 7 | |
| 8 | |

| Time | Students That Know About the Rumor |
|---|---|
| Hours | Students |
| 9 | |
| 10 | |
| 11 | |
| 12 | |
| 13 | |
| 14 | |
| 15 | |
| 16 | |
| 17 | |

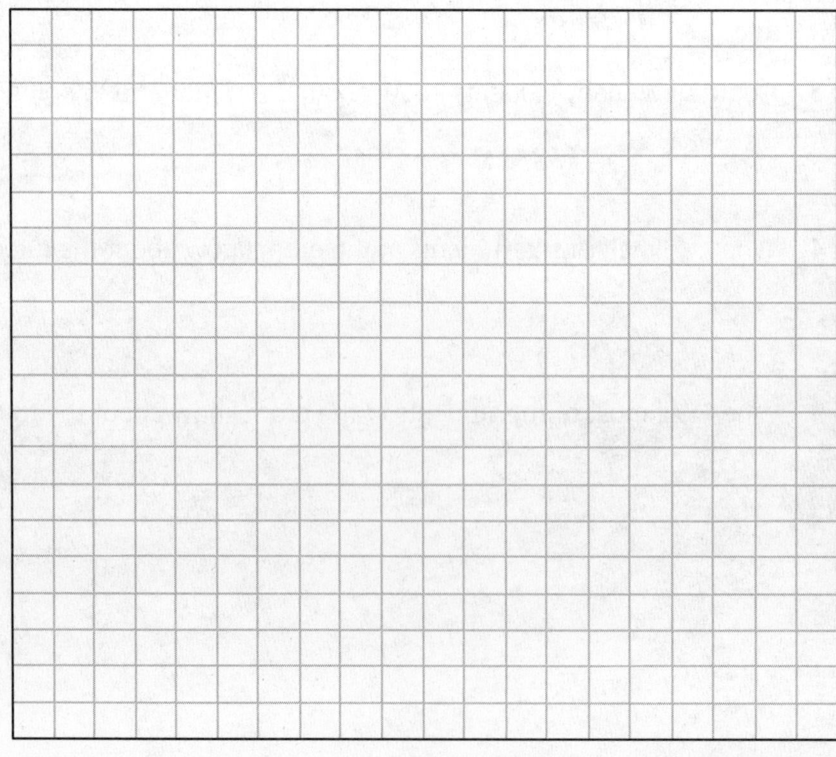

process: summary

## I Heard a Rumor

7. Describe the key behaviors or characteristics of this function and explain how they relate to the problem situation.

   a. What are the domain and range of the function?

   b. What are the intercepts of the function?

   c. How can you describe the rate of change of the function?

   d. What are the maxima and minima?

   e. What are the intervals of increase or decrease?

process: summary

## I Heard a Rumor

f. Describe the end behavior of the function.

g. Is the function continuous or discontinuous? If it is discontinuous, what the locations of discontinuity?

h. List any asymptotes for the function.

8. For this logistic function identify the initial growth, exponential growth, dampened growth, and equilibrium stages and the carrying capacity.

   a. initial growth stage

   b. exponential growth stage

   c. dampened growth stage

   d. equilibrium stage

   e. carrying capacity

# Unit 3: Functions, Relations, and Inverses

## Contents

## Functions and Relations

input

A **relation** is any correspondence between members of the domain and members of the range. A **function** is a special type of relation in which every member of the domain is associated with exactly one member of the range. In other words, a function is a relationship in which each input value has a unique output value.

If you are selling candy bars for a fundraiser, the relationship between the number of candy bars sold and the resulting income is a function. For any amount of candy bars sold (the input or domain), there is a unique value for the income (the output or range).

If you are counting how much money you have based on the number of coins in your pocket, the relationship is not a function. The amount of money may vary depending on the types of coins you have. While this is not an example of a function, it is still a relation.

## Functions and Relations

analysis

1. Every holder of a social security card in the United States is assigned a nine-digit social security number.

   a. Let the domain be all assigned social security numbers and the range be the names of all social security card-holders. Is this relationship a function? Why or why not?

   b. Consider reversing the situation. Let the domain be the names of all social security card-holders, and the range be all assigned social security numbers. Is this reverse relationship a function? Why or why not?

analysis

## Functions and Relations

2. Every member on the school football team is assigned a number.
   a. Let the domain be the names of all the football players on the school team and the range be all of the assigned player numbers. Is this relationship a function? Why or why not?

   b. Reverse the situation. Let the domain be all assigned player numbers and the range be all of the football players on the school team. Is this reverse relationship a function? Why or why not?

3. Each person has an astrological sun sign (e.g., Gemini, Libra) based on when the person was born.
   a. Let the domain be all people in the world and let the range be the astrological sun signs. Is this relationship a function? Why or why not?

   b. Reverse the situation so that the domain is the astrological sun signs and the range is all people in the world. Is the reverse relationship a function? Why or why not?

## Put it in Reverse

input

In the previous section, you considered general relationships and their reverse relationships and decided whether or not they were functions. Now you will look at more specific situations and determine their reverse. In other words, you will determine how to "undo" the situation. "Undoing," working backwards, or retracing steps to return to an original value or position is referred to as finding the **inverse operations**.

## Put it in Reverse

process: evaluate

Write a phrase, expression, or sentence for the inverse of each given action.

1. Open a door.

2. Turn on a light.

3. Add 6 to a number.

4. Divide a number by 4.

5. Jump in the deep end of a pool and swim to the shallow side.

6. Walk 2 blocks east and then 3 blocks south.

process: evaluate

## Put it in Reverse

7. Multiply a number by 3 and subtract 5.

8. Load a truck at the warehouse, drive to a dock, and put the cargo on a barge.

9. Enter a building, walk down the hall, and ride an elevator up six floors.

10. Add 2 to a number, multiply the result by 3, and subtract 1.

analysis

## Put it in Reverse

11. For question 7 above, Jan gave the inverse as "Divide by 3 and add 5." Marcus gave the inverse as "Add 5 and divide by 3." Who is correct? Why?

analysis

## Put it in Reverse

12. When asked for the inverse of "Walk two blocks north then one block east," Javon and Denise gave different answers. Javon said, "Walk one block west and then two blocks south." Denise said, "Walk two blocks south and then one block west." Who is correct? Why?

13. Does every situation have an inverse? Explain your answer.

process: summary

## Put it in Reverse

14. Write a short paragraph explaining how to determine the inverse of a situation.

## Change for a Dollar

input

You are planning a trip to Australia and you decide to exchange some American money for Australian money before you go. The current exchange rate is listed at 1.50, meaning that for every American dollar, you will get 1.5 Australian dollars. When you return from your Australian vacation, you will need to exchange any remaining Australian dollars back to American dollars. Assume the exchange rate is constant.

## Change for a Dollar

process: model

1. Complete the tables of values for exchanging from American to Australian dollars, and back from Australian to American dollars.

| American Currency | Australian Currency |
|---|---|
| dollars | dollars |
| 200 | |
| 400 | |
| 600 | |
| 800 | |
| 1000 | |
| 1200 | |
| 1400 | |
| 1600 | |
| 1800 | |

| Australian Currency | American Currency |
|---|---|
| dollars | dollars |
| 200 | |
| 400 | |
| 600 | |
| 800 | |
| 1000 | |
| 1200 | |
| 1400 | |
| 1600 | |
| 1800 | |

2. Write an algebraic function for the conversion from American dollars to Australian dollars.

3. Write an algebraic function for the conversion from Australian dollars to American dollars.

process: model

## Change for a Dollar

4. Create a graph for each of the exchange rate situations on the grid below. Also graph the function $y = x$ using a dotted line for that function. Use the same intervals on both the x- and y-axes.

analysis

## Change for a Dollar

Converting from American to Australian dollars and converting from Australian to American dollars are inverses of each other.

5. Write a few sentences to explain how the problem situation indicates that these situations are inverses.

## Change for a Dollar

analysis

6. How do the values in your tables indicate that the situations are inverses?

7. How do the algebraic functions indicate that the situations are inverses?

8. Consider your graph for this situation.

   a. Choose a point on the graph of your first function.

   b. Reflect that point over the line $y = x$. What is the new point? Is it on the graph of your second function?

   c. Choose a different point on the graph of your first function.

   d. Reflect that point over the line $y = x$. What is the new point? Is it on the graph of your second function?

   e. What do the results of the reflections indicate about the relationship of the graphs for the functions?

   f. How can the graphs help determine whether or not two functions are inverses?

## A Square Deal

process: model

1. You can create the growing pattern of squares below by adding equally sized toothpicks to the pattern. Complete the table by describing each part of the sequence in terms of the number of squares created and the number of toothpicks used.

| Number of Squares | Number of Toothpicks |
|---|---|
| 1 | 4 |
| 2 | |
| 3 | |
| 4 | |
| | |
| | |

2. Write an algebraic function for finding the number of toothpicks needed based on the number of squares.

3. Draw a graph of the function on the grid below. Use the same intervals on both the x- and y-axes.

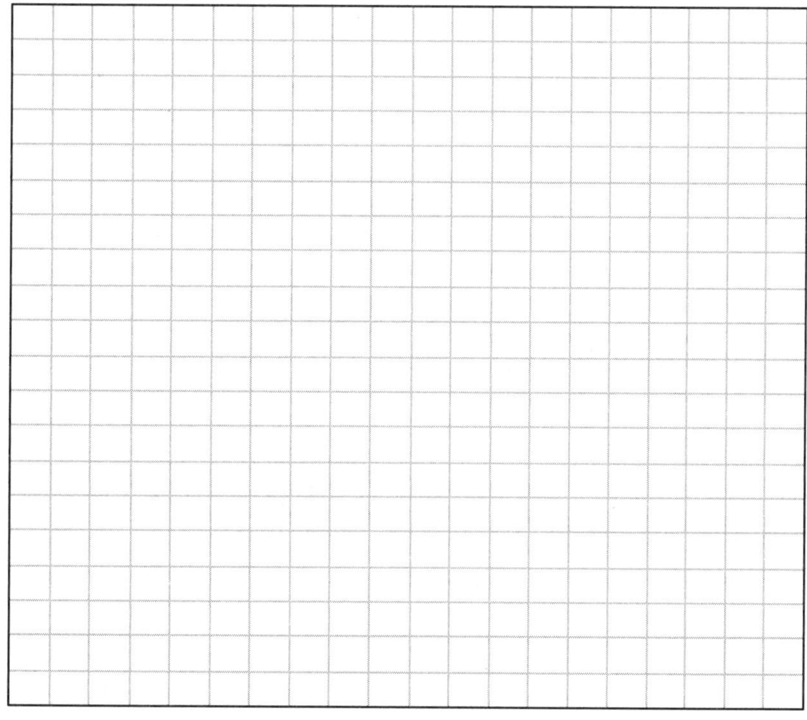

process: model

## A Square Deal

4. Suppose you want to create a pattern of squares using the toothpicks you can find. How many squares could you make with:

   a. 22 toothpicks?

   b. 55 toothpicks?

5. Write an algebraic function for finding the number of squares that can be created based on the number of toothpicks.

6. Complete the table of values.

| Number of Toothpicks | Number of Squares |
|---|---|
| 4 | |
| 7 | |
| 10 | |
| 13 | |
| | |
| | |

7. Graph the function on the grid you used on the previous page. Also graph $y = x$.

analysis

## A Square Deal

8. Do the relationships between toothpicks and squares represent inverse functions? Explain using evidence from the problem situation, tables, equations, and graph.

## Identities and Inverses

input

An **inverse element** within a given operation is an element that "undoes" another element.

## Identities and Inverses

process: evaluate

1. What *addition* will "undo" each of the following?

   a. Add -5

   b. Add 3

   c. Add 125

   d. Add $-\frac{3}{4}$

2. What is the inverse element for addition associated with each of the following?

   a. -5

   b. 3

   c. 125

   d. $-\frac{3}{4}$

3. Fill in the missing addend in each of the following.

   a. $-5+\square=0$

   b. $3+\square=0$

   c. $\square+125=0$

   d. $-\frac{3}{4}+\square=0$

analysis

## Identities and Inverses

4. What happens when a number and its additive inverse are combined by adding them together? Why?

process: evaluate

## Identities and Inverses

5. What *multiplication* will "undo" each of the following?

a. Multiply by $\frac{1}{2}$

b. Multiply by $-\frac{1}{3}$

c. Multiply by $\frac{4}{3}$

d. Multiply by 4

6. What is the inverse element for multiplication associated with each of the following?

a. $\frac{1}{2}$

b. $-\frac{1}{3}$

c. $\frac{4}{3}$

d. 4

7. Fill in the missing factor in each of the following.

a. $\frac{1}{2} \cdot \square = 1$

b. $-\frac{1}{3} \cdot \square = 1$

c. $\square \cdot \frac{4}{3} = 1$

d. $4 \cdot \square = 1$

analysis

## Identities and Inverses

8. What happens when a number and its multiplicative inverse are combined by multiplying them together? Why?

input

## Identities and Inverses

An **identity element** for a given operation is the element that produces no change when combined with other elements.

analysis

## Identities and Inverses

9. What is the identity element for addition? Why?

10. What is the identity element for multiplication? Why?

11. What are other names for additive and multiplicative inverses that you have used before?

## Identities and Inverses

analysis

12. Explain why $f(x) = x$ is called the **identity function.**

13. In *Change for a Dollar*, you identified the functions $f(x) = 1.5x$ and $g(x) = \frac{x}{1.5}$ as inverse functions. What should happen when these two functions are combined?

14. In *A Square Deal*, you identified the functions $f(x) = 3x + 1$ and $g(x) = \frac{x-1}{3}$ as inverse functions. What should happen when these two functions are combined?

input

## Motions of the Square (optional)

**Symmetries** or **motions of a figure** are transformations of the figure that preserve its size, shape, and apparent orientation.

A square has many motions. For example, if the square below is reflected over the dotted line, it maintains its size, shape, and apparent orientation. While we know the square has been "turned over," this would not be apparent without the labels on the vertices.

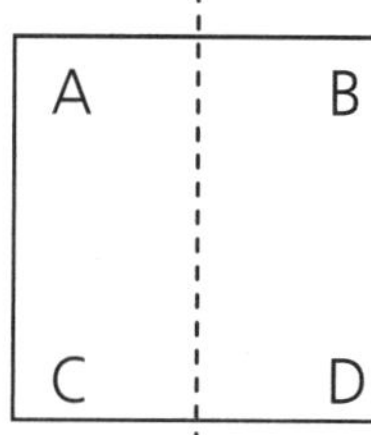


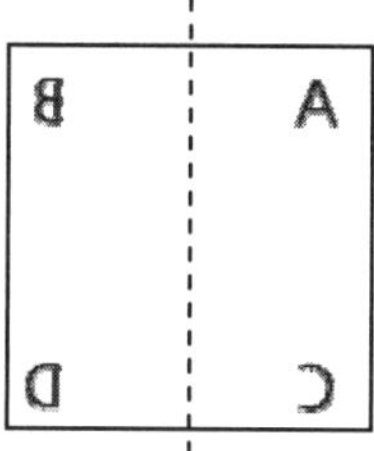


Another motion of the square is a rotation of 90 degrees counter-clockwise.

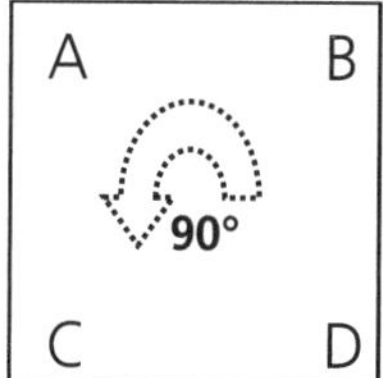


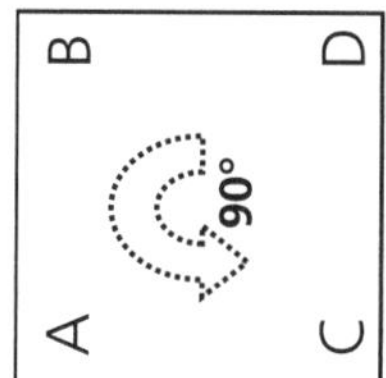


While it may seem to be different, a rotation of 270 degrees clockwise is not considered a new motion for the square because the end result is the same as the rotation of 90 degrees counter-clockwise. That means the vertices are in the same location as with the transformation above, so it is considered the same motion.

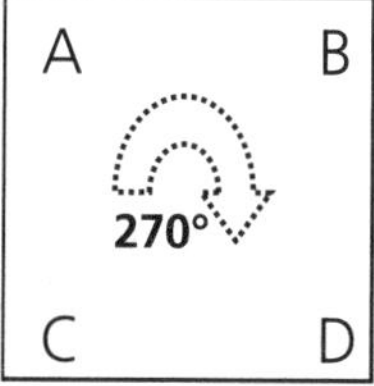


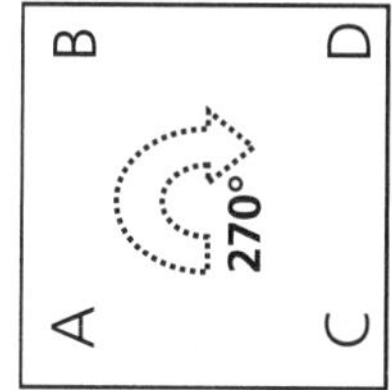

process: evaluate

## Motions of the Square (optional)

1. Make a complete list of the motions of the square. Be careful not to list equivalent motions. Either describe or draw the motion.

analysis

## Motions of the Square (optional)

2. How could you have predicted the number of motions of a square before finding them? Explain why or why not.

## Motions of the Square (optional)

process: evaluate

3. In your group, create symbols for each of the motions of the square. Consider what makes a good symbol.

4. Explain why you chose your symbols. What makes them good choices?

input

## Motions of the Square (optional)

A **Cayley table** shows the result of using an operation on two elements. For example, a multiplication table is a Cayley Table that shows what happens when two numbers are multiplied.

| * | 0 | 1 | 2 | 3 | ... |
|---|---|---|---|---|---|
| 0 | 0 | 0 | 0 | 0 | |
| 1 | 0 | 1 | 2 | 3 | |
| 2 | 0 | 2 | 4 | 6 | |
| 3 | 0 | 3 | 6 | 9 | |
| ⋮ | | | | | |

Process: model

## Motions of the Square (optional)

5. To begin creating a Cayley table for the motions of a square, choose a symbol for your operation and place it in the upper left cell of the table. Then add symbols for the motions of the square along the top row and left column of the table.

| | | | | | | | | |
|---|---|---|---|---|---|---|---|---|
| | | | | | | | | |
| | | | | | | | | |
| | | | | | | | | |
| | | | | | | | | |
| | | | | | | | | |
| | | | | | | | | |
| | | | | | | | | |
| | | | | | | | | |

## Motions of the Square (optional)

input

Call the operation on the motions of the square "followed by." The result of the operation will be the single motion that produces the same result as the two motions combined. For example, the motion *rotate 90 degrees counter-clockwise* followed by the motion *reflect over vertical line through midpoint of AB and CD* yields the same result as the motion *reflect over line through B and C.*

Use this definition of the operation on the motions of the square to complete your Cayley table on the previous page.

## Motions of the Square (optional)

analysis

6. Does the operation "followed by" demonstrate **closure** for the motions of a square? That is, will the result always be one of the elements (motions of a square)?

7. What is the identity element for the operation "followed by" on the motions of a square? How do you know it is an identity?

8. What are the inverse elements for each motion of the square under the operation of "followed by?" How do you know what elements are inverses?

analysis

## Motions of the Square (optional)

9. Does the associative property apply to the operation "followed by" on the motions of the square? Explain.

10. Does the commutative property apply to the operation "followed by" on the motions of the square? Explain.

extension

## Motions of the Square (optional)

The motions of a square form a **group**. A group is a set of elements under an operation that have the following properties:

- The operation is closed for the elements.
- There is an identity element.
- Each element has a unique inverse.
- The associative property applies.

## Motions of the Square (optional)

extension

11. For each of the following, decide whether it forms a group. Explain your answer.
    a. Addition with integers.
    b. Multiplication with real numbers.
    c. Division with integers.
    d. Subtraction with real numbers.
    e. Addition with natural numbers.

## Picture Perfect

input

You work for Picture Perfect, a company that creates digital pictures from old photographs and stores the photos on CD-ROM for the customer. When people bring in old photos to be digitized, they are charged a flat rate of $10.00 plus a fee of $1.50 for each photo.

## Picture Perfect

process: evaluate

1. Calculate the cost to digitize photos for a customer with:
   a. 12 photos.
   b. 24 photos.
   c. 36 photos.
2. How did you calculate the cost?
3. Write an algebraic function, $f(x)$, for determining the cost based on the number of photos.

input

## Picture Perfect

When a customer picks up their CD, the clerk must verify that all original photos are being returned. As a check, the clerk must determine how many photos should be returned based on the cost.

process: evaluate

## Picture Perfect

4. Calculate the number of photos for a customer who is charged:
   a. $28.00
   b. $46.00
   c. $64.00
5. How did you calculate the number of photos?
6. Write an algebraic function, $g(x)$, for the determining the number of photos from the cost.

process :model

## Picture Perfect

7. On the grid below, graph the function, y = x using a dotted line. Then graph both of the functions, f(x) and g(x) from this problem on the same grid. Use the same intervals on both the x- and y-axes.

analysis

## Picture Perfect

8. What do you notice about the graphs of *f*(*x*) and *g*(*x*)?

analysis

## Picture Perfect

9. How are questions 1 and 4 related?

10. How does your answer for question 2 compare to your answer for question 5?

11. a. Use 17 as the value of $x$. Compute $f(17)$.

    b. Use the result of part a in the function $g(x)$. What is the result?

12. Another way to ask the previous question is, "What is $g(\ f(17)\ )$?" Explain how the questions are the same.

13. If any value is used for $x$ in your function, $f(x)$, from question 3, and then the result is entered into the function, $g(x)$, from question 6, will you be able to predict the final result without doing calculations for each function? In other words, can you predict the result of $g(f\ (x)\ )$? Why or why not?

input

## Picture Perfect

The Picture Perfect situation is modeled by a function for calculating the cost based on the number of photos, and its inverse function for calculating the number of photos from the cost.

The input and output values for the first function, $f(x)$, were reversed for the second function, $g(x)$. That is, the function $f(x)$ consists of the ordered pairs, $(x, y)$ or $(x, f(x))$, while the function $g(x)$ consists of the ordered pairs, $(y, x)$. The roles of the independent and dependent variables have been switched. This was evident in your answers to questions 1 and 4, and from your graph of the functions.

The inverse of a function is denoted by $f^{-1}(x)$ and is read "*f* inverse of *x*." In this problem, $f^{-1}(x) = g(x)$.

You can switch the roles of the variables through a series of algebraic manipulations. Find the inverse for $f(x) = 1.5x + 10$ to verify your work in this problem.

Step 1. Replace $f(x)$ with $y$.

Step 2. Switch the $x$ and $y$ variables in the equation.

Step 3. Solve for $y$.

Step 4. Replace $y$ with $f^{-1}(x)$.

process: summary

## Picture Perfect

14. So far, you have explored several different methods to represent inverse functions. Now explain each method in a brief paragraph. Reference the Picture Perfect situation in your explanations as needed.

    a. Explain inverse functions as a process of "undoing."

    b. Explain inverse functions as reflected graphs.

    c. Explain the relationship of inverse functions when the results of one function are used as the inputs of the other function.

    d. Explain inverse functions as interchanging the domain and range.

15. If f and g are inverse functions, what is the result of $g(f(x))$? What is the result of $f(g(x))$?

process: practice

## Inverses of Linear Functions

Use the four-step process described in the previous problem to determine the inverse for the given functions algebraically.

1. Find $f^{-1}(x)$ for $f(x) = 12x$.

2. Find $g^{-1}(x)$ for $g(x) = 3x - 5$.

3. Find $h^{-1}(x)$ for $h(x) = -2x + 7$

4. Find $j^{-1}(x)$ for $j(x) = \dfrac{x+2}{3}$

## Inverses of Linear Functions

process: evaluate

5. Using the function $f(x) = 12x$ from question 1, evaluate:

   a. $f(3)$ b. $f^{-1}(36)$ c. $f(f^{-1}(3))$

   d. $f^{-1}(-24)$ e. $f(-2)$ f. $f^{-1}(f(-24))$

6. Using the function $g(x) = 3x - 5$ from question 2, evaluate:

   a. $g(3.5)$ b. $g^{-1}(5.5)$ c. $g(g^{-1}(3.5))$

   d. $g^{-1}(25)$ e. $g(10)$ f. $g^{-1}(g(25))$

7. Using the function $h(x) = -2x + 7$ from question 3, evaluate:

   a. $h(-3)$ b. $h^{-1}(13)$ c. $h(h^{-1}(-3))$

   d. $h^{-1}(3)$ e. $h(2)$ f. $h(h^{-1}(3))$

8. Using the function $j(x) = \dfrac{x+2}{3}$ from question 4, evaluate:

   a. $j(4)$ b. $j^{-1}(2)$ c. $j(j^{-1}(4))$

   d. $j^{-1}(8)$ e. $j(22)$ f. $j(j^{-1}(8))$

## Inverses of Linear Functions

analysis

9. What do you notice about the functions and their inverses from questions 5 and 6?

10. If a function is evaluated for an input value and its inverse is evaluated with the resulting output, what is the final result? Why?

11. Verify that $f(x)$ and $f^{-1}(x)$ from problem 1 are inverses by evaluating:

    a. $f(f^{-1}(x))$

    b. $f^{-1}(f(x))$

12. Verify that $g(x)$ and $g^{-1}(x)$ from problem 2 are inverses by evaluating:

    a. $g(g^{-1}(x))$

    b. $g^{-1}(g(x))$

## Inverses of Linear Functions

analysis

13. Verify that $h(x)$ and $h^{-1}(x)$ from problem 3 are inverses by evaluating:
    a. $h(h^{-1}(x))$

    b. $h^{-1}(h(x))$

14. Verify that j(x) and $j^{-1}(x)$ from problem 4 are inverses by evaluating:
    a. $j(j^{-1}(x))$

    b. $j^{-1}(j(x))$

15. How do your results show that the functions are inverses?

## Inverses of Linear Functions

process: model

16. Graph $g(x)$, $g^{-1}(x)$, and the identity function, $g(x) = x$ on the grid below.

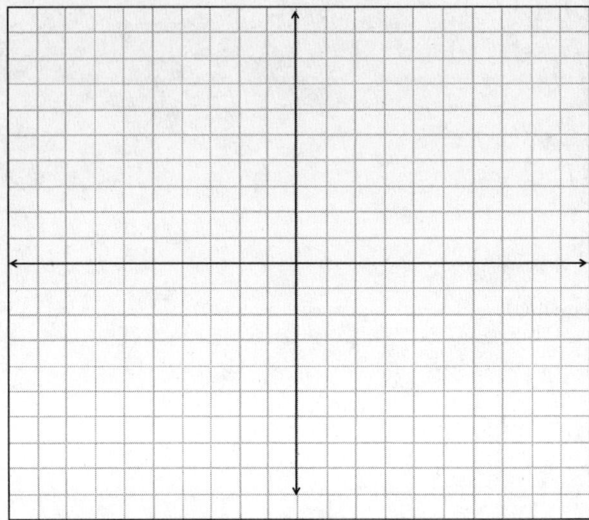

17. Graph $h(x)$, $h^{-1}(x)$, and the identity function, $h(x) = x$ on the grid below.

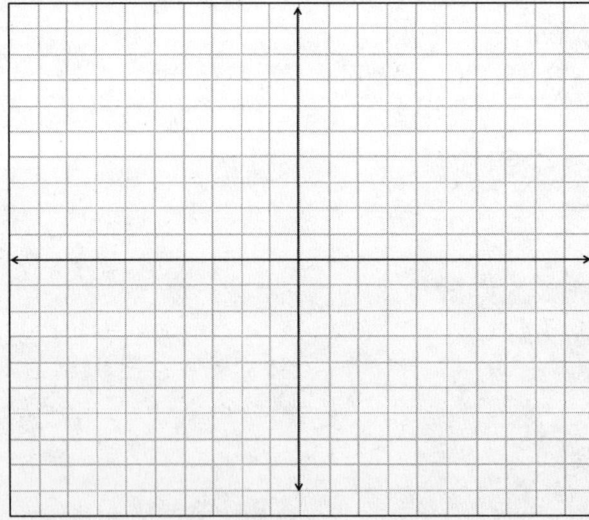

18. How do your results show that the functions are inverses?

process: evaluate

## Inverses of Linear Functions

19. In a bowling tournament, participants are assigned a handicap. The handicap is an adjustment to the score in each game to account for differences in bowlers' abilities. A handicap is based on a bowler's average score per game. One tournament used the function, $h(a) = 0.8(225 - a)$, to determine each bowler's handicap.

    On the day of the tournament, the director could not find the list of averages, but he did find the list of handicaps. What function can be used to determine averages based on handicaps?

process: evaluate

## Inverses of Linear Functions

20. The amount of wood in a tree can be estimated based on the radius of the tree trunk. However, it is difficult to measure the radius of a tree trunk without damaging the tree. It is easier to measure the circumference around the tree. You know that the circumference of a circle can be found with the function $C = 2\pi r$. What function can determine the radius of the circle given the circumference?

## Inverses of Linear Functions

analysis

21. Is every linear relation a function? Why or why not?

22. Is the inverse of every linear function also a function? Why or why not?

23. Can a linear function and its inverse be the same function? If so, list the linear functions that demonstrate this. If not, explain why not.

## Classified Ad

input

To determine the cost of a classified ad, a local newspaper charges $0.50 per line of text plus a flat fee of $5.00 to place the ad.

## Classified Ad

process: model

1. Write a sentence to describe the inverse of the problem situation above.

2. Represent the situation and its inverse with algebraic functions. Use appropriate function and inverse function notation.

3. Complete the tables of values for the function and its inverse function.

| Lines of Text | Cost for Ad |
|---|---|
| text lines | dollars |
| | |
| | |
| | |
| | |
| | |

| Cost for Ad | Lines of Text |
|---|---|
| dollars | text lines |
| | |
| | |
| | |
| | |
| | |

process: model

## Classified Ad

4. Create a graph of the function and its inverse on the grid provided. Use the same intervals on both the x- and y-axes.

process: evaluate

## Classified Ad

5. Evaluate $f(5)$.

   a. What does $f(5)$ represent in terms of the problem situation?

6. Evaluate $f^{-1}(10)$.

   a. What does $f^{-1}(10)$ represent in terms of the problem situation?

## Classified Ad

process: evaluate

7. Evaluate $f^{-1}(5)$.

   a. What does $f^{-1}(5)$ represent in terms of the problem situation?

8. Evaluate the following functions.

   a. $f(f^{-1}(8))$

   b. $f^{-1}(f(8))$

   c. $f(f^{-1}(20))$

   d. $f^{-1}(f(20))$

9. Evaluate $f(f^{-1}(x))$

10. Evaluate $f^{-1}(f(x))$

## Selling Appliances

input

You are starting a new job at Crazy Joe's World of Appliances. Crazy Joe agrees to pay you $500 per week. He also offers a commission as an incentive for you to sell more appliances. As an example of what you could earn in a week, he shows you the following table posted in the employee lounge.

| Weekly Sales ($) | Weekly Earnings ($) |
|---|---|
| 0 | 500 |
| 1000 | 540 |
| 2000 | 580 |
| 3000 | 620 |
| 4000 | 660 |
| 5000 | 700 |

## Selling Appliances

process: model

1. Crazy Joe's table is a little old and beat up. He wants you to make a new table for him to post. However, he has decided that the new table should be the inverse of the original, showing how much an employee must sell based on how much they want to earn. Create this table for Crazy Joe.

| | |
|---|---|
| | |
| | |
| | |
| | |
| | |
| | |

process: model

## Selling Appliances

2. Create a graph of the function and its inverse on the grid below. Use the same intervals on both the x- and y-axes.

process: model

## Selling Appliances

3. Write an algebraic function, $f(x)$, that represents the information in the original table.

4. Write an algebraic function, $f^{-1}(x)$, for the inverse function.

process: model

## Selling Appliances

5. Describe the original function in a sentence.

6. Describe the inverse function in a sentence.

process: evaluate

## Selling Appliances

7. Evaluate $f(2500)$ and explain what it means in the problem situation.

8. Evaluate $f^{-1}(750)$ and explain what it means in the problem situation.

9. Evaluate $f(f^{-1}(x))$.

10. Evaluate $f^{-1}(f(x))$.

## Burning Calories

input

A newspaper article about exercise showed the following graph as an example of the calories an average adult could burn by bicycling.

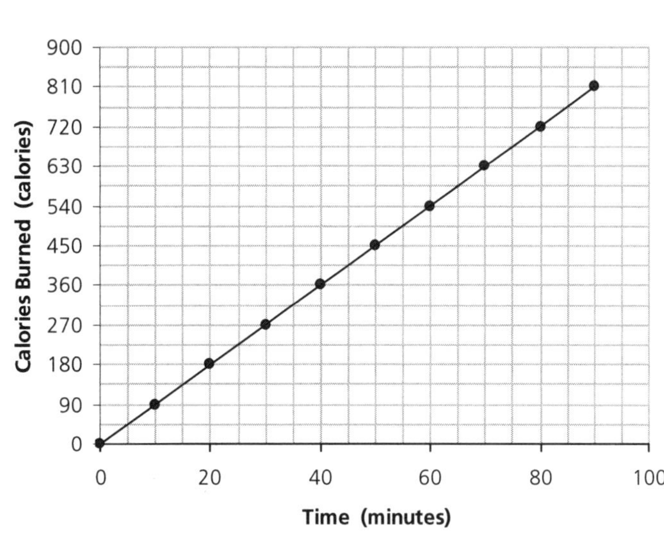


## Burning Calories

process: model

1. Create a graph showing the inverse of the graph from the newspaper.

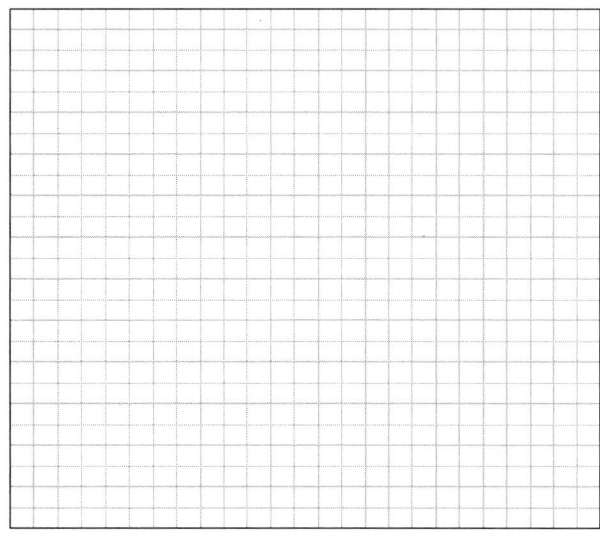

process: model

## Burning Calories

2. Complete the tables of values for the function and its inverse function.

| Time | Calories Burned |
|---|---|
| Minutes | Calories |
| | |
| | |
| | |
| | |
| | |
| | |

| Calories Burned | Time |
|---|---|
| Calories | Minutes |
| | |
| | |
| | |
| | |
| | |
| | |

3. Describe the original function in a sentence.

4. Describe the inverse function in a sentence.

5. Represent the situation and its inverse with algebraic functions.

process: evaluate

## Burning Calories

6. Evaluate $f(45)$ and explain what it means in the problem situation.

7. Evaluate $f(-10)$ and explain what it means in the problem situation.

8. Evaluate $f^{-1}(300)$ and explain what it means in the problem situation.

9. Evaluate $f^{-1}(0)$ and explain what it means in the problem situation.

10. Evaluate $f(f^{-1}(x))$.

11. Evaluate $f^{-1}(f(x))$.

## Inverses of Power Functions

process: model

1. Graph $y = x^2$ and $y = x$ on the grid below. Use the same intervals on both the x- and y-axis.

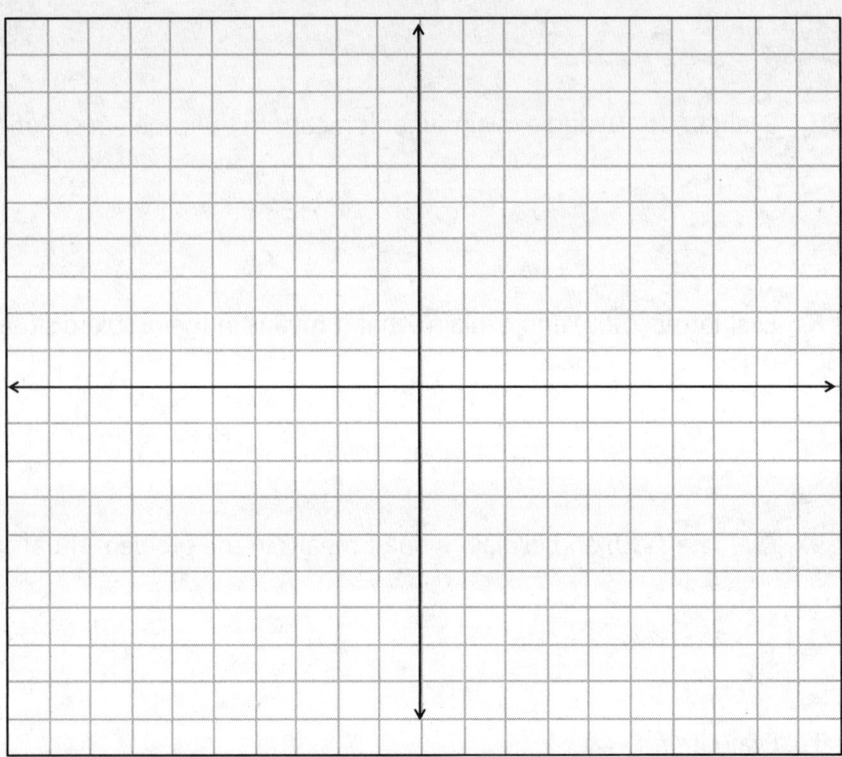

2. Sketch the reflection of the graph of $y = x^2$ over the line $y = x$.

## Inverses of Power Functions

analysis

3. Are the graphs of $y = x^2$ and its reflection over $y = x$ inverses of each other? Explain your answer.

analysis

## Inverses of Power Functions

4. Do the graphs of $y = x^2$ and its reflection over $y = x$ represent functions? Explain your answer.

process: model

## Inverses of Power Functions

5. Find the inverse of the function $y = x^2$ algebraically.

6. Use your calculator to graph $y=x^2$ and the inverse you found in question 5. Compare the graphs on your calculator with your graphs from questions 1 and 2. Do they match? If not, what is different and why?

input

## Inverses of Power Functions

It is a mathematical convention for $\sqrt{x}$ to represent only the positive square root of *x*, and $-\sqrt{x}$ to represent the negative square root of *x*. Restricting the values in this way avoids the problems associated with relations that are not functions. There is no longer confusion over which value in the range should be associated with each value in the domain.

Most of the time, $y = \sqrt{x}$ is treated as a function with a domain and range restricted to non-negative real numbers.

process: model

## Inverses of Power Functions

7. Graph $y = x^3$ and $y = x$ on the grid below. Use the same intervals on both the x- and y-axis.

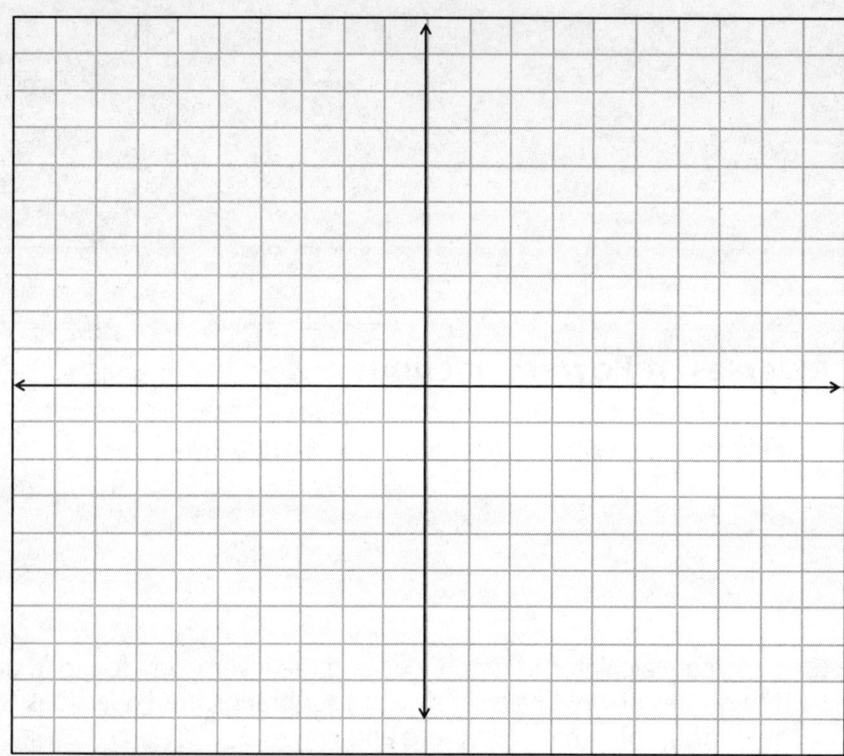

8. Sketch the reflection of the graph of $y = x^3$ over the line $y = x$.

analysis

## Inverses of Power Functions

9. Are the graphs of $y = x^3$ and its reflection over $y = x$ inverses of each other? Explain your answer.

analysis

## Inverses of Power Functions

10. Do the graphs of $y = x^3$ and its reflection over $y = x$ represent functions? Explain your answer.

process: model

## Inverses of Power Functions

11. Find the inverse of the function $y = x^3$ algebraically.

12. Use your calculator to graph $y = x^3$ and the inverse you found in question 11. Compare the graphs on your calculator with your graphs from questions 1 and 2. Do they match? If not, what is different and why?

input

## Inverses of Power Functions

Functions with inverses that are also functions are called **one-to-one** functions. These are functions for which each element of the domain corresponds to exactly one element in the range, and for which each element of the range corresponds to exactly one element in the domain.

If a function, $f(x)$ is one-to-one, its inverse will also be a function. If a function is not one-to-one, its inverse will not be a function and it is not appropriate to represent the inverse with functional notation.

## Inverses of Power Functions

analysis

13. Can you predict if other power functions, like $y = x^4$ or $y = x^5$, are one-to-one? If so, how? Hint: graphing the functions with your calculator might help.

## Inverses of Power Functions

process: practice

Determine the inverses of the following functions algebraically.

14. $f(x) = x^2 + 4$

15. $g(x) = 2x^2 - 3$

16. $h(x) = 4x^4 - 2$

17. $j(x) = 6x^5 + 2$

process: practice

## Inverses of Power Functions

Determine the inverses of the following functions algebraically.

18. $k(x) = x^3 + 7$

19. $m(x) = x^2 + 6x$

20. $n(x) = x^2 - x$

21. $p(x) = x^2 + 4x + 1$

analysis

## Inverses of Power Functions

22. Which of the functions in questions 14 to 21 are one-to-one? How do you know?

## Designing Cologne Bottles

input

You work for a company that produces cologne to be sold worldwide. The development team has created a new fragrance for both men and women called *Confession*. As part of the design team, you design bottles for the products. The team has decided on simple designs. *Confession for Men* will be sold in a cube-shaped container. *Confession for Women* will be sold in a sphere-shaped container placed on a small stand.

As part of your job, you need to determine the sizes for bottles of cologne. Knowing that one cubic centimeter will hold one milliliter of cologne, you consider the sizes of various bottles. The volume of the cube based on the length of a side can be determined with the function $f(x) = x^3$, and the volume of the sphere based on the diameter can be determined with the function $g(x) = \frac{\pi x^3}{6}$.

## Designing Cologne Bottles

process evaluate

1. How much *Confession for Men* will fill a bottle with a side measuring:
   a. 2 cm?
   b. 3 cm?
   c. 4 cm?

process evaluate

## Designing Cologne Bottles

2. How much *Confession for Women* will fill a bottle with a side measuring:
   a. 2 cm?
   b. 3 cm?
   c. 4 cm?

process model

## Designing Cologne Bottles

3. Find the inverse of the function $f(x) = x^3$ algebraically.

4. In a sentence, explain what the inverse function, $f^{-1}(x)$, represents in the problem.

5. Find the inverse of the function $g(x) = \frac{\pi x^3}{6}$ algebraically.

6. In a sentence, explain what the inverse function, $g^{-1}(x)$, represents in the problem.

process evaluate

## Designing Cologne Bottles

7. Colognes are frequently sold in 30 ml and 40 ml bottles. Determine how large the bottles must be for *Confession for Men* colognes sold in these sizes.

8. Determine the size of the bottle for 30 ml and 40 ml sizes of *Confession for Women.*

extension

## Designing Cologne Bottles

9. If the design of the bottle changed to a rectangular prism with a square base, and a height that is double the length of a side of the square base, what are the dimensions for a 30 ml and 40 ml bottle?

## The Square Root Function

input

The function, $f(x) = \sqrt{x}$, is the **square root function**. The function is only defined for non-negative values of *x*.

The graph of the parent square root function is shown below.

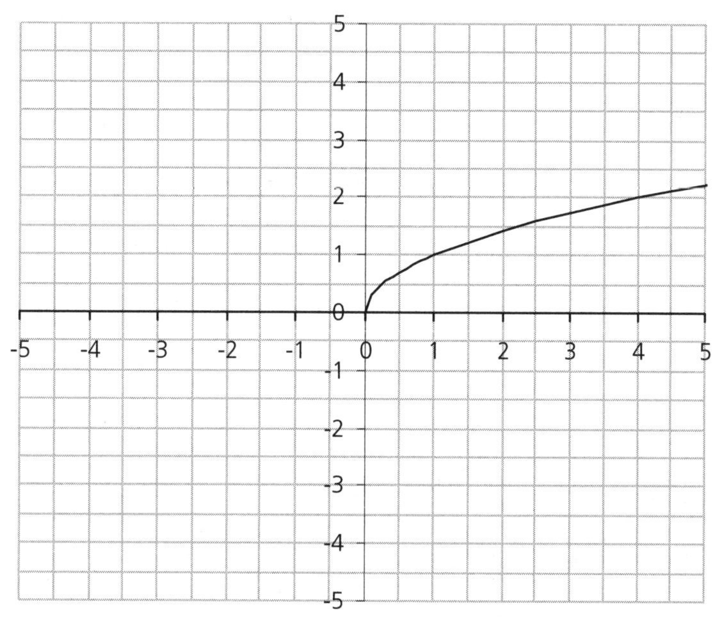


## The Square Root Function

analysis

Describe the parent square root function by answering the following questions.

1. What are the domain and range of the square root function?

2. What are the x- and y-intercepts?

3. What are the intervals of increase and decrease?

analysis

## The Square Root Function

4. What is the end behavior?

5. Does the function have any asymptotes? If so, list them.

6. Is the graph continuous or discontinuous? If it is discontinuous, list any locations of discontinuity.

process: model

## The Square Root Function

Sketch the graph of each square root function below by transforming the parent square root function. Write a description of the transformations used and describe the domain and range of the transformed function.

7. $f(x) = -\sqrt{x}$

process: model

## The Square Root Function

8. $g(x) = 2\sqrt{x}$

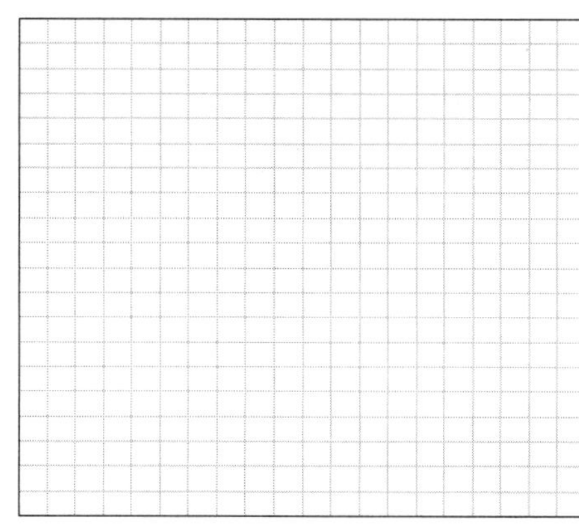

9. $h(x) = \sqrt{x} - 5$

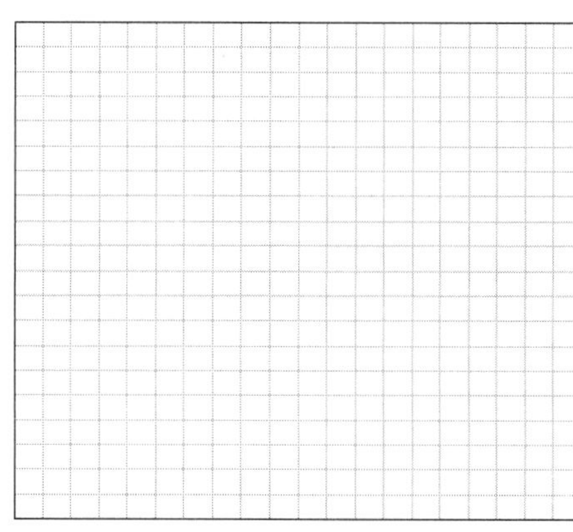

10. $j(x) = \sqrt{x+2}$

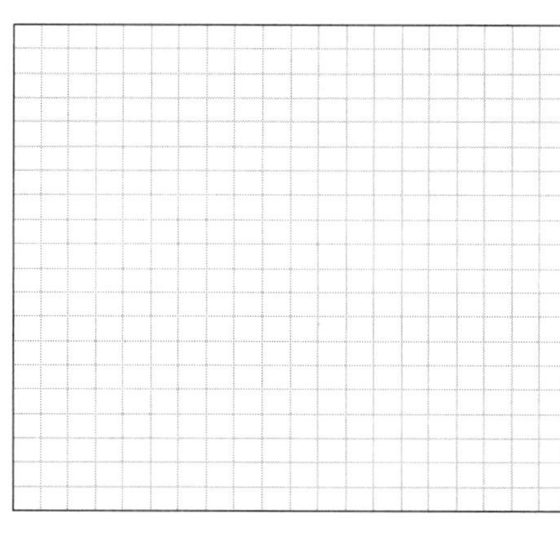

## The Square Root Function

process: model

11. $k(x) = -\frac{1}{2}\sqrt{x+4}$

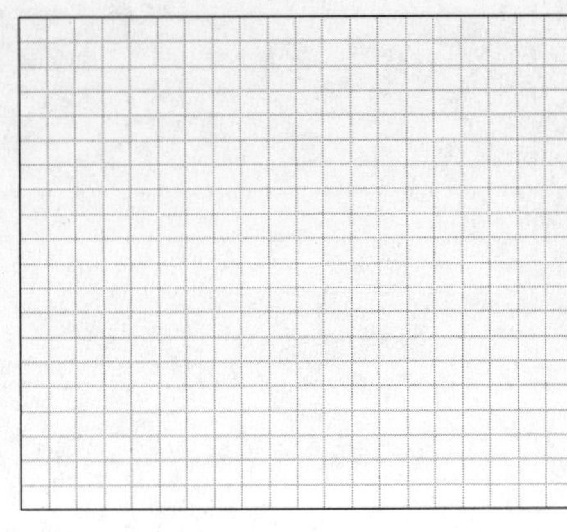

12. $m(x) = \sqrt{x-1} + 3$

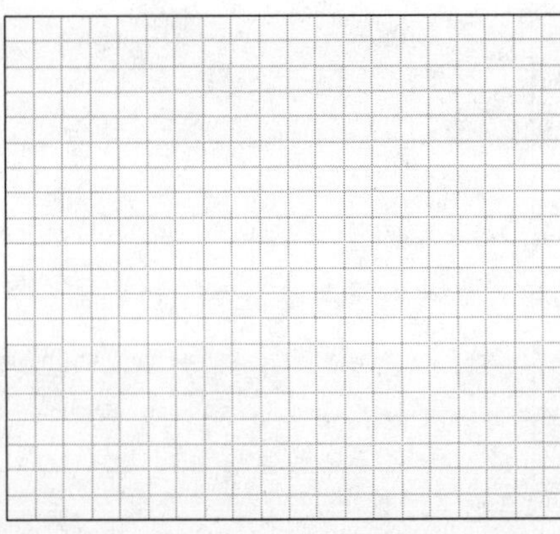

13. $n(x) = 4\sqrt{x} + 2.5$

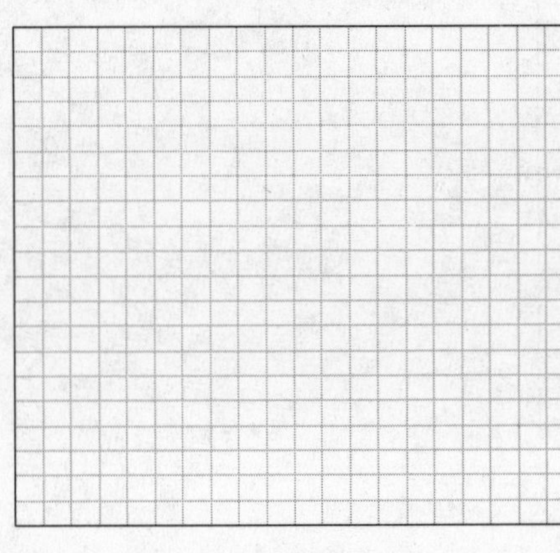

## The Square Root Function

process: evaluate

14. What is the inverse of the function $f(x) = \sqrt{x}$ ? Remember to consider an appropriate domain.

15. What is the inverse of the function $j(x) = \sqrt{x+2}$ ?

16. What is the inverse of the function $h(x) = \sqrt{x} - 5$ ?

17. What is the inverse of the function $n(x) = 4\sqrt{x} + 2.5$ ?

analysis

## The Square Root Function

18. How did you determine the domain for each of the inverse functions in questions 14 through 17?

process: evaluate

## The Square Root Function

19. Verify that $j(x)$ and $j^{-1}(x)$ are inverses by evaluating:

    a. $j(j^{-1}(x))$

    b. $j^{-1}(j(x))$

## The Square Root Function

process: evaluate

20. Verify that $h(x)$ and $h^{-1}(x)$ are inverses by evaluating:

    a. $h(h^{-1}(x))$

    b. $h^{-1}(h(x))$

## The Rotor

input

The Rotor is a popular amusement park ride shaped like a cylindrical room. Riders stand against the circular wall of the room while the room spins. When The Rotor reaches the necessary speed, the floor drops out and centripetal force leaves riders pinned up against the wall.

The minimum speed (measured in meters per second) required to keep a person pinned against the wall during the ride can be determined with the function $s(r) = 4.95\sqrt{r}$, where $r$ is the radius of The Rotor measured in meters.

## The Rotor

process: evaluate

1. An amusement park designed a rotor ride with a radius of 2 meters. At what speed does it need to spin?

2. The same park decided to build a larger rotor ride with a radius of 4 meters. At what speed does it need to spin?

3. Designers at another park have a motor that could spin a rotor ride at 6 meters per second. How big can they make the ride?

process: model

## The Rotor

4. Algebraically determine the inverse of the function used to determine the speed of The Rotor.

5. Create a graph of the function and its inverse on the grid below.

## The Rotor

process: model

6. Write a paragraph describing the initial function, $s(r) = 4.95\sqrt{r}$ . Be sure to include information about the parent function, domain, range, and behavior of the function.

7. Write a paragraph describing the inverse function. Be sure to include information about the parent function, domain, range, and behavior of the function.

## The Rotor

process: evaluate

8. The designers estimate that they must allow 60 cm (0.6 meters) of space along the wall for each person on the ride. For each of the designs from questions 1 and 2, determine the maximum number of people who should be allowed on the ride at one time.

process: model

## The Rotor

9. Write an algebraic function for determining the number of people a rotor ride can hold based on its radius.

10. Find the inverse of your function from question 9 algebraically.

11. Graph your functions from questions 9 and 10 on the grid below.

process: evaluate

## The Rotor

12. The ride designers at an amusement park want to create a rotor ride that can accommodate 25 riders at a time.

    a. Would you use the function given at the start of the problem or its inverse to determine the size of the ride? Why?

    b. How big must they make the ride?

process: model

## The Rotor

13. Write a paragraph describing your function from question 9. Be sure to include information about the parent function, domain, range, and behavior of the function.

14. Write a paragraph describing the inverse function from question 10. Be sure to include information about the parent function, domain, range, and behavior of the function.

## Exponentials and Logarithms

input

Previously you leaned about exponential and logarithmic functions. Now you will further investigate the relationship between these functions.

Recall that an exponential function with base $b$ can be written in the form $y = b^x$. The related logarithmic function is the inverse of the exponential function and can be written as $y = \log_b x$, which is read, "$y$ equals log base b of $x$."

Common logarithms have a base of 10 and are usually written without a base. For example, $y = \log_{10} x$ is usually written as $y = \log x$.

Natural logarithms have a base of e and are usually written without a base and using "ln" instead of "log." For example, $y = \log_e x$ is usually written as $y = \ln x$.

Most scientific and graphing calculators have buttons for common and natural logarithms.

## Exponentials and Logarithms

process: model

1. Find the inverse of the exponential function $f(x) = 10^x$.

2. Complete the tables of values below to show the relationship between $f(x)$ and $f^{-1}(x)$.

| $x$ | $f(x)$ |
|---|---|
| -2 | |
| -1 | |
| 0 | |
| 1 | |
| 2 | |

| $x$ | $f^{-1}(x)$ |
|---|---|
| | |
| | |
| | |
| | |
| | |

process: model

## Exponentials and Logarithms

3. Create a graph showing $f(x)$, $f^{-1}(x)$, and the identity function.

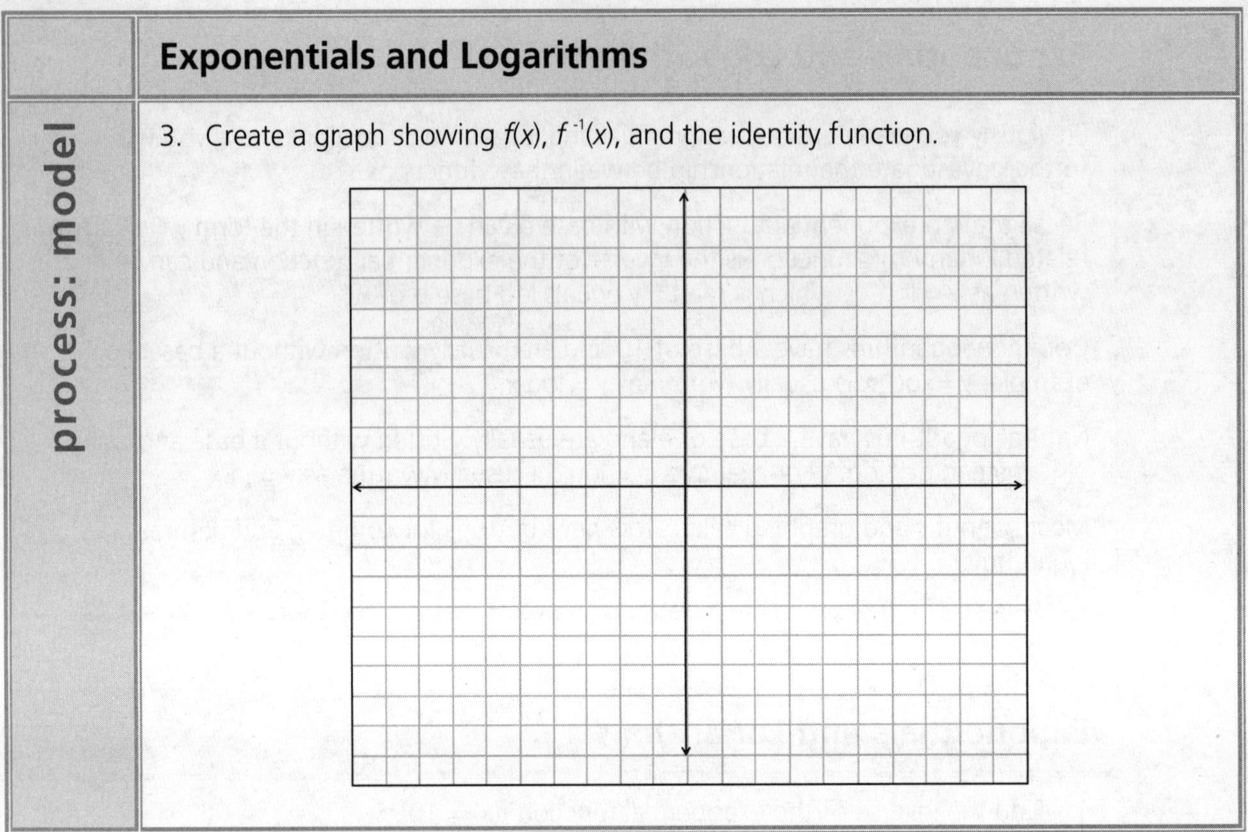

analysis

## Exponentials and Logarithms

4. What are the intercepts of $f(x) = 10^x$? What are the intercepts of $f^{-1}(x)$?

5. Is this consistent with what you know about inverses? Explain.

analysis

## Exponentials and Logarithms

6. What is the asymptote for $f(x) = 10^x$? What is the asymptote for $f^{-1}(x)$?

7. Is this consistent with what you know about inverses? Explain.

8. Write a paragraph describing similarities and differences of the graphs of the exponential and logarithmic functions.

process: model

## Exponentials and Logarithms

9. Find the inverse of the exponential function $g(x) = e^x$.

## Exponentials and Logarithms

process: model

10. Create a graph showing $g(x)$, $g^{-1}(x)$, and the identity function.

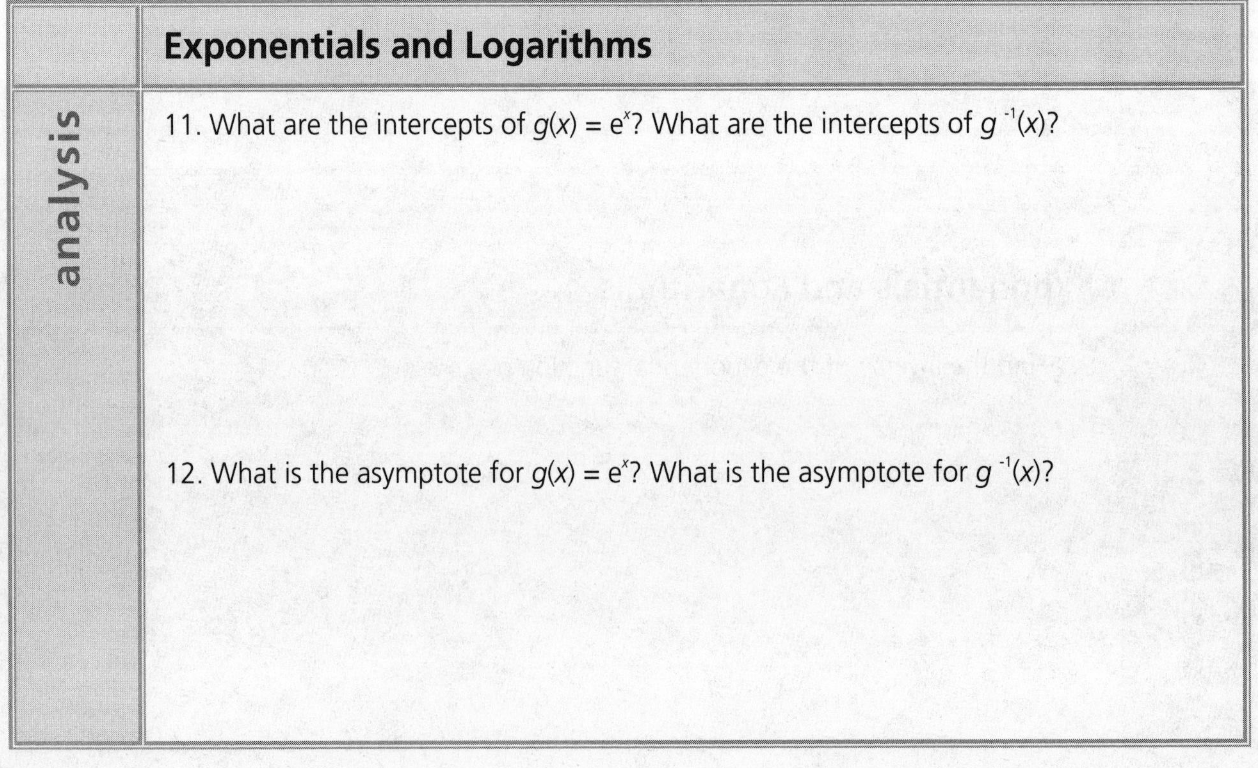

## Exponentials and Logarithms

analysis

11. What are the intercepts of $g(x) = e^x$? What are the intercepts of $g^{-1}(x)$?

12. What is the asymptote for $g(x) = e^x$? What is the asymptote for $g^{-1}(x)$?

## Exponentials and Logarithms

analysis

13. Describe the similarities and differences for the graphs of $f(x)$ and $g(x)$.

14. Describe the similarities and differences for the graphs of $f^{-1}(x)$ and $g^{-1}(x)$.

15. Describe the relationship between an exponential function and its inverse in general.

## Becoming More Profitable

input

Your manager holds a meeting with all of the company's employees. He announces that even though the new company has just finished its first year, the company has made a profit. The profit is only $100, but your manager assures everyone that if the company doubles its annual profit each year, he will be satisfied.

## Becoming More Profitable

process: model

1. Assuming the annual profit doubles each year, complete the table below for the problem situation.

| Number of Years | Profit ($) |
|---|---|
| 1 | |
| 2 | |
| 3 | |
| 4 | |
| 5 | |
| 6 | |

2. Write an algebraic function that models the annual profit based on the number of years.

process: evaluate

## Becoming More Profitable

3. By doubling the profit each year, what annual profit is expected after 8 years?

4. By doubling the profit each year, what annual profit is expected after 10 years?

process: model

## Becoming More Profitable

5. Find an algebraic function that models the inverse of the function.

6. What does the inverse function represent in terms of the problem situation?

## Becoming More Profitable

process: model

7. If the company can continue to double its profit each year, how long will it take for the company to have an annual profit of at least $100,000?

8. If the company can continue to double its profit each year, how long will it take for the company to have an annual profit of a quarter million dollars?

9. If the company can continue to double its profit each year, how long will it take for the company to have an annual profit of one million dollars?

## E. Coli

input

*Escherichia coli* (E. coli) is a bacterium that is a common inhabitant of the stomach of warm-blooded animals, including humans. Most strains of E. coli are harmless. However, some strains such as *E. coli* O157:H7 can cause illness. Harmful E. coli bacteria can be transmitted to humans through contaminated food. Good hygiene and thorough cooking of food help prevent the spread of harmful bacteria.

A microbiologist studying E. coli starts a culture with only one E. coli bacterium. Under ideal conditions for growth (proper temperature and available nutrients), the E. coli bacteria population can quadruple in one hour.

## E. Coli

process: model

1. Complete the table below for the problem situation.

| Time | Number of Bacteria |
|---|---|
| Hours | Bacteria |
| 0 | 1 |
| | |
| | |
| | |
| | |
| | |

2. Write an algebraic function to model the situation.

process: model

## E. Coli

3. Create a graph of the function for the situation.

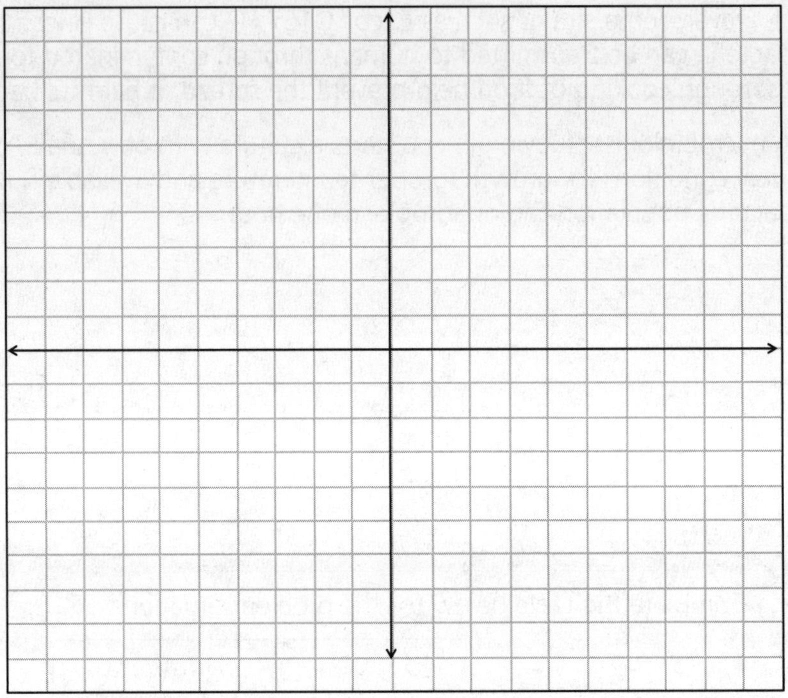

process: evaluate

## E. Coli

4. How many E. Coli bacteria will be present after 8 hours?

5. How many E. Coli bacteria will be present after 2 hours and 30 minutes?

6. How many E. Coli bacteria will be present after 4.25 hours?

| | E. Coli |
|---|---|
| process: model | 7. Find an algebraic function that models the inverse of this situation. |

| | E. Coli |
|---|---|
| process: evaluate | 8. How long would it take before there were 16,384 bacteria?<br><br>9. How long would it take before there were 100,000 bacteria?<br><br>10. How long would it take before there were one million bacteria? |

extension

## E. Coli

To create a culture, bacteria can be added to a fresh medium for growth. This problem only considered the exponential phase; however, a bacterial population typically goes through four stages:

- Lag phase
- Exponential (or logarithmic) phase
- Stationary phase
- Death (or decline) phase

Stationary Phase

Exponential Phase

Death Phase

Lag Phase

Bacterial Population

Time

extension

## E. Coli

11. Describe the graph showing the four phases of a bacterial culture population growth.

## E. Coli

extension

12. What do you think accounts for the different stages in the bacterial culture population growth?
    a. Lag phase

    b. Exponential phase

    c. Stationary phase

    d. Death phase

## Inverses of Rational Functions

process: model

1. Graph the function, $f(x)=\dfrac{3}{x+2}$.

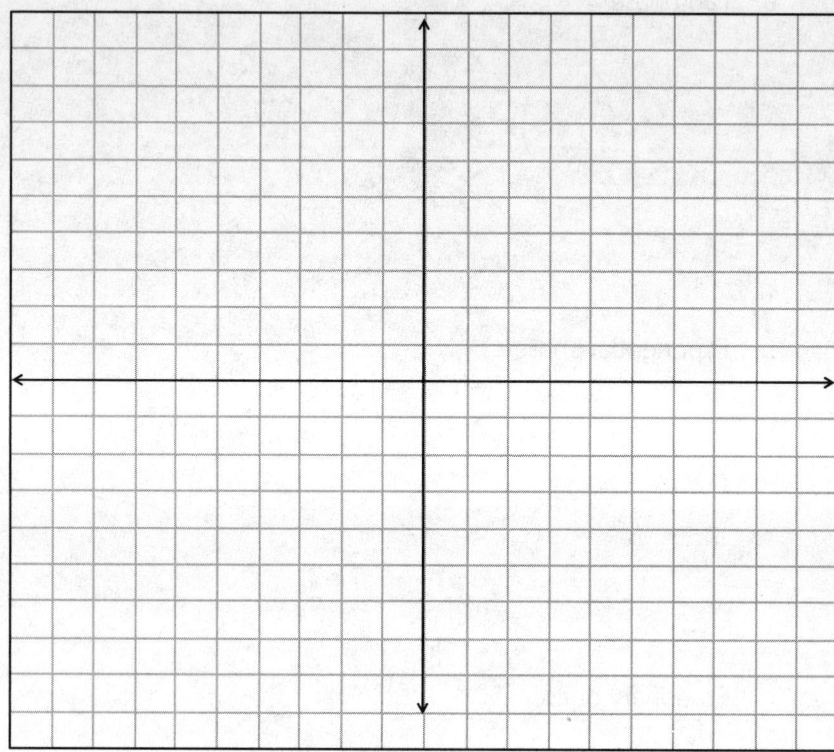

2. For the function, $f(x)=\dfrac{3}{x+2}$:

   a. What is the domain?

   b. What is the range?

   c. What are the intercepts?

   d. What are the asymptotes?

process: model

## Inverses of Rational Functions

3. Find the inverse of $f(x) = \dfrac{3}{x+2}$ algebraically.

4. Graph $f^{-1}(x)$ on the same grid as $f(x)$.

5. For the inverse function, $f^{-1}(x)$:
   a. What is the domain?

   b. What is the range?

   c. What are the intercepts?

   d. What are the asymptotes?

analysis

## Inverses of Rational Functions

6. Compare your answers to questions 2 and 5. What do you notice?

process: model

## Inverses of Rational Functions

7. Graph the function, $g(x) = \dfrac{x+1}{x-5}$.

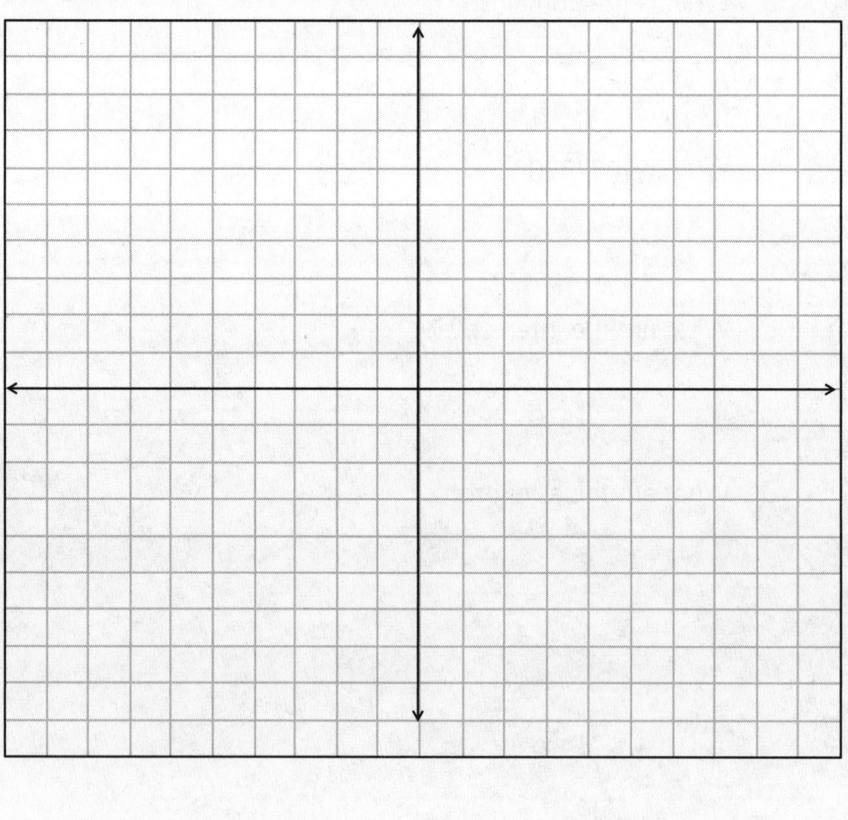

## Inverses of Rational Functions

process: model

8. For the function, $g(x) = \dfrac{x+1}{x-5}$:

   a. What is the domain?

   b. What is the range?

   c. What are the intercepts?

   d. What are the asymptotes?

9. Without finding the inverse of g(x), answer the following questions about $g^{-1}(x)$.

   a. What is the domain?

   b. What is the range?

   c. What are the intercepts?

   d. What are the asymptotes?

10. Sketch the graph of $g^{-1}(x)$ on the same grid as $g(x)$.

process: model

## Inverses of Rational Functions

11. Find $g^{-1}(x)$ algebraically.

analysis

## Inverses of Rational Functions

12. How does the description of the graph of a rational function relate to the description of its inverse?

process: practice

## Inverses of Rational Functions

13. Sketch a graph of the inverse of each rational function graphed below. Sketch the inverse on the same grid as its function.

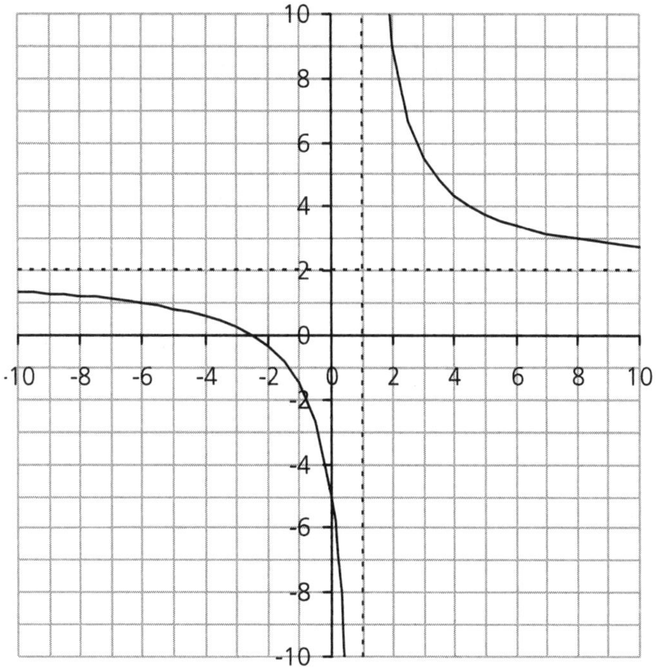

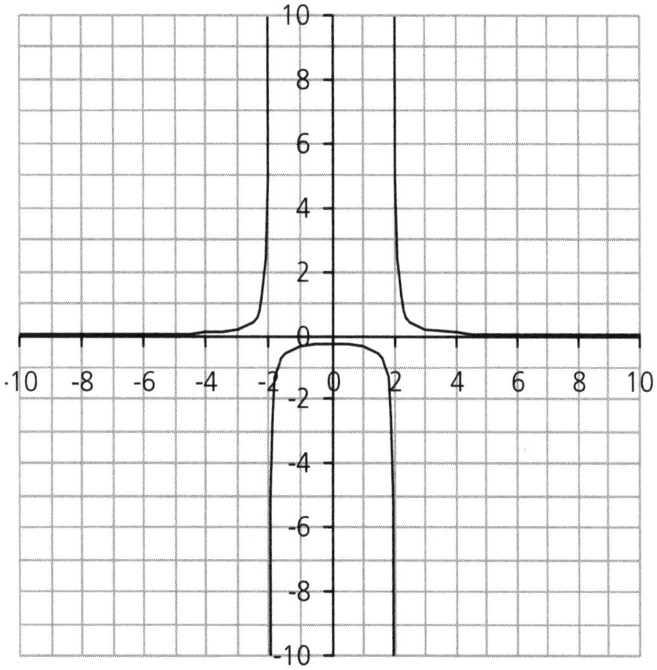

## Inverses of Rational Functions

process: practice

14. Determine the inverse of each of the following functions algebraically.

a. $f(x) = \dfrac{4}{x-3}$

b. b. $g(x) = \dfrac{1}{2x}$

c. $h(x) = \dfrac{x}{x+2}$

d. $j(x) = \dfrac{2x+5}{x-2}$

extension

## Inverses of Rational Functions

Sometimes, determining the inverse of a rational function algebraically can be more difficult. Consider the following example.

Find the inverse of $f(x)=\dfrac{4x^2}{x+3}$.

$$x=\frac{4y^2}{y+3}$$
$$x(y+3)=4y^2$$
$$xy+3x=4y^2$$
$$0=4y^2-xy-3x$$

Solving for $y$ using the quadratic formula, $a=4$, $b=-x$, and $c=-3x$.

$$y=\frac{x\pm\sqrt{(-x)^2-4(4)(-3x)}}{2(4)}$$
$$y=\frac{x\pm\sqrt{x^2+48x}}{8}$$

extension

## Inverses of Rational Functions

15. Find the inverses of the following functions algebraically.

a. $f(x)=\dfrac{3x}{2x^2-1}$

b. $g(x)=\dfrac{x^2+1}{x}$

## Drama Club Fundraiser

input

For a fundraising project, your drama club is selling posters from Broadway shows. The cost of the digital images and the rights to use them is $600. In addition to this one-time cost, the unit cost to print each poster is $1.50.

You head the committee that must decide how many posters to order for sale and how much to charge for the posters.

## Drama Club Fundraiser

process: model

1. Write an algebraic function that models the average cost per poster based on the number of posters printed.

2. Create a table of values based on your algebraic model

| Number of Posters | Average Cost ($) |
|---|---|
| | |
| | |
| | |
| | |
| | |
| | |

## Drama Club Fundraiser

process: model

3. Create a graphic model of your algebraic function.

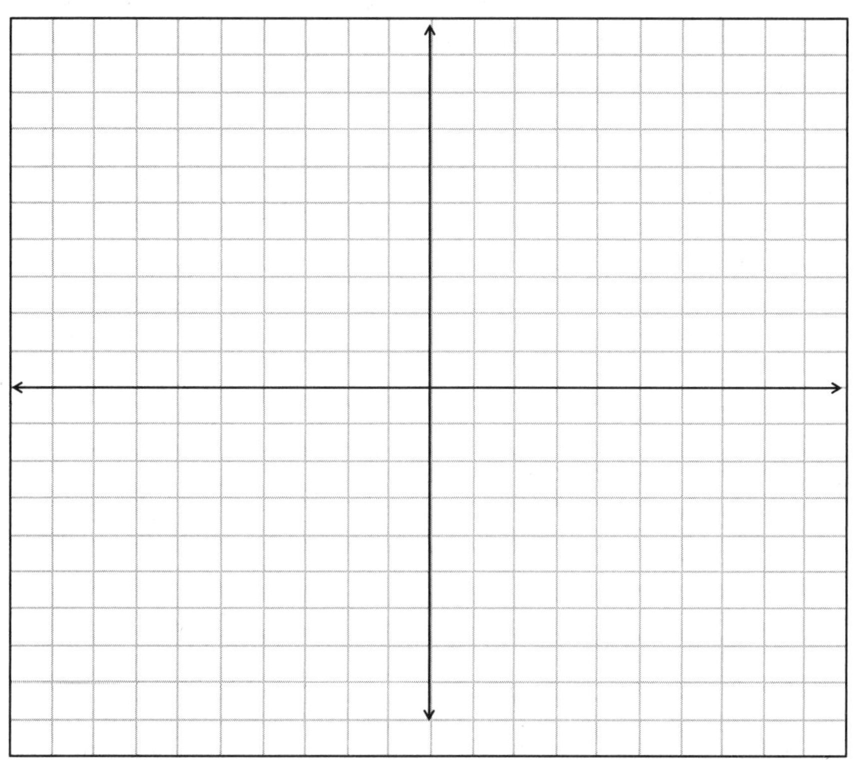

4. Describe the function in terms of domain and range, intercepts, asymptotes, and continuity. Differentiate between the mathematical model and the real problem situation.

process: model

## Drama Club Fundraiser

5. Find the inverse of the function that models the average cost per poster based on the number of posters printed.

process: evaluate

## Drama Club Fundraiser

6. Determine how many posters need to be printed to bring the average cost per poster down to:

   a. $5.00

   b. $4.00

   c. $2.00

## Drama Club Fundraiser

analysis

7. There are 40 members of the drama club and about 600 students in your high school. Consider how many posters you think the members of the club can sell. Make a recommendation about how many posters to print and how much to charge per poster. Write a paragraph explaining your recommendation. Use mathematics to support your decision. Include estimates of your club's profits from the sale.

# Unit 4: Polynomial and Rational Functions

## Contents

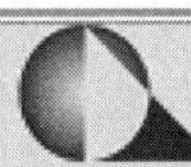

## Adding and Subtracting Polynomials

preview

Total attendance per year at American League baseball games can be modeled by the following polynomial function.

$$y = 8350x^3 - 386{,}040x^2 + 2{,}423{,}160x + 24{,}414{,}310$$

Total attendance per year at National League baseball games can be modeled by the following polynomial function.

$$y = 19{,}320x^3 - 266{,}310x^2 + 1.195.580x + 22{,}655{,}750$$

In both models, $x$ represents the time since 1985 in years and $y$ represents the total attendance for the year.

1. According to the model, what was the total attendance in 1990 for the American League and the National League? What was the total Major League attendance in 1990?

2. According to the model, what was the total attendance in 1995 for the American League and the National League? What was the total Major League attendance in 1995?

3. Write a polynomial function to compute the total Major League attendance.

## Adding and Subtracting Polynomials

input

Polynomials are added or subtracted by combining like terms. Two terms are considered **like terms** if they involve the same variables raised to the same exponents.

Consider the sum of polynomials $(3x^2 + 5x + 1) + (x^2 + 2x + 4)$

You can use the Commutative Property to rewrite the expression so like terms are grouped together.

$(3x^2 + x^2) + (5x + 2x) + (1 + 4)$

Like terms are combined by adding or subtracting the constants for each like term while holding the variable fixed.

$4x^2 + 7x + 5$

The sum of the polynomials $(3x^2 + 5x + 1) + (x^2 + 2x + 4)$ is equal to $4x^2 + 7x + 5$.

When subtracting polynomials, it is often easier to distribute the negative sign prior to grouping and combing like terms. Consider the difference of polynomials $(2x^2 - 7x + 3) - (4x^2 + x - 1)$.

Distributing the negative sign gives the following result.

$2x^2 - 7x + 3 - 4x^2 - x + 1$

As with addition of polynomials, you can use the Commutative property to rewrite the expression so like terms are grouped together.

$(2x^2 - 4x^2) + (-7x - x) + (3 + 1)$

Like terms are combined by adding or subtracting the constants for each like term while holding the variable fixed.

$-2x^2 - 8x + 4$

The difference of the polynomials $(2x^2 - 7x + 3) - (4x^2 + x - 1)$ is equal to $-2x^2 - 8x + 4$.

## Adding and Subtracting Polynomials

process: practice

4. Simplify each expression by finding the sum or difference.

   a. $(-3x^4 + 2x^2 - 6) + (x^3 - x^2 + 4)$

   b. $(3r^2 - 4rs + s^2) - (2r^2 + rs - 3s^2)$

   c. $(-3x^4 - x^3 + 6x^2) - (-3x^3 - x^2 + 6x)$

   d. $(x^2 + xy + 3x^2y - 3y^2) + (2x^2 - 6xy^2 + y^2 - 4xy)$

   e. $(3x^2 + 6x) + (-5x^2 + 3x) - (5x^2 + 5x)$

   f. $(-3x^4 - x^3 + 6x^2) + (3x^4 + 2x^3) - (-x^3 + 6x^2 - 6x) + (3x^4 + 7x^3 + 3x^2)$

   g. $(-x^3 - 2x^2) - (3x^4 - 2x^3 + 6x^2) - (-7x^3 - 3x) - (9x^4 - 3x^3 + x^2)$

process: practice

## Adding and Subtracting Polynomials

The number of male students participating in high school sports can be modeled as

$$y = 172.28x^3 - 4982.27x^2 + 2224.53x + 3{,}926{,}420.22$$

The number of female students participating in high school sports can be modeled as

$$y = -38.88x^4 + 2809.59x^3 - 67{,}574.84x^2 + 644{,}262.77x - 209{,}594.90$$

In both models, $x$ is the time since 1970 in years and $y$ is the number of students participating in high school sports.

5. Write a function to compute the total number of male and female students participating in high school sports.

6. According to the models, how many males participated in high school sports in 1975? How many females? How many total students?

7. According to the models, how many males participated in high school sports in 1995? How many females? How many total students?

8. According to the models, how many males will participate in high school sports in 2005? How many females? How many total students?

## Multiplying Polynomials

input

You can multiply polynomials using the Distributive Law and properties of exponents.

Consider the product of a monomial and a polynomial, such as $2x^2(x^3 - 5x^2 - 3x + 1)$.

To multiply, distribute the monomial $2x^2$ over the polynomial $x^3 - 5x^2 - 3x + 1$.

$$2x^2(x^3) + 2x^2(-5x^2) + 2x^2(-3x) + 2x^2(1)$$

By applying properties of exponents, you can simplify the expression to

$$2x^5 - 10x^4 - 6x^3 + 2x^2.$$

## Multiplying Polynomials

process: practice

1. Compute each product and simplify.

   a. $3x(2x^5 - 3x^3 - 2x + 5)$

   b. $-2ab(a^2 + 3a^2b - ab + 5ab^2 - 4b^2)$

   c. $a^2b(4a^2 - 2ab - 8ab^2 + 3b^2)$

input

## Multiplying Polynomials

Consider the product of a polynomial and a polynomial, such as $(2x^2 + 1)(3x^2 - x + 2)$.

To multiply, distribute the binomial $2x^2 + 1$ over the polynomial $3x^2 - x + 2$.

$$(2x^2 + 1)(3x^2) + (2x^2 + 1)(-x) + (2x^2 + 1)(2)$$

You must then apply a second distribution.

$$(6x^4 + 3x^2) + (-2x^3 - x) + (4x^2 + 2)$$

Combine the like terms.

$$6x^4 - 2x^3 + 7x^2 - x + 2$$

process: practice

## Multiplying Polynomials

2. Compute each product and simplify.
   a. $(3x + 1)(x^2 + 2x - 4)$

   b. $(a + b)(a^2 - 2ab + 3b^2)$

   c. $(3x^2 - 2x - 2)(x^2 + x - 3)$

input

## Multiplying Polynomials

Consider a polynomial raised to a power, such as $\left(3x^2 - 2x + 1\right)^2$.

By applying the definition of an exponent, you can rewrite the expression as

$(3x^2 - 2x + 1)(3x^2 - 2x + 1)$.

To multiply, distribute the polynomial $3x^2 - 2x + 1$ over the polynomial $3x^2 - 2x + 1$.

$(3x^2 - 2x + 1)(3x^2) + (3x^2 - 2x + 1)(-2x) + (3x^2 - 2x + 1)(1)$

You must then apply a second distribution.

$(9x^4 - 6x^3 + 3x^2) + (-6x^3 + 4x^2 - 2x) + (3x^2 - 2x + 1)$

Combine the like terms.

$9x^4 - 12x^3 + 10x^2 - 4x + 1$

process: practice

## Multiplying Polynomials

3. Simplify each expression.

   a. $\left(2x + 1\right)^2$

   b. $\left(x^2 - x - 1\right)^2$

   c. $\left(x + 1\right)^3$

process: practice

## Multiplying Polynomials

4. Simplify each expression.

   a. $(x^2 - y^2)(x - y)$

   b. $(-3x + 2)^2(-2x + 5)$

   c. $(3x - y)^3$

process: practice

## Multiplying Polynomials

d. $(3x^2 + 4x - 1)(x^3 - 2x^2 - 4x + 7)$

e. $(x+1)^4$

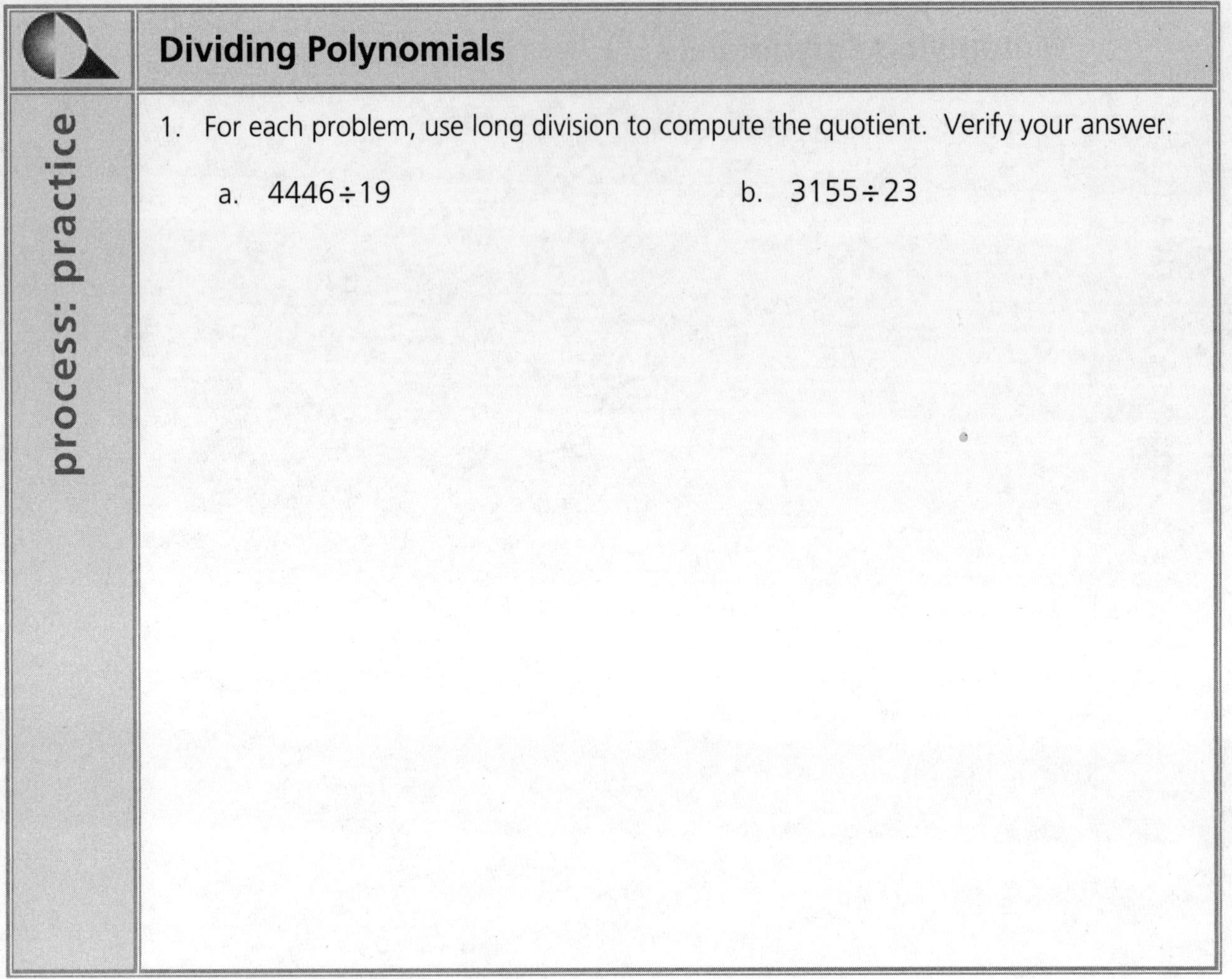

## Dividing Polynomials

process: practice

1. For each problem, use long division to compute the quotient. Verify your answer.

   a. $4446 \div 19$

   b. $3155 \div 23$

## Dividing Polynomials

analysis

2. You are volunteering at an after school tutor center. One of the students is beginning to learn about long division. How would you explain how to solve each of the two problems above?

input

## Dividing Polynomials

You can divide polynomials using long division as well.

Consider the problem $(x^2+4x+3)\div(x+3)$ where $(x^2 + 4x + 3)$ is the dividend and $(x + 3)$ is the divisor. You can rewrite the problem as

$$x+3\overline{)x^2+4x+3}$$

What term can you multiply with the first term of the divisor, $x$, to get the first term of the dividend, $x^2$? This term is placed above the $x$ term of the dividend.

$$\begin{array}{r} \boldsymbol{x}\phantom{+4x+3} \\ \boldsymbol{x}+3\overline{)\boldsymbol{x^2}+4x+3} \end{array}$$

Multiply the x in the quotient by the divisor. Place the result, $x^2+3x$, under the dividend and subtract. The next term of the dividend, 3, is brought down.

$$\begin{array}{r} x\phantom{+4x+3} \\ x+3\overline{)x^2+4x+3} \\ \underline{-(x^2+3x)}\downarrow \\ x+3 \end{array}$$

Repeat the process for the remaining portion of the dividend, x+3. The 1 is placed above the constant term of the divisor. The result of multiplying the 1 in the quotient by the divisor is $(x+3)$. Subtracting results in zero.

$$\begin{array}{r} x+1 \\ x+3\overline{)x^2+4x+3} \\ \underline{-(x^2+3x)}\phantom{+3} \\ x+3 \\ \underline{-(x+3)} \\ 0 \end{array}$$

You can check the quotient by performing multiplication.

$$\underbrace{(x+1)}_{\text{Quotient}}\underbrace{(x+3)}_{\text{Divisor}}=\underbrace{x^2+4x+3}_{\text{Dividend}}$$

process: practice

## Dividing Polynomials

3. Use long division to compute each quotient. Verify your answer.

   a. $(2x^2 - 5x - 3) \div (x - 3)$

   b. $(9x^2 - 4) \div (3x + 2)$

## Dividing Polynomials

process: practice

c. $\left(3x^3 - 8x^2 + 7x - 2\right) \div \left(x^2 - 2x + 1\right)$

d. $\left(x^3 + 3x^2y + 3xy^2 + y^3\right) \div \left(x + y\right)$

input

## Dividing Polynomials

Consider the problem $(4x^3+2x^2+3)\div(2x-1)$. You can rewrite the problem as

$$2x-1\overline{)4x^3+2x^2+0x+3}$$

What term can you multiply with the first term of the divisor, $2x$, to get the first term of the dividend, $4x^3$? This term is placed above the $x^2$ term of the dividend. Multiply the $2x^2$ in the quotient by the divisor. Place the result, $4x^3-2x^2$, under the dividend and subtract. The next term of the dividend, $0x$, is brought down.

$$\begin{array}{r} 2x^2\phantom{+2x^2+0x+3} \\ 2x-1\overline{)4x^3+2x^2+0x+3} \\ \underline{-(4x^3-2x^2)}\downarrow\phantom{+3} \\ 4x^2+0x\phantom{+3} \end{array}$$

Repeat the process to complete the problem.

$$\begin{array}{r} 2x^2+2x+1 \\ 2x-1\overline{)4x^3+2x^2+0x+3} \\ \underline{-(4x^3-2x^2)}\downarrow\phantom{+0x+3} \\ 4x^2+0x\phantom{+3} \\ \underline{-(4x^2-2x)}\downarrow \\ 2x+3 \\ \underline{-(2x-1)} \\ 4 \end{array}$$

The quotient is written as $2x^2+2x+1+\dfrac{4}{2x-1}$

You can check the quotient by performing multiplication.

$$\underbrace{(2x^2+2x+1)}_{\text{Quotient}}\underbrace{(2x-1)}_{\text{Divisor}}+\underbrace{4}_{\text{Remainder}}=\underbrace{4x^3+2x^2+3}_{\text{Dividend}}$$

process: practice

## Dividing Polynomials

4. Use long division to compute each quotient. Verify your answer.

   a. $(x^2 - x - 4) \div (x + 1)$

   b. $(6x^3 - x^2 + x - 3) \div (3x - 2)$

process: practice

## Dividing Polynomials

c. $\left(4x^4 + x^3 - 12x^2 + x + 5\right) \div \left(4x + 1\right)$

d. $\left(x^3 - 1\right) \div \left(x + 1\right)$

## U.S. Shirts Revisited

input

As a member of the student council, you must help organize the annual fund drive to raise money for the prom. This year, the council decides to sell t-shirts with the school's mascot on the front.

A local T-shirt shop, U.S. Shirts, charges $8 per shirt plus a one-time charge of $15 to set up the design.

## U.S. Shirts Revisited

process: model

1. Complete the table below.

| Number of Shirts Sold | Total Cost of All Shirts | Average Cost of a Shirt |
|---|---|---|
| Shirts | Dollars | Dollars/Shirt |
| 50 | | |
| 100 | | |
| 150 | | |
| 200 | | |
| 250 | | |
| 300 | | |
| 350 | | |
| 400 | | |
| 450 | | |
| 500 | | |
| 550 | | |
| 600 | | |
| 650 | | |
| 700 | | |
| 750 | | |

process: model

## U.S. Shirts Revisited

2. Write an algebraic equation to calculate the total cost of the shirts.

3. Write an algebraic equation to calculate the average cost of a shirt.

analysis

## U.S. Shirts Revisited

4. What is similar about the two equations you wrote?

5. What is different about the two equations you wrote?

## U.S. Shirts Revisited

process: model

6. Create a graph for the total cost of the shirts.

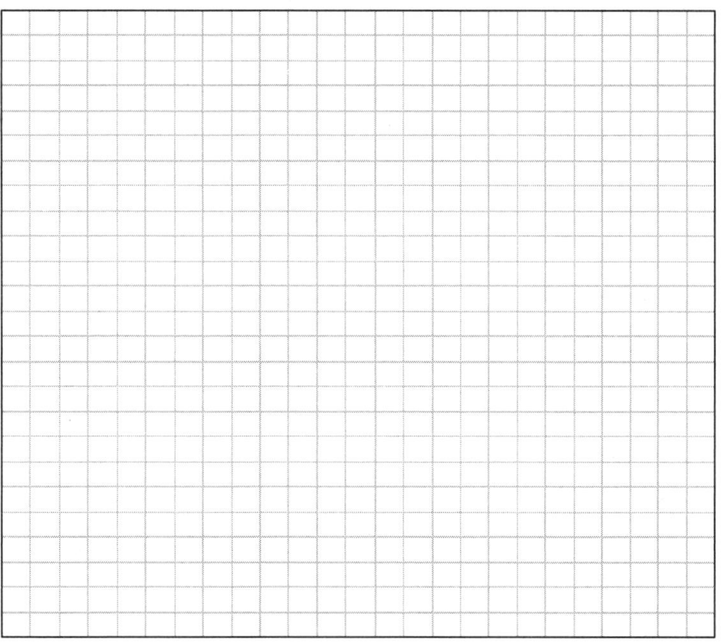

7. Create a graph for the average cost of a shirt.

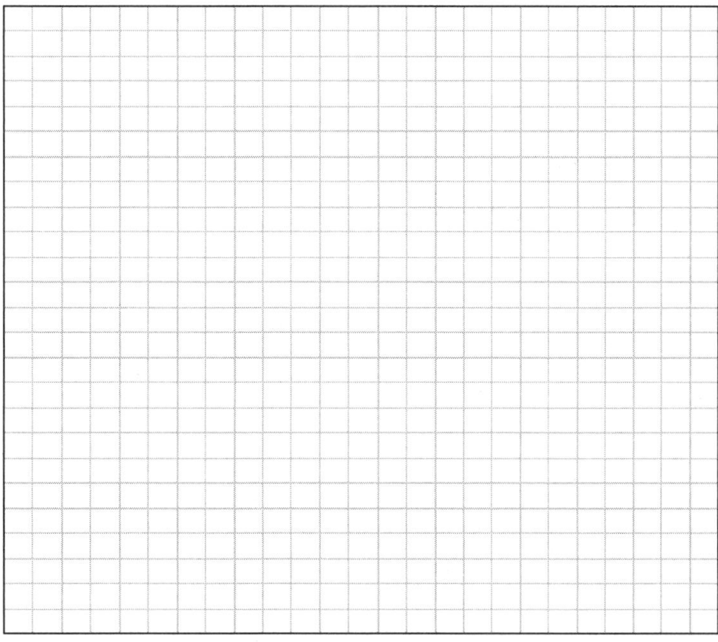

## U.S. Shirts Revisited

analysis

8. What is similar about the two graphs?

9. What is different about the two graphs?

10. What is the average cost of a shirt if you order zero shirts? Use the equation to explain why this is the case.

11. What is the lowest average cost you can get? Use your graph to explain why this is the case.

## Simplifying Rational Expressions

input

The function to describe average cost is one example of a rational function. A **rational function** is defined as the quotient of two polynomial expressions.

Operations on rational functions, such as simplifying, addition, subtraction, multiplication and division, are very similar to operations on rational numbers or fractions.

## Simplifying Rational Expressions

analysis

1. Based on the definition, are the following rational expressions? Explain why or why not.

a. $\frac{x^2+2x+6}{x-5}$

d. $\frac{4x^2-3x+7}{2x^3}$

b. $\frac{1}{x+3}$

e. $\frac{2^x}{x^2-4}$

c. $\frac{\sqrt{x}}{2}$

f. $\frac{(x-3)(x+2)}{(x-5)(x+4)}$

process: practice

## Simplifying Rational Expressions

2. Simplify each fraction below.

a. $\frac{2}{4}$

c. $\frac{44}{121}$

b. $\frac{12}{18}$

d. $\frac{40}{72}$

analysis

## Simplifying Rational Expressions

3. Explain the procedure for simplifying a fraction. Provide an example to demonstrate.

## Simplifying Rational Expressions

input

To simplify a rational expression such as $\frac{4x^2-10x-6}{6x^2-12x-18}$, factor both the numerator and denominator completely.

The first step in factoring involves factoring the greatest common factor (GCF) of the numerator and denominator. Rewrite the rational expression after factoring the GCF.

The second step in factoring involves factoring the remaining trinomials in the numerator and denominator. Rewrite the rational expression after factoring the remaining trinomials in the numerator and denominator.

After factoring both the numerator and denominator completely, cancel common factors from the numerator and denominator.

What common factors appear in both the numerator and denominator?

Rewrite the rational expression after canceling the common factors from the numerator and denominator.

## Simplifying Rational Expressions

process: practice

4. Factor the numerator and denominator of each rational expression and simplify completely.

a. $\dfrac{3xy^2}{9x^2y}$

b. $\dfrac{2x^3+2x^2}{x^2+x}$

c. $\dfrac{x^2+2x-3}{2x^2-3x+1}$

d. $\dfrac{x^2-9}{2x-6}$

e. $\dfrac{3x^3+x^2-10x}{3x^2-8x+5}$

f. $\dfrac{4x^4-2x^3-6x^2}{16x^3-36x}$

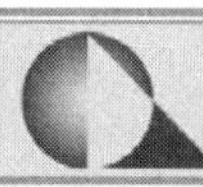

## Multiplying Rational Expressions

process: practice

1. Compute each product and simplify.

a. $\frac{2}{3} \cdot \frac{1}{4}$

c. $\frac{49}{100} \cdot \frac{4}{7}$

b. $\frac{24}{35} \cdot \frac{7}{8}$

d. $\frac{105}{144} \cdot \frac{18}{35}$

## Multiplying Rational Expressions

analysis

2. Explain the procedure for multiplying two fractions. Provide an example to demonstrate.

## Multiplying Rational Expressions

input

To multiply two rational expressions such as $\frac{2x+12}{5x^2+25x-30} \bullet \frac{5x^2+40x-45}{4x+12}$, factor the numerator and denominator of both fractions completely.

Rewrite both rational expressions in factored form by factoring the greatest common factor.

Rewrite both rational expressions in factored form by factoring the remaining trinomial.

Cancel all possible factors from the numerator and denominator of both fractions.

Perform the multiplication by multiplying the numerators and multiplying the denominators. Write the resulting product.

## Multiplying Rational Expressions

process: practice

3. Compute each product and simplify completely.

a. $\dfrac{2x+1}{3} \bullet \dfrac{6x}{4x^2-1}$

b. $\dfrac{d-9}{d^2-15d+56} \bullet \dfrac{d^2-d+56}{2d+14}$

c. $\dfrac{3x+36}{4x^2-36x} \bullet \dfrac{8x^3-72x^2}{5x+60}$

d. $\dfrac{4w+1}{5w+10} \bullet \dfrac{30w+60}{2w-2}$

e. $\dfrac{6x^2+30x+36}{x^2+7x+12} \bullet \dfrac{x^2+9x+20}{3x^2+21x+30}$

## Dividing Rational Expressions

process: practice

1. Compute each quotient and simplify.

a. $\frac{2}{3} \div \frac{4}{9}$

c. $\frac{56}{121} \div \frac{7}{11}$

b. $\frac{7}{8} \div \frac{35}{64}$

d. $\frac{105}{180} \div \frac{75}{40}$

## Dividing Rational Expressions

analysis

2. Explain the procedure for dividing two fractions. Provide an example to demonstrate.

input

## Dividing Rational Expressions

To divide two rational expressions such as $\frac{x^2+8x+16}{x+2} \div \frac{x^2+2x-8}{x^2-4}$, multiply the first rational expression by the inverse of the second rational expression. Rewrite the original problem after performing this step.

After rewriting the expression, the product can now be found using the procedure for multiplying two rational expressions.

Complete the problem below.

process: practice

## Dividing Rational Expressions

3. Compute the quotient and simplify completely.

   a. $\frac{x^2+6x-7}{3x^2} \div \frac{x+7}{6x}$

process: practice

## Dividing Rational Expressions

b. $\frac{x^2-14x+48}{x^2-6x} \div (3x-24)$

c. $\frac{2x^2-12x}{x^2-7x+6} \div \frac{2x}{3x-3}$

d. $\frac{3x+36}{4x^2-36x} \div \frac{5x+60}{8x^3-72x^2}$

e. $\frac{5x^2+10x}{x^2-x-6} \div \frac{15x^3+45x^2}{x^2-9}$

f. $\frac{5x}{3x-12} \div \frac{x^2-2x}{x^2-6x+8}$

## Adding and Subtracting Rational Expressions

process: practice

1. Add or subtract each fraction and simplify.

a. $\frac{2}{4}+\frac{1}{4}$

b. $\frac{2}{7}+\frac{4}{7}$

c. $\frac{7}{12}-\frac{4}{12}$

d. $\frac{13}{24}-\frac{11}{24}$

## Adding and Subtracting Rational Expressions

analysis

2. Explain the procedure for adding or subtracting two fractions with a common denominator. Provide an example to demonstrate.

## Adding and Subtracting Rational Expressions

input

To add or subtract two rational expressions such as $\frac{x^2}{x^2-x-2}-\frac{2x+3}{x^2-x-2}$, keep the common denominator and add or subtract the numerators.

Write the difference of the two rational expressions.

Distribute the subtraction over the quantity $2x+3$. Rewrite the rational expression after the distribution.

The final step involves simplifying the sum or difference. Recall that simplifying involves factoring both the numerator and denominator completely and canceling any common factors.

Factor the numerator and denominator of the rational expression.

Rewrite the rational expression after canceling the common factors from the numerator and denominator.

## Adding and Subtracting Rational Expressions

process: practice

3. Compute the sum or difference and simplify completely.

   a. $\frac{4x^2}{9}+\frac{2x^2}{9}$

   b. $\frac{2a+5}{a+1}+\frac{a-4}{a+1}$

   c. $\frac{4s^2}{2s+1}-\frac{1}{2s+1}$

   d. $\frac{2t+1}{t-1}-\frac{t+2}{t-1}$

   e. $\frac{6}{5x^3}+\frac{4}{5x^3}$

   f. $\frac{3c+2}{c+4}-\frac{c-6}{c+4}$

process: practice

## Adding and Subtracting Rational Expressions

4. Add or subtract each fraction and simplify.

a. $\frac{1}{2}+\frac{1}{4}$

c. $\frac{3}{8}-\frac{5}{12}$

b. $\frac{2}{3}+\frac{3}{4}$

d. $\frac{11}{16}-\frac{5}{24}$

analysis

## Adding and Subtracting Rational Expressions

5. Explain the procedure for adding or subtracting two fractions without a common denominator. Provide an example to demonstrate.

input

## Adding and Subtracting Rational Expressions

To add or subtract two rational expressions without a common denominator such as $\frac{3}{x-1}-\frac{2}{x+2}$, determine the common denominator. What is the common denominator in this problem?

What form of one do you need to multiply the first rational expression by?

What form of one do you need to multiply the second rational expression by?

Rewrite the difference after multiplying each rational expression by a form of one.

Write the difference of the two rational expressions.

Perform distribution in the numerator. Rewrite the rational expression after the distribution.

Simplify the numerator.

Always attempt to simplify the result if possible. Is it possible to simplify the difference?

process: practice

## Adding and Subtracting Rational Expressions

6. Compute the sum or difference and simplify completely.

a. $\frac{7}{3p}+\frac{2}{5p^4}$

b. $\frac{4}{t+1}+\frac{t+2}{t+5}$

c. $\frac{4}{d+5}-\frac{3}{d-5}$

d. $\frac{1}{g+2}+\frac{1}{g^2+3g+2}$

e. $\frac{2}{x^2+4x+3}-\frac{1}{x^2+3x+2}$

preview

## Solving Rational Equations

You were always told that if you start counting when you see a bolt of lightning and stop counting when you hear the thunder, you could figure out how far the storm is from you. Each second represents a mile's distance between you and the storm.

In science class, however, you learned that the time between seeing the lightning and hearing thunder was not only a function of distance, but also of temperature.

The time ($t$) between seeing the lightning and hearing the thunder = $\frac{d}{1.09T + 1050}$.

In this equation, $d$ is the distance you are from the lightening in feet and $T$ is the temperature in Fahrenheit.

Suppose you know $t = 10$ seconds and $d = 2$ miles. What is the temperature in Fahrenheit?

input

## Solving Rational Equations

In general, solving a rational equation is similar to solving any polynomial equation. You want to find a value for the variable such that the equation is true. In the case of a rational equation, you will need to check the solution(s) to make sure they make sense in the problem. What would cause a solution to fail?

process: evaluate

## Solving Rational Equations

1. Using your knowledge of solving equations involving ratios, show how you would solve the rational equation $\frac{1}{a+7}=\frac{2}{a+2}$.

2. Use your knowledge of combining rational expressions and solving equations to solve the rational equation $\frac{1}{a+7}-\frac{7}{a^2+9a+14}=\frac{-2}{a+2}$.

## Solving Rational Equations

input

The two methods used to solve rational equations are formalized below.

Method 1

- Find the least common denominator (LCD).
- Multiply both sides of the equation by the LCD (each and every term) to transform the rational equations to polynomial form (clearing the denominator).
- Solve for the unknown using appropriate techniques.

Method 2

- Combine expressions as required on the left and right sides by applying the rules for combining expressions and writing equivalent expressions.
- Multiply by the reciprocal of the denominator on both sides to transform the rational equations to polynomial form (clearing the denominator).
- Solve for the unknown using appropriate techniques.

## Solving Rational Equations

process: practice

3. Solve the following rational equations for the variable and check the solutions.

$$\frac{4}{s-2} = 4$$

## Solving Rational Equations

process: practice

$$\frac{4}{x}+\frac{5}{2}=-\frac{11}{x}$$

$$\frac{2x}{x+3}=\frac{4}{x+1}$$

$$\frac{2}{y}+\frac{2y-6}{y-6}=\frac{2y-4}{y}$$

## Winning Percentage

input

School spirit is at an all time high as the school's baseball team makes its run for the state championship. Currently, the team has a record of 7 wins and 8 losses. The good news is that two key players have returned to the lineup after being injured and a number of the upcoming opponents have losing records. The coach believes that the team can win all of the remaining games on its schedule.

## Winning Percentage

process: evaluate

1. What is the team's current winning percentage?

2. What will be the team's winning percentage if it wins the next game?

3. What will be the team's winning percentage if it wins the next 2 games?

4. What will be the team's winning percentage if it wins the next 5 games?

5. What will be the team's winning percentage if it wins the next 10 games?

## Winning Percentage

process: model

6. Complete the table below.

| Number of additional games won in a row | Winning percentage |
|---|---|
| **Games** | **Percent** |
| 0 | |
| 1 | |
| 2 | |
| 3 | |
| 4 | |
| 5 | |
| 6 | |
| 7 | |
| 8 | |
| 9 | |
| 10 | |

7. Write an equation to calculate the team's winning percentage from the number of additional games won in a row.

## Winning Percentage

process: model

8. Graph the function below.

## Winning Percentage

process: evaluate

9. How many additional games would the team need to win in a row to have a winning percentage of 75%? Solve algebraically.

## Winning Percentage

process: evaluate

10. How many additional games would the team need to win in a row to have a winning percentage of 80%? Solve algebraically.

11. How many additional games would the team need to win in a row to have a winning percentage of 95%? Solve algebraically.

## Take Your Medicine

input

There are many "rules of thumb" to calculate the dosage of medicine for a child from the known dosage for an adult.

One rule is known as Young's Rule and is modeled by the equation

$$c = \frac{ad}{a+12}$$

Another rule, known as Webster's Rule, is modeled by the equation

$$c = \frac{d(a+1)}{a+7}$$

Both models use the following.

- $a$ is the child's age in years
- $d$ is the adult dosage
- $c$ is the corresponding child's dosage

Pain-B-Gone pain reliever recommends an adult dosage of 200 milligrams.

## Take Your Medicine

process: model

1. Write an equation to calculate the recommended child dosage of Pain-B-Gone using Young's Rule.

2. Write an equation to calculate the recommended child dosage of Pain-B-Gone using Webster's Rule.

process: model

## Take Your Medicine

3. Graph both equations on the same set of axes.

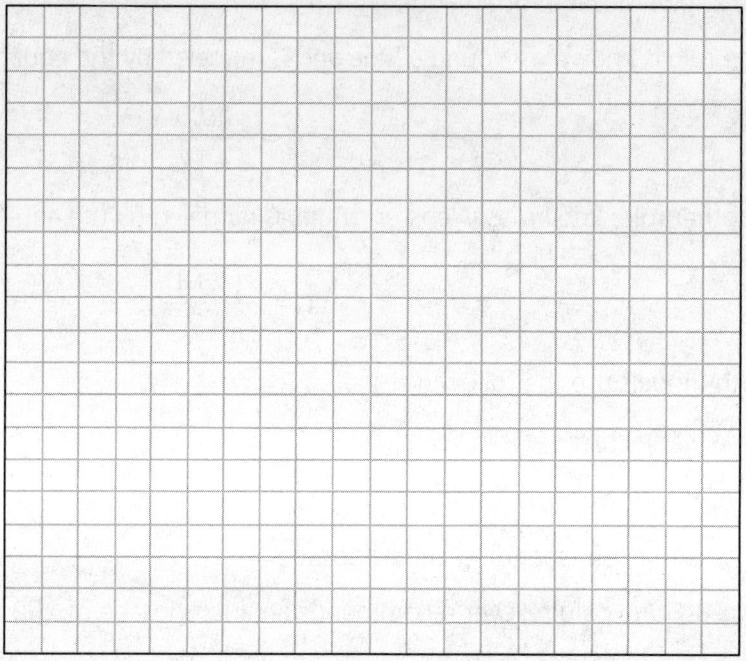

process: evaluate

## Take Your Medicine

4. How many milligrams of Pain-B-Gone should a 5-year old receive using Young's Rule? Webster's Rule?

process: evaluate

## Take Your Medicine

5. How many milligrams of Pain-B-Gone should a 13-year old receive using Young's Rule? Webster's Rule?

6. What age child should receive 75 milligrams of Pain-B-Gone according to Young's Rule? According to Webster's Rule? Do these answers make sense in terms of the problem situation?

process: evaluate

## Take Your Medicine

7. What age child should receive 150 milligrams of Pain-B-Gone according to Young's Rule? According to Webster's Rule? Do these answers make sense in terms of the problem situation?

8. Will Young's Rule and Webster's Rule ever recommend the same dosage for a child of a given age?

# Unit 5: Statistics and Probability

## Contents

## The Five-Number Summary

input

In a previous course you learned about box and whisker plots and the five statistics used to create a box and whisker plot for a set of data.

The **median** is the middle value of a set of data. Half of the values in the data set are greater than the median and half of the values are less than the median. To determine the median of a given set of data, order the values from least to greatest. If there are an odd number of values, then the middle score is the median. If there are an even number of scores, the median is halfway between the two scores closest to the middle value. The median is also called the **second quartile** or the **50$^{th}$ percentile**.

The **first quartile ($Q_1$)**, is the value at the **25$^{th}$ percentile** for a set of data meaning that 25% of the values are less than the first quartile. To determine the first quartile, consider only the values less than the median in a set of data. The median of this subset of values is the first quartile.

The **third quartile ($Q_3$)**, is the value at the **75$^{th}$ percentile** for a set of data meaning that 75% of the values are less than the third quartile. To determine the third quartile, consider only the values greater than the median in a set of data. The median of this subset of values is the third quartile.

The **minimum value** is the lowest value in a set of data.

The **maximum value** is the greatest value in a set of data.

A box and whiskers plot is useful to compare the relative positions of different data sets.

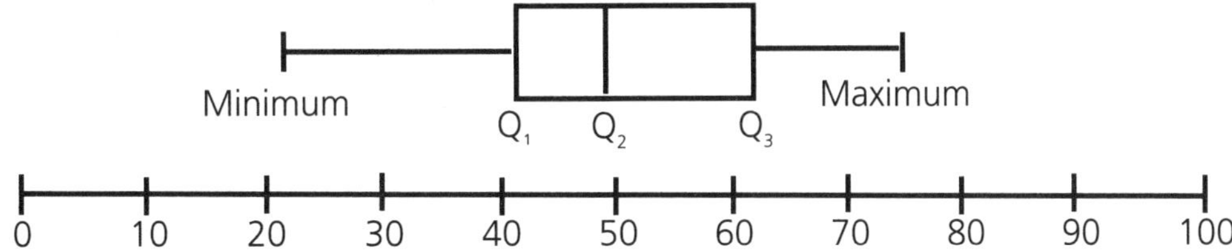


To create a box and whisker plot:

- draw a number line representing the full range of values
- mark the median, minimum, maximum, first quartile, and third quartile values
- draw a box between the first and third quartile
- Mark the median with a vertical line in the box
- Draw whiskers from the first quartile to the minimum and from the third quartile to the maximum

analysis

## The Five-Number Summary

1. Compare and contrast the two box and whisker plots below. What can you say about the data represented by the plots?

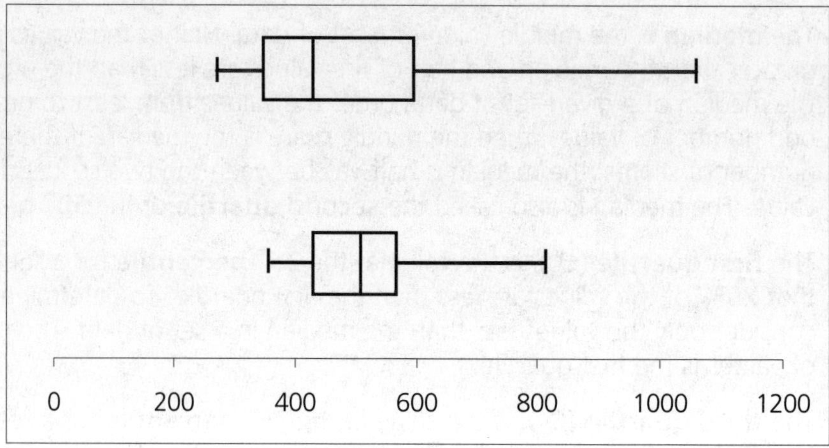


2. Answer the following questions based on the graph above.

   a. Approximately what is the median of each of the data sets represented in the graph?

   b. What are the approximate minimum and maximum values?

   c. What are the approximate interquartile ranges (IQR)?

| | **The Five-Number Summary** |
|---|---|
| input | The box and whisker plots on the previous page are based on calorie counts for sandwiches from Burger King, McDonald's, and Wendy's. Data was gathered from the restaurants' web sites in April 2004. The top plot represents calorie counts for hamburger sandwiches. The bottom plot represents calorie counts for chicken sandwiches. |

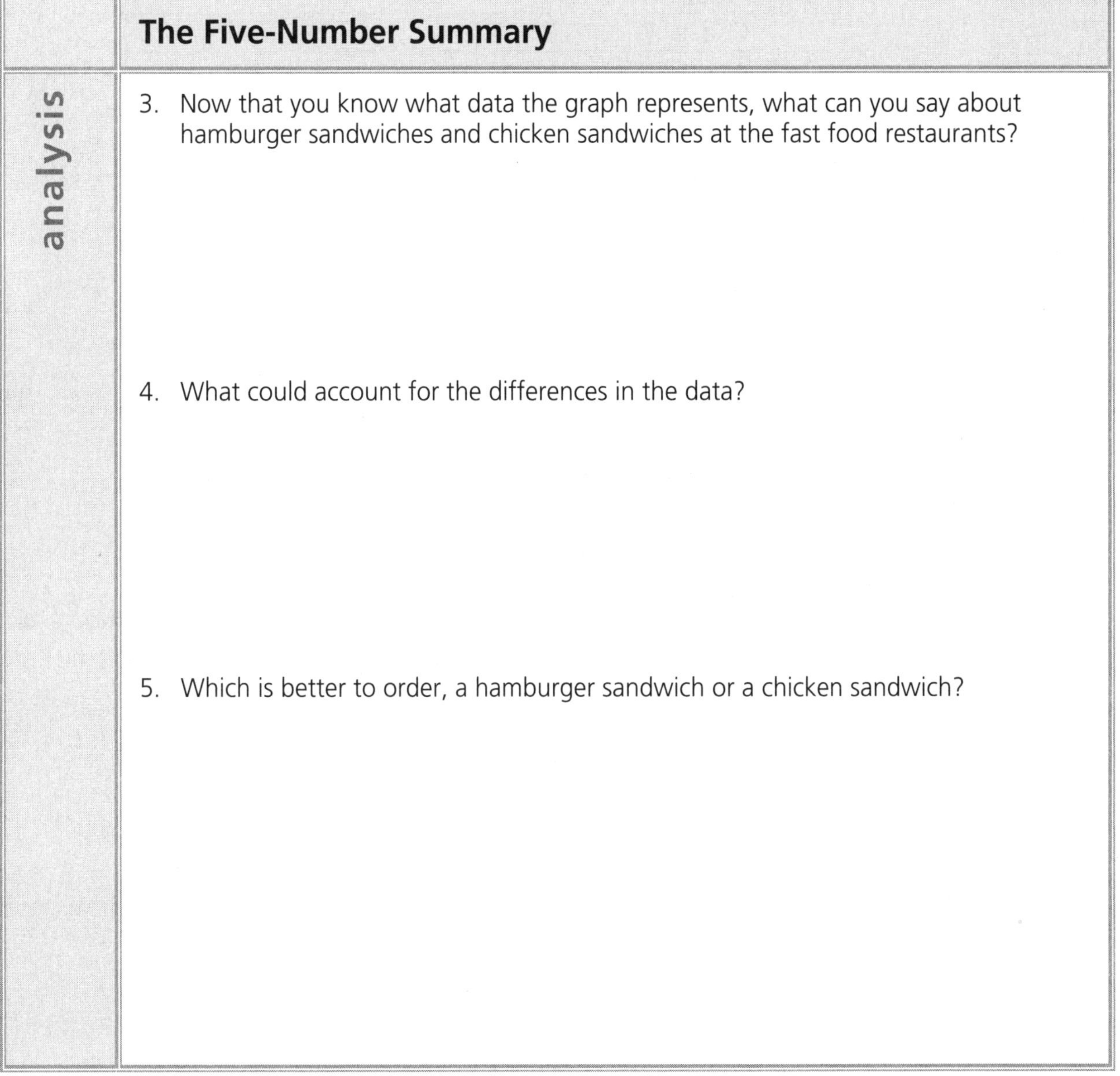

| | **The Five-Number Summary** |
|---|---|
| analysis | 3. Now that you know what data the graph represents, what can you say about hamburger sandwiches and chicken sandwiches at the fast food restaurants?<br><br>4. What could account for the differences in the data?<br><br>5. Which is better to order, a hamburger sandwich or a chicken sandwich? |

input

## The Five-Number Summary

The calorie counts for hamburger and chicken sandwiches from Burger King, gathered from their web site in April 2004 are listed below. Information is from the Nutrition Information for Menu Items download. Sandwiches not included in that document are not listed.

| Burger King Sandwich | Calorie Count |
|---|---|
| Original WHOPPER® | 700 |
| Original WHOPPER® With Cheese | 800 |
| Original DOUBLE WHOPPER® | 970 |
| Original DOUBLE WHOPPER® With Cheese | 1060 |
| Original WHOPPER JR.® | 390 |
| Original WHOPPER JR.® With Cheese | 430 |
| Hamburger | 310 |
| Cheeseburger | 350 |
| Double Hamburger | 440 |
| Double Cheeseburger | 530 |
| Bacon Cheeseburger | 390 |
| Bacon Double Cheeseburger | 570 |
| Chicken WHOPPER® | 570 |
| Original Chicken | 560 |
| TenderCrisp Chicken | 810 |
| Spicy TenderCrisp Chicken | 750 |

process: evaluate

## The Five-Number Summary

6. For the Burger King data above, determine the minimum, maximum, median, first quartile, and third quartile values.

input

## The Five-Number Summary

The calorie counts for hamburger and chicken sandwiches from Wendy's gathered from their web site in April 2004 are listed below. Information is from the Complete Nutrition Guide download. Sandwiches not included in that document are not listed.

| Wendy's Sandwich | Calorie Count |
|---|---|
| Jr. Hamburger | 270 |
| Jr. Cheeseburger | 310 |
| Jr. Cheeseburger Deluxe™ | 350 |
| Jr. Bacon Cheeseburger | 380 |
| Classic Single with Everything | 410 |
| Big Bacon Classic® | 580 |
| Ultimate Chicken Grill | 360 |
| Spicy Chicken Fillet | 510 |
| Homestyle Chicken Fillet | 540 |

process: evaluate

## The Five-Number Summary

7. For the Wendy's data above, determine the minimum, maximum, median, first quartile, and third quartile values.

input

## The Five-Number Summary

The calorie counts for hamburger and chicken sandwiches from McDonald's gathered from their web site in April 2004 are listed below. Information is from the Nutrition Facts download. Sandwiches not included in that document are not listed.

| McDonald's Sandwich | Calorie Count |
|---|---|
| Hamburger | 280 |
| Cheeseburger | 330 |
| Double Cheeseburger | 490 |
| Quarter Pounder® | 430 |
| Quarter Pounder® with Cheese | 540 |
| Double Quarter Pounder® with Cheese | 770 |
| Big Mac® | 600 |
| Big N' Tasty® | 540 |
| Big N' Tasty® with Cheese | 590 |
| Chicken McGrill® | 400 |
| Crispy Chicken | 510 |
| McChicken® | 430 |
| Hot 'n Spicy McChicken® | 450 |

process: evaluate

## The Five-Number Summary

8. For the McDonald's data above, determine the minimum, maximum, median, first quartile, and third quartile values.

process: model

## The Five-Number Summary

9. Create box and whisker plots for the calorie count data from each of the restaurants. Put all three plots on the same graph.

analysis

## The Five-Number Summary

10. What can you conclude about the calorie count of the sandwiches from the three restaurants based on your graph?

## Mean and Standard Deviation

input

In a previous course, you learned about a measure of central tendency called the mean and a measure of spread called the standard deviation. The combination of the mean and standard deviation is the most common way to describe a set of numerical data.

The **mean**, $\overline{x}$ (read "x – bar"), represents the sum of all values in a data set divided by the number of values. Note that $\overline{x}$ represents a sample mean; μ (the greek letter, mu) will be used to represent the mean of a whole population.

$$\overline{x} = \frac{\text{the sum of the observations}}{\text{the number of observations}} = \frac{\sum x}{n}$$

The **standard deviation**, σ, measures the average distance of individual observations from their mean. To calculate a standard deviation, find the distance of each data point from the mean and square the distances. Find the average of the distances (the variance) by dividing their sum by *n*. The square root of the variance is the standard deviation. To get a sample standard deviation, *s*, we divide by (*n* – 1) to adjust for error.

$$\sigma = \sqrt{\frac{\sum_{i=1}^{n}(x_i - \overline{x})^2}{n}} \qquad s = \sqrt{\frac{\sum_{i=1}^{n}(x_i - \overline{x})^2}{(n-1)}}$$

## Mean and Standard Deviation

process: evaluate

1. The number of goals scored by Wayne Gretzky in his first 10 seasons in the National Hockey League is 51, 55, 92, 71, 87, 73, 52, 62, 40, and 54. Find the mean and standard deviation of the data above.

2. System Bank just opened a new ATM at the mall. The first 8 withdrawals were each for $100. Find the mean and standard deviation of those withdrawals.

process: evaluate

## Mean and Standard Deviation

3. Graduation rates from 1998 are given for each state and the District of Columbia in the following chart. Determine the mean graduation rate and the standard deviation for the data. Also determine the median of the data.

| State | Graduation Rate | State | Graduation Rate | State | Graduation Rate |
|---|---|---|---|---|---|
| AL | 62% | KY | 71% | ND | 87% |
| AK | 70% | LA | 66% | OH | 78% |
| AZ | 60% | ME | 77% | OK | 75% |
| AR | 71% | MD | 79% | OR | 67% |
| CA | 73% | MA | 80% | PA | 85% |
| CO | 70% | MI | 77% | RI | 77% |
| CT | 81% | MN | 84% | SC | 72% |
| DE | 75% | MS | 60% | SD | 78% |
| DC | 60% | MO | 77% | TN | 59% |
| FL | 63% | MT | 80% | TX | 68% |
| GA | 57% | NE | 85% | UT | 77% |
| HI | 72% | NV | 63% | VT | 85% |
| ID | 75% | NH | 74% | VA | 76% |
| IL | 82% | NJ | 80% | WA | 72% |
| IN | 74% | NM | 63% | WV | 78% |
| IA | 93% | NY | 74% | WI | 87% |
| KS | 76% | NC | 66% | WY | 76% |

process: model

## Mean and Standard Deviation

4. Create a stem-and-leaf plot of the data.

analysis

## Mean and Standard Deviation

5. Describe the distribution of the graduation rate data.

6. Standard deviation, *s*, measures the spread about the mean, $\bar{x}$.

   a. What does a standard deviation of zero indicate about data?

   b. If a set of data has a standard deviation that is large, what do you know about the data?

input

## Mean and Standard Deviation

During the 2003-2004 basketball season, the Los Angeles Lakers players' annual salaries were as follows.

| Player | Salary |
|---|---|
| Shaquille O'Neal | $26.5 million |
| Kobe Bryant | $13.5 million |
| Gary Payton | $4.9 million |
| Rick Fox | $4.6 million |
| Devean George | $4.5 million |
| Derek Fisher | $3.0 million |
| Karl Malone | $1.5 million |
| Stanislav Medvedenko | $1.5 million |

| Player | Salary |
|---|---|
| Kareem Rush | $1.1 million |
| Bryon Russell | $1.1 million |
| Horace Grant | $1.1 million |
| Brian Cook | $0.8 million |
| Jamal Sampson | $0.6 million |
| Jannero Pargo | $0.6 million |
| Luke Walton | $0.4 million |

process: evaluate

## Mean and Standard Deviation

7. Determine the mean and the median for the Laker's salary data.

process: model

## Mean and Standard Deviation

8. Create a stem-and-leaf plot of the data.

9. Describe the distribution of the salary data.

analysis

## Mean and Standard Deviation

10. Why is the mean of the Laker's salaries so much higher than the median?

11. If Shaquille O'Neal made $30 million dollars, how would the mean of the data change? How would the median change?

12. Which is a better measure of central tendency, the mean or the median?

input

## Mean and Standard Deviation

Two common descriptors of the center and spread of data are the five-number summary and the mean and standard deviation. The five-number summary uses the median to describe the center and the quartiles, maximum, and minimum values to describe spread. The standard deviation, when used with mean as the measure of center, measures spread as a type of average of distances from the mean.

A skewed distribution or a distribution with outliers is usually better described with the five-number summary. The mean and standard deviation are good statistics to describe distributions that are reasonably symmetric and free of outliers.

The mean and standard deviation are important for describing a type of symmetric distribution called the normal distribution which you will study later in this unit.

## Binomial Probabilities

preview

1. Suppose you are taking a true/false quiz with a total of five questions. How likely is it for you to get at least four of the questions correct just by guessing? Explain how you determined your answer.

## Binomial Probabilities

process: evaluate

2. Consider the five question true/false quiz. List all of the possible ways to get:

| 0 correct | 1 correct | 2 correct | 3 correct | 4 correct | 5 correct |
|---|---|---|---|---|---|
| | | | | | |

process: evaluate

## Binomial Probabilities

3. How many possible outcomes are there on the five question true/false test?

4. How many possible ways are there to get:
   a. 0 correct answers
   b. 1 correct answer
   c. 2 correct answers
   d. 3 correct answers
   e. 4 correct answers
   f. 5 correct answers

5. Write out at least the first six rows of Pascal's Triangle.

6. Calculate:
   a. ${}_5C_0$
   b. ${}_5C_1$
   c. ${}_5C_2$
   d. ${}_5C_3$
   e. ${}_5C_4$
   f. ${}_5C_5$

7. Expand $(x + y)^5$ and write the terms in order of descending degree for $x$.

## Binomial Probabilities

analysis

8. How do your answers to question 3 relate to Pascal's Triangle, the combinations in question 5, and the binomial expansion in question 6? How can you account for the relationship?

## Binomial Probabilities

Input

The binomial expansion of $(x + y)^5$ can help determine how many ways there are to get a given number of correct answers by guessing on a five question true/false quiz. A situation like this is called a **binomial experiment**.

Binomial experiments have the following features.

1. The situation involves repeated trials.
2. Every trial has exactly two possible outcomes (success or failure).
3. The trials are independent.
4. Each trial has the same probability of success.

In a binomial experiment, the number of ways to achieve exactly *k* successes among *h* trials is ${}_hC_k$. For example, the number of ways to get exactly 3 correct answers (successes) on the true/false quiz with 5 questions (trials) is ${}_5C_3$.

Input

## Binomial Probabilities

Calculating a binomial probability involves determining the probability for each trial and multiplying by the number of ways to get that outcome. For instance, to find the probability of exactly 3 correct answers, multiply the probability of each trial outcome by the number of ways to get exactly 3 correct answers. In this case, there are 10 combinations that give 3 correct answers. The probability for each of the three correct answers is 0.5 and the probability for each of the two incorrect answers is 0.5.

$$\begin{aligned} P(\text{exactly 3 correct answers}) &= {}_5C_3(0.5)^3(0.5)^2 \\ &= 10(0.5)^3(0.5)^2 \\ &= .3125 \end{aligned}$$

In general, you can determine a complete binomial probability distribution by expanding $(p+q)^n$ where $p$ = probability of success, $q$ = probability of failure, and $n$ = the number of trials. For example, if the teacher gave a multiple choice quiz with five options for each of four questions, the following binomial probability distribution would result.

$$\begin{array}{lcccccccccc} & & \text{4 correct} & & \text{3 correct} & & \text{2 correct} & & \text{1 correct} & & \text{0 correct} \\ (p+q)^4 & = & 1p^4 & + & 4p^3q & + & 6p^2q^2 & + & 4pq^3 & + & 1q^4 \\ & = & (0.2)^4 & + & 4(0.2)^3(0.8) & + & 6(0.2)^2(0.8)^2 & + & 4(0.2)(0.8)^3 & + & (0.8)^4 \\ & = & 0.0016 & + & 0.0256 & + & 0.1536 & + & 0.4096 & + & 0.4096 \end{array}$$

process: evaluate

## Binomial Probabilities

9. About 13% of the general population is left handed.

   a. Out of 20 randomly selected people, about how many would you expect to be left handed?

   b. What is the probability that exactly 2 of 20 randomly selected people are left handed?

## Binomial Probabilities

process: evaluate

c. A baseball team has 20 pitchers on its roster. Six of the pitchers are left handed. How likely is it that this is by chance? Justify your answer.

10. A drug company claims that its medication for treating allergy symptoms produces mild side effects in 5% of patients. A doctor notes that 4 of the 30 patients for whom he has prescribed the medication have complained of side effects. What is the probability of this happening due to chance?

11. A basketball player's free throw shooting percentage is 75%. This can be interpreted that the probability that he will make a free throw is 75% or ¾. It also means that the probability he will miss a free throw is 25% or ¼. If the player gets four free throw attempts in a game, calculate the following probabilities.

    a. The player makes exactly three of the four free throws.

    b. The player makes at least three of the four free throws.

## Binomial Probability Distributions

input

Recall that a binomial experiment is one in which there are exactly two possible outcomes (success and failure), and for which there is a fixed number of independent trials. A **binomial distribution function** is a function used to calculate the probability of getting exactly $k$ successes in $n$ trials of a binomial experiment. The binomial distribution function can be modeled $B(k) = {}_nC_k(p)^k(1-p)^{n-k}$ where $p$ represents the probability of success. You can determine the values of a binomial distribution function and graph them as a histogram.

For example, suppose that the five day weather forecast calls for a 30% chance of rain each day. You can calculate the probability distribution for the number of days with rain over the next five days and graph the distribution as a histogram.

$$B(k) = {}_5C_k(0.3)^k(0.7)^{8-k}$$

Evaluating the function for each value of $k$ generates the numbers in the following table. The data are graphed in the histogram with bars centered on the value of the random variable.

| $k$ | $B(k)$ |
|---|---|
| 0 | $1\,(0.3)^0\,(0.7)^5 = 0.16807$ |
| 1 | $5\,(0.3)^1\,(0.7)^4 = 0.36015$ |
| 2 | $10\,(0.3)^2\,(0.7)^3 = 0.3087$ |
| 3 | $10\,(0.3)^3\,(0.7)^2 = 0.1323$ |
| 4 | $5\,(0.3)^4\,(0.7)^1 = 0.02835$ |
| 5 | $1\,(0.3)^5\,(0.7)^0 = 0.00243$ |

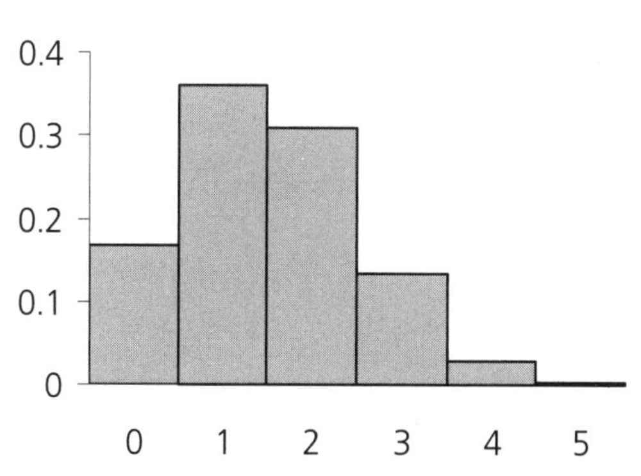


## Binomial Probability Distributions

analysis

1. If the probability of rain in the example above was changed to 50% or 80%, how do you think the probability distribution and graph would change?

## Binomial Probability Distributions

process: model

2. Graph the probability distributions for the following binomial experiments.

   a. 6 trials ($n = 6$) and a probability of success of 0.1 ($p = 0.1$)

   b. 6 trials ($n = 6$) and a probability of success of 0.25 ($p = 0.25$)

   c. 6 trials ($n = 6$) and a probability of success of 0.5 ($p = 0.5$)

process: model

## Binomial Probability Distributions

d. 6 trials ($n = 6$) and a probability of success of 0.75 ($p = 0.75$)

e. 6 trials ($n = 6$) and a probability of success of 0.9 ($p = 0.9$)

analysis

## Binomial Probability Distributions

3. Compare and contrast the graphs you created in question 2.

   a. How are the graphs in part (a) and part (e) related? Why do you think this happens?

## Binomial Probability Distributions

analysis

b. How are the graphs in part (b) and part (d) related? Why do you think this happens?

c. How is the graph in part (c) different than the other graphs?

4. What would the graph of a probability distribution with $p = 0$ look like?

5. What would the graph of a probability distribution with $p = 1$ look like?

6. What is the mode of each of the distributions you graphed?

input

## Binomial Probability Distributions

You can measure the center and spread of a binomial probability distribution given the number of trials, $n$, the probability of success, $p$, and the probability of failure, $q$.

The mean, $\mu$, of a binomial probability distribution can be calculated using the formula $\mu = np$

The standard deviation, $\sigma$, can be calculated using the formula $\sigma = \sqrt{npq}$

process: evaluate

## Binomial Probability Distributions

7. Calculate the mean and standard deviations for each of the distributions from question 2.

8. A candidate up for election is expected to win with about 75% of the vote.

   a. If 80 voters are chosen at random, how many people do you expect would vote for the candidate?

   b. Calculate the mean and standard deviation for a binomial probability distribution with $n = 80$ and $p = 0.75$.

process: evaluate

## Binomial Probability Distributions

9. Clinical trials show that a new treatment cures an illness is 40% of patients.
   a. If 50 patients who received the treatment are chosen at random, how many do you expect would have been cured?

   b. Calculate the mean and standard deviation for a binomial probability distribution with $n = 50$ and $p = 0.40$.

10. An airline claims that 85% of its flights arrive on time.
    a. If 120 of that airline's flights are chosen at random, how many do you expect would have arrived on time?

    b. Calculate the mean and standard deviation for a binomial probability distribution with $n = 50$ and $p = 0.40$.

analysis

## Binomial Probability Distributions

11. An independent research group did a study of the airline in question 10. They randomly chose 120 flights and found that 96 of them were on time. They claim the airline lied about their percentage of on time flights. Did you think the airline lied? Justify your answer.

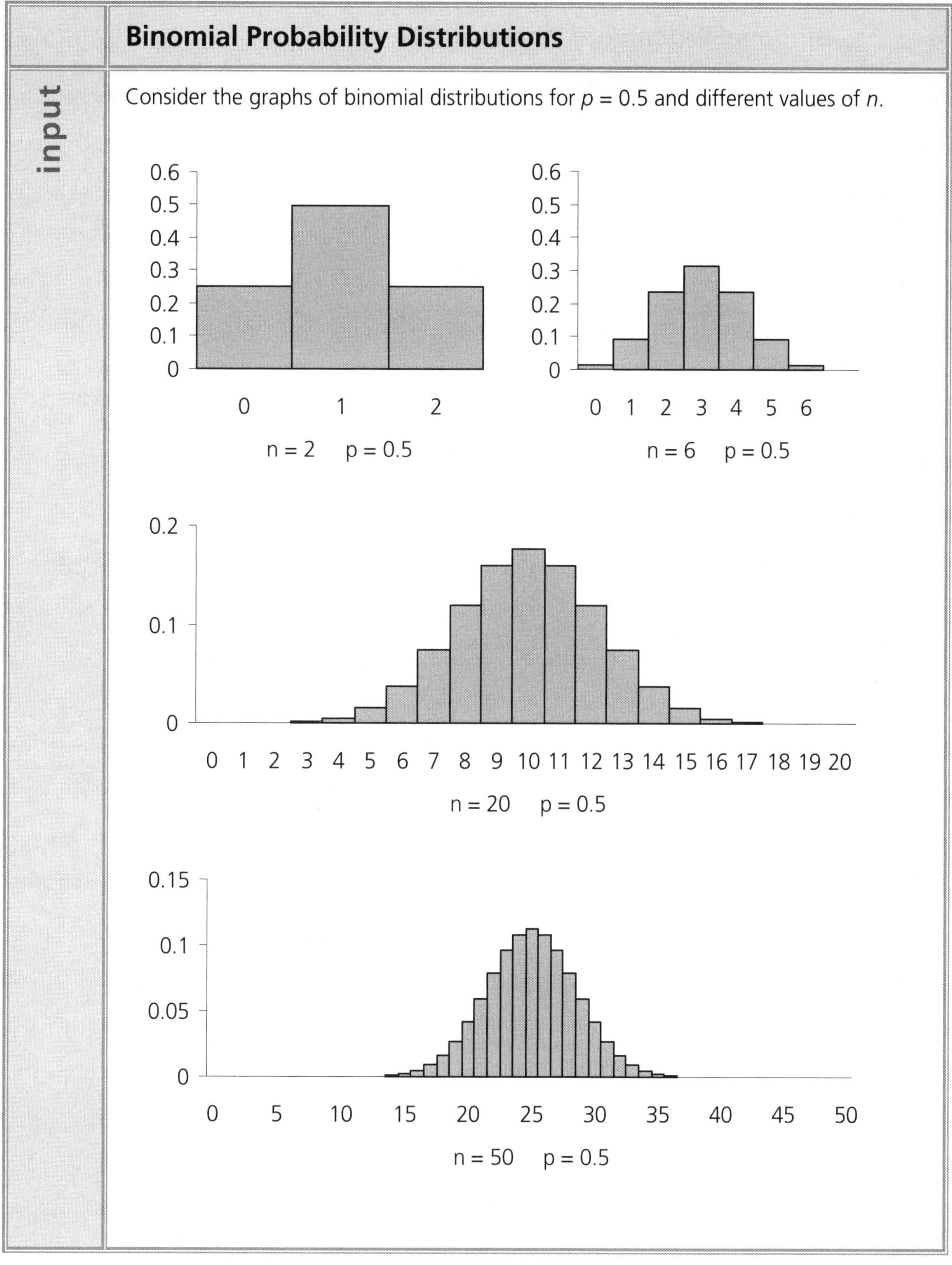

Binomial Probability Distributions
input
Consider the graphs of binomial distributions for p = 0.5 and different values of n.
0.6
0.5
0.4
0.3
0.2
0.1
0
0 1 2
n = 2 p = 0.5
0 1 2 3 4 5 6
n = 6 p = 0.5
0.2
0.1
0 1 2 3 4 5 6 7 8 9 10 11 12 13 14 15 16 17 18 19 20
n = 20 p = 0.5
0.15
0.1
0.05
0 5 10 15 20 25 30 35 40 45 50
n = 50 p = 0.5

analysis

## Binomial Probability Distributions

12. Determine the mean and standard deviation of each of the distributions graphed on the previous page.

13. Describe the effect of increasing the value of $n$ for probability distributions with $p = 0.5$.

    a. What happens to the mode of the distribution? How can you determine the mode given $n$?

    b. What happens to the line of symmetry? How can you determine the line of symmetry given $n$?

    c. What happens to the spread of the distribution?

    d. What happens to the range of the distribution? What happens to the maximum value?

14. What do you notice about the means and standard deviations for these distributions with $p = 0.5$?

## The Normal Curve

input

In the previous section you saw that when the probability of success in a binomial experiment is fixed (for example, $p = 0.5$), then as the number of trials increases, the graph of the probability distribution becomes more like a bell-shaped curve. We can use a **density curve** to picture the overall shape of such distributions.

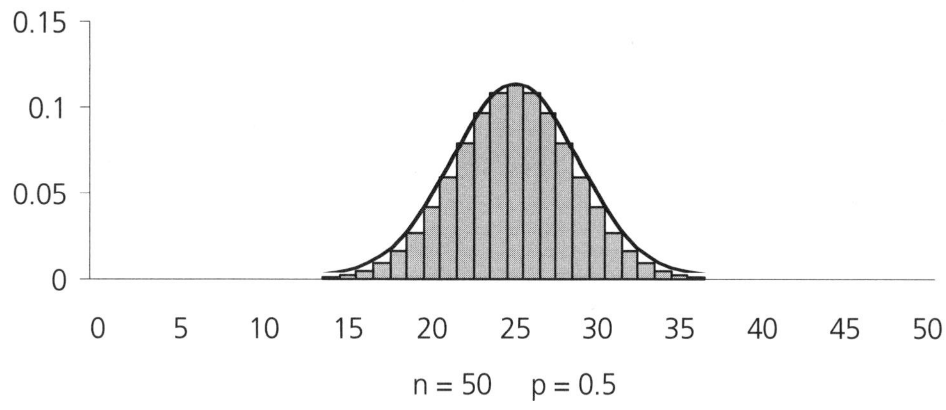


The histograms you used to show the probability distributions indicated probability with the heights of the bars. Density curves, like the one above, indicated the proportion of observations in a region by the area under the curve. Since a density curve indicates probabilities, the total area under the density curve must equal one.

## The Normal Curve

analysis

1. Describe the density curve above.
   a. Is the curve symmetric? If so, where is the line of symmetry?

   b. What is the domain and range of the curve?

   c. Is the curve continuous or discontinuous? If it is discontinuous, list any points of discontinuity

## The Normal Curve

analysis

d. Does the curve represent a function? How do you know?

e. Does the curve have any asymptotes? If so, where?

2. What is the area under the curve to the left of the line of symmetry? How do you know?

3. What does the area under the curve to the left of the line of symmetry represent?

## The Normal Curve

input

Graphs of symmetric density curves represent special distributions called **normal distributions**. The graphs of normal distributions are called **normal curves**.

Normal curves can be completely described by their mean and standard deviation. The mean determines the center of the distribution. It is located at the center of symmetry. The standard deviation determines the shape of the normal curve. It is the distance from the mean to the change of curvature points on either side of the mean.

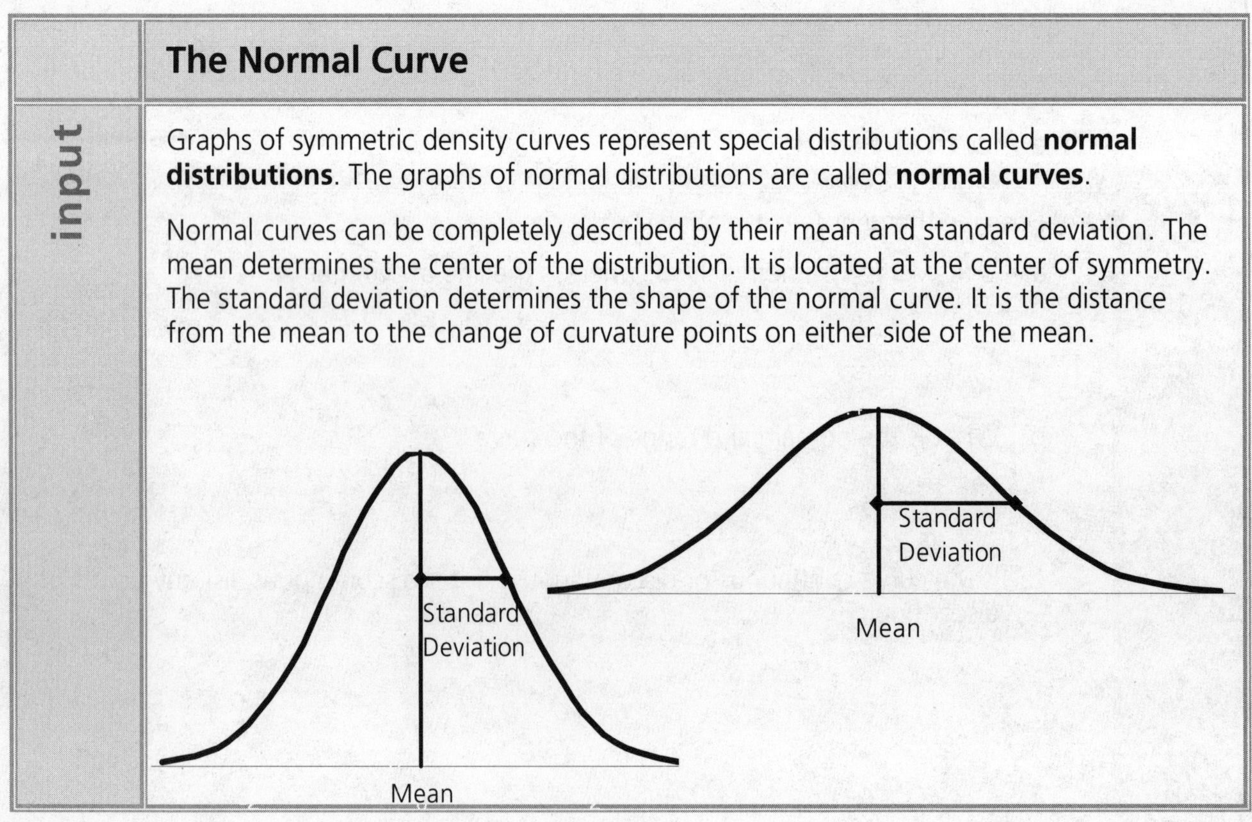

process: evaluate

## The Normal Curve

4. Estimate the mean and standard deviation of each of the normal curves below.

a.

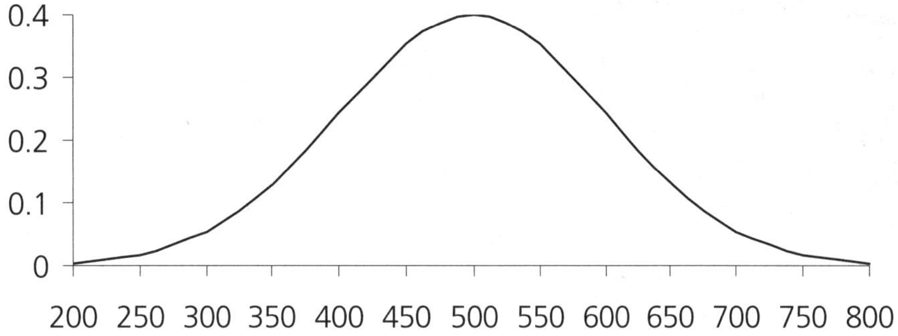

b.

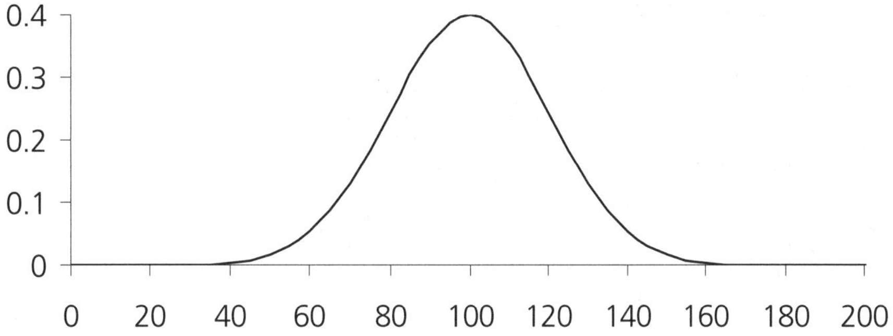

c.

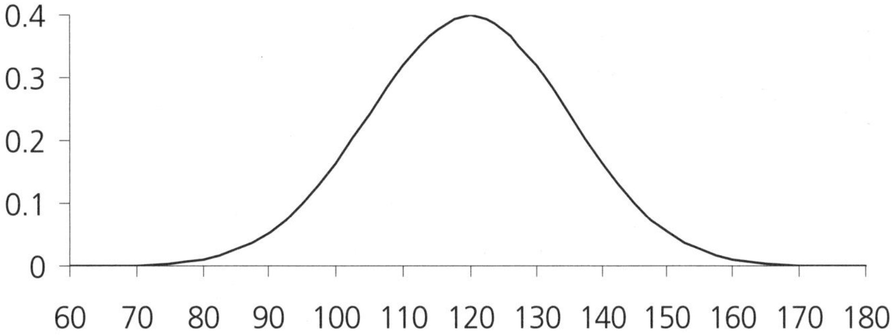

d.

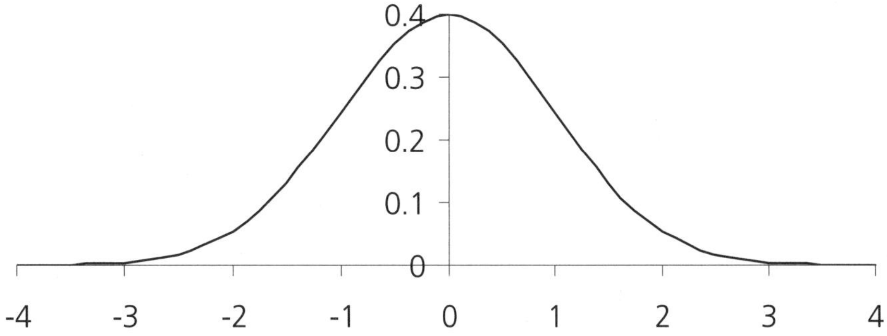

## Standard Scores

input

The **standard normal curve** is a normal distribution with a mean equal to zero and a standard deviation equal to one.

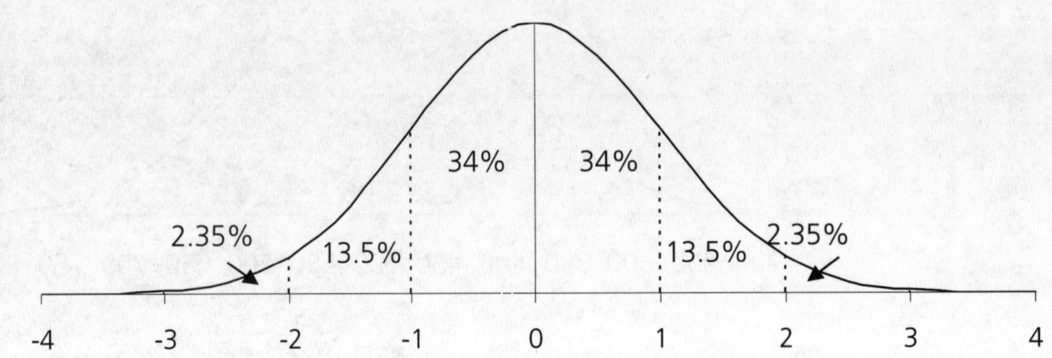


**The 68-95-99.7 Rule**

In any normal distribution, approximately
68% of the observations fall within one standard deviation of the mean.
95% of the observations fall within two standard deviations of the mean.
99.7% of the observations fall within three standard deviations of the mean.

No set of data is perfectly normal. However, remembering the 68-95-99.7 rule will allow you to think about normal distributions without making lots of detailed calculations.

## Standard Scores

process: model

1. The mean height for adult men is approximately 176 cm with a standard deviation of about 7 cm. Sketch a normal curve modeling the height of adult men.

## Standard Scores

analysis

2. How many standard deviations from the mean is an adult man with a height of:
   a. 183 cm?

   b. 162 cm?

   c. 204 cm?

   d. 186.5 cm?

3. What percentage of adult men are:
   a. shorter than 176 cm?

   b. between 169 and 183 cm in height?

   c. shorter than 190 cm?

   d. between 176 and 190 cm in height?

## Standard Scores

input

Observations stated in terms of standard deviations are called **standard scores** because the standard deviation is the natural measurement for a normal distribution.

The standard score for any observation can be computed as follows.

$$\textbf{standard score} = \frac{\textbf{observation - mean}}{\textbf{standard deviation}}$$

Standard scores are often called **z-scores**.

analysis

## Standard Scores

4. On the mathematics portion of the Scholastic Aptitude Test (SAT), the mean score is approximately 500 and the standard deviation is approximately 100.
   a. If you score 640 on the math SAT, what is your standard score (z-score)?

   b. What does that standard score mean?

5. Your friend takes the American College Testing (ACT) test instead of the SAT. On the mathematics portion of the ACT, the mean score is approximately 18 and the standard deviation is approximately 3. Your friend scored 22 on the ACT.
   a. If your friend scored 22 on the ACT, what is her standard score (z-score)?

   b. What does that standard score mean?

6. Assuming that both the SAT and ACT mathematics tests measure the same type of ability, who scored better, you or your friend? Explain.

7. Another friend of yours has decided to take the SAT again next time it is offered because he only scored 450 on the mathematics test. What is his standard score and what does it mean?

8. Why do you suppose the SAT and ACT use scores other than standard scores to report results?

## Standard Scores

analysis

9. Consider a mean SAT math score of 500.
   a. What is the standard score (z-score) associated with it?

   b. What is the percentile score? What percent of scores are lower than 500?

   c. What is the probability that a randomly selected test taker scored lower than 500? Higher than 500?

10. Consider an SAT math score of 400.
    a. What is the standard score (z-score) associated with it?

    b. What is the percentile score?

    c. What is the probability that a randomly selected test taker scored lower than 400? Higher than 400?

11. Consider an SAT math score of 700.
    a. What is the standard score (z-score) associated with it?

    b. What is the percentile score?

    c. What is the probability that a randomly selected test taker scored lower than 700? Higher than 700?

input

## Standard Scores

The 68-95-97.7 rule allows you to estimate probabilities with the standard normal distribution. To find more specific probabilities with the normal distribution, you can use a table of probabilities. A table of standard normal probabilities for z-scores from -3 to 3 can be found at the end of this unit.

Using the partial table below, you can see how specific probabilities and percentiles can be determined.

**Standard Normal Probabilities**

| z | 0 | 1 | 2 | 3 | 4 | 5 | 6 | 7 | 8 | 9 |
|---|---|---|---|---|---|---|---|---|---|---|
| +0.9 | .8159 | .8186 | .8212 | .8238 | .8264 | .8289 | .8315 | .8340 | .8365 | .8389 |
| +1.0 | **.8413** | .8438 | .8461 | .8485 | .8508 | .8531 | .8554 | .8577 | .8599 | .8621 |
| +1.1 | .8643 | .8665 | .8686 | .8708 | .8729 | .8749 | .8770 | .8790 | .8810 | .8830 |
| +1.2 | .8849 | .8869 | .8888 | .8907 | .8925 | .8944 | .8962 | .8980 | .8997 | .9015 |
| +1.3 | .9032 | .9049 | .9066 | .9082 | .9099 | .9115 | .9131 | .9147 | .9162 | .9177 |
| +1.4 | .9192 | .9207 | .9222 | .9236 | .9251 | .9265 | .9279 | .9292 | .9306 | .9319 |
| +1.5 | .9332 | .9345 | .9357 | .9370 | .9382 | .9394 | .9406 | .9418 | .9429 | .9441 |
| +1.6 | .9452 | .9463 | .9474 | .9484 | .9495 | .9505 | .9515 | .9525 | .9535 | .9545 |
| +1.7 | .9554 | .9564 | .9573 | .9582 | .9591 | .9599 | .9608 | .9616 | .9625 | .9633 |

You estimated that a standard score of 1 correlated to the 84$^{th}$ percentile. From the bold value in the table, you can see that a more precise percentile value is 84.13. You can also calculate values for other standard scores. For example, the boxed value in the table indicates that the percentile associated with a standard score of 1.27 is approximately 89.80. You can also determine a standard score required for a certain probability or percentile. The shaded values in the table indicate that the 95$^{th}$ percentile is associated with a standard score between 1.64 and 1.65.

process: evaluate

## Standard Scores

12. In Great Falls, MN, the mean January temperature is approximately 23°F with a standard deviation of about 8°F.

    a. Estimate the probability that a January temperature in Great Falls will be less than freezing, 32°F.

    b. Estimate the probability that a January temperature will be less than 0°F.

process: evaluate

## Standard Scores

13. Many IQ tests are designed to have a mean score of 100 and a standard deviation of 15.

   a. Determine the probability that a randomly selected person would score 110 or less on an IQ test.

   b. Determine the probability that a randomly selected person would score 90 or less on an IQ test.

   c. Determine the probability that a randomly selected person would score between 90 and 110 on an IQ test.

   d. What IQ score places a person in the 70$^{th}$ percentile?

14. A study of human birth weights at one hospital showed that the mean weight at birth was 8.08 pounds with a standard deviation of 1.12 pounds.

   a. What is the probability of a birth weight at that hospital between 7.5 and 8 pounds?

   b. A baby born at that hospital weighing 8.6 pounds is at what percentile of birth weights?

## Confidence Intervals

input

Before elections, news organization will often report poll results as an indication of how likely it is that a candidate will be elected. The day before one election, the TV news runs a story in which they state, "In a poll of 1100 registered voters, 594 respondents or 54% said they would vote for Shondra Wilson in the upcoming mayoral election. The margin of error for this poll is ± 3 percentage points."

## Confidence Intervals

analysis

1. Does the information about the poll mean that Shondra Wilson will get 54% of the vote? Explain your answer.

2. What exactly does the 54% represent?

3. Based on the poll, can you say that Shondra Wilson will win the election? Explain your answer.

input

## Confidence Intervals

You note the source of the poll listed on the screen and look up the source on the internet. The results there are stated differently. On the polling company's web site, the results read, "With 95% confidence, our poll indicates that Shondra Wilson will receive between 51% and 57% of the votes in the mayoral election tomorrow." To understand the real results, you must understand the language of sampling distributions and confidence intervals.

Numbers that describe a population (like voters in the election) are called **parameters**. Since it is too cumbersome in most situations to count the entire population, you estimate population parameters based on a sample of the population. The numbers calculated from the sample (like the 54% of voters favoring Shondra Wilson) are called **statistics**.

A **95% confidence interval** is an interval calculated from sample data that is guaranteed to contain the true population parameter in 95% of all samples. Most statistics reported on the news without a confidence interval stated are calculated with a 95% confidence interval.

analysis

## Confidence Intervals

4. If the poll was conducted again with a new sample of 1100 registered voters, do you think 594 respondents would report that they will vote for Shondra Wilson? Explain your answer.

5. Is it possible for fewer than 50% of respondents in a new sample to say they will vote for Shondra Wilson? Is it likely? Explain.

input

## Confidence Intervals

If you were to repeat the experiment of conducting a poll of voters many times, the sample proportion, $\hat{p}$ (p-hat), would not always be the same. In the poll cited in this problem, $\hat{p} = 0.54$ (54%). While additional samples would not likely yield exactly the same results, the samples would, in the long run form a pattern. The pattern is well described by the normal curve.

The **sampling distribution** of a statistic is the distribution of values for the statistic in all possible samples of the same size from the same population. If $\hat{p}$ is the sample proportion of successes and the sample is large enough:

- The sampling distribution for $\hat{p}$ will be approximately normal.
- The mean of the sampling distribution with be $\hat{p}$.
- The standard deviation of the sampling distribution will be $\sqrt{\dfrac{p(1-p)}{n}}$.

process: evaluate

## Confidence Intervals

6. Consider the election poll from this activity again. 594 out of 1100 people polled said they would vote for Shondra Wilson.
   a. What is $\hat{p}$, the sample proportion of successes? That is, what proportion of people said they would vote for Shonda Wilson?

   b. What is the standard deviation of the sampling distribution?

7. From the 68-95-97.7 rule, you know that 95% of all sample outcomes will fall within two standard deviations of the mean.
   a. What is the mean of the sample distribution?

   b. What values are two standard deviations below and above the mean of the sample distribution?

analysis

## Confidence Intervals

8. Based on your answers to questions 6 and 7, explain the statement, "With 95% confidence, our poll indicates that Shondra Wilson will receive between 51% and 57% of the votes in the mayoral election tomorrow."

input

## Confidence Intervals

A **confidence interval** for a parameter has two parts.

- An **interval** that is calculated from the data (e.g. 0.54 ± 0.015)
- A **confidence level**, *C*, which tells the probability that the interval contains the true parameter value in repeated samples (e.g. with 95% confidence)

Confidence levels other than 95% can be used. The table below indicates the **critical values**; the number of standard deviations needed to achieve various confidence intervals.

| Confidence Level C | Critical Value z* |
|---|---|
| 50% | 0.67 |
| 60% | 0.84 |
| 70% | 1.04 |
| 80% | 1.28 |

| Confidence Level C | Critical Value z* |
|---|---|
| 90% | 1.64 |
| 95% | 1.96 |
| 99% | 2.58 |
| 99.9% | 3.29 |

Notice that the 95% confidence level is actually 1.96 standard deviations. Previously, you used an approximation of 2 standard deviations from the 68-95-99.7 rule. The value in the table is more exact.

In general, an approximate confidence interval for a simple random sample of size *n* from a population in which $\hat{p}$ is the proportion of successes from the sample, and $z^*$ is the critical value for probability *C* can be calculated as

$$\hat{p} \pm z^* \sqrt{\frac{\hat{p}(1-\hat{p})}{n}}$$

## Confidence Intervals

process: evaluate

9. A poll of 993 adults in the United States asked, "Do you feel that executing people who commit murder deters others from committing murder, or do you think such executions don't have much effect?" Of the sample surveyed, 407 believed the death penalty has a deterrent effect.

   a. What is the population proportion for this poll?

   b. Find a 95% confidence interval for *p*.

   c. Find a 99% confidence interval for *p*.

10. 2201 adults in the United States were asked if they believe in the survival of the soul after death. Of those polled, 1849 responded that they did believe the soul survives after death.

    a. What is the population proportion for this poll?

    b. Find a 95% confidence interval for *p*.

    c. Find a 90% confidence interval for *p*.

## Confidence Intervals

analysis

11. The report of a sample survey of 1500 teenagers says, "With 95% confidence, between 23% and 29% of all teenagers believe that bullying is the most serious problem for students in school." A friend who knows nothing about statistics asks what the phrase "95% confidence" means. Write your explanation.

12. An opinion poll finds that 40% of its sample of registered voters believes that the economy is the most important issue in the presidential election. Give a 95% confidence interval for the proportion of registered voters who believe this if the sample size is:

    a. 500

    b. 1000

    c. 2000

13. What can you conclude about the effect of increasing the size of the sample?

## Confidence Intervals

analysis

14. An opinion poll finds that 70% of residents in the city favor legislation to ban smoking in all restaurants and bars. For a sample size of 1000, give:
    a. an 80% confidence interval
    b. a 90% confidence interval
    c. a 95% confidence interval

15. What can you conclude about the effect of increasing the confidence level?

## Confidence Intervals

input

You can use **statistical inference** to draw conclusions about a population based on data from a sample of the population. Since the entire population is not surveyed, the conclusions you make will have some uncertainty.

A **confidence interval** estimates a parameter for a population and tells how uncertain the estimate is. The interval indicates how closely the parameter is estimated. The **confidence level** is a probability that indicates the likelihood that a random sample from the population will fall in that interval.

## Hypothesis Testing

input

Commercials running in New York City for Whatta Water's bottled water claim that it tastes better than tap water. A skeptic claims that people wouldn't be able to tell the difference without the packaging. The skeptic sets up an experiment to test his claim. Each of 120 randomly selected subjects tastes New York City tap water and Whatta Water from unmarked cups. They are asked to select which tastes better. Of the 120 participants, 54 choose the New York City tap water as tasting better.

## Hypothesis Testing

process: evaluate

1. If people could not tell the difference between the taste of Whatta Water and tap water they would be randomly selecting one or the other. What percentage of the subjects would you expect to choose the tap water is this case?

2. If the taste of Whatta Water is actually better than tap water, what would you expect to be true about the percentage of the subjects choosing it?

3. In the binomial experiment set up by the skeptic,

   a. What is the value of $n$?

   b. If people can't taste the difference, what is the expected value of $p$?

   c. What is the mean and standard deviation for a binomial probability distribution with the $n$ and $p$ values you identified?

4. Of the 120 participants in the skeptic's experiment, 66 choose Whatta Water as tasting better. What is the z-score associated with this sample result?

input

## Hypothesis Testing

A graph of the information shows the standard normal curve and the section between the expected z-score of zero and the actual sample z-score.

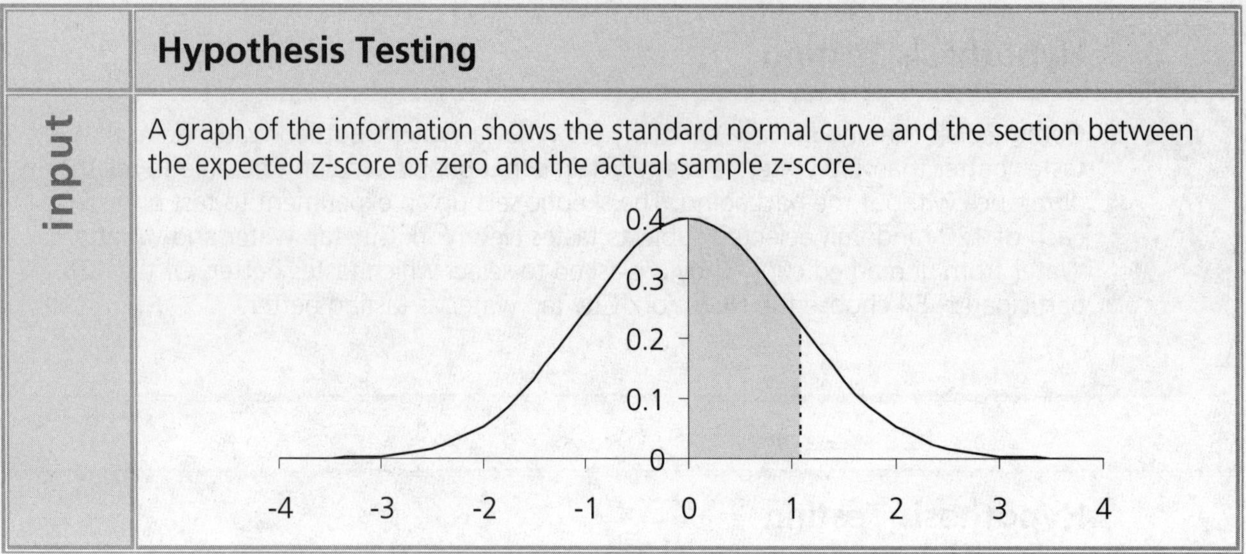


analysis

## Hypothesis Testing

5 What is the probability of a sample value with a z-score that is greater than the expected z-score of zero?

6 What is the probability of a z-score between the expected value of zero and the actual value you computed in question 4? Use the probability tables at the end of the unit to help you.

7 What is the probability of a value of a z-score at least as large as the observed z-score?

8 Based on your results, do you think that Whatta Water can support their claim to be better tasting than tap water? Explain your response.

input

## Hypothesis Testing

**Hypothesis testing** is a way to decide if the evidence supporting a claim is convincing. In practice, a hypothesis test indicates whether the effect observed in a sample could be by chance or if it is evidence that the effect is really present in the whole population.

Hypothesis testing is generally composed of four steps.

| **Formulate the Hypothesis** | |
|---|---|
| The **null hypothesis, $H_0$** (read H-naught) is usually a statement of no effect or no difference. It states that the observations are purely the result of chance. | In the Whatta Water problem, the null hypothesis ($H_0$) was that there is no difference between the taste of Whatta Water and tap water, so the preferred water would be chosen randomly ($p$ = 0.5). |
| The **alternate hypothesis, $H_\alpha$** (read H-alpha) is usually a statement that there is a real effect. It states that the observations are a result of a real effect (plus chance variation). | In the Whatta Water problem, the alternate hypothesis ($H_\alpha$) was that Whatta Water tastes better ($p > 0.5$). |
| **Choose the test statistic** | |
| The test statistic identifies a statistic that will assess evidence against $H_0$ and in support of $H_\alpha$. | In the Whatta Water problem, the test statistic is the binomial random variable with $p = 0.5$ and $n = 120$. |
| **Determine the p-value** | |
| Determining the p-value is developing a probability statement responding to the question, "If $H_0$ is true, what is the probability of observing a test statistic at least as extreme as the one observed." | In the Whatta Water problem, the p-value was 0.1357. It represented the probability of getting at least 66 of the 120 respondents to choose Whatta Water by chance. |
| **Compare the p-value to a predetermined significance level** | |
| Generally, a significance level is chosen before the experiment is conducted. The significance level, α, serves as a cut off value below which we agree that an effect is statistically significant (not likely to happen by chance). If the p-value is less than α, the null hypothesis is rejected and we accept that the observed effect is real. | In the Whatta Water problem, you did not choose a significance level in advance to decide if the claim of better tasting water is true. Often, a significance level of $\alpha = 0.05$ or $\alpha = 0.01$ is used. In either case, the p-value for the Whatta Water experiment (0.1357) is greater and we must reject the alternative hypothesis (rejecting) the company's claim and keep the null hypothesis. |

input

## Hypothesis Testing

A local organization runs a weekly bingo. One option for players when they purchase their bingo cards is something called the x-factor. If a player who has purchased an x-factor ticket for the evening wins a bingo, a number between 2 and 5 is drawn. The player's winnings for that game are multiplied by the x-factor number. The organization says there is an equal chance of any number from 2 through 5 being drawn as the x-factor value.

One bingo player claims that the x-factor drawing is not fair. He claims that 5 is not drawn as much as it should be drawn indicating that the numbers are not equally likely to come up. He asks you to help validate his claim. You decide to do a hypothesis test with a significance level of $\alpha = 0.05$

process: model

## Hypothesis Testing

9. Formulate the hypothesis for the experiment.

   a. What is the null hypothesis, $H_0$?

   b. What is the alternate hypothesis, $H_\alpha$?

10. In one bingo session, 24 games are played. You monitor the x-factor drawing for all of the games.

    a. Describe your test statistic including the value for *p* and *n*.

    b. Determine the mean and standard deviation for the binomial experiment.

## Hypothesis Testing

process: model

11. During the bingo session, the x-factor value of 5 is drawn only two times.

    a. Determine the standard score for the sample statistic.

    b. Find the p-value, the probability of getting a result at least as extreme as the observed result.

## Hypothesis Testing

analysis

12. Compare your calculated p-value to the predetermined significance level of $\alpha = 0.05$. What conclusion can you draw about the bingo player's claim?

13. If we had predetermined a significance level of $\alpha = 0.01$, would your interpretation of the results be the same?

input

## Hypothesis Testing

Hypothesis tests like the ones used for the Whatta Water and bingo claims are called **one sided tests** because the p-value comes from the area under one tail of the normal curve. In each of those situations, you were only concerned about the probability of getting a result either higher (Whatta Water) or lower (bingo) than expected.

Sometimes, you will encounter situations for which the direction of the result is not relevant. When a hypothesis test is concerned with observations in either direction, it is a **two-sided test**.

input

## Hypothesis Testing

It is estimated that the average young adult gets about 6 colds per year. A manufacturer of a natural supplement called Rutitol wants to see if use of the supplement affects the number of colds that young adults get. They will work with a significance level of $\alpha = 0.05$

process: evaluate

## Hypothesis Testing

14. Formulate the hypothesis for the experiment.
    a. What is the null hypothesis, $H_0$?

    b. What is the alternate hypothesis, $H_\alpha$?

## Hypothesis Testing

process: evaluate

15. Over a year, 160 young adults using Rutitol were monitored. On average, they had 5.6 colds with a standard deviation of 3.7. Determine the z-score for the sample statistic.

16. The original study was devised to find out if Rutitol affected the number of colds, not how it might affect the colds. Sketch a normal curve showing the area in *either* direction representing results at least as extreme as the sample result.

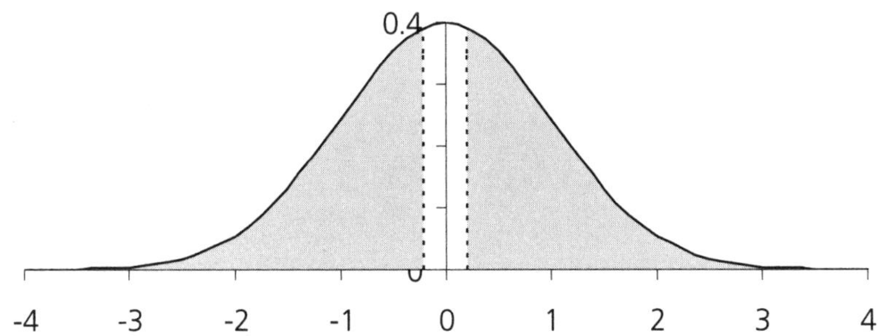


17. Calculate the p-value for the experiment including the area from both tails of the normal curve.
    a. What is the probability of a result at least as extreme as the sample statistic in the left tail of the normal curve?
    b. What is the probability of a result at least as extreme as the sample statistic in the right tail of the normal curve? Hint: consider the symmetry of the curve.
    c. What is the p-value?

## Hypothesis Testing

analysis

18. Compare your calculated p-value to the predetermined significance level of $\alpha = 0.05$. What conclusion can you draw about the use of Rutitol as a supplement?

## The Other Side of the Coin

input

Coin flips are used in situations in which an equal chance of two outcomes is desired. For example, a coin is flipped to decide which team will get the ball first in an overtime situation in NFL football games. When a coin flip is used, there is an assumption that the coin is fair, providing an equal chance of either outcome.

While one-Euro coins are all the same diameter and weight, each country in the European Union has its own design on the "heads" side of the coin. In Belgium, the heads side of the coin is a depiction of King Albert II. Statistics students in Belgium decided to check their one-Euro coin for fairness. They flipped the coin 250 times. Of the 250 flips, 140 of them came up heads. Knowing that 56% of the coin flips came up heads, is the coin fair?

## The Other Side of the Coin

process: evaluate

1. Formulate the hypothesis for the experiment.
   a. What is $H_0$?

   b. What is $H_\alpha$?

2. Determine the mean and standard deviation for the binomial experiment.

3. Determine the z-score for the sample statistic.

4. Find the p-value.

## The Other Side of the Coin

analysis

5. At a significance level of $\alpha = 0.05$, what conclusion can you make about the Euro based on the experiment?

6. If the experiment with the Belgian Euro was conducted with 500 coin flips and 56% of the flips were heads, how would the results be different?

   a. How many of the coin flips would have come up heads?

   b. What would be the mean and standard deviation of the modified experiment?

   c. What would be the z-score for the sample statistic?

   d. What would be the p-value?

   e. What conclusion could you draw about the fairness of the coin at the 0.05 significance level?

analysis

## The Other Side of the Coin

7. If the experiment with the Belgian Euro was conducted with 50 coin flips and 56% of the flips were heads, how would the results be different?
   a. How many of the coin flips would have come up heads?
   b. What would be the mean and standard deviation of the modified experiment?
   c. What would be the z-score for the sample statistic?
   d. What would be the p-value?
   e. What conclusion could you draw about the fairness of the coin at the 0.05 significance level?

8. Considering the effect of sample size, is it appropriate to make a judgment based only on a sample proportion ( $\hat{p}$ )? For example, could you conclude that a six sided die is unfair based on the knowledge that in an experiment the proportion of ones rolled is 0.5?

## Statistics Project

input

You now know enough about statistics to design and conduct your own experiment. It is time to do such a project. For the project you will be responsible for the following.

**Choose a topic**
What will you study? How will you measure it? For example, the students in Belgium studied the fairness of a coin. They measured it by flipping the coin and counting the number of heads.

**Collect and organize the data**
Decide what method you will use to collect the data. Ensure you have all materials you need to gather the data. Create an appropriate graphical display of the data you gather. For example, if you are measuring free throw shooting, you will need access to a basketball and hoop. If you are conducting a survey, you will need to design the survey and get a random sample.

**Perform a statistical analysis of your data.**
Provide the statistics that represent your data. What is the mean, standard deviation, sample size, etc.?

**Perform a hypothesis test.**
Be sure to indicate your chosen significance level, the null hypothesis and the alternate hypotheses. Describe the test statistic. Sketch the normal curve representing your work. Find the p-value and state your conclusion.

**Write up your results including**:

- what you decided to study and why you chose to study it
- your numerical data and an appropriate graphic display of the data
- your statistical analysis
- your hypothesis test including the significance level, hypotheses, test statistics, a sketch of the curve, and the p-value
- your conclusion based on the hypothesis test

| Table of Standard Normal Probabilities | | | | | | | | | | |
|---|---|---|---|---|---|---|---|---|---|---|
| *z* | **0** | **1** | **2** | **3** | **4** | **5** | **6** | **7** | **8** | **9** |
| **-3.0** | .0013 | .0013 | .0013 | .0012 | .0012 | .0011 | .0011 | .0011 | .0010 | .0010 |
| **-2.9** | .0019 | .0018 | .0018 | .0017 | .0016 | .0016 | .0015 | .0015 | .0014 | .0014 |
| **-2.8** | .0026 | .0025 | .0024 | .0023 | .0023 | .0022 | .0021 | .0021 | .0020 | .0019 |
| **-2.7** | .0035 | .0034 | .0033 | .0032 | .0031 | .0030 | .0029 | .0028 | .0027 | .0026 |
| **-2.6** | .0047 | .0045 | .0044 | .0043 | .0041 | .0040 | .0039 | .0038 | .0037 | .0036 |
| **-2.5** | .0062 | .0060 | .0059 | .0057 | .0055 | .0054 | .0052 | .0051 | .0049 | .0048 |
| **-2.4** | .0082 | .0080 | .0078 | .0075 | .0073 | .0071 | .0069 | .0068 | .0066 | .0064 |
| **-2.3** | .0107 | .0104 | .0102 | .0099 | .0096 | .0094 | .0091 | .0089 | .0087 | .0084 |
| **-2.2** | .0139 | .0136 | .0132 | .0129 | .0125 | .0122 | .0119 | .0116 | .0113 | .0110 |
| **-2.1** | .0179 | .0174 | .0170 | .0166 | .0162 | .0158 | .0154 | .0150 | .0146 | .0143 |
| **-2.0** | .0228 | .0222 | .0217 | .0212 | .0207 | .0202 | .0197 | .0192 | .0188 | .0183 |
| **-1.9** | .0287 | .0281 | .0274 | .0268 | .0262 | .0256 | .0250 | .0244 | .0239 | .0233 |
| **-1.8** | .0359 | .0351 | .0344 | .0336 | .0329 | .0322 | .0314 | .0307 | .0301 | .0294 |
| **-1.7** | .0446 | .0436 | .0427 | .0418 | .0409 | .0401 | .0392 | .0384 | .0375 | .0367 |
| **-1.6** | .0548 | .0537 | .0526 | .0516 | .0505 | .0495 | .0485 | .0475 | .0465 | .0455 |
| **-1.5** | .0668 | .0655 | .0643 | .0630 | .0618 | .0606 | .0594 | .0582 | .0571 | .0559 |
| **-1.4** | .0808 | .0793 | .0778 | .0764 | .0749 | .0735 | .0721 | .0708 | .0694 | .0681 |
| **-1.3** | .0968 | .0951 | .0934 | .0918 | .0901 | .0885 | .0869 | .0853 | .0838 | .0823 |
| **-1.2** | .1151 | .1131 | .1112 | .1093 | .1075 | .1056 | .1038 | .1020 | .1003 | .0985 |
| **-1.1** | .1357 | .1335 | .1314 | .1292 | .1271 | .1251 | .1230 | .1210 | .1190 | .1170 |
| **-1.0** | .1587 | .1562 | .1539 | .1515 | .1492 | .1469 | .1446 | .1423 | .1401 | .1379 |
| **-0.9** | .1841 | .1814 | .1788 | .1762 | .1736 | .1711 | .1685 | .1660 | .1635 | .1611 |
| **-0.8** | .2119 | .2090 | .2061 | .2033 | .2005 | .1977 | .1949 | .1922 | .1894 | .1867 |
| **-0.7** | .2420 | .2389 | .2358 | .2327 | .2296 | .2266 | .2236 | .2206 | .2177 | .2148 |
| **-0.6** | .2743 | .2709 | .2676 | .2643 | .2611 | .2578 | .2546 | .2514 | .2483 | .2451 |
| **-0.5** | .3085 | .3050 | .3015 | .2981 | .2946 | .2912 | .2877 | .2843 | .2810 | .2776 |
| **-0.4** | .3446 | .3409 | .3372 | .3336 | .3300 | .3264 | .3228 | .3192 | .3156 | .3121 |
| **-0.3** | .3821 | .3783 | .3745 | .3707 | .3669 | .3632 | .3594 | .3557 | .3520 | .3483 |
| **-0.2** | .4207 | .4168 | .4129 | .4090 | .4052 | .4013 | .3974 | .3936 | .3897 | .3859 |
| **-0.1** | .4602 | .4562 | .4522 | .4483 | .4443 | .4404 | .4364 | .4325 | .4286 | .4247 |
| **-0.0** | .5000 | .4960 | .4920 | .4880 | .4840 | .4801 | .4761 | .4721 | .4681 | .4641 |

## Table of Standard Normal Probabilities

| z | 0 | 1 | 2 | 3 | 4 | 5 | 6 | 7 | 8 | 9 |
|---|---|---|---|---|---|---|---|---|---|---|
| **+0.0** | .5000 | .5040 | .5080 | .5120 | .5160 | .5199 | .5239 | .5279 | .5319 | .5359 |
| **+0.1** | .5398 | .5438 | .5478 | .5517 | .5557 | .5596 | .5636 | .5675 | .5714 | .5753 |
| **+0.2** | .5793 | .5832 | .5871 | .5910 | .5948 | .5987 | .6026 | .6064 | .6103 | .6141 |
| **+0.3** | .6179 | .6217 | .6255 | .6293 | .6331 | .6368 | .6406 | .6443 | .6480 | .6517 |
| **+0.4** | .6554 | .6591 | .6628 | .6664 | .6700 | .6736 | .6772 | .6808 | .6844 | .6879 |
| **+0.5** | .6915 | .6950 | .6985 | .7019 | .7054 | .7088 | .7123 | .7157 | .7190 | .7224 |
| **+0.6** | .7257 | .7291 | .7324 | .7357 | .7389 | .7422 | .7454 | .7486 | .7517 | .7549 |
| **+0.7** | .7580 | .7611 | .7642 | .7673 | .7704 | .7734 | .7764 | .7794 | .7823 | .7852 |
| **+0.8** | .7881 | .7910 | .7939 | .7967 | .7995 | .8023 | .8051 | .8079 | .8106 | .8133 |
| **+0.9** | .8159 | .8186 | .8212 | .8238 | .8264 | .8289 | .8315 | .8340 | .8365 | .8389 |
| **+1.0** | .8413 | .8438 | .8461 | .8485 | .8508 | .8531 | .8554 | .8577 | .8599 | .8621 |
| **+1.1** | .8643 | .8665 | .8686 | .8708 | .8729 | .8749 | .8770 | .8790 | .8810 | .8830 |
| **+1.2** | .8849 | .8869 | .8888 | .8907 | .8925 | .8944 | .8962 | .8980 | .8997 | .9015 |
| **+1.3** | .9032 | .9049 | .9066 | .9082 | .9099 | .9115 | .9131 | .9147 | .9162 | .9177 |
| **+1.4** | .9192 | .9207 | .9222 | .9236 | .9251 | .9265 | .9279 | .9292 | .9306 | .9319 |
| **+1.5** | .9332 | .9345 | .9357 | .9370 | .9382 | .9394 | .9406 | .9418 | .9429 | .9441 |
| **+1.6** | .9452 | .9463 | .9474 | .9484 | .9495 | .9505 | .9515 | .9525 | .9535 | .9545 |
| **+1.7** | .9554 | .9564 | .9573 | .9582 | .9591 | .9599 | .9608 | .9616 | .9625 | .9633 |
| **+1.8** | .9641 | .9649 | .9656 | .9664 | .9671 | .9678 | .9686 | .9693 | .9699 | .9706 |
| **+1.9** | .9713 | .9719 | .9726 | .9732 | .9738 | .9744 | .9750 | .9756 | .9761 | .9767 |
| **+2.0** | .9773 | .9778 | .9783 | .9788 | .9793 | .9798 | .9803 | .9808 | .9812 | .9817 |
| **+2.1** | .9821 | .9826 | .9830 | .9834 | .9838 | .9842 | .9846 | .9850 | .9854 | .9857 |
| **+2.2** | .9861 | .9864 | .9868 | .9871 | .9875 | .9878 | .9881 | .9884 | .9887 | .9890 |
| **+2.3** | .9893 | .9896 | .9898 | .9901 | .9904 | .9906 | .9909 | .9911 | .9913 | .9916 |
| **+2.4** | .9918 | .9920 | .9922 | .9925 | .9927 | .9929 | .9931 | .9932 | .9934 | .9936 |
| **+2.5** | .9938 | .9940 | .9941 | .9943 | .9945 | .9946 | .9948 | .9949 | .9951 | .9952 |
| **+2.6** | .9953 | .9955 | .9956 | .9957 | .9959 | .9960 | .9961 | .9962 | .9963 | .9964 |
| **+2.7** | .9965 | .9966 | .9967 | .9968 | .9969 | .9970 | .9971 | .9972 | .9973 | .9974 |
| **+2.8** | .9974 | .9975 | .9976 | .9977 | .9977 | .9978 | .9979 | .9979 | .9980 | .9981 |
| **+2.9** | .9981 | .9982 | .9983 | .9983 | .9984 | .9984 | .9985 | .9985 | .9986 | .9986 |
| **+3.0** | .9987 | .9987 | .9987 | .9988 | .9988 | .9989 | .9989 | .9989 | .9990 | .9990 |

# Unit 6: Circles

## Contents

## Making a Dog Pen

preview

You are the proud owner of a new dog. Unfortunately, you live on a very busy street and need to confine your dog to the back yard. It would be nice to fence the entire yard, but because of the expense, you cannot. A friend has given you a piece of fencing that measures 4 feet in height and 30 feet in length, so you decide to make the most of it.

Design the area to be fenced. Include measurements on your drawing and remember you want to enclose the largest possible area.

1. What is the total area of the dog pen? Show your work and explain your calculations.

2. If you doubled the length of the fence, how would that impact the area available to your dog?

3. If you tripled the length of fence, how would that affect the area?

## Circles

process: model

1. In the middle of the space below, draw a point and label it *B*.

2. Use a ruler to locate and draw another point that is exactly 5 cm from point *B*. Label this point A.

3. Locate another point that is exactly 5 cm from point B. Label this point C.

4. Repeat this procedure until you have drawn at least ten points that are each exactly 5 cm from point *B*.

## Circles

analysis

5. What do you notice about the collection of points you drew in parts 2 through 4?

6. How many other points could be located exactly 5 cm from point *B*?

## Circles

process: summary

7. What special name applies to point *B* in relation to all the other points you drew?

8. What special name applies to the distance from point *B* to any other point you drew?

9. What special name applies to twice the distance from point *B* to any of the other points?

process: model

## Circles

10. Describe the position of the other points in this figure in relation to point $C$

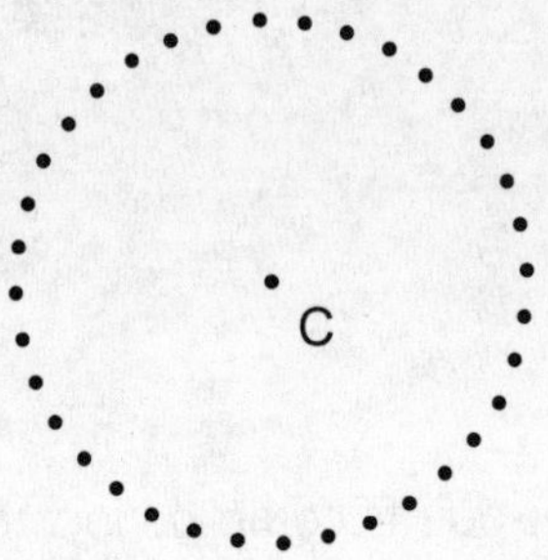

C

a. Select a point and measure the distance from the selected point to point $C$. Do this again with some of the other points. What can you conclude after making these measurements?

11. What special name applies to the figure below?

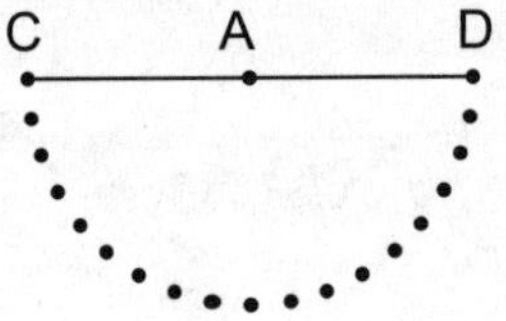

C A D

analysis

## Circles

12. Based on what you've done in questions 1 through 10, write a definition for the word **circle**.

process: model

## Parts of a Circle

1. In the figures below, identify as many examples of the terms listed as possible. Use symbols to list the examples next to each term.

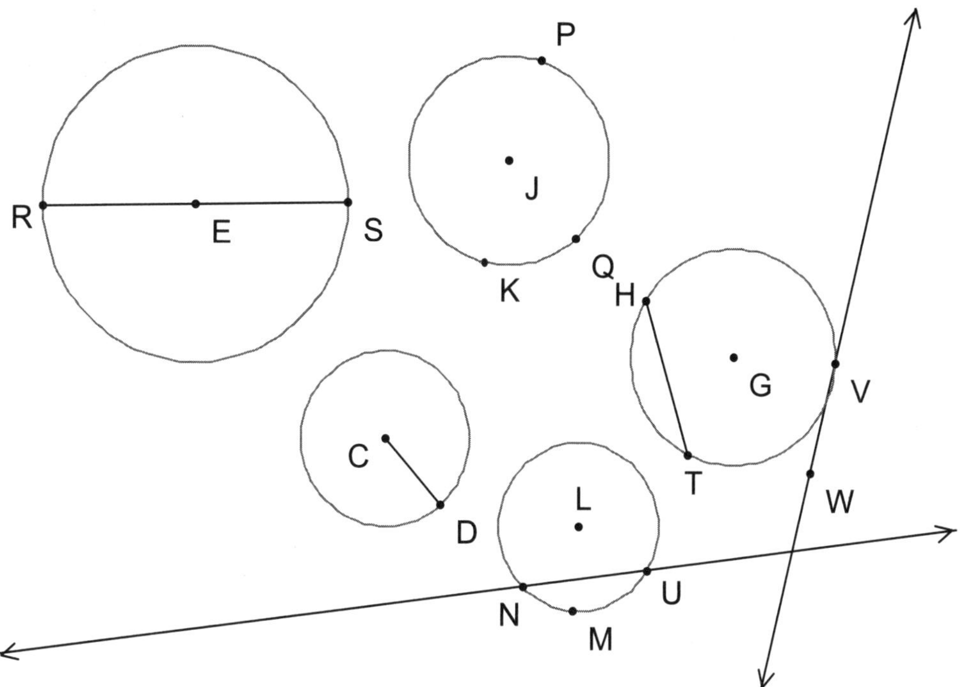


a. Chord:

b. Radius:

c. Diameter:

d. Tangent:

e. Arc: :

f. Major Arc:

g. Minor Arc:

h. Semicircle:

i. Secant:

## Parts of a Circle

process: model

2. Write a definition and draw a picture for each of the following terms.

   a. Arc:

   b. Chord:

   c. Diameter:

   d. Major Arc:

   e. Minor Arc:

process: model

## Parts of a Circle

f. Radius:

g. Secant:

h. Semicircle:

i. Tangent:

## Angles of a Circle

input

∠*GEF* is a *central angle* of circle *E*, and arc *FG* is its *intercepted arc*.
∠*JHK* is a *central angle* of circle *H*, and arc *JK* is its *intercepted arc*.
∠*PHK* is a *central angle* of circle *H*, and arc *PK* is its *intercepted arc*.

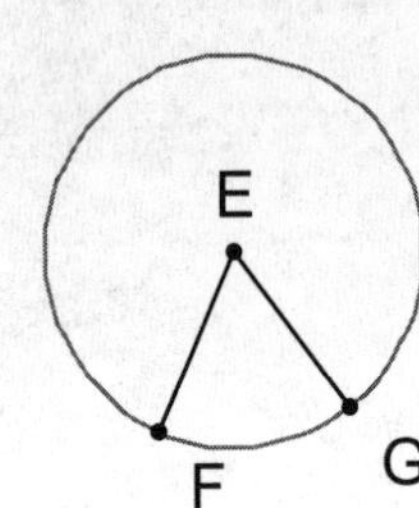


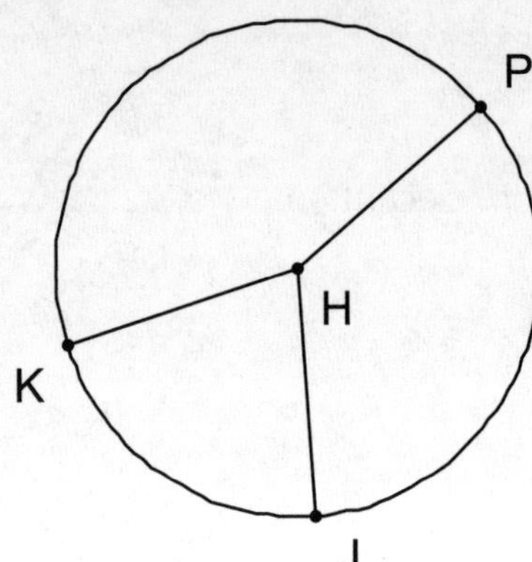


## Angles of a Circle

analysis

1. Describe what appears to be true about the position of every central angle of a circle.

2. How can you locate an intercepted arc from the central angle?

process: model

## Angles of a Circle

3. Draw a circle. Draw a central angle of your circle and shade its intercepted arc.

4. Write definitions for a **central angle** of a circle and its **intercepted arc**.

input

## Angles of a Circle

The following statement describes a property of any central angle of a circle and its intercepted arc:

*The measure of a central angle is equal to the measure of its intercepted arc.*

process: practice

## Angles of a Circle

Refer to circle E and circle H below to answer each of the following questions.

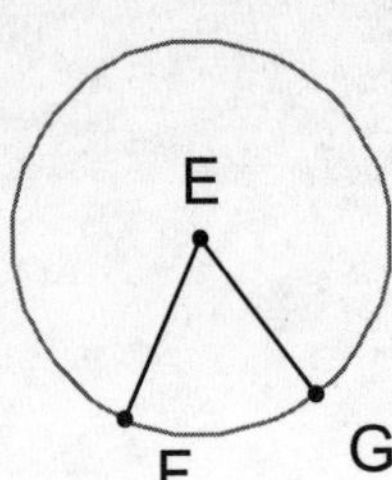


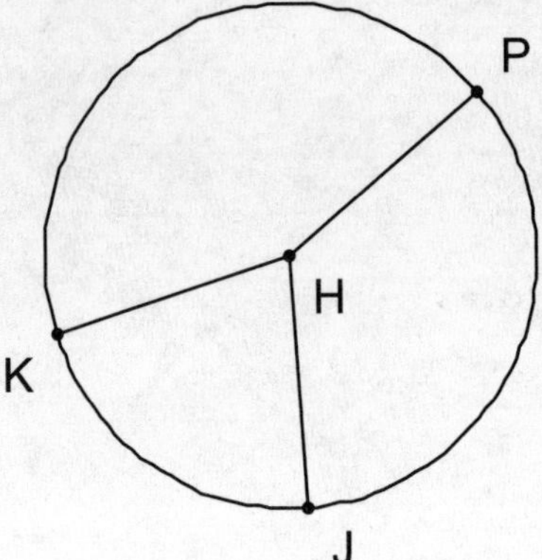


5. If the measure of ∠*FEG* is 52°, what is the measure of its intercepted arc *FG*?

6. If the measure of arc *JK* is 49°, what is the measure of ∠*KHJ*?

7. If the measure of ∠*KHP* is 134°, what is the measure of its intercepted arc *PK*?

8. Draw a circle with a central angle that measures 90°. Shade the arc intercepted by the central angle.

input

## Angles of a Circle

In the figure below, ∠*FRG* is an **inscribed angle** in circle *E*, and arc *FG* is its **intercepted arc**.

∠*PQK* is an **inscribed angle** in circle *H*, and arc *PK* is its **intercepted arc**.

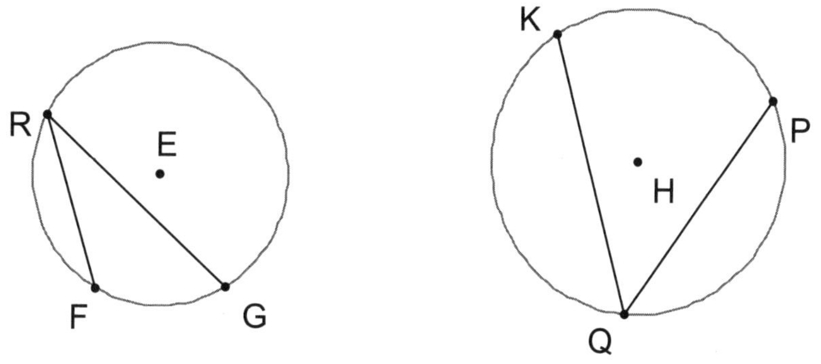

R E F G K P H Q

analysis

## Angles of a Circle

9. Describe what appears to be true about the position of inscribed angles.

process: model

## Angles of a Circle

10. Draw a circle. In your circle, draw an inscribed angle and shade its intercepted arc. Describe how you located the intercepted arc.

process: model

## Angles of a Circle

11. Draw a circle with an inscribed angle whose measure is 90°.

12. Write definitions for an **inscribed angle** in a circle and its **intercepted arc**.

process: practice

## Angles of a Circle

Refer to circle A, pictured below, to answer each of the following questions.

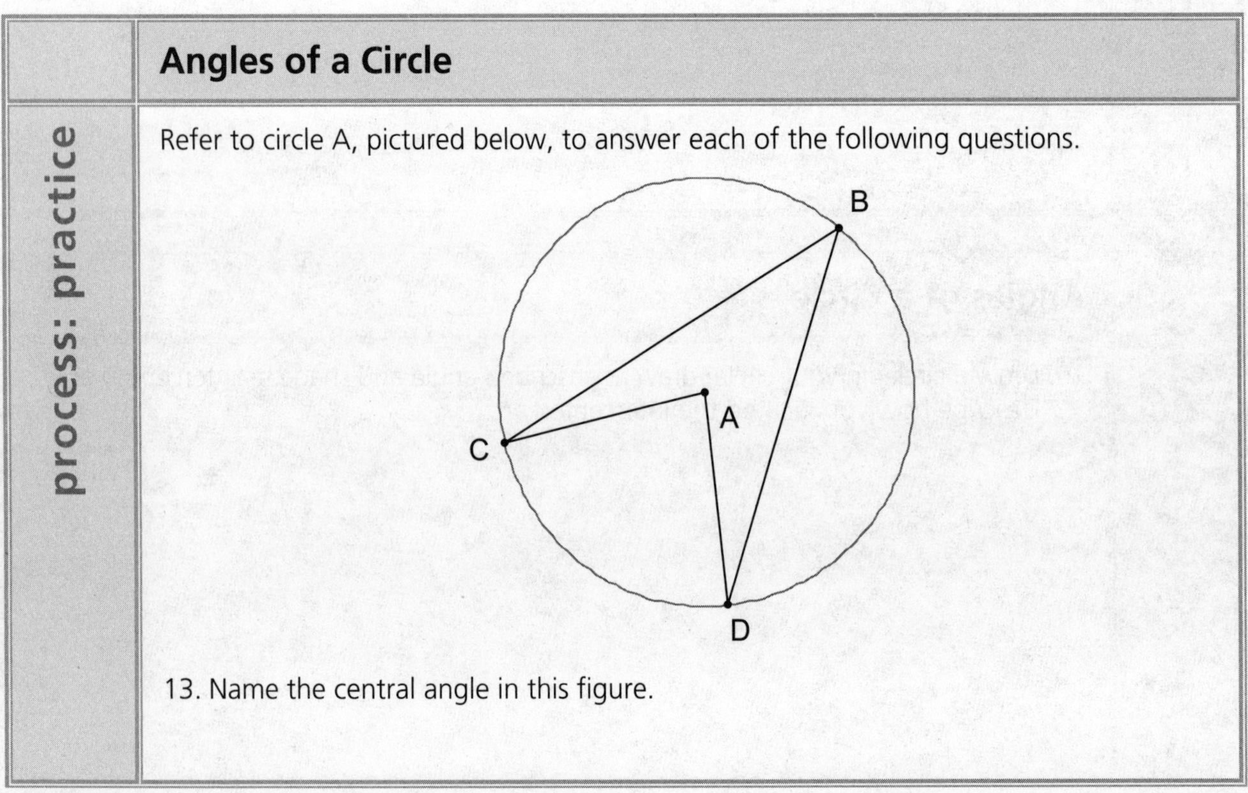


13. Name the central angle in this figure.

## Angles of a Circle

process: practice

14. What is the name of the arc intercepted by the central angle?

15. What is the inscribed angle in this figure?

16. What is the name of the arc intercepted by the inscribed angle?

17. Measure the central angle. Record your measurement on the figure.

18. What is the measure of the intercepted arc?

19. Measure the inscribed angle. Record your measurement on the figure.

## Angles of a Circle

analysis

20. What is the relationship between the measure of an inscribed angle and the measure of its intercepted arc?

process practice

## Angles of a Circle

Refer to the circles E and H below to answer the following questions.

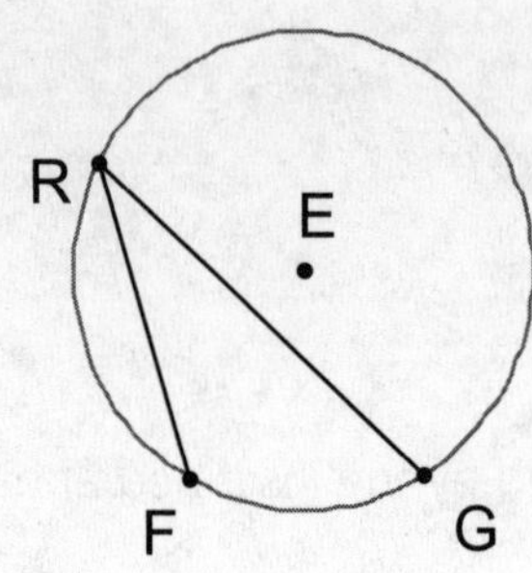


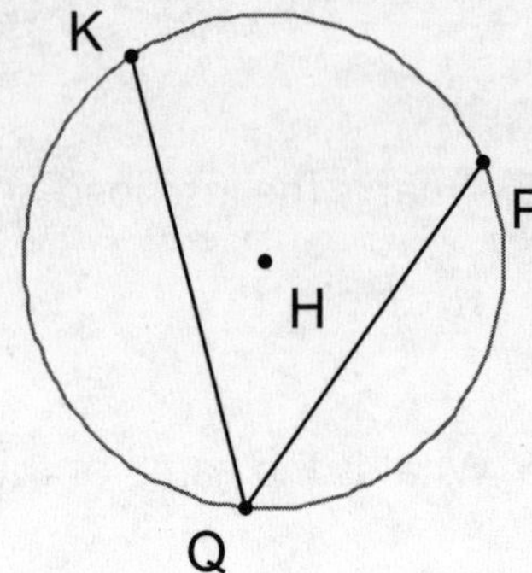


21. If the measure of $\angle FRG$ is 38°, what is the measure of arc *FG*?

22. If the measure of arc *KP* is 124°, what is the measure of $\angle KQP$?

23. Measure $\angle EAC$ in circle A. Use what you know to determine the measure of $\angle CDE$, and explain how you found this measure.

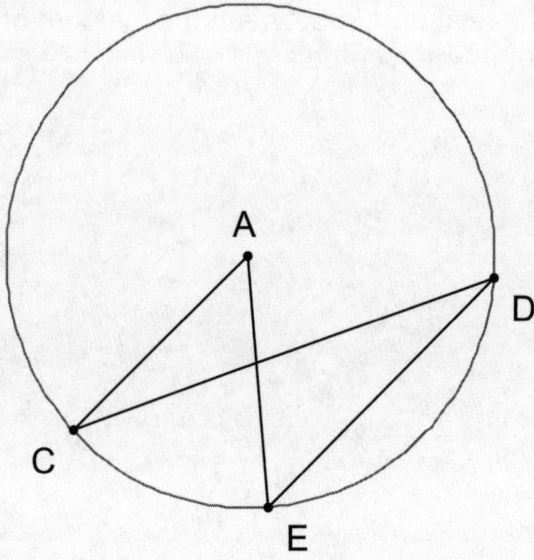

## Interior Angles of a Circle (Optional)

input

In the previous section, you learned about angles whose vertices were at the center of the circle or on the circle. What are the names given to these angle types?

In this section, you will find the measure of angles whose vertices are located somewhere inside a circle—not at the center and not on the circle itself. If the vertex of the angle is inside the circle, the angles created are called **interior angles**.

Consider circle *A* and find a way to calculate the measure of angle $\angle FGB$.

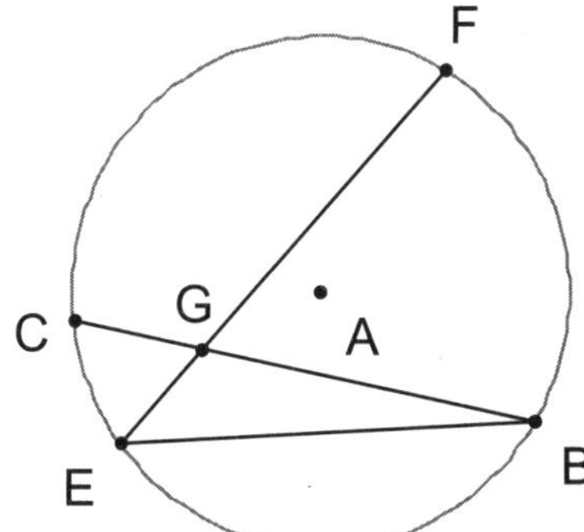


$m\overset{\frown}{CE}$ on ⊙AB = 30°

$m\overset{\frown}{FB}$ on ⊙AB = 90°

## Interior Angles of a Circle (Optional)

process: evaluate

1. Use what you learned in the previous section to find $m\angle FEB$ and $m\angle CBE$. Record these answers on the figure.

2. Use what you know about triangles and angles to find $m\angle FGB$.

3. Find the measure of $\angle CGF$ and the measure of $\angle CGE$.

## Interior Angles of a Circle (Optional)

analysis

4. Explain in complete sentences how you used the measures of arcs CE and FB to find the measure of ∠*FGB*.

## Interior Angles of a Circle (Optional)

process: evaluate

Refer to the figure below to answer each of the following questions.

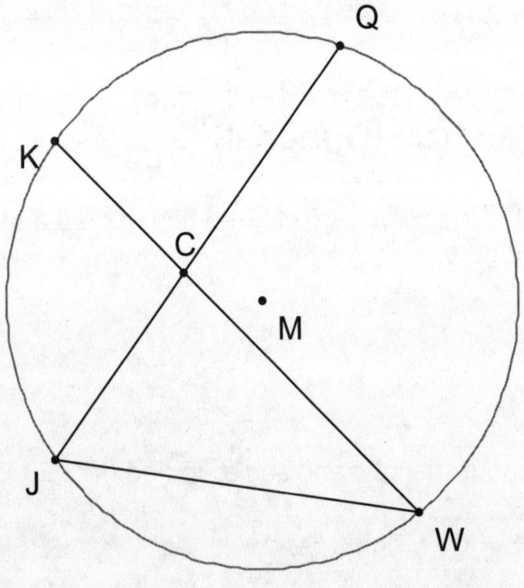

Q

K

C

M

J

W

m $\overset{\frown}{QW}$ on ⊙MW = 124°
m $\overset{\frown}{KJ}$ on ⊙MW = 72°

5. Find the measure of ∠*KCJ* in circle *M*, using the information given.

6. Find the measure of ∠*KCQ* and ∠*JCW*.

analysis

## Interior Angles of a Circle (Optional)

7. Explain in complete sentences how you could use the measures of arcs *KJ* and *QW* to find the measure of angle $\angle KCJ$.

process: practice

## Interior Angles of a Circle (Optional)

8. Find the measure of $\angle KCQ$.

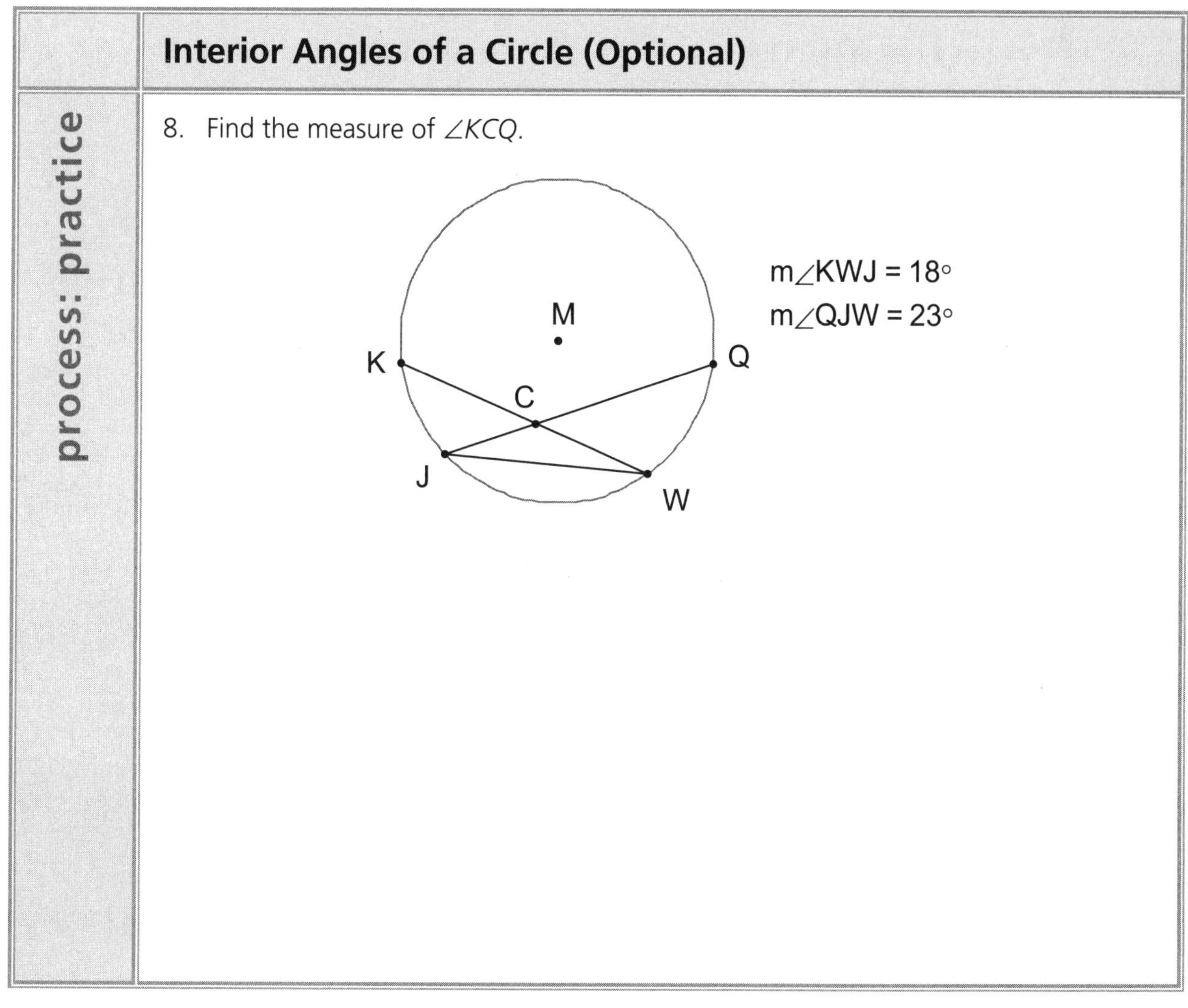

process: practice

## Interior Angles of a Circle (Optional)

9. Find the measure of ∠*KCQ*.

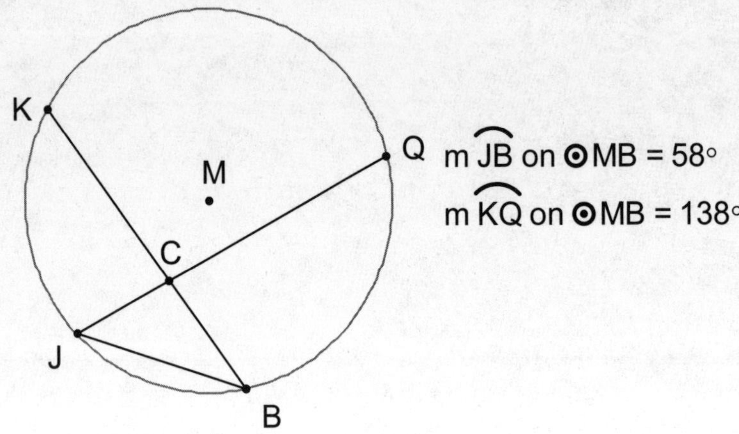


10. Find the measure of ∠*JCK*.

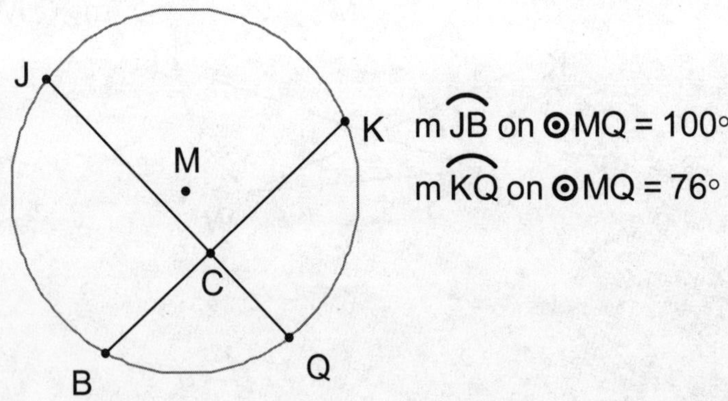

## Interior Angles of a Circle (Optional)

process: practice

11. Find the measure of $\angle KCQ$.

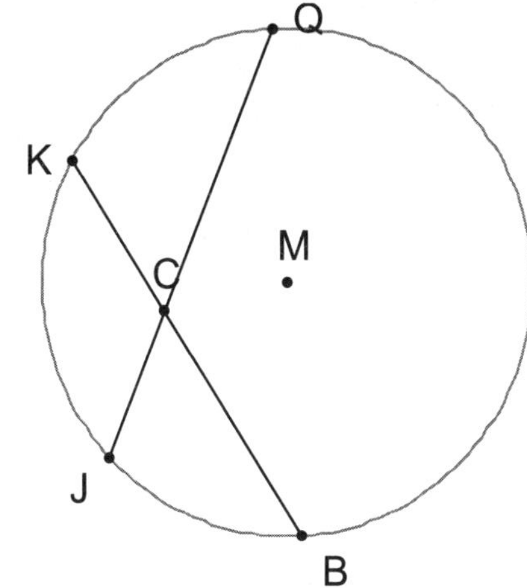


m $\overset{\frown}{JB}$ on ⊙MB = 50°

m $\overset{\frown}{KQ}$ on ⊙MB = 58°

## Interior Angles of a Circle (Optional)

analysis

12. How does m$\angle KCQ$ relate to the measure of a central angle that intercepts arc KQ? (You can draw $\overline{KM}$ and $\overline{MQ}$ to create $\angle KMQ$ if needed.)

13. How does m$\angle KCQ$ relate to the measure of an inscribed angle that intercepts arc KQ? (You can draw $\overline{KJ}$ or $\overline{QB}$ if you have not done so already.)

process: model

## Interior Angles of a Circle (Optional)

14.

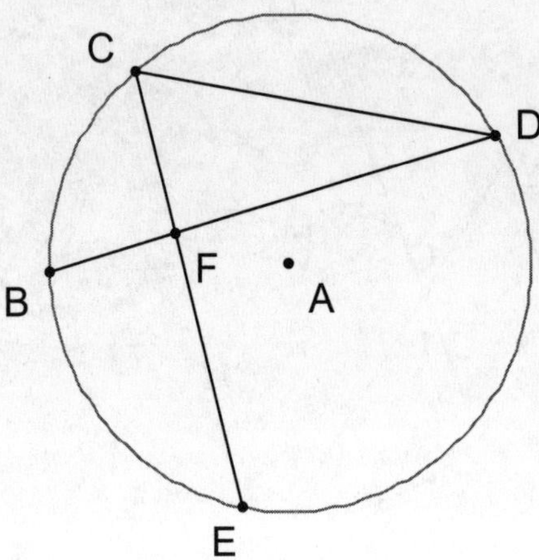


a. If the measure of arc BC = x°, find m∠*BDC*.

b. If the measure of arc DE = y°, find m∠*DCE*.

c. Use these two answers to find the m∠*DFE*.

d. How does the measure of an interior angle relate to the measures of the two arcs intercepted by it and its vertical angle?

e. Does this make sense? Why or why not?

process: summary

## Interior Angles of a Circle (Optional)

15. Design your own problem and give it to your neighbor to solve. Make sure you have provided enough information to find the measure of an interior angle of the circle. Refer to circle *A* to construct your problem.

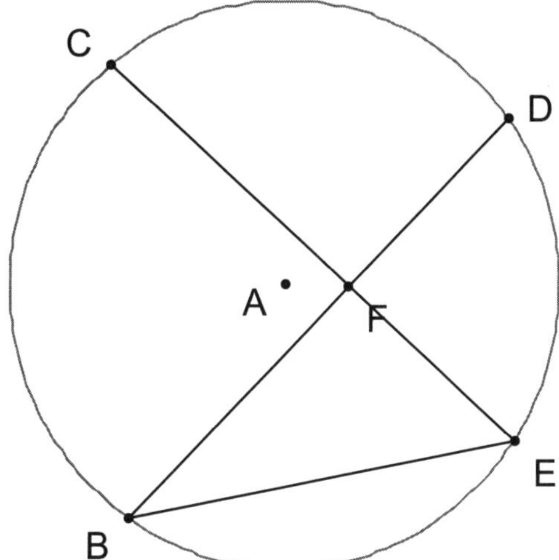


Given information:

Find the measure of interior angle __________:

## Exterior Angles of a Circle (Optional)

input

So far, you have learned to find the measure of angles whose vertices are at the center, inside the circle, or on the circle. How do we find the measure of angles whose vertices are located *outside* of a circle? **Exterior Angles** of a circle are angles whose vertices are outside the circle.

## Exterior Angles of a Circle (Optional)

process: evaluate

Consider circle *A* and find the measure of ∠*C*, located outside the circle.

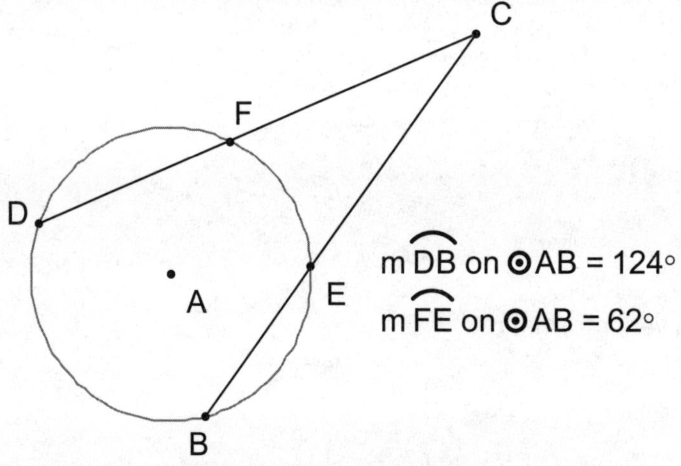


1. First, connect point D to E with $\overline{DE}$, and find m∠*DEB*.

2. What is the measure of ∠*CDE*? (Note that ∠*CDE* is the same as ∠*FDE*.)

3. What is the measure of ∠*C*?

## Exterior Angles of a Circle (Optional)

analysis

4. Explain in complete sentences how you used the measure of arcs FE and DB to find the measure of ∠C.

## Exterior Angles of a Circle (Optional)

process: practice

5. Find the measure of ∠W.

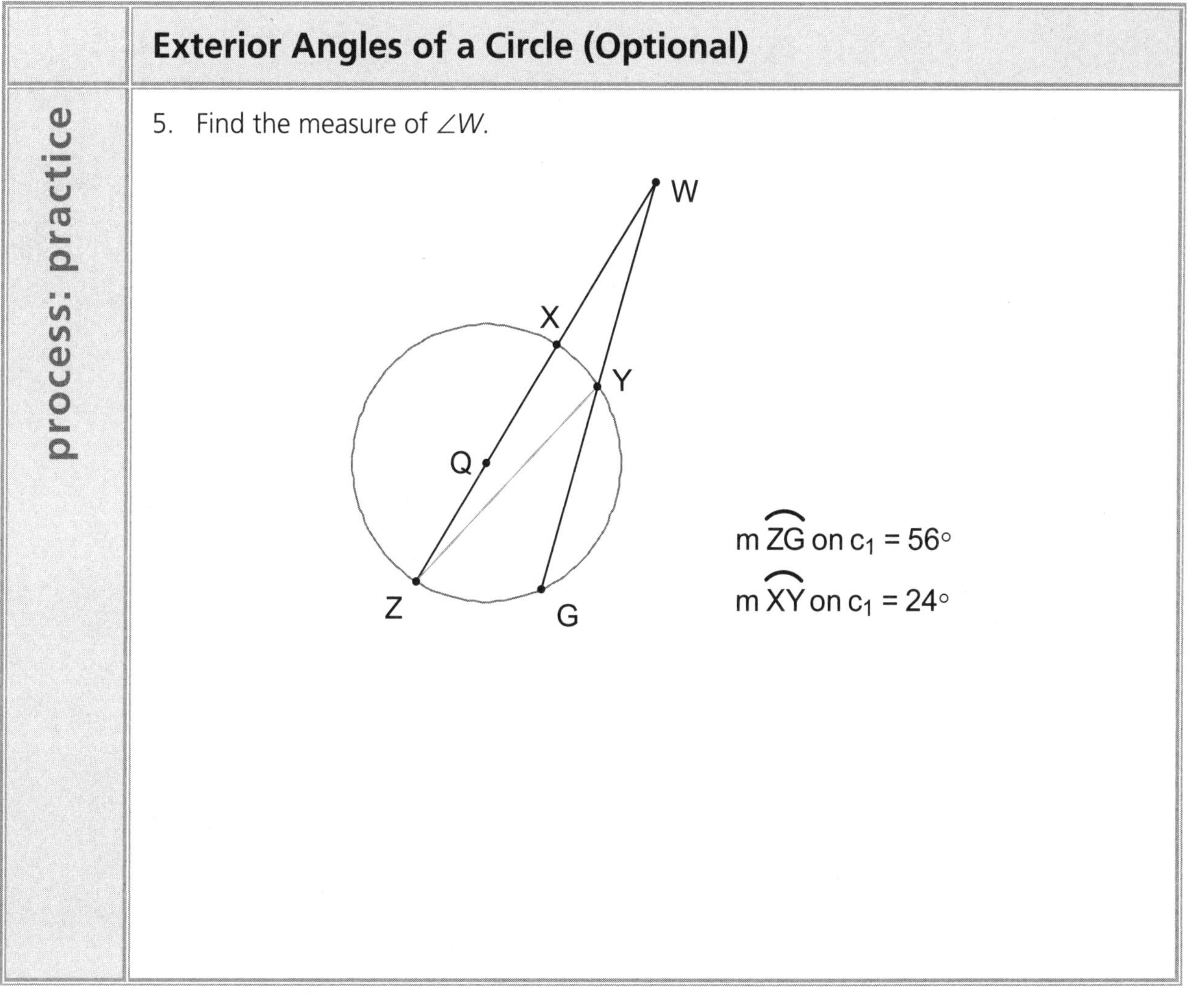

## Exterior Angles of a Circle (Optional)

process: practice

6. Find the measure of $\angle W$. Hint: Create $\overline{YZ}$ or $\overline{XG}$ first.

$m\widehat{ZG}$ on $c_1 = 114°$

$m\widehat{XY}$ on $c_1 = 44°$

## Exterior Angles of a Circle (Optional)

process: evaluate

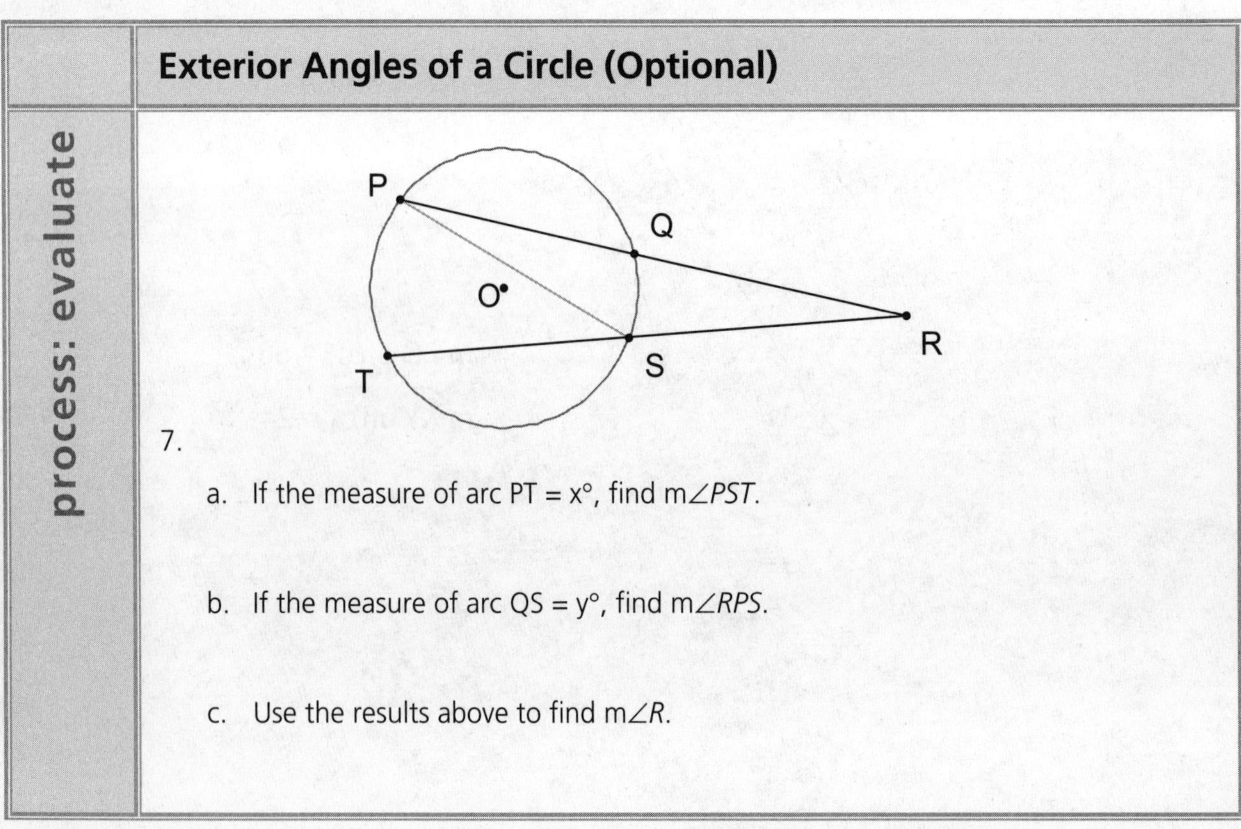


7.

a. If the measure of arc PT = x°, find $m\angle PST$.

b. If the measure of arc QS = y°, find $m\angle RPS$.

c. Use the results above to find $m\angle R$.

process: model

## Exterior Angles of a Circle (Optional)

8. How does the measure of an exterior angle relate to the measures of the two arcs it intercepts?

9. Does that make sense? Why or why not?

process: practice

## Exterior Angles of a Circle (Optional)

10. Find the measure of arc *GZ*.

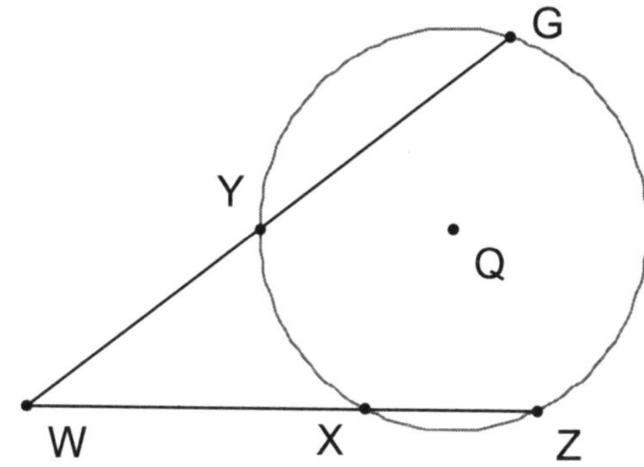


$m\angle ZWG = 37°$

$m\overset{\frown}{YX}$ on ⊙QG = 63°

11. $\overline{WZ}$ is a tangent segment. Find the measure of ∠*W* in circle *Q*.

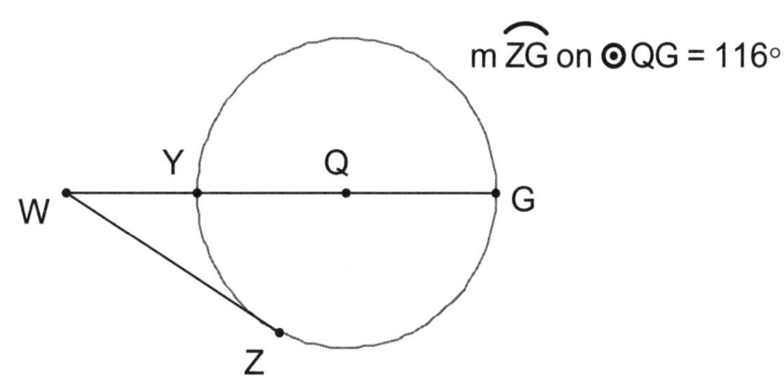

process: summary

## Exterior Angles of a Circle (Optional)

12. Design your own problem and give it to your neighbor to solve. Make sure you have provided enough information to find the measure of an exterior angle of the circle. Refer to this circle to construct your problem.

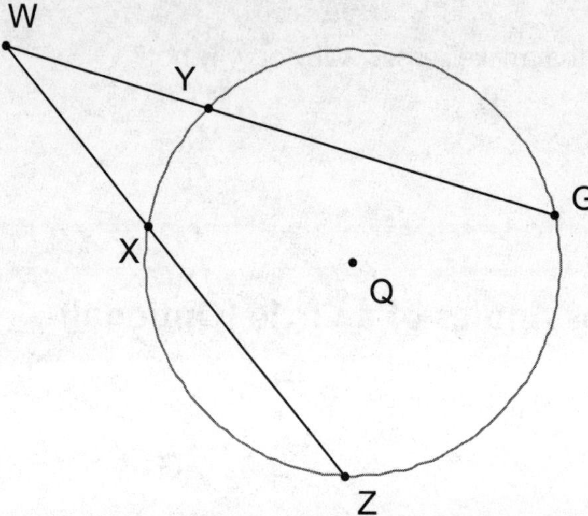


Given information:

Find the measure of $\angle W$:

process: summary

## Exterior Angles of a Circle (Optional)

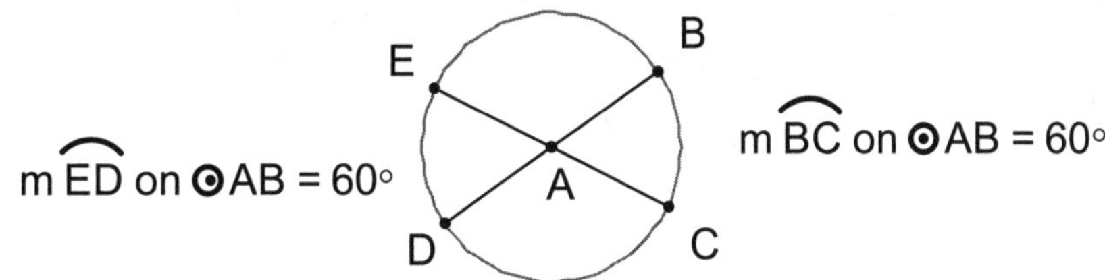


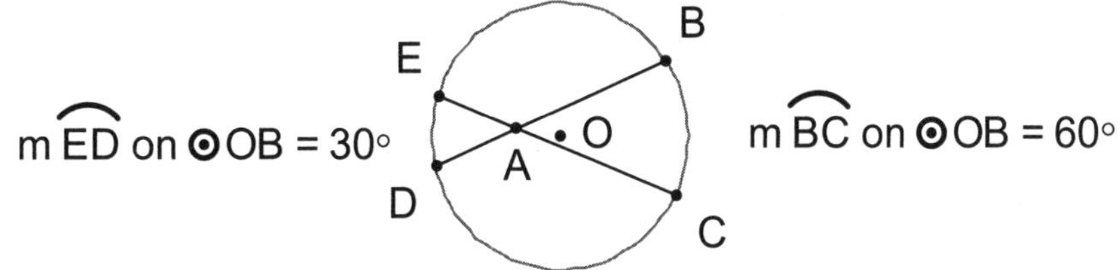


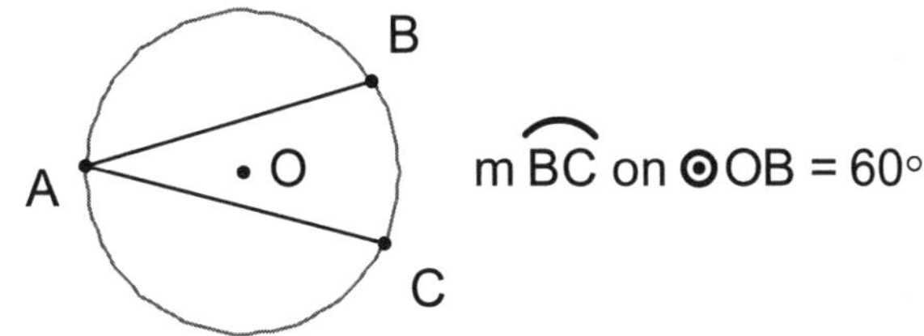


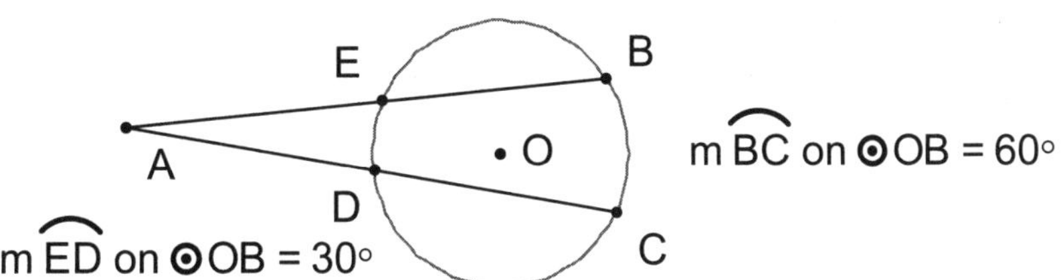


Pictured above are four circles showing the same arc BC with a measure of 60°. However, one shows a central angle, one an interior angle of a circle, the next an inscribed angle, and the last an exterior angle of a circle.

13. Calculate the m∠BAC and record it on the figure.

   a. What do you observe about the measure of ∠BAC as you work from one figure to the next?

   b. Explain why this observation is or is not in line with your previous observations about angles in circles.

## Arcs of a Circle

input

Previously, you saw that arcs of circles will determine certain angles, like central angles or inscribed angles. A **chord** also determines an arc. In circle *F*, chord *AB* determines arc *AB*.

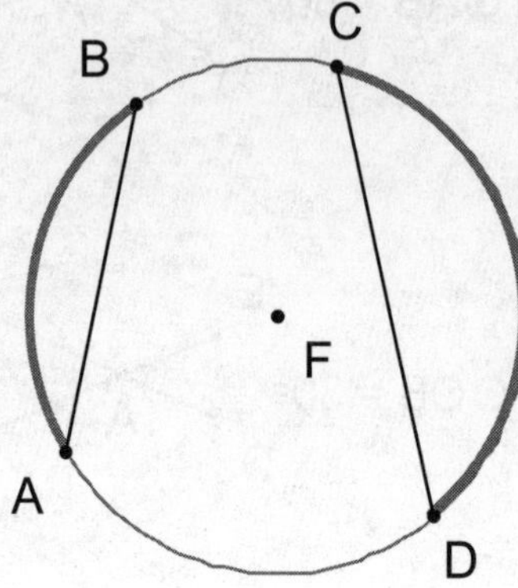


## Arcs of a Circle

process: model

1. Shade the arc determined by chord *CD*.
2. Which arc appears to be larger, arc *AB* or arc *CD?*

3. Which chord appears to be longer, chord *AB* or chord *CD*?

4. What is the relationship between the length of a chord and the size of the arc determined by that chord?

analysis

## Arcs of a Circle

5. Theorize about whether **congruent arcs** determine **congruent chords.** Test your theory in circle *H* by constructing two arcs, each measuring 90°. Draw the chords that determine those arcs. Measure the chords. Write a statement describing your conclusion.

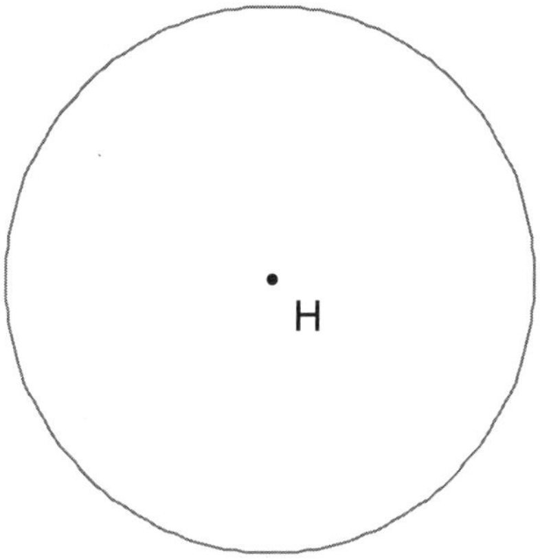

H

process: model

## Arcs of a Circle

6. Two points located on a circle divide the entire circle into two separate arcs. One is called a **major arc** and the other is called a **minor arc**.

   a. Shade the minor arc in circle *H* below.

   b. Using a different color or darker shading, shade the major arc in circle *H* below.

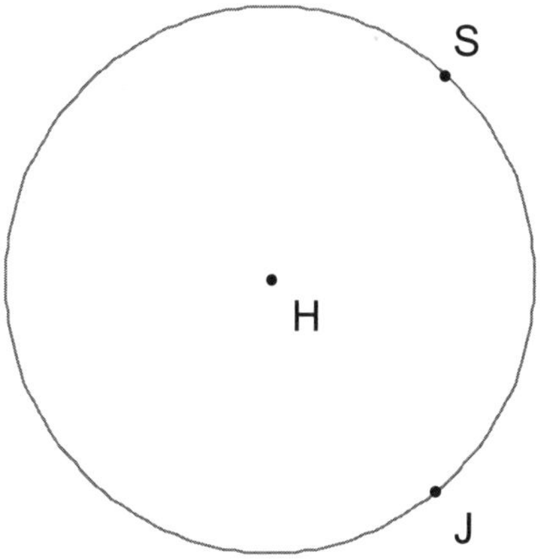

S

H

J

analysis

## Arcs of a Circle

7. Based on the names minor arc and major arc and based upon the circle pictured on the previous page, what is the difference between a major arc and a minor arc?

8. In circle *M*, is arc *PQ* a major arc or a minor arc? Explain your answer.

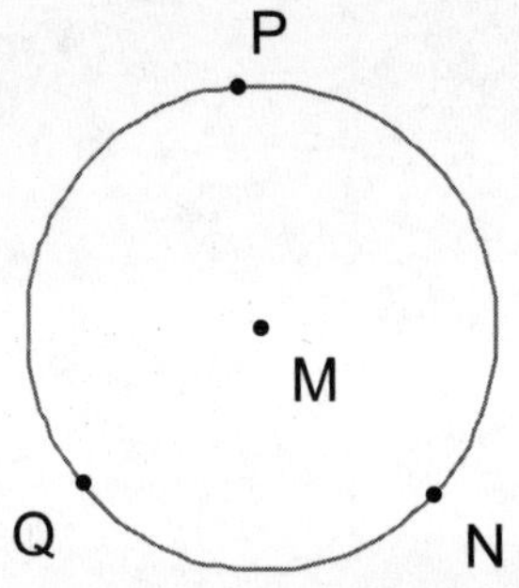

P

M

Q

N

9. Can you suggest a method for naming arcs that would help to make it clear whether the arc being named is a major arc or a minor arc?

10. What is a better name for the major arc PQ?

process: practice

## Arcs of a Circle

Refer to circle *C* to answer each of the following questions.

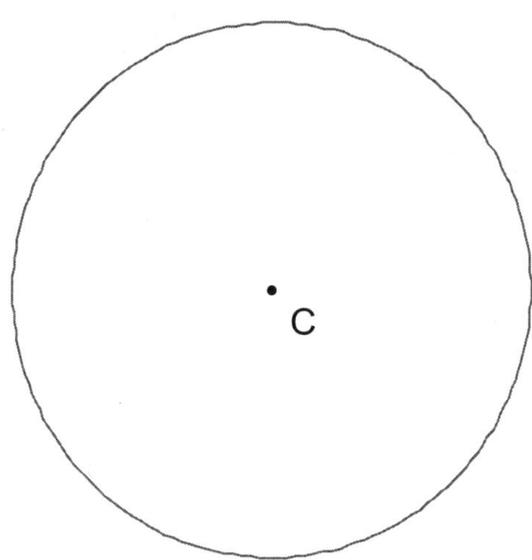


11. Locate points that determine a major arc measuring 200°. Use as few points as possible, and label your points. You may use any tools you choose.
12. What is the name of the major arc?
13. What is the name of the minor arc that completes the circle when added to the major arc?
14. What is the measure of this minor arc?
15. Are there any other minor arcs in this circle? If so, what are their names? If not, why not?
16. Are there any other major arcs in this circle? If so, what are their names? If not, why not?

## Arc Length

input

**Arc Length** is the length of an arc measured along the circumference of a circle. Imagine outlining a part of the circle with string and then stretching the string along a ruler. This length, measured in linear units such as centimeters or inches, is the arc length.

## Arc Length

process: evaluate

1. Find the circumference of circle *A*.

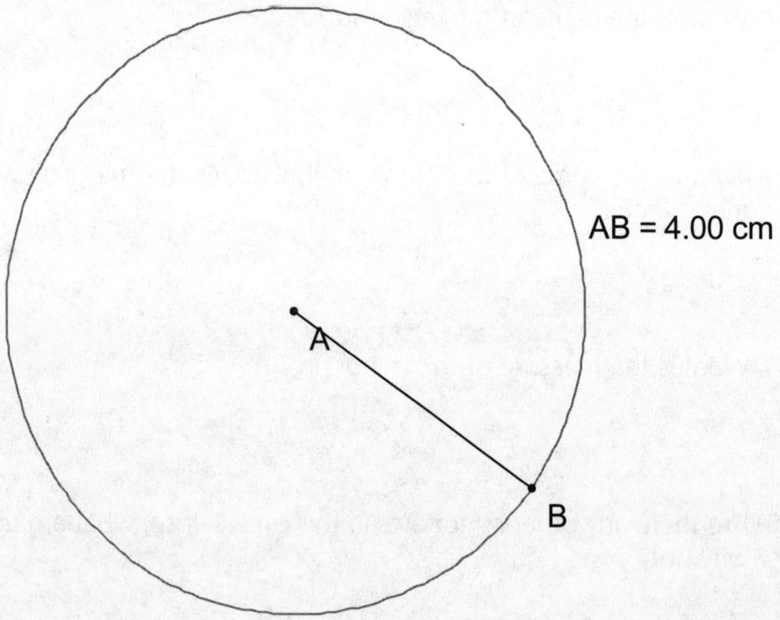

process: evaluate

## Arc Length

2. This circle is identical to the circle in question 1, with an additional radius. If arc *CDB* is a semicircle, what is the *length* of arc *CDB*?

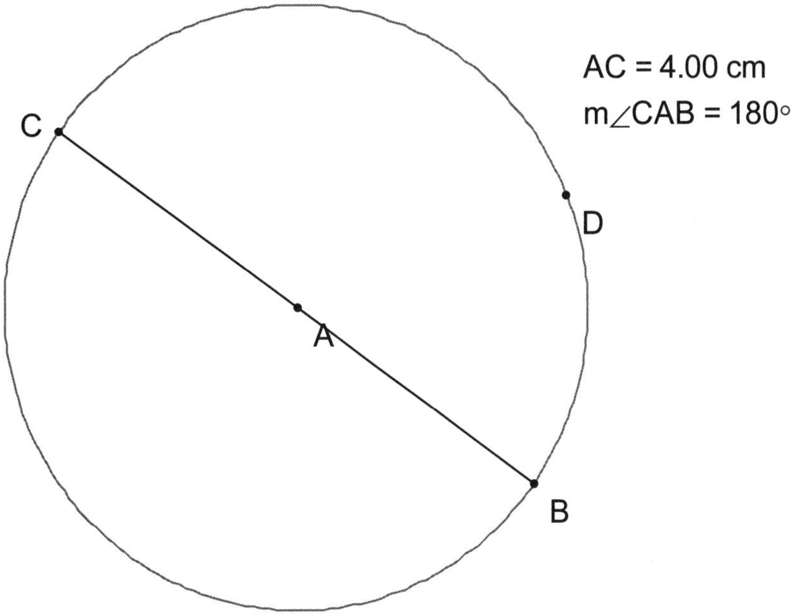

AC = 4.00 cm
m∠CAB = 180°

C
D
A
B

analysis

## Arc Length

a. Explain how you found the length of the arc. How is this method different than one you might have used to find the measure of arc *CDB?*

process: practice

## Arc Length

3. In this problem the measure of $\angle CAB$ is 90°. What is the length of arc *CB*?

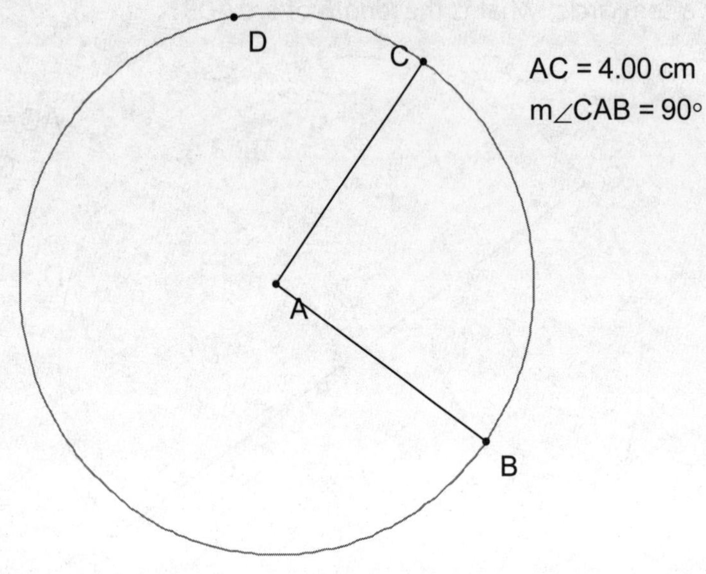


4. In this problem the measure of $\angle CAB$ is 40°. What is the length of arc *CB*?

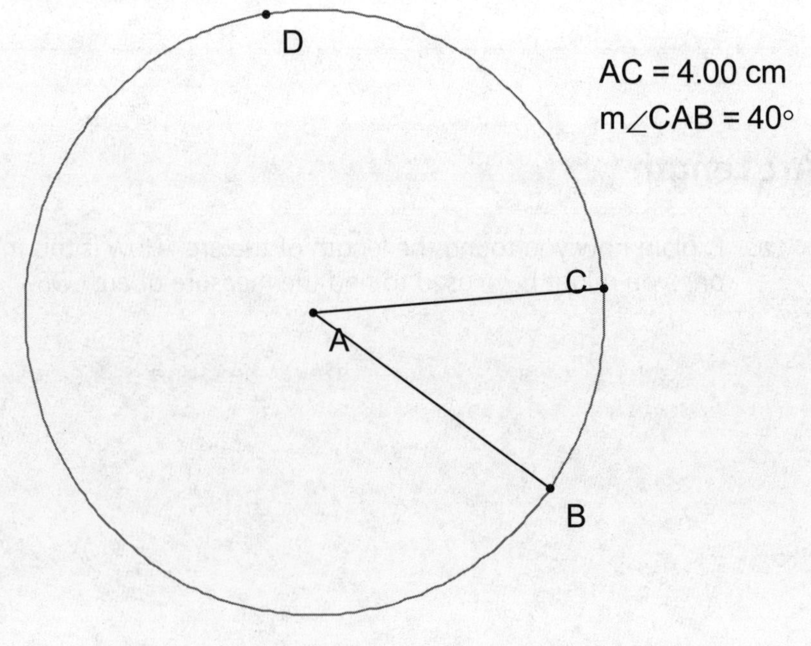

process: practice

## Arc Length

5. Given circle $O$ with $\angle MOB$, find the length of arc $MB$ *and* the measure of arc $MB$.

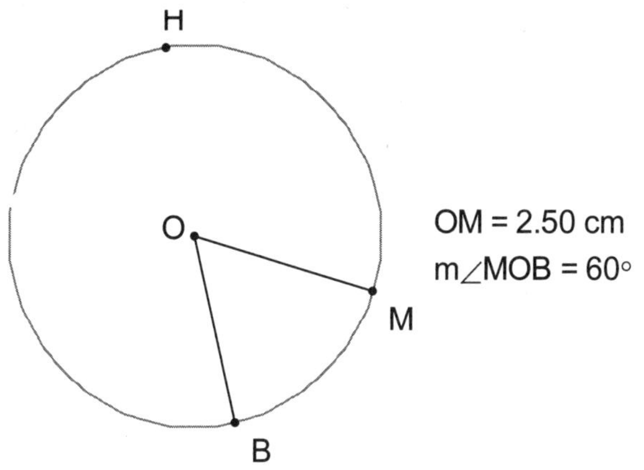

H
O
M
B
OM = 2.50 cm
m∠MOB = 60°

analysis

## Arc Length

6. Explain how you found the arc lengths in the previous three problems.

7. Describe the relationship between the degree measure of an arc and its arc length.

8. In general, explain the difference between the length of an arc and the measure of an arc.

process: summary

## Arc Length

9. Write clear, concise, and accurate instructions for finding arc length if you are given the measure of the arc and the length of the radius of the circle. Make up your own problem to illustrate your explanation.

## Sectors of a Circle

input

A sector of a circle is the region bounded by an arc and two radii of the circle.

In circle *A*, $\overline{AB}$, $\overline{AC}$, and arc *BC* are the borders of a sector. A sector looks like a slice of pie.

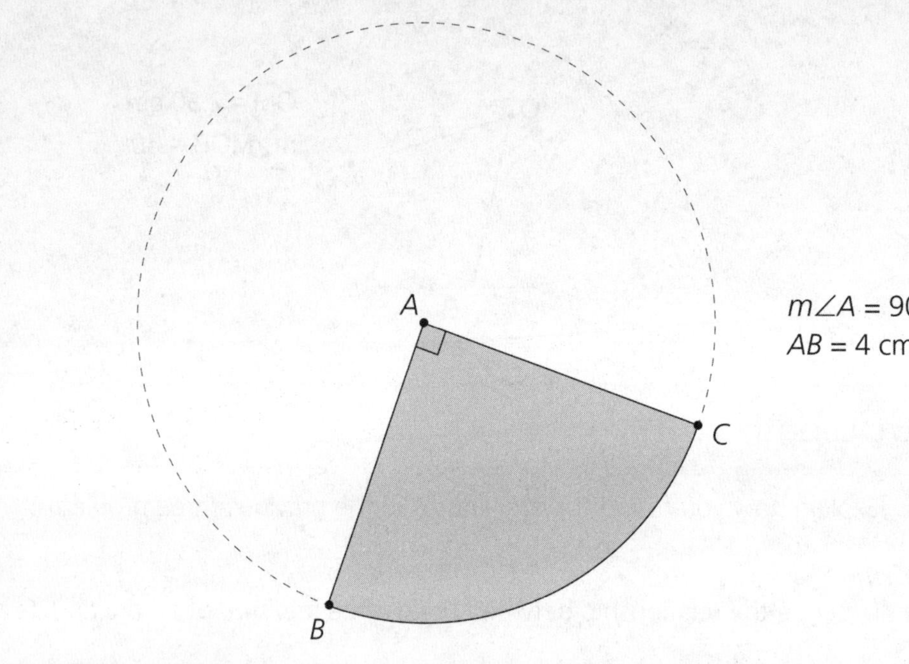


## Sectors of a Circle

process: evaluate

1. Given the radius in the circle above is 4 cm, what is the area of circle *A*?

2. Knowing the area of circle *A*, find the area of the given sector. Explain your reasoning.

process: evaluate

## Sectors of a Circle

3. Find the area of the given sector of circle $A$, by first finding the area of the circle. Explain your reasoning.

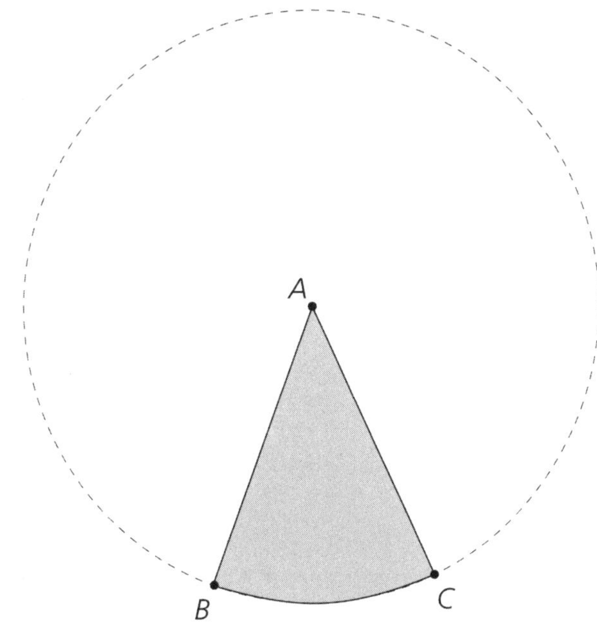


$AB = 4$ cm
$m\angle BAC = 45°$

4. Find the area of the given sector of circle $X$, by first finding the area of the circle. Explain your reasoning.

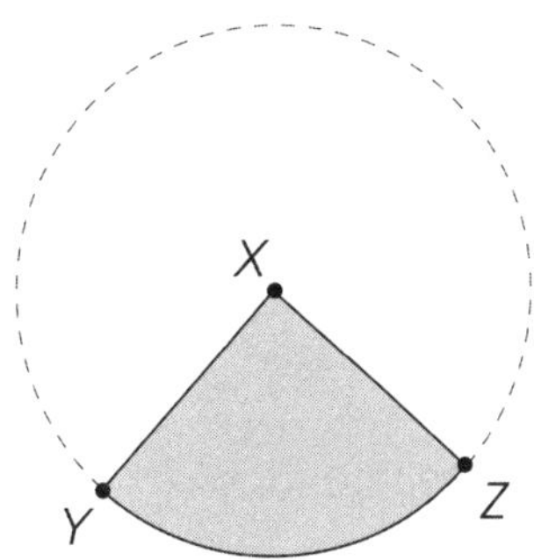


$XY = 2$ cm
$m\angle YXZ = 90°$

process: practice

## Sectors of a Circle

5. The following are measures of central angles that define sectors of a circle. Find the fraction of a circle represented by the sector.

a. What fraction of the interior of a circle is determined by a 90° central angle?

b. Given a 60° central angle, what fraction of the interior of a circle is the sector?

c. What fraction of the interior of a circle is the sector created by a 15° central angle?

d. If the central angle is 100°, what fraction of the interior of a circle is the sector?

e. What fraction of the interior of a circle is the sector determined by a 36° central angle?

process: practice

## Sectors of a Circle

6. Suppose the area of a circle is $24\pi$ square units. Find the area of the sectors of this circle that correspond to the given central angle. Show your work.

   a. 45°

   b. 15°

   c. 100°

   d. 36°

analysis

## Sectors of a Circle

7. How does the area of a sector of a circle relate to the area of the circle? Explain how you know this.

8. Use a compass to create a circle with a radius of your choosing. Find the area of your circle.

   a. Within your circle, create a central angle that measures 50°. What portion or fraction of a circle is determined by your 50° central angle?

   b. Explain how you use that fraction and the area of the circle to find the area of the sector?

9. If the radius of a circle is *r* units, what is the area of the circle? (Create a sketch.)

   a. If the measure of a central angle in the same circle is *m* degrees, what portion of the circle is determined by the central angle?

   b. Use your results to the last two questions to create a formula for finding the area of a sector.

## Sectors of a Circle

process: practice

10. If the radius of a circle is 15 cm and the central angle that determines a sector measures 30°, find the area of the sector. Show your work.

11. If the circumference of a circle is $14\pi$ units and the central angle that determines a sector measures 40°, find the area of the sector. Show your work.

12. If the area of a sector is $6\pi$ square units and the central angle that determines that sector measures 15°, find the area of the circle. Show your work.

process: practice

## Sectors of a Circle

13. If the area of a circle is $30\pi$ square units and the area of a sector of that circle is $5\pi$ square units, what fractional portion of the interior of the circle does the sector represent? Show your work.

14. If the central angle that determines a sector measures 180°, what can you conclude about the area of the sector? Draw an illustration of this situation.

process: summary

## Sectors of a Circle

15. Design your own problem about the sector of a circle. Solve it here first. Then write the problem on another sheet of paper and give it to a neighbor to solve.

## Segments of a Circle (Optional)

A **segment of a circle** is the region bounded by an arc of the circle and its chord.

In circle *A*, $\overline{BC}$ and arc *BC* are the borders of a segment of the circle.

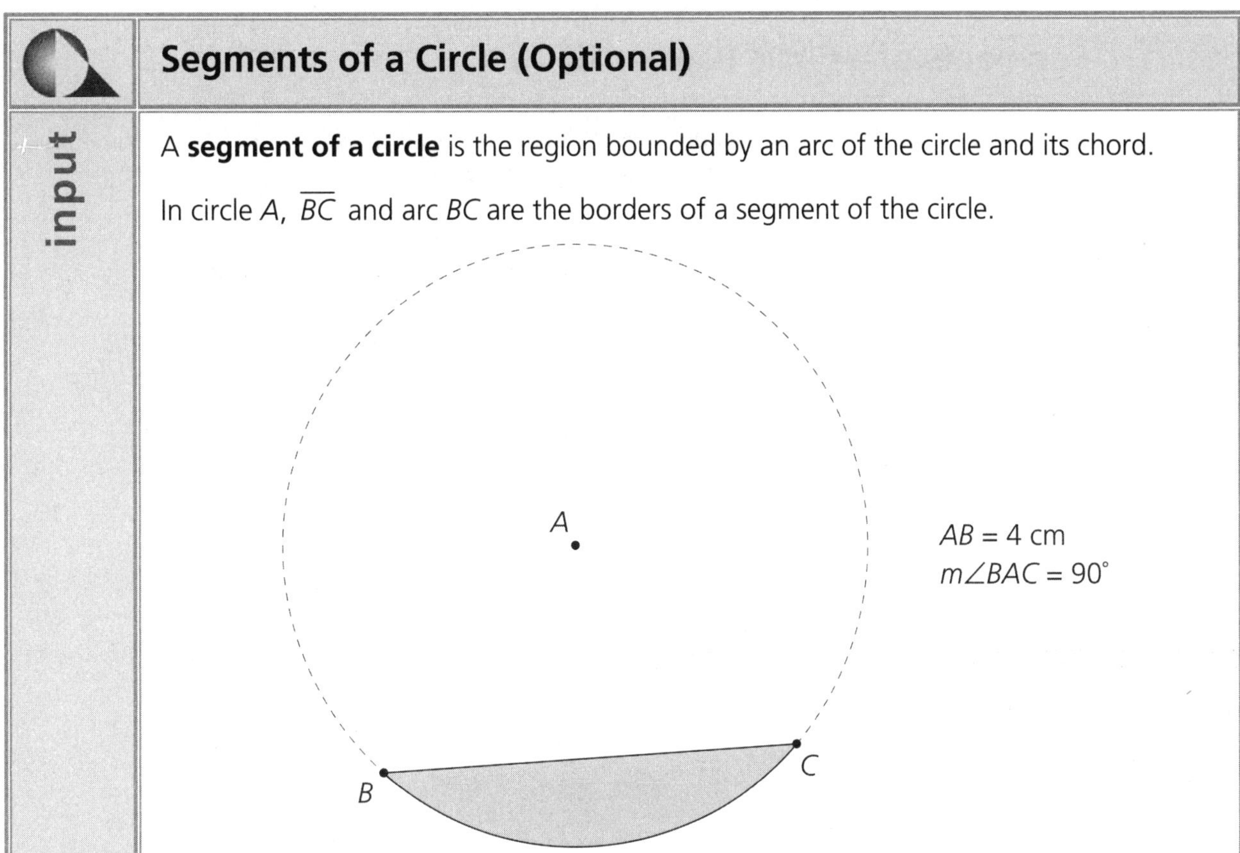


## Segments of a Circle (Optional)

1. Get together with your group and brainstorm ways to find the area of a segment of a circle. Refer to the segment of circle *A* in your discussion. Write a brief description of your method for finding the area of a segment of a circle. (Hint: Add one or more line segments to the figure, and use what you have learned about sectors.)

## Segments of a Circle (Optional)

process: evaluate

2. Try using the three steps listed below to find the area of the segment of circle *A* bounded by $\overline{BC}$ and arc *BC*. Work with a friend, if you like. Show your work.

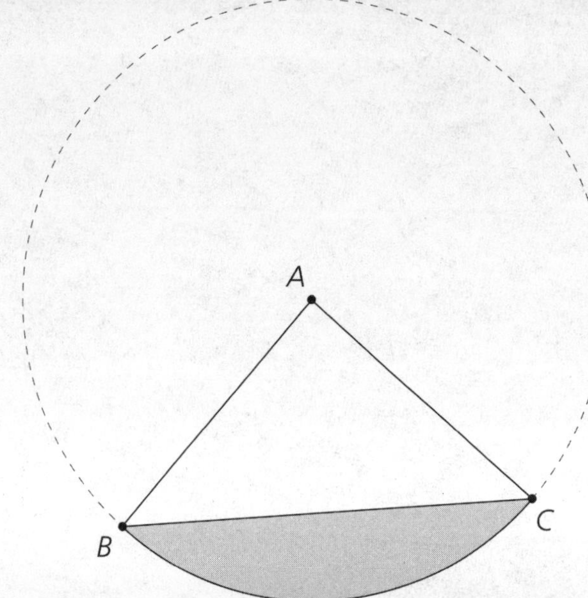


$AB = 4$ cm
$m\angle BAC = 90^\circ$

a. Find the area of the sector.

b. Find the area of $\Delta ABC$.

c. Find the area of the segment using the answers above.

process: practice

## Segments of a Circle (Optional)

3. Refer to the information given in this figure to find the area of the segment of circle *A* bounded by $\overline{CB}$ and arc *CB*. Show your work. (Hint: Use what you know about special right triangles or trigonometric ratios to find the height of the triangle.)

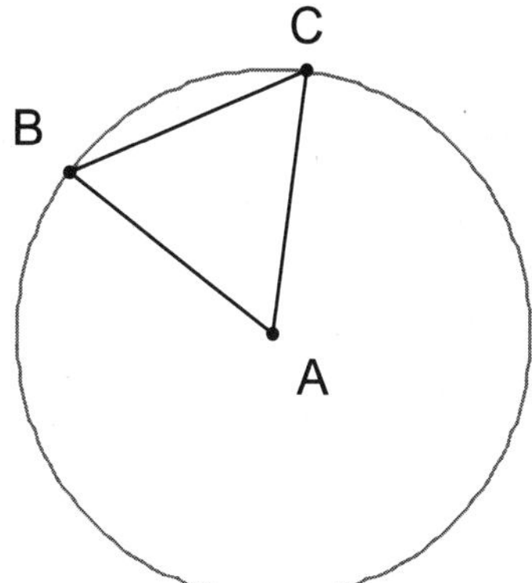


CA = 2.23 cm
CB = 2.23 cm

## Chords of a Circle

process: model

1. Using a ruler, draw a chord of length 5 cm in circle C. Create a segment that will measure the distance from the center of the circle to the chord.

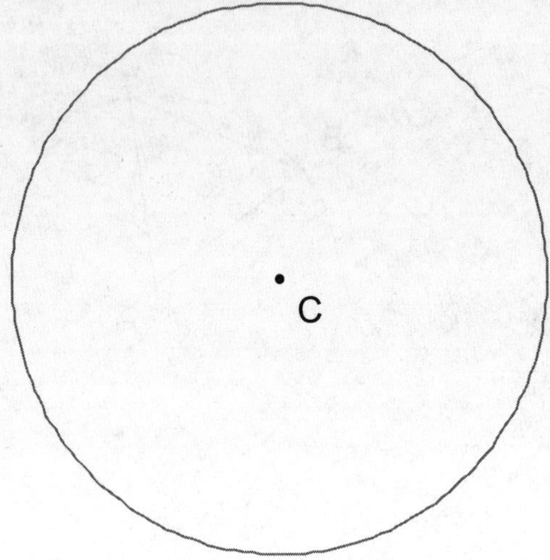


   a. Can you create another segment measuring the distance from the center of the circle to the chord that is the shortest one possible? If so, create it. If not, why not?

   b. How far is the chord from the center?

   c. What is the relationship between the chord and the segment measuring its distance from the center?

2. Refer to circle C in question 1. Using a ruler, draw a second chord of length 5 cm. Create the shortest segment possible to measure the distance of this chord from the center.

   a. How far is this second chord from the center of the circle?

## Chords of a Circle

process: model

3. If it is possible to create another chord of 5 cm in circle C, use your ruler to draw it. Measure its distance from the center and record it here.

   a. How many more 5 cm chords can be drawn in this circle?

## Chords of a Circle

analysis

4. What observations can you make about congruent chords and their distance from the center of a circle?

## Chords of a Circle

process: model

5. Use your ruler to draw the longest chord possible in circle *C*.

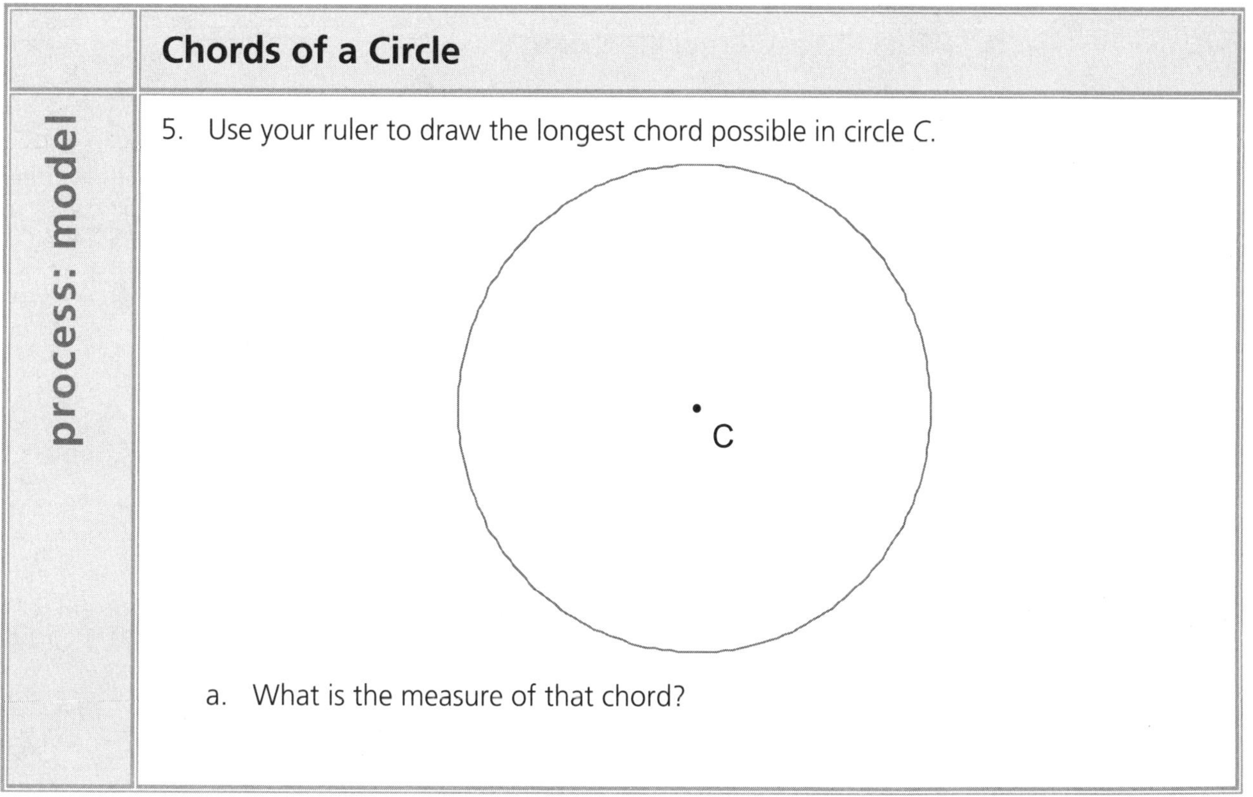


   a. What is the measure of that chord?

## Chords of a Circle

process: model

b. How do you know that the chord you drew in question 5 is the longest possible chord?

c. How many other chords of the same length can be drawn in this circle?

d. What do all of these chords have in common?

e. How far are these chords from the center of the circle?

6. Refer to circle *C* in question 5. Create the shortest chord you can in circle *C*.

a. What is the measure of this chord?

b. How many other chords of the same measure can be drawn?

c. What do all of these chords have in common?

d. How far are these chords from the center of the circle?

process: model

## Chords of a Circle

7. Create a chord in circle C that is not a diameter. Measure its length and record the length on the chord.

C

a. Create the perpendicular bisector of that chord.

b. Create a second chord in circle C. Measure its length and record the length.

c. Create the perpendicular bisector of the second chord.

analysis

## Chords of a Circle

d. Where do the two perpendicular bisectors intersect?

e. Create a third chord along with its perpendicular bisector. Are the three chords you created concurrent (i.e., do they intersect at the same point)?

## Chords of a Circle

analysis

f. Create a hypothesis to explain why all of the perpendicular bisectors of the chords in a circle intersect at the center?

8. Locate the center of this circle by drawing chords and their perpendicular bisectors.

## Tangents of a Circle

process: model

1. In terms of the number of intersection points, a line and a circle can be related in three different ways. Draw a line in each figure to illustrate these different relationships. Next to each figure, write a statement describing the relationship between the circle and the line.

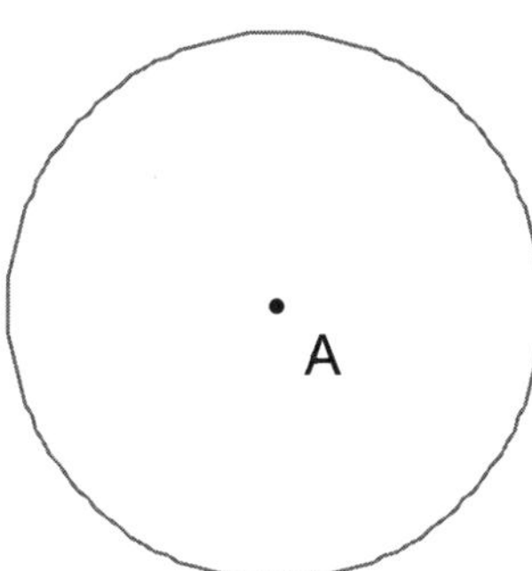


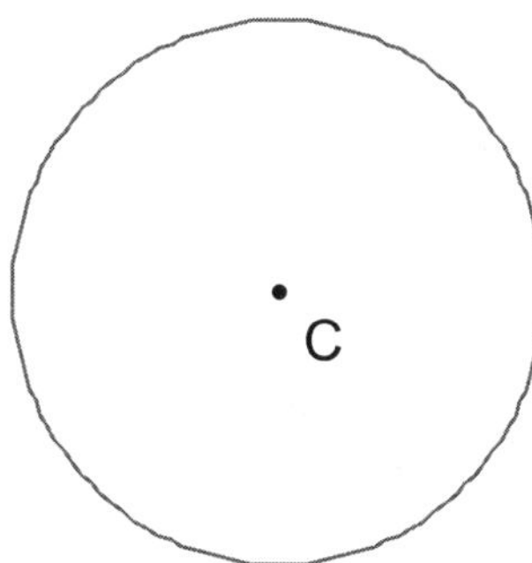


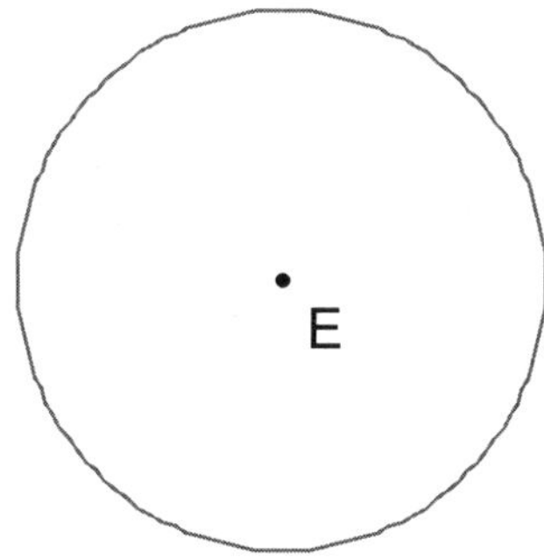

input

## Tangents of a Circle

A tangent to a circle is a line that intersects the circle at exactly one point. Line DT in this figure is a tangent to circle Z. Point T is called the point of tangency.

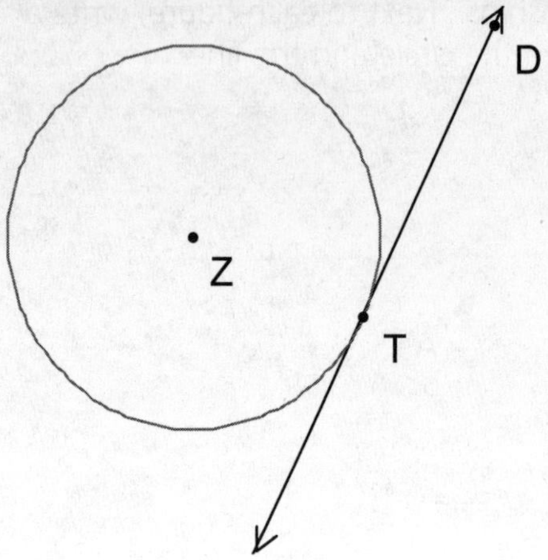

analysis

## Tangents of a Circle

2. How many different tangents to one circle can be drawn?

3. How many different tangent lines can be drawn to point T on circle Z above?

process: model

## Tangents of a Circle

4. If $K$, $D$, and $E$ on circle $P$ are points of tangency, draw the tangent lines.

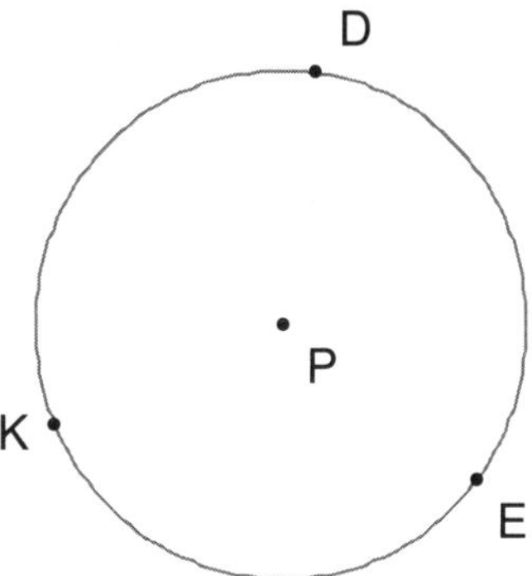


   a. Use a ruler or straightedge to draw the radii $\overline{PE}$, $\overline{PD}$, and $\overline{PK}$.

   b. Measure the angles formed by the radii and the tangents at points P, E, and K. What do you notice about the measures of these angles?

5. Use these circles to verify or dispute the special relationship you found.

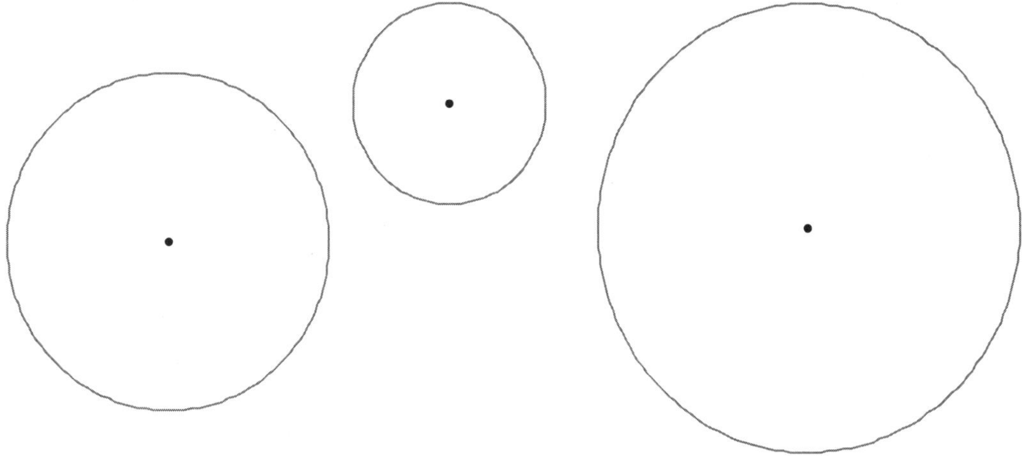

   a. Draw a radius in each circle and a tangent to each circle. Make sure one endpoint of the radius is also the point of tangency. Measure the angle formed by the radius and the tangent.

   b. Does the relationship you discovered between the tangent and the radius drawn to the point of tangency hold true?

## Tangents of a Circle

process: model

6. How many different lines can be drawn through point P that are tangent to circle A?

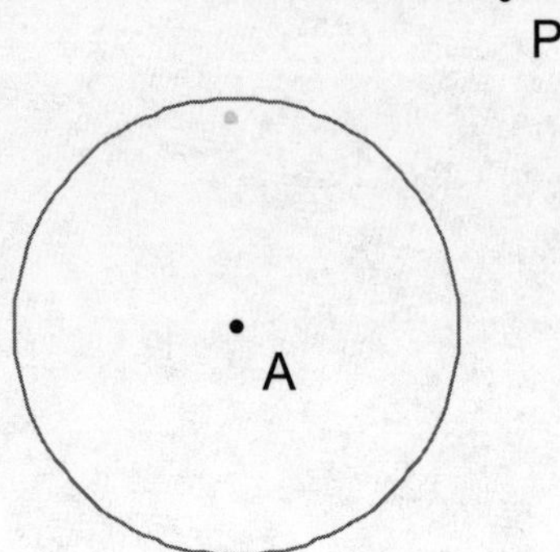


   a. Draw the possible tangent(s) and label the point(s) of tangency.

7. Locate a point anywhere outside the circle and label it.

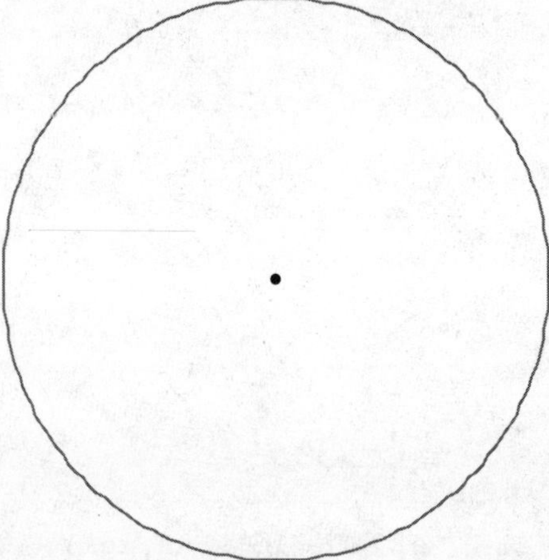

   a. How many different lines that are tangent to the circle can be drawn through the outside point you created?

   b. Draw the possible tangent(s) and label the point(s) of tangency.

## Tangents of a Circle

input

Point *P* is located outside of circle *A*. Two lines, *PG* and *PB*, are drawn through point *P* and tangent to circle *A*.

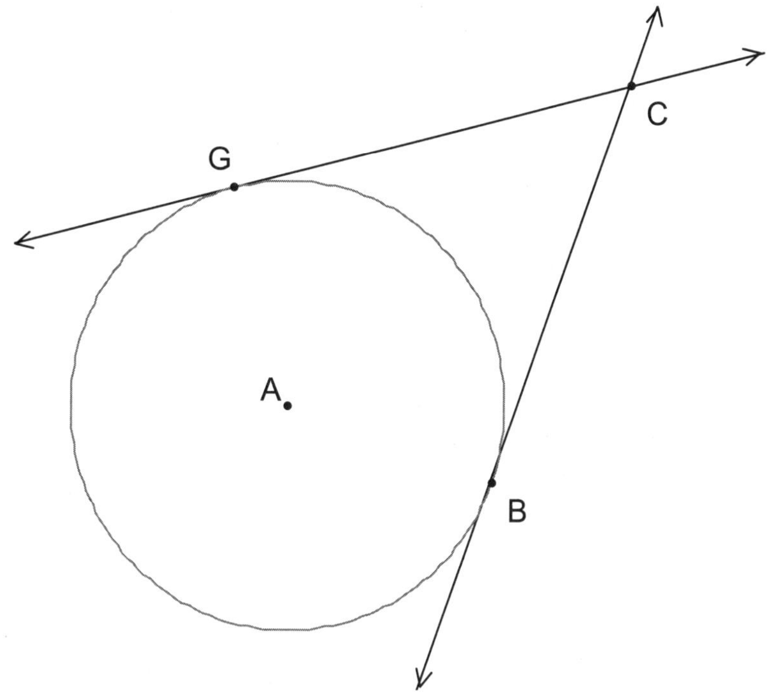


Line segment *PG* and line segment *PB* are called **tangent segments**. A tangent segment is a segment that is located on a tangent line with one endpoint at the point of tangency.

## Tangents of a Circle

analysis

8. Two tangents are associated with each point located outside a circle. Why?

 a. Measure the tangent segments in your circles for questions 6 and 7, and measure $\overline{PB}$ and $\overline{PG}$. What appears to be true about the lengths of the two tangent segments drawn from the same point outside the circle?

## Tangents of a Circle

process: model

9. Refer to these circles to verify what you have found.

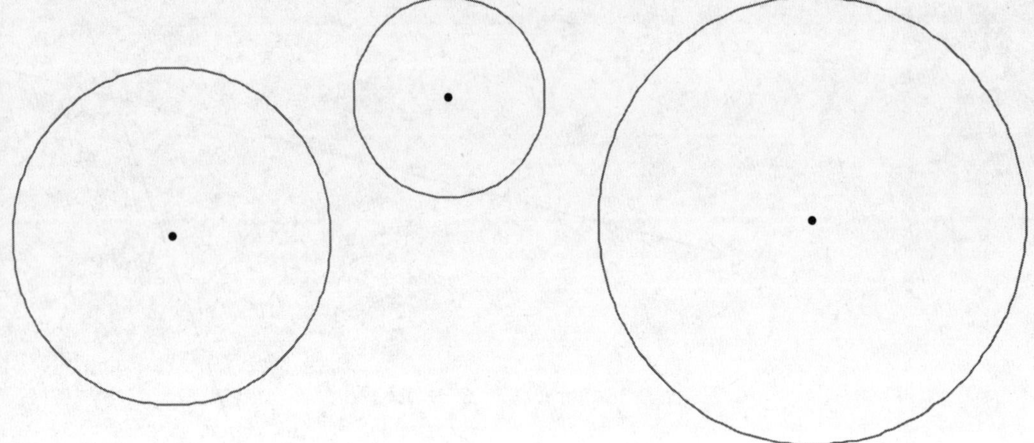

a. Locate one point outside each circle, and draw the tangent lines to each circle.

b. Measure each pair of tangent segments.

c. Does your conclusion regarding pairs of tangent segments hold true?

## Tangents of a Circle

process: model

10. Think about the different ways two circles can intersect each other. Create at least two drawings showing the different ways circles can intersect one another.

    a. Is it possible for two circles to intersect each other at exactly one point? If you have not done so above, create at least one drawing showing two circles that intersect at exactly one point. If the circles cannot intersect at one point, explain why.

    b. In how many different ways can you draw circles intersecting each other at exactly one point? If there is more than one way for circles to intersect in exactly one point, then create drawings of the additional way(s). If only one way is possible, explain why.

input

## Tangents of a Circle

Tangent circles are circles that intersect at exactly one point. One of these figures illustrates circles that are internally tangent. The other figure illustrates circles that are externally tangent. Can you tell which one is which and why?

process: model

## Tangents of a Circle

11. Return to your drawing(s) of circles that intersect in exactly one point. Label the drawing(s) as internally tangent or externally tangent.

process: summary

## Tangents of a Circle

12. Summarize your findings about tangent lines, tangent segments, radii, and circles.

extension

## Tangents of a Circle

13. Lines *TJ* and *TH* are drawn tangent to circle *A*. Find the measure of ∠*JTH*, *using only the given information*, and explain your reasoning.

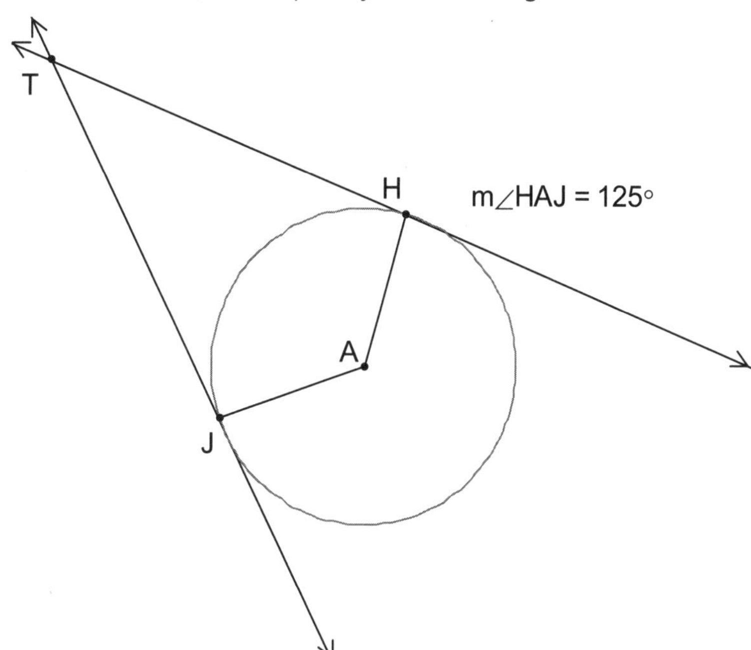


14. If the measure of ∠*HAJ* is changed to 62.5° (half its size in question 12), will the measure of ∠*JTH* also shrink to half the size it is in question 12? Explain your answer.

extension

## Tangents of a Circle

15. Lines *KP* and *KG* are drawn tangent to circle *W*. Find the measure of ∠*KPG*, and explain your reasoning.

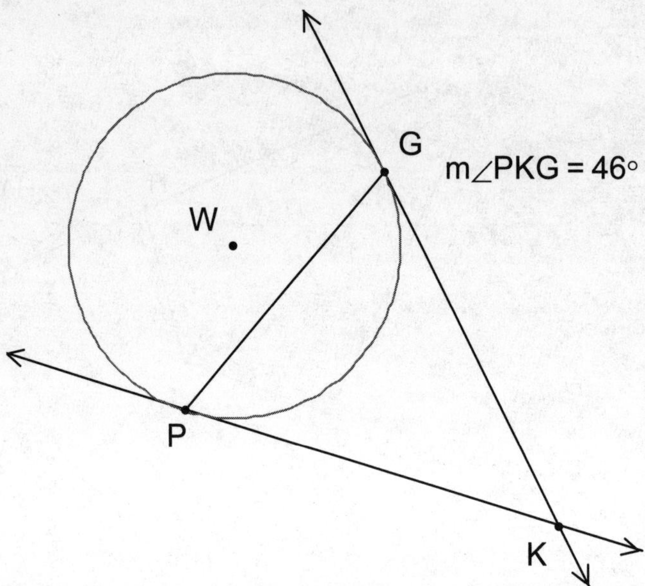


16. Explain why ∠*KPG* and ∠*KGP* in problem 14 are not right angles.

extension

## Tangents of a Circle

17. Line *PS* is drawn tangent to circle *M*. Find the measure of $\angle MPS$, and explain your reasoning.

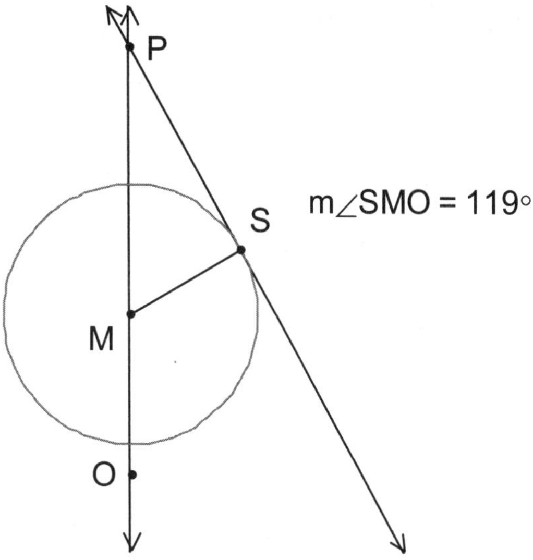


18. Line *XZ* is drawn tangent to circles *T* and *V*. If the measure of $\angle T$ is 8° more than the measure of $\angle V$, what can you conclude? Explain your reasoning.

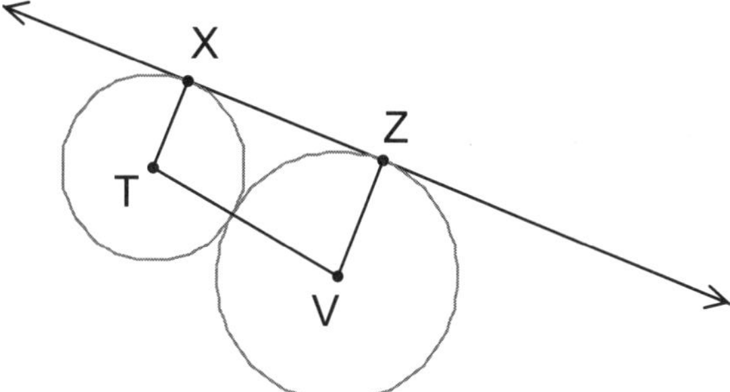

# Unit 7: Periodic Functions

## Contents

## You're Getting Warmer

input

A thermostat controls the temperature in many homes, keeping the house warm in the winter and cool in the summer. A programmable thermostat allows the home owner to set the temperature of the home over periods of time so that energy is not wasted when nobody is at home.

In winter, a home owner sets the thermostat as follows:

| | | |
|---|---|---|
| Wake | 5:00 a.m. | 68°F |
| Leave | 8:00 a.m. | 62°F |
| Return | 5:00 p.m. | 68°F |
| Sleep | 10:00 p.m. | 62°F |

These settings maintain an overnight temperature of 62°F. The temperature begins to rise at 5:00 a.m. to 68°F. The temperature stays at 68°F until 8:00 a.m., when it goes back down to 62°F. The temperature stays at 62°F during the day and begins to rise again at 5:00 p.m. for an evening temperature of 68°F. This temperature is maintained until 10:00 p.m., when the temperature drops again to its overnight level of 62°F.

## You're Getting Warmer

process: model

1. Sketch a graph showing the temperature of the house in winter as a function of time. Show the temperature over at least three days.

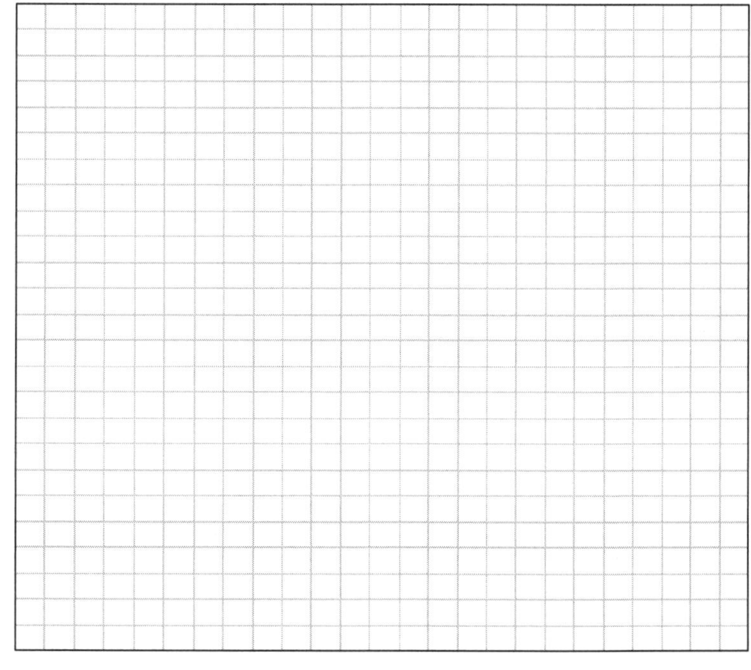

## You're Getting Warmer

analysis

2. Using sentences, describe the graph you created.

3. What is different about this graph when compared to many other graphs you have previously created?

4. Some programmable thermostats work differently. They gradually change the temperature so that the desired temperature is reached at the time set on the thermostat instead of starting the temperature change at that time. In this example, that means that the house begins warming up before 5:00, and the temperature reaches 68°F at 5:00. If the thermostat reached the desired temperature at the time programmed, how would your graph change?

## Tower Drop

input

The Tower Drop is a new thrill ride at Oaks Point amusement park. This ride's car climbs 200 feet from the station straight up a tower over a period of 16 seconds. When it reaches the tower, the car pauses for 4 seconds and then drops towards the ground. Riders feel as if they will crash into the ground below, but at the last second, the ride slows and returns the riders safely to the starting point. The descent takes about 4 seconds. When one set of thrill seekers disembarks, a new set of riders enters the car. They are strapped in for safety, and 40 seconds later, the Tower Drop starts to climb again.

## Tower Drop

process: model

1. Create a graph of the height of the ride's car over time for at least three repeats of the ride.

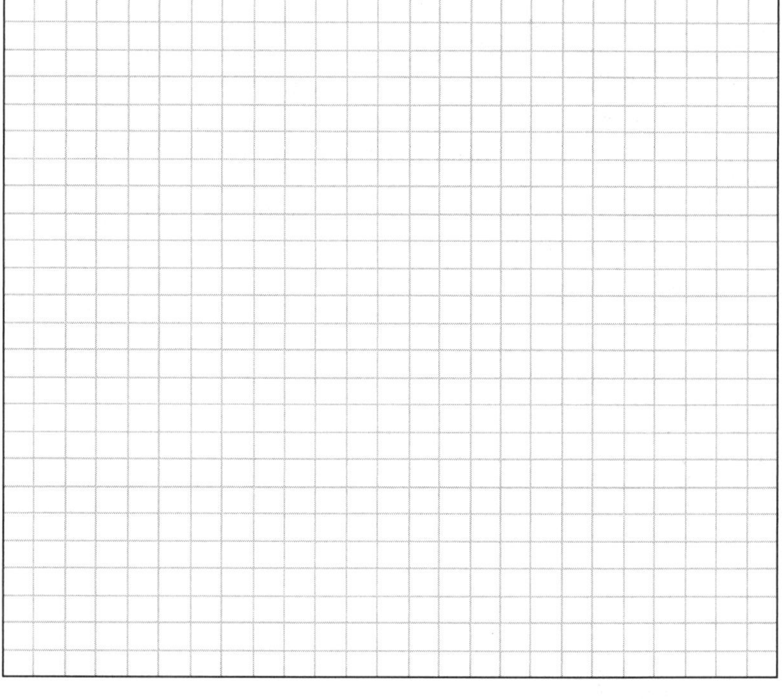

## Tower Drop

analysis

2. Compare and contrast your graphs for *You're Getting Warmer* and *Tower Drop*.

## Tower Drop

input

You may have noticed that the graphs you created for *You're Getting Warmer* and *Tower Drop* demonstrate a repetitive behavior. The temperature in one problem and the height of the ride in the other problem rise and fall in a regular pattern. Graphs that demonstrate this type of repetition are called **periodic**. The functions modeling these problem situations are **periodic functions**.

The **period** of a function is the interval over which one cycle or repetition of the function pattern occurs.

## Tower Drop

analysis

3. For *Tower Drop*, over what period of time does the height pattern repeat itself? In other words, what is the period of the function?

4. For *You're Getting Warmer*, over what period of time does the temperature pattern repeat itself? What is the period of the function?

analysis

## Tower Drop

5. Does the graph for *Tower Drop* have one absolute maximum or absolute minimum value like a parabola?

6. Does the graph for *Tower Drop* have relative maxima or minima? If so, what are the maxima and minima?

process: evaluate

## Tower Drop

7. Every periodic function has relative maxima and relative minima.
   a. What is the maximum $y$-value for the Tower Drop problem?
   b. What is the minimum $x$-value for the Tower Drop problem?
   c. What is the difference between the maximum and minimum values? In other words, what is the range of possible values for this periodic function?

8. The absolute value of half of the difference between the values of the maxima and minima is called the **amplitude** of a periodic function. What is the amplitude of the *Tower Drop* function?

9. What is the range and amplitude of the *You're Getting Warmer* function?

## Gum on a Bicycle Wheel

input

While riding to school on your bicycle, you ran over a piece of gum which stuck to the wheel. It went around and around the wheel for the remainder of the trip.

After seeing the gum stick to the wheel, you began watching the big wad of pink gum move around on the bicycle wheel. You found yourself wondering about the movement of the sticky gum with respect to the road on which you were riding. Later at school, you pull out some paper from your notebook and being to analyze the situation using the fact that the diameter of your bicycle wheel is 28 inches.

## Gum on a Bicycle Wheel

process: model

1. Create a rough sketch of what you think the position of the gum with respect to the turn of the bike wheel will look like.

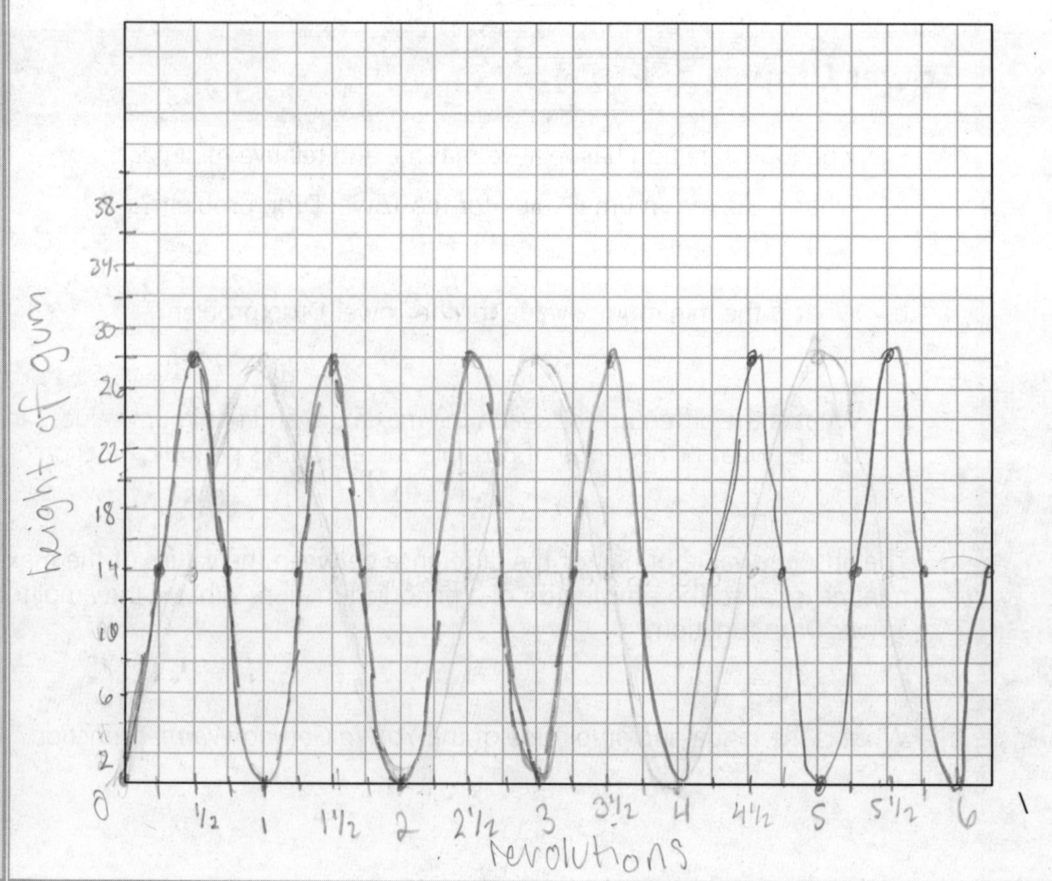

## Gum on a Bicycle Wheel

process: model

2. Create a table of values and a graph of the distance of the gum above the ground with respect to the amount of turn of the bicycle wheel.

| Amount of Turn of the Bicycle Wheel | Height of Gum |
|---|---|
| Revolutions | Inches |
| 0 | 0 |
| ¼ | 14 |
| ½ | 28 |
| ¾ | 14 |
| 1 | 0 |
| 1 ¼ | 14 |
| 1 ½ | 28 |
| 1 ¾ | 14 |
| 2 | 0 |
| 2 ½ | 14 |
| 3 | 28 |

3. Often, the amount of turn is measured in degrees rather than revolutions.

a. How many degrees are in one revolution?

360° in one revolution.

b. How many degrees are in two revolutions?

720° in two revolutions.

c. How many degrees are in ½ revolution?

180° in 1/2 a revolution

d. How many degrees are in 1¼ revolutions?

450° in 1 and 1/4 a revolutions.

process: model

## Gum on a Bicycle Wheel

4. Add a column to your table that shows the amount of turn being measured in degrees.

| Amount of Turn of the Bicycle Wheel | | Height of Gum |
|---|---|---|
| Revolutions | Degrees | Inches |
| 0 | 0 | 0 |
| ¼ | 90 | 14 |
| ½ | 180 | 28 |
| ¾ | 270 | 14 |
| 1 | 360 | 0 |
| 1 ¼ | 450 | 14 |
| 1 ½ | 540 | 28 |
| 1 ¾ | 630 | 14 |
| 2 | 720 | 0 |
| 2 ½ | 900 | 28 |
| 3 | 1080 | 0 |

5. Write a sentence describing how to determine the number of degrees given the number of revolutions.

there are 360° in one revolution, so ½ a revolution would be 180°, when cut into 4th it's 90° portions so each ¼, add 90°.

6. Write an algebraic function for determining the number of degrees given the number of revolutions.

degrees = 360r

(r=revolutions) d = 360r

7. Write the algebraic inverse of the function from question 6.

$r = \frac{d}{360}$

8. In a sentence, explain what the inverse function represents.

the inverse shows degrees divided by 360° to represent the number of revolutions

## Gum on a Bicycle Wheel

**process: evaluate**

9. Determine the number of degrees represented by the given amount of revolution.

   a. $\frac{1}{12}$ of a revolution

   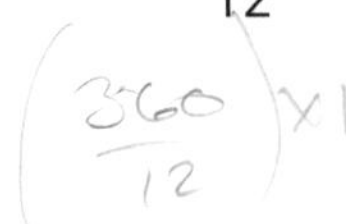

   b. $\frac{1}{8}$ of a revolution

   c. $\frac{5}{12}$ of a revolution

   d. $\frac{2}{3}$ of a revolution

   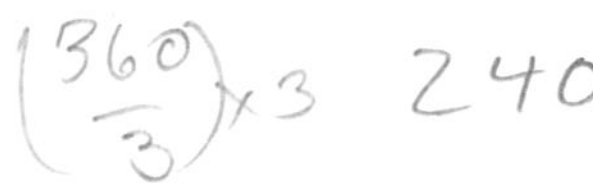

   e. $\frac{5}{6}$ of a revolution

   f. $\frac{7}{8}$ of a revolution

10. Determine the amount of revolution represented by the given number of degrees.

    a. 60°

    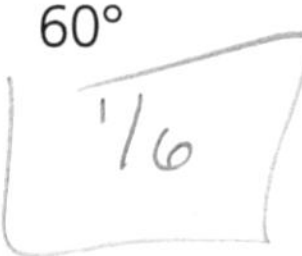

    b. 120°

    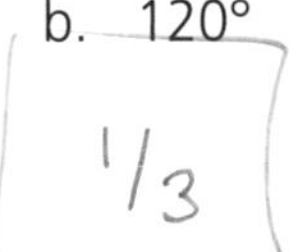

    c. 135°

    d. 210°

    e. 225°

    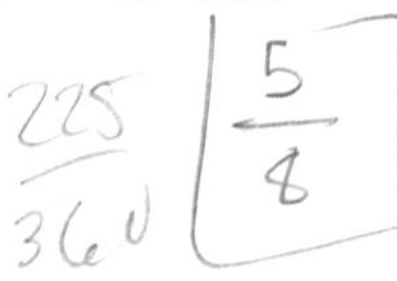

    f. 330°

## Gum on a Bicycle Wheel

process: model

11. The circle in the drawing below represents a bicycle wheel, and the line segment represents the ground.

   a. Mark a point where the wheel meets the ground to represent where the gum first sticks to the wheel. Label this point A.

   b. Mark a point to represent the position of the gum after a turn of 45°.Label this point B.

   c. Draw radii to each of the points you have marked. Label the center point C.

   d. Create a right triangle by drawing a segment from point B perpendicular to segment AC (and parallel to the ground). Label the intersection of AC and the perpendicular point D.

C 45 D B A

## Gum on a Bicycle Wheel

process: evaluate

12. Which special right triangle have you formed?

the 45-45-90 triangle

13. Determine the length of each side of the special right triangle. Remember that the radius of the bicycle wheel is 14 inches.

$\frac{14}{\sqrt{2}}$ = 9.9 inches

hypotenuse = 14 inches

each other sides = 9.9 inches

14. What is the height of the gum after the wheel has turned 45°?

14.0 − 9.9

4.1 inches height

process: model

## Gum on a Bicycle Wheel

15. The circle in the drawing below represents a bicycle wheel, and the line segment represents the ground.
    a. Mark a point where the wheel meets the ground to represent where the gum first sticks to the wheel. Label this point A.
    b. Mark a point to represent the position of the gum after $\frac{1}{6}$ of a revolution. Label this point B.
    c. Draw radii to each of the points you have marked. Label the center point C.
    d. Create a right triangle by drawing a segment from point B perpendicular to segment AC (and parallel to the ground). Label the intersection of AC and the perpendicular point D.

C
$7\sqrt{3}$
14
7
B

process: evaluate

## Gum on a Bicycle Wheel

16. Which special right triangle have you formed?

30-60-90 TRIANGLE

17. Determine the length of each side of the special right triangle. Remember that the radius of the bicycle wheel is 14 inches.

60
30

18. What is the height of the gum after the wheel has turned $\frac{1}{6}$ of a revolution?

process: model

## Gum on a Bicycle Wheel

19. Use your work from the previous two pages to complete the table below. You may need to make some additional computations.

| Amount of Turn of the Bicycle Wheel | Height of Gum |
|---|---|
| Degrees | Inches |
| 0 | 0 inch |
| 30 | inch |
| 45 | inch |
| 60 | inch |
| 90 | inch |
| 120 | inch |
| 135 | |
| 150 | |
| 180 | 14 |
| 210 | |
| 225 | |
| 240 | |
| 270 | |
| 300 | |
| 315 | |
| 330 | |
| 360 | 0 |
| 450 | |
| 540 | |
| 630 | |
| 720 | |
| 900 | |
| 1080 | |

## The Ferris Wheel

**input**

At the amusement park, there is a Ferris wheel with a diameter of 50 feet. It takes a total of two minutes to complete one revolution. Today, you will examine how the rider's distance from the ground depends on the time since the Ferris wheel started. To do this, you will construct a model of the Ferris wheel on the grid below by drawing a circle. Then you will find the distance off the ground at various points on the circle.

## The Ferris Wheel

**process: model**

1. Use a compass to construct a circle in the middle of the grid so that it has a radius of 25. Why should the radius be 25 units?

**process: model**

## The Ferris Wheel

2. Divide the circle into twelve equal parts by following the directions below.
    a. Draw a horizontal diameter of the circle.
    b. Draw a vertical diameter of the circle (perpendicular to the horizontal diameter).
    c. Use a protractor to divide the newly formed 90° angles into 30° angles.
3. You should now have 6 diameters drawn in your circle that divide it into 12 equal pie-shaped pieces. Label the points at the ends of the diameters with the letters A through L. Label the lowest point on the circle "A" and continue labeling the points alphabetically in a counter-clockwise direction.

**process: model**

## The Ferris Wheel

4. The circle you drew represents the Ferris wh[illegible]el. Determine when the rider will be at the various labeled points if the rider starts a[illegible] point A. Do this for three complete revolutions of the Ferris wheel by completing the first three columns of the table

| | Time when the Ferris wheel reach[illegible]s the point | | | Distance from the ground |
|---|---|---|---|---|
| | 1st revolution | 2nd revolution | 3[illegible]d revolution | |
| | seconds | seconds | seconds | feet |
| A | 0 | 120 | 240 | 0 |
| B | 10 | 130 | [illegible]50 | 3.5 |
| C | 20 | 140 | [illegible]60 | 12.5 |
| D | 30 | 150 | 2[illegible]0 | 25 |
| E | 40 | 160 | 2[illegible]0 | 37.5 |
| F | 50 | 170 | 29[illegible] | 46.7 |
| G | 60 | 180 | 300 | 50 |
| H | 70 | 190 | 310 | 46.7 |
| I | 80 | 200 | 320 | 37.5 |
| J | 90 | 210 | 330 | 25 |
| K | 100 | 220 | 340 | 12.5 |
| L | 110 | 230 | 350 | 3.5 |

## The Ferris Wheel

**process: model**

5. Using the model you created for the Ferris wheel, determine the height of the rider above the ground at each of the points A through L by counting gridlines. Record your answers in the last column of the table.

6. Construct a graph of the height of the rider above the ground with respect to the time since the ride started.

   a. Label each of the axes and scale your graph appropriately.

   b. Plot the points from your table.

   c. Connect these points with a smooth curve.

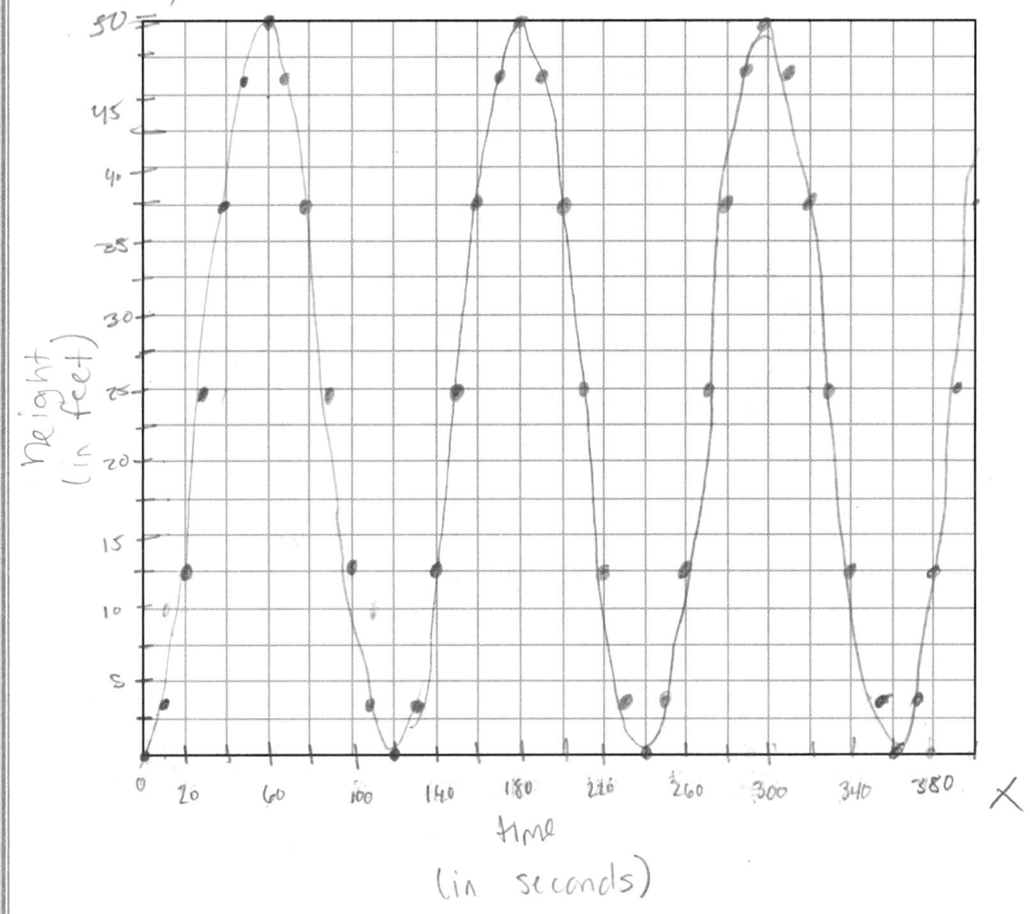

## The Ferris Wheel

**analysis**

7. Answer the following questions based on the function you graphed.

   a. What are the domain and range of the function?

   x∈ℝ
   the domain is x is all real numbers from -∞ to ∞.
   the range is between 0 and 50 inclusive.

   b. What are the x- and y-intercepts of the function, and what do they represent in the problem situation?

   The x-int. are at every multiple of (120, 0) *including (0,0)
   they represent when the wheel starts a new cycle from the ground (they happen at every multiple of 120)
   The y-int is at (0,0) where the wheels first begins (initial starting point)

   c. What are the maxima and minima of the function?

   the maxima is 50 feet and
   the minima is 0 feet.
   the maxima occur every 60 seconds added by a mult. of 120. (60 + 120x)
   the minima occur every 0 seconds added by a mult. of 120. (0 + 120x)

   d. What are the intervals of increase or decrease for the function?

   the intervals of increase happen every first 60 seconds of a cycle. The intervals of decrease happen the last 60 seconds of a cycle.
   (increase: 120x to (120x + 60)
   decrease from 60 to 120 ...)

   e. Is this function continuous or discontinuous? If it is discontinuous, what are the points of discontinuity?

   This is a continuous function. there are no points of discontinuity.

   f. What is the period of the function?

   The period of the function is 120 seconds

   g. What is the amplitude of the function?

   The amplitude of the function is 25 feet.

## Radians

input

Another way to measure angles is called **radian** measure. The values you found for the lengths of the intercepted arcs in the unit circle are also the radian measures of their central angles. In other words, the radian measure of an angle is equal to the length of its intercepted arc in the unit circle.

## Radians

process: evaluate

1. Every angle can be measured in revolutions, degrees, or radians. Complete the table below for equivalent measures of angles.

| Angle Measure | | |
|---|---|---|
| **revolutions** | **degrees** | **radians** |
| $\frac{1}{12}$ | | |
| $\frac{1}{8}$ | | |
| | 60 | |
| | | $\frac{\pi}{2}$ |
| | | $\pi$ |
| $\frac{5}{8}$ | | |
| | 360 | |
| 2 | | |

process: model

## Radians

2. This exercise helps you to better understand radians by creating a way to measure them. To do this, you will need a cylindrical object (e.g., a soup can) and some string.

   a. Measure the diameter of your cylindrical object and calculate its radius. On the string, mark off units equal in length to the radius of the cylinder to create a number line. Mark off at least seven units of one radius length.

   b. Wrap the string around your cylindrical object. How many radii are needed to go around the cylinder one time? Is this consistent with the arc length measure you got for a 360° angle in the unit circle?

   c. Place the cylindrical object on a sheet of paper and trace around it to draw a circle. Leave the cylinder on the paper and wrap the string around the bottom of the cylinder. Mark off an arc of the circle equal to one radius unit. This is also called one **radian**.

   d. Locate the center of your circle. (Hint: Paper folding can help you do this.) Draw a central angle that intercepts your arc of one radius unit. This angle will measure one radian.

   e. Use a protractor to measure the angle in degrees. How many degrees is the angle? How many revolutions? How many radians?

input

## Radians

The radian measure of a central angle is formally defined as the number of radius units in the length of its intercepted arc. The relationship with the radius is how radian measure got its name.

You now know three ways to measure an angle: revolutions, degrees, and radians. For example, all of the following are equivalent measures of an angle: one-fourth of a revolution, 45 degrees, and $\frac{\pi}{4}$ radians.

## Sine Language

analysis

3. How do the equations of the functions you graphed relate to the similarities and differences in the graphs?

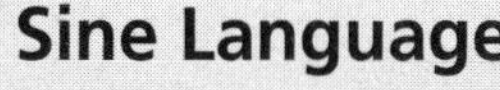

## Sine Language

process: model

4. Graph the following functions on your calculator (set to radian measure), and sketch them on the grid below. Show the parent function over at least two periods.

   a. $y = \sin \theta$  b. $y = \sin 2\theta$  c. $y = \sin \frac{1}{2}\theta$

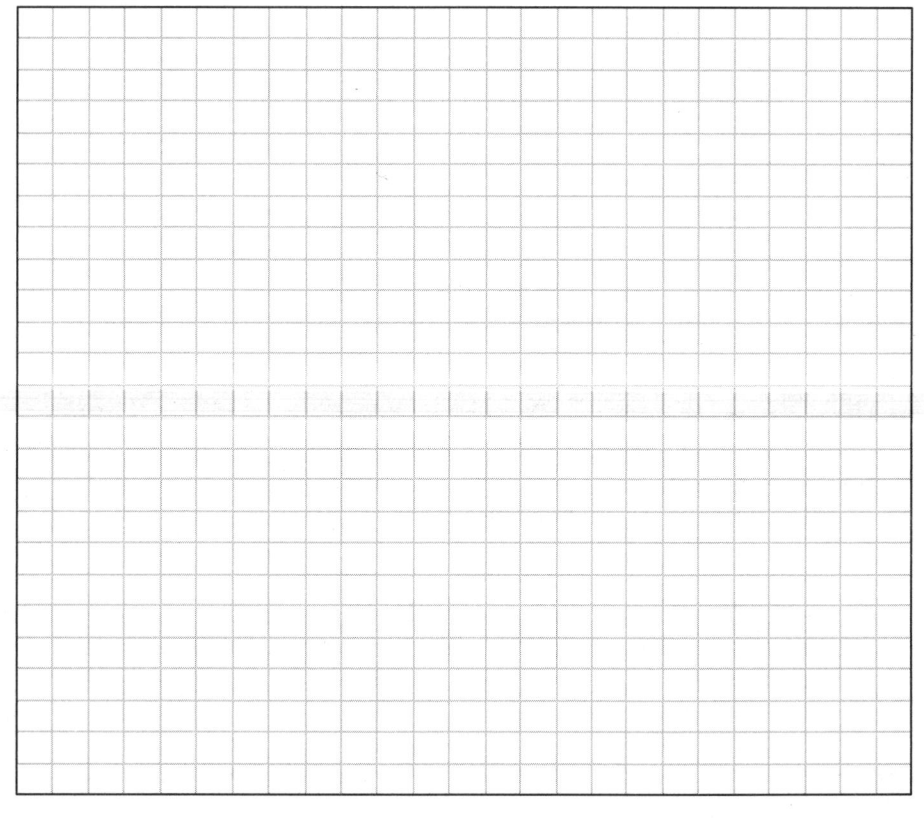

analysis

## Sine Language

5. What similarities and differences do you notice about the three functions?

6. How do the equations of the functions you graphed relate to the similarities and differences in the graphs?

input

## Sine Language

Each of the transformations on the sine function that you have examined so far in this section is a type of dilation. The first transformation affected the **amplitude** of the function, resulting in a **vertical stretch or shrink.** This is a dilation with respect to the x-axis. Note that the coefficient of a sine function is equal to the amplitude of the sine function.

The second transformation affected the **period** of the function, resulting in a **horizontal stretch or shrink**. This is a dilation with respect to the y-axis. Note that the coefficient of the dependent variable, $\theta$, is *not* the same as the period. Instead, it represents the number of repetitions of the function between 0 and 2π.

**Frequency** is related to the period of the function. The frequency is the reciprocal of the period and specifies the number of repetitions of the graph of a periodic function per unit (in this case, radians). For example, the function $y = \sin 2\theta$ has a period of π radians because it has two repetitions of the function between 0 and 2π. The frequency becomes the reciprocal of $\frac{1}{\pi}$, meaning the function completes $\frac{1}{\pi}$ cycles over one radian.

## Sine Language

process: model

7. Graph the following functions on your calculator (set to radian measure), and sketch them on the grid below. Show the parent function over at least two periods.

a. $y = \sin\ \theta$ b. $y = \sin\ \theta + 2$ c. $y = \sin\ \theta\ - 3$

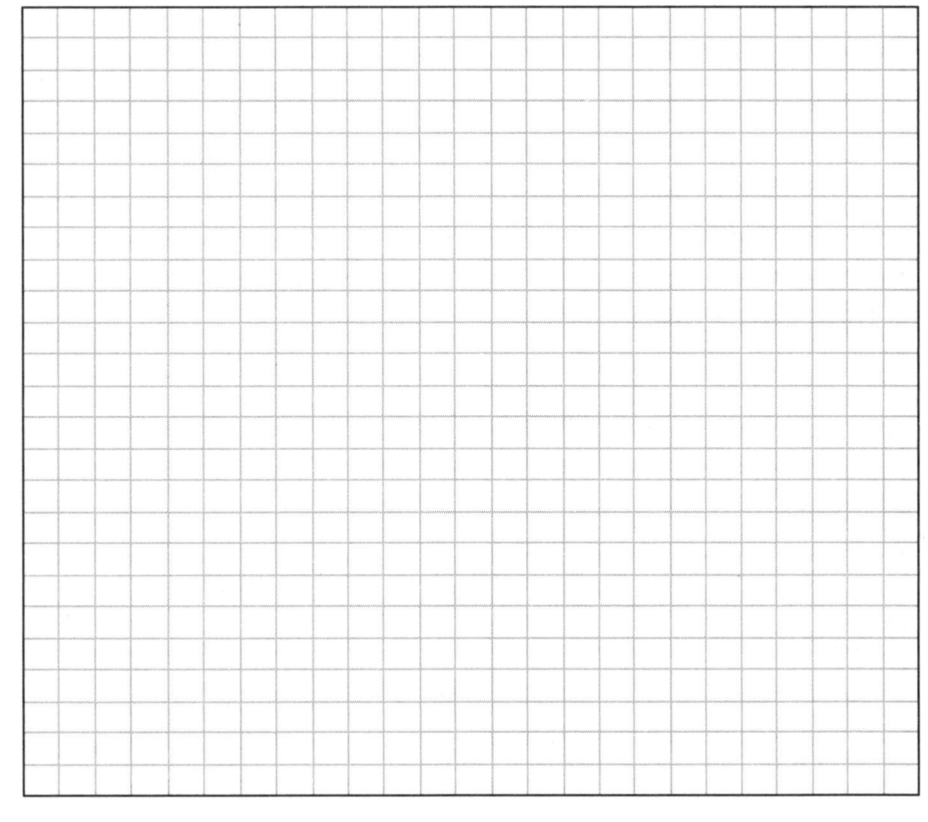

## Sine Language

analysis

8. What similarities and differences do you notice about the three functions?

## Sine Language

analysis

9. How do the equations of the functions you graphed relate to the similarities and differences in the graphs?

## Sine Language

process: model

10. Graph the following functions on your calculator (set to radian measure), and sketch them on the grid below. Show the parent function over at least two periods.

a. $y = \sin\ \theta$ b. $y = \sin\ (\theta + \frac{\pi}{2})$ c. $y = \sin\ (\theta\ -\ \pi)$

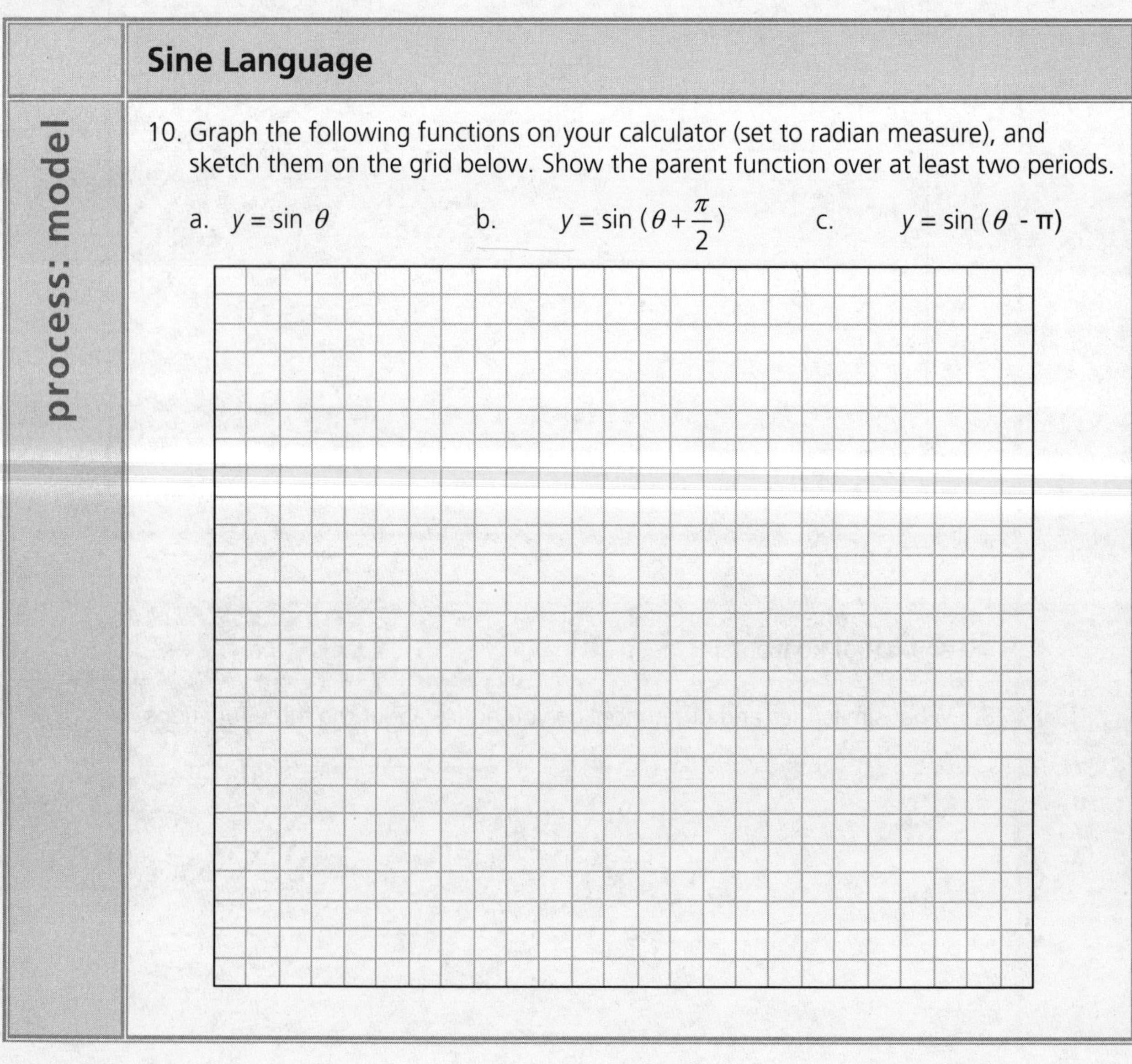

## Sine Language

analysis

11. What similarities and differences do you notice about the three functions?

12. How do the equations of the functions you graphed relate to the similarities and differences in the graphs?

## Sine Language

input

The last two transformations on the sine function that you examined in this section are translations. The first transformation, adding a number to the entire function, resulted in a vertical translation. The second transformation, adding a number to the dependent variable, resulted in a horizontal translation. These transformations act just as they have on other functions you studied.

For periodic functions, horizontal translations are called **phase shifts**. The phase shift is the amount of horizontal translation of a new sinusoidal graph from the parent function of $y = \sin\theta$. For example, the function $y = \sin\left(\theta + \frac{\pi}{2}\right)$ represents a phase shift of $-\frac{\pi}{2}$ units, and the function $y = \sin(\theta - \pi)$ represents a phase shift of $\pi$ units.

## What's Your Sine?

process: practice

Sketch the graph of each sinusoidal function by transforming the parent sine function. Write a description of the transformations from the parent function, $y = \sin\,\theta$.

1. $y = 2\sin\,\theta + 1$

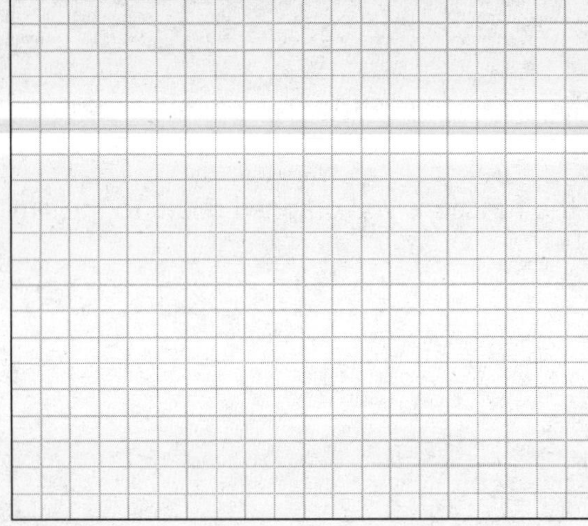

2. $y = 0.5\sin\left(\theta + \frac{\pi}{2}\right)$

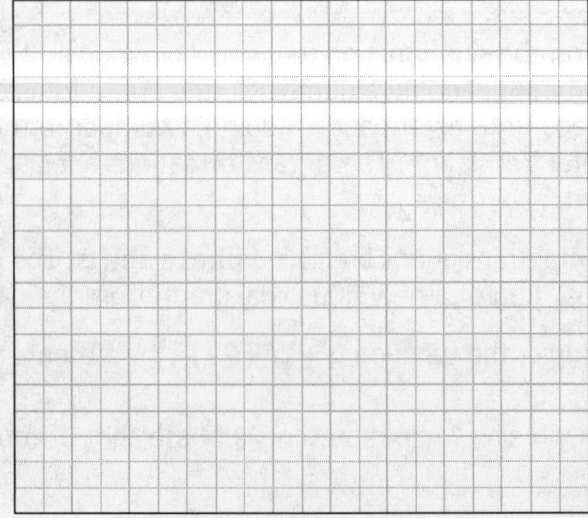

process: practice

## What's Your Sine?

Sketch the graph of each sinusoidal function by transforming the parent sine function. Write a description of the transformations from the parent function, $y = \sin\ \theta$.

3. $y = 3\sin\left(\frac{1}{2}\theta\right) + 5$

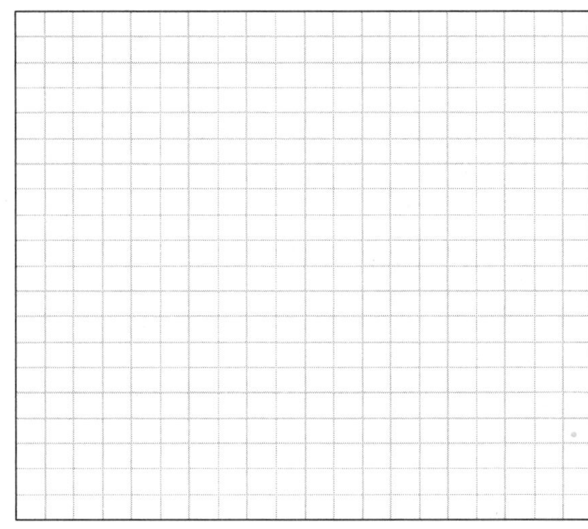

4. $y = -\sin 2\theta$

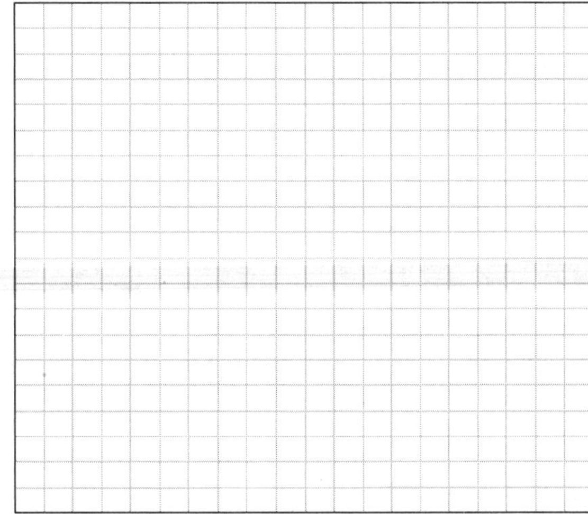

process: practice

## What's Your Sine?

Write the algebraic function for each graph shown based on the transformations performed on the parent sine function. Also, write a description of the transformation from the parent function, $y = \sin\theta$.

5.

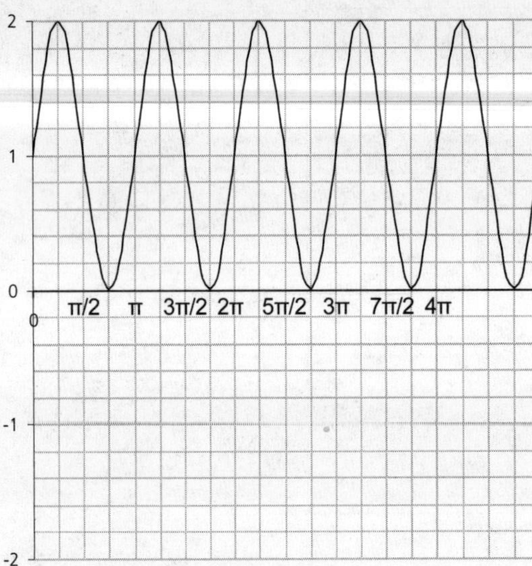


6.

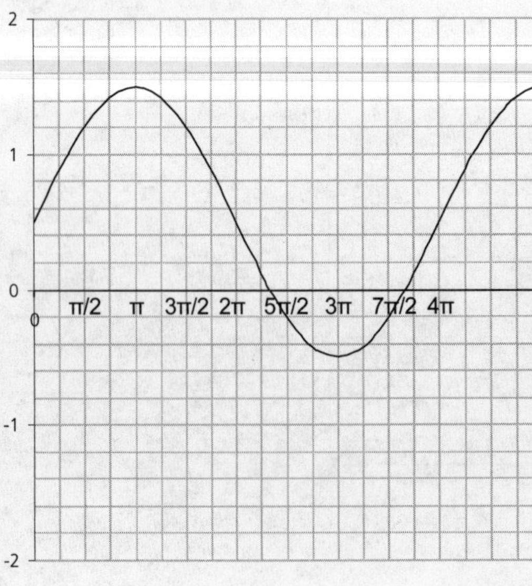

process: practice

## What's Your Sine?

Write the algebraic function for each graph shown based on the transformations performed on the parent sine function. Also, write a description of the transformation from the parent function, $y = \sin\theta$.

7.

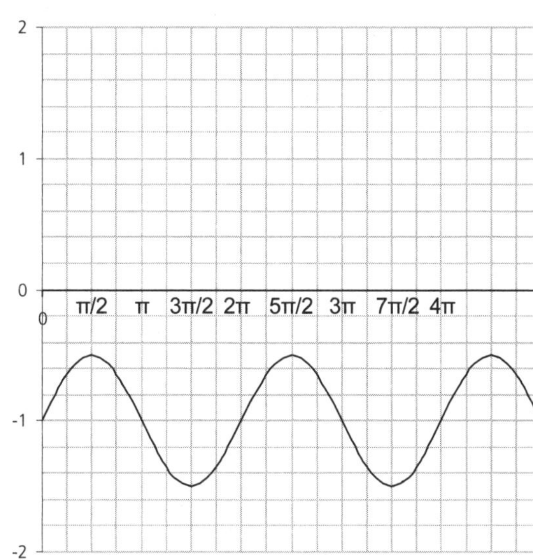


8.

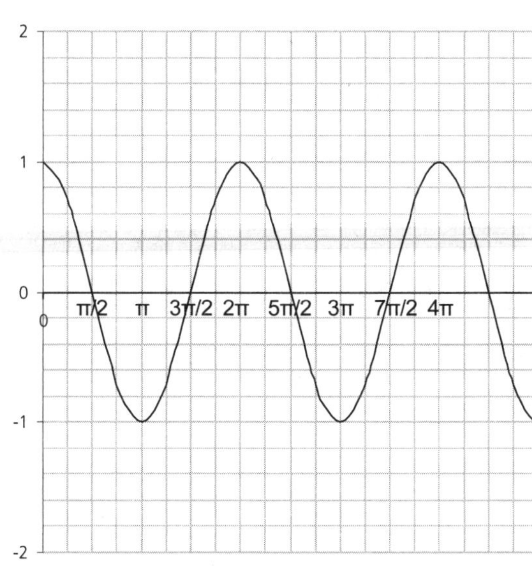

process: model

## Transforming Between Functions

1. Sketch a graph of the function, $y = \cos\,\theta$ on the grid below.

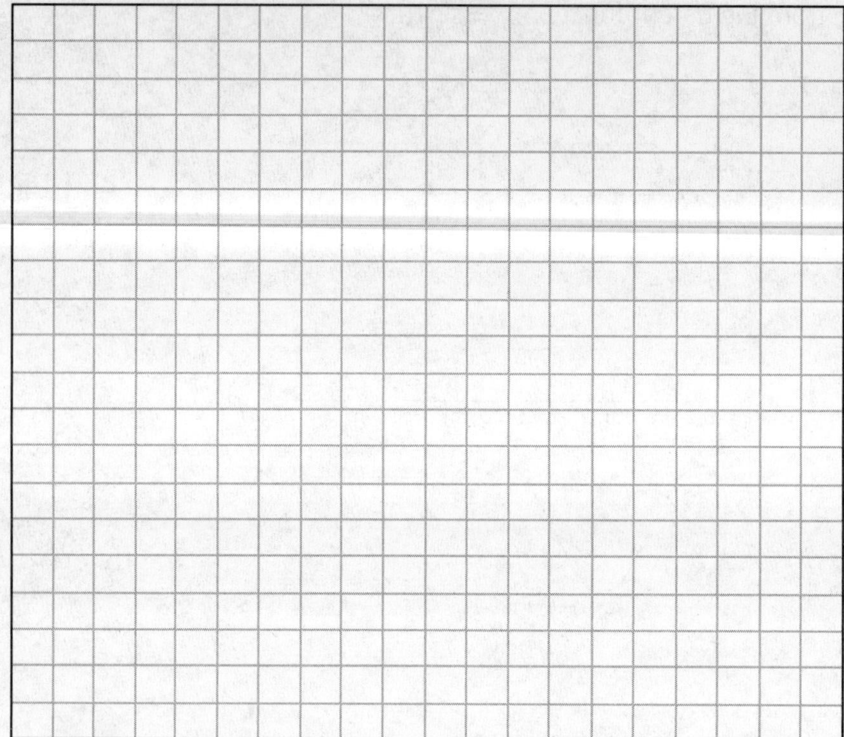

2. Consider how the graph represents a transformation on the sine function. Write an algebraic function for the graph using the sine function.

3. How can the function $y = \cos\,\theta + 1$ be written using the sine function? Sketch the graph on paper or on your calculator if it will help.

4. How can the function $y = 2\cos\left(\theta + \frac{\pi}{2}\right)$ be written using the sine function? Sketch the graph on paper or on your calculator if it will help.

process: model

## Transforming Between Functions

5. Sketch a graph of the function, $y = \sin\ \theta$ on the grid below.

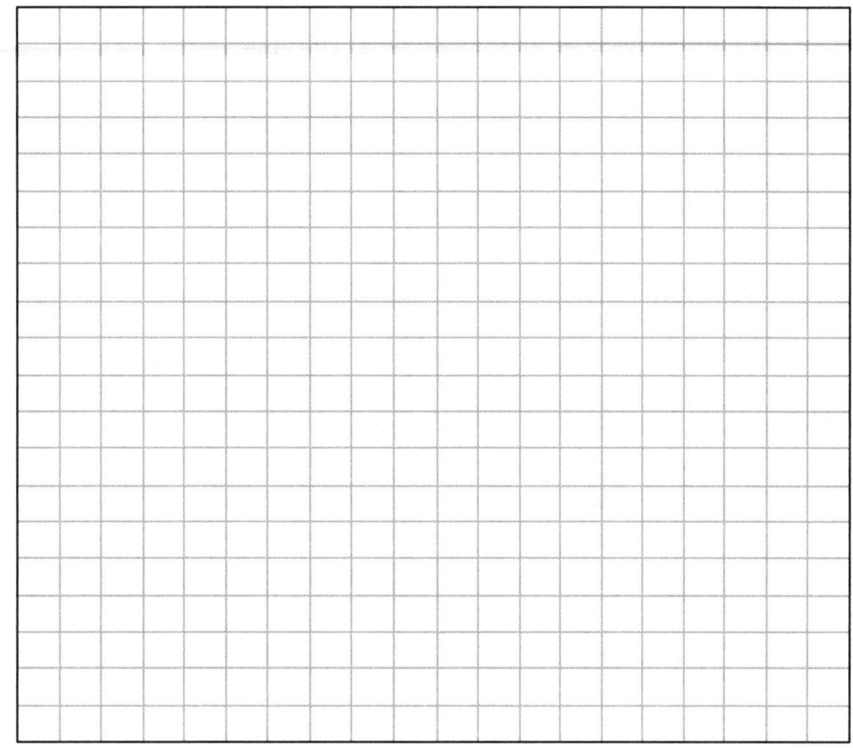

6. Consider how the graph represents a transformation on the cosine function. Write an algebraic function for the graph using the cosine function.

7. How can the function $y = 3\cos\ \theta + 1$ be written using the sine function? Sketch the graph on paper or on your calculator if it will help.

8. How can the function $y = \frac{1}{2}\cos(\theta + \frac{\pi}{2})$ be written using the sine function? Sketch the graph on paper or on your calculator if it will help.

## Rabbit Population

input

The rabbit population in a national park rises and falls throughout the year. The population is at its approximate minimum of 6000 rabbits in December. As the weather gets warmer and food becomes more available, the population grows to its approximate maximum of 16,000 rabbits in June.

The function describing the rabbit population is $f(x) = 5000\sin(\frac{\pi}{6}x - \frac{\pi}{2}) + 11000$ where $x$ is the time in months and $f(x)$ is the rabbit population.

## Rabbit Population

process: model

1. Complete the table showing the rabbit population through one year.

| Month | Time | Rabbit Population |
|---|---|---|
| | months | rabbits |
| December | 0 | 6000 |
| January | 1 | |
| February | | |
| March | | |
| April | | |
| May | | |
| June | | |
| July | | |
| August | | |
| September | | |
| October | | |
| November | | |
| December | | |

process: model

## Rabbit Population

2. Sketch a graph of the function showing at least two periods of the function.

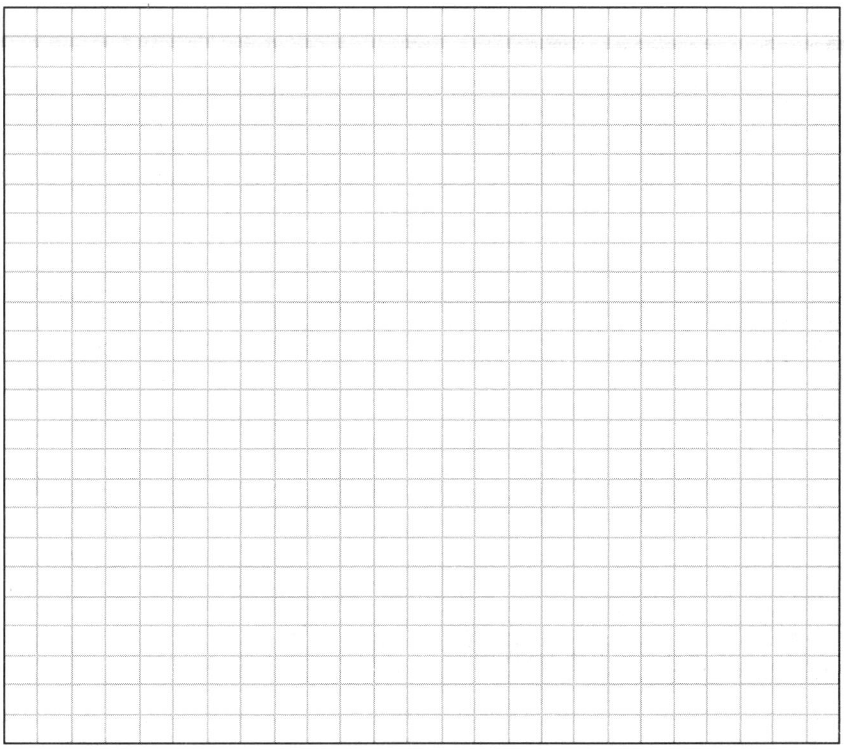

process: evaluate

## Rabbit Population

3. How has the function been translated vertically from the parent sine function?

4. What is the amplitude of the function?

5. What is the period of the function?

6. What is the phase shift of the function?

## Rabbit Population

analysis

7. How is the vertical translation related to the algebraic function? What does it represent in terms of the problem situation?

8. How is the amplitude related to the algebraic function? What does it represent in terms of the problem situation?

9. How is the period related to the algebraic function? What does it represent in terms of the problem situation?

10. How is the phase shift related to the algebraic function? What does it represent in terms of the problem situation?

11. If the rabbit population cycle occurred over six months instead of one year, how would the equation and graph change?

12. If the rabbit population has a minimum of 4000 and a maximum of 20000, how would the equation and graph change?

summary

## Rabbit Population

Think about your work on the Rabbit Population problem as you answer the following questions about the general sine function, $f(x) = A \sin (Bx + C) + D$.

13. What does the value of *A* represent in the general sine function?

14. What does the value of *B* represent in the general sine function?

15. What does the value of *C* represent in the general sine function?

16. What does the value of *D* represent in the general sine function?

## Seasonal Affective Disorder

input

Seasonal changes can affect a person's mood. You may have heard someone say that they have "the winter blues." Some people are so strongly affected by a lack of daylight that they experience a severe depression when the hours of daylight are shortest. This condition is called seasonal affective disorder (SAD).

A person with SAD experiences depression for several months when daylight hours are short. In general, the occurrence of SAD increases as the location gets further from the equator. Symptoms of SAD include disruption of sleep patterns, feeling fatigued, overeating, craving carbohydrates, depression, and avoidance of social activity.

Patterns of daylight are related to SAD. The amount of daylight varies in a periodic manner and can be modeled by a sine function. The following table shows the number of approximate daylight hours in Chicago, Illinois, which has a latitude of 42° N.

| Date | Day | Daylight Hours |
|---|---|---|
| Dec. 31 | 0 | 9.2 |
| Jan. 10 | 10 | 9.3 |
| Jan. 20 | 20 | 9.6 |
| Jan. 30 | 30 | 9.9 |
| Feb. 9 | 40 | 10.3 |
| Feb. 19 | 50 | 10.7 |
| Mar. 1 | 60 | 11.4 |
| Mar. 11 | 70 | 11.7 |
| Mar. 21 | 80 | 12.2 |
| Mar. 31 | 90 | 12.7 |
| Apr. 10 | 100 | 13.1 |
| Apr. 20 | 110 | 13.6 |
| Apr. 30 | 120 | 14 |
| May 10 | 130 | 14.4 |
| May 20 | 140 | 14.7 |
| May 30 | 150 | 15 |
| June 9 | 160 | 15.2 |
| June 19 | 170 | 15.2 |
| June 29 | 180 | 15.2 |

| Date | Day | Daylight Hours |
|---|---|---|
| July 9 | 190 | 15.1 |
| July 19 | 200 | 14.8 |
| July 29 | 210 | 14.5 |
| Aug. 8 | 220 | 14.2 |
| Aug. 18 | 230 | 13.7 |
| Aug. 28 | 240 | 13.3 |
| Sept. 7 | 250 | 12.9 |
| Sept. 17 | 260 | 12.4 |
| Sept. 27 | 270 | 12 |
| Oct. 7 | 280 | 11.5 |
| Oct. 17 | 290 | 11 |
| Oct. 27 | 300 | 10.6 |
| Nov. 6 | 310 | 10.2 |
| Nov. 16 | 320 | 9.8 |
| Nov. 26 | 330 | 9.5 |
| Dec. 6 | 340 | 9.2 |
| Dec. 16 | 350 | 9.2 |
| Dec. 26 | 360 | 9.1 |

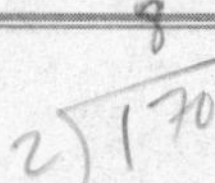

process: model

## Seasonal Affective Disorder

8. Enter the data from the table into your calculator. Use the calculator to perform a sinusoidal regression for this data. Write the regression equation from the calculator. How does it compare to your equation?

analysis

## Seasonal Affective Disorder

9. Seasonal affective disorder appears to vary according to latitude. The farther a location is from the equator, the more prevalent cases of SAD become. Why might this happen?

10. Anchorage, Alaska is located at a latitude of 61° N. This is considerably farther north than Chicago. If we created a graph to model the daylight hours in Anchorage, how do you think it would compare to the graph for daylight hours in Chicago? In what ways would it be the same? In what ways would it be different?

11. In locations like Chicago and Anchorage, SAD is most likely to occur around the month of January. In locations in the southern hemisphere, like Santiago, Chile, (latitude 33.5° S.), SAD occurs around the month of July. Why does this happen?

analysis

## Seasonal Affective Disorder

12. If we created a graph to model the daylight hours in Santiago, Chile, how would it compare to the graph for Chicago?

input

## Seasonal Affective Disorder

The following tables list daylight hours in Anchorage, Alaska (latitude 61° N).

| Day | Daylight Hours |
|---|---|
| 0 | 5.6 |
| 10 | 6 |
| 20 | 6.7 |
| 30 | 7.5 |
| 40 | 8.4 |
| 50 | 9.3 |
| 60 | 10.4 |
| 70 | 11.3 |
| 80 | 12.3 |
| 90 | 13.3 |
| 100 | 14.2 |
| 110 | 15.2 |
| 120 | 16.1 |

| Day | Daylight Hours |
|---|---|
| 130 | 17 |
| 140 | 17.9 |
| 150 | 18.6 |
| 160 | 19.1 |
| 170 | 19.4 |
| 180 | 19.3 |
| 190 | 18.8 |
| 200 | 18.2 |
| 210 | 17.4 |
| 220 | 16.5 |
| 230 | 15.5 |
| 240 | 14.6 |

| Day | Daylight Hours |
|---|---|
| 250 | 13.7 |
| 260 | 12.8 |
| 270 | 11.8 |
| 280 | 10.9 |
| 290 | 9.9 |
| 300 | 9 |
| 310 | 8.1 |
| 320 | 7.2 |
| 330 | 6.5 |
| 340 | 5.9 |
| 350 | 5.5 |
| 360 | 5.5 |

input

## Seasonal Affective Disorder

Below is data regarding daylight hours in Santiago, Chile, latitude 33.5° S.

| Day | Daylight Hours |
|---|---|
| 0 | 14.3 |
| 10 | 14 |
| 20 | 14 |
| 30 | 13.8 |
| 40 | 13.5 |
| 50 | 13.2 |
| 60 | 12.8 |
| 70 | 12.5 |
| 80 | 12.1 |
| 90 | 11.8 |
| 100 | 11.4 |
| 110 | 11.1 |
| 120 | 10.8 |

| Day | Daylight Hours |
|---|---|
| 130 | 10.4 |
| 140 | 10.3 |
| 150 | 10.1 |
| 160 | 10 |
| 170 | 10 |
| 180 | 10 |
| 190 | 10.1 |
| 200 | 10.3 |
| 210 | 10.5 |
| 220 | 10.7 |
| 230 | 11 |
| 240 | 11.3 |

| Day | Daylight Hours |
|---|---|
| 250 | 11.6 |
| 260 | 11.9 |
| 270 | 12.3 |
| 280 | 12.6 |
| 290 | 13 |
| 300 | 13 |
| 310 | 13.3 |
| 320 | 13.6 |
| 330 | 14.1 |
| 340 | 14.2 |
| 350 | 14.3 |
| 360 | 14.3 |

process: model

## Seasonal Affective Disorder

13. Enter the data for daylight hours in Anchorage and Santiago in your calculator. Use the calculator to do a sinusoidal regression for each of these locations, and record the equations below.

    a. Regression equation for daylight hours in Anchorage

    b. Regression equation for daylight hours in Santiago

analysis

## Seasonal Affective Disorder

14. Compare the regression equations for daylight hours in each of the three cities.

    a. How do the amplitudes compare? What might account for any similarities or differences?

    b. How do the periods compare? What might account for any similarities or differences?

    c. How do the phase shifts compare? What might account for any similarities or differences?

    d. How do the vertical shifts compare? What might account for any similarities or differences?

## Off on a Tangent

input

In an earlier course, you learned that the tangent ratio represents the ratio of the lengths of the opposite side and the adjacent side in a right triangle. You will use this definition and the unit circle to begin exploring the tangent function. The tangent function is usually abbreviated *tan*.

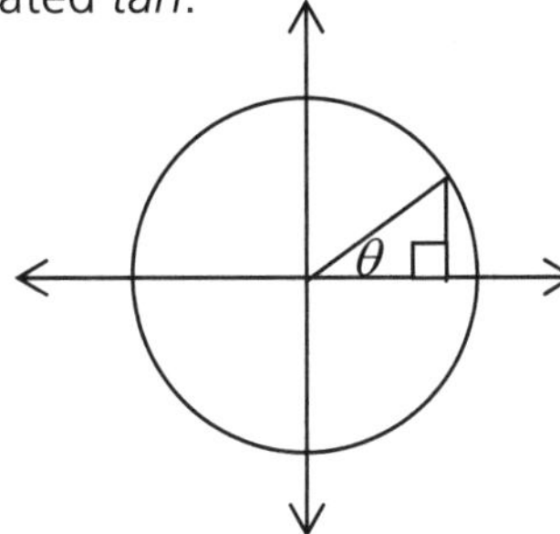


Recall that the radius of the unit circle is one unit in length, so the hypotenuse of the right triangle is one unit in length.

## Off on a Tangent

process: evaluate

1. If the measure of $\theta$ is $\frac{\pi}{6}$ radians (30 degrees), what are the lengths of the opposite and adjacent sides of the right triangle?

2. Evaluate $\sin\frac{\pi}{6}$, $\cos\frac{\pi}{6}$, and $\tan\frac{\pi}{6}$.

3. What is the slope of the terminal ray of an angle measuring $\frac{\pi}{6}$ radians? (Hint: Think about slope as "rise over run.")

process: evaluate

## Off on a Tangent

4. If the measure of $\theta$ is $\frac{\pi}{4}$ radians (45 degrees), what are the lengths of the opposite and adjacent sides of the right triangle?

5. Evaluate $\sin\frac{\pi}{4}$, $\cos\frac{\pi}{4}$, and $\tan\frac{\pi}{4}$.

6. What is the slope of the terminal ray of an angle measuring $\frac{\pi}{4}$ radians?

7. If the measure of $\theta$ is $\frac{\pi}{3}$ radians (60 degrees), what are the lengths of the opposite and adjacent sides of the right triangle?

8. Evaluate $\sin\frac{\pi}{3}$, $\cos\frac{\pi}{3}$, and $\tan\frac{\pi}{3}$.

9. What is the slope of the terminal ray of an angle measuring $\frac{\pi}{3}$ radians?

## Off on a Tangent

analysis

10. How is the tangent value related to the values for sine and cosine?

11. How is the tangent value related to the slope of the terminal ray of the central angle?

12. What is the slope of the terminal ray of an angle measuring 0 radians? What is the value of tan 0?

13. What is the slope of the terminal ray of an angle measuring $\frac{\pi}{2}$ radians? What is the value of $\tan\frac{\pi}{2}$?

extension

## Off on a Tangent

You learned about tangents to a circle in a previous unit of this text. Sometimes mathematical terms are used differently in different contexts. A tangent to a circle is a line that intersects the circle at exactly one point. In trigonometry, a tangent is a ratio of the opposite and adjacent sides of a right triangle.

extension

## Off on a Tangent

14. Draw a tangent to the circle at the point where the terminal ray of the central angle intersects the circle.

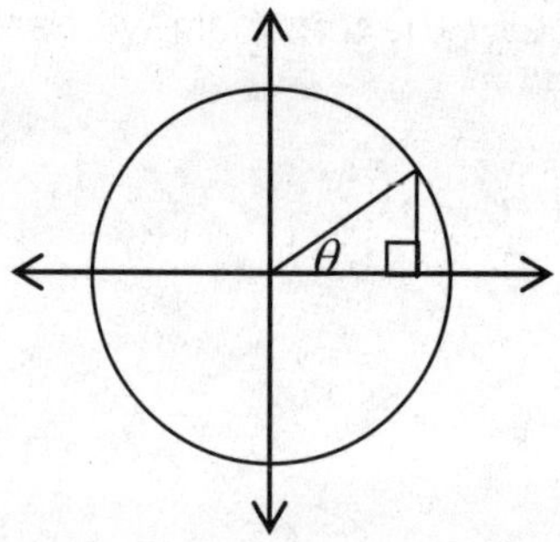

$\theta$

15. How is the terminal ray of the central angle related to the line tangent to the circle?

16. How is the slope of the terminal ray related to the slope of the line tangent to the circle?

extension

## Off on a Tangent

17. What is the slope of the line tangent to the circle at the point where the terminal ray of a central angle measuring $\frac{\pi}{6}$ radians intersects the circle?

18. What is the slope of the line tangent to the circle at the point where the terminal ray of a central angle measuring $\frac{\pi}{4}$ radians intersects the circle?

19. If $\tan\theta = -\frac{2}{3}$, what is the slope of the line tangent to the circle where the terminal ray of the angle intersects the circle?

20. If the tangent line to the circle has a slope of zero, what is the measure of the central angle with the terminal ray intersecting the circle at the tangent point?

## The Tangent Function

input

The tangent function, like the sine and cosine functions, is a trigonometric function and is periodic. Creating and analyzing a graph of the tangent function will help you understand the nature of the tangent function.

## The Tangent Function

process: model

1. Complete the table of values below for the tangent function. It may help to look at the table for sine and cosine that you completed in *The Sine and Cosine Functions* section.

| **Angle measure ($\theta$)** | **tan$\theta$** |
|---|---|
| **radians** | |
| 0 | |
| $\frac{\pi}{6}$ | |
| $\frac{\pi}{4}$ | |
| $\frac{\pi}{3}$ | |
| $\frac{\pi}{2}$ | |
| $\frac{2\pi}{3}$ | |
| $\frac{3\pi}{4}$ | |
| $\frac{5\pi}{6}$ | |

| **Angle measure ($\theta$)** | **tan$\theta$** |
|---|---|
| **radians** | |
| $\pi$ | |
| $\frac{7\pi}{6}$ | |
| $\frac{5\pi}{4}$ | |
| $\frac{4\pi}{3}$ | |
| $\frac{3\pi}{2}$ | |
| $\frac{5\pi}{3}$ | |
| $\frac{7\pi}{4}$ | |
| $\frac{11\pi}{6}$ | |
| $2\pi$ | |

process: model

## The Tangent Function

2. Plot the points from your table on the coordinate grid below.

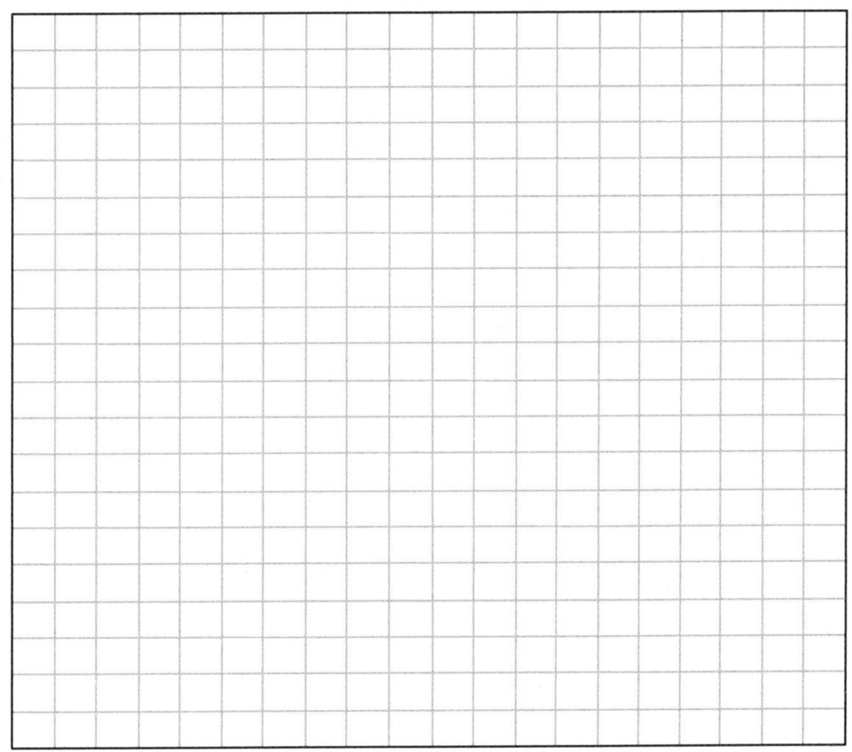

3. How does the slope of the terminal ray change as the central angle of the unit circle changes? It may help to consider the changes through each quadrant of the coordinate plane.

4. Use the points you plotted and your answer to question 3 to sketch the graph of the tangent function on the grid.

## The Tangent Function

analysis

5. Answer the following questions based on the function you graphed.
   a. What are the domain and range of the function?

   b. What are the x- and y-intercepts of the function?

   c. What are the maxima and minima of the function?

   d. What are the intervals of increase or decrease for the function?

   e. Is this function continuous or discontinuous? If it is discontinuous, what are the points of discontinuity?

   f. What is the period of the function?

   g. What is the amplitude of the function?

process: model

## The Tangent Function

6. Graph the following functions on your calculator (set to radian measure), and sketch them on the grid provided. Show the parent function over at least two periods.

   a. $y = \tan\ \theta$ b. $y = 2\tan\ \theta$ c. $y = \frac{1}{2}\tan\theta$

analysis

## The Tangent Function

7. What similarities and differences do you notice about the three functions? What in the algebraic function accounts for the similarities and differences?

## The Tangent Function

analysis

8. The coefficient on the sine and cosine functions affect the amplitude of the functions. The coefficient on the tangent function does not change the amplitude. Why is this?

## The Tangent Function

process: model

9. Graph the following functions on your calculator (set to radian measure), and sketch them on the grid provided. Show the parent function over at least two periods.

a. $y = \tan \theta$ b. $y = \tan 2\theta$ c. $y = \tan \frac{1}{2}\theta$

## The Tangent Function

analysis

10. What similarities and differences do you notice about the three functions? What in the algebraic function accounts for the similarities and differences?

## The Tangent Function

process: model

11. Graph the following functions on your calculator (set to radian measure), and sketch them on the grid provided. Show the parent function over at least two periods.

a. $y = \tan \theta$  b. $y = \tan \theta + 2$  c. $y = \tan \theta - 3$

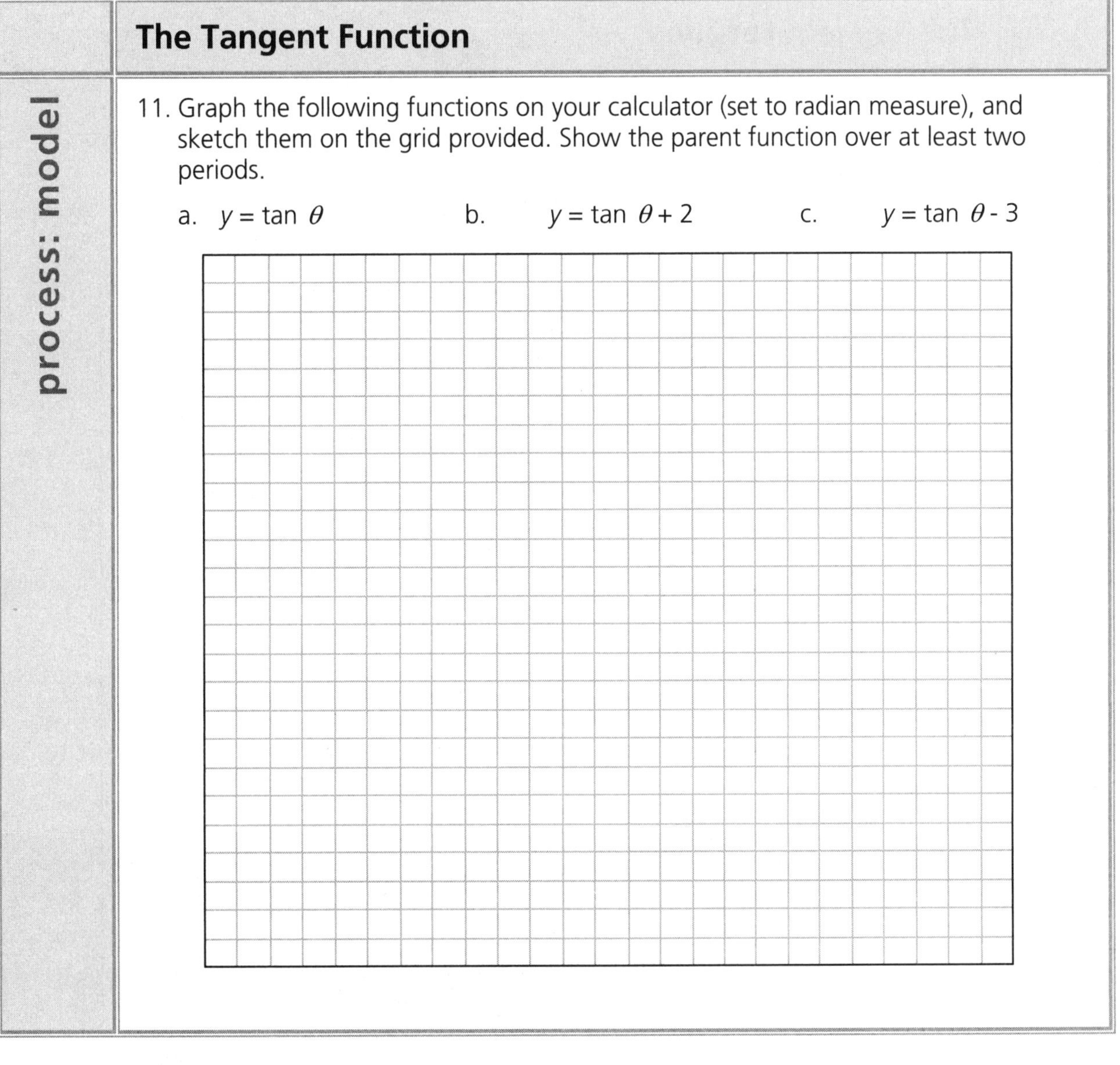

analysis

## The Tangent Function

12. What similarities and differences do you notice about the three functions? What in the algebraic function accounts for the similarities and differences?

process: model

## The Tangent Function

13. Graph the following functions on your calculator (set to radian measure), and sketch them on the grid provided. Show the parent function over at least two periods.

a. $y = \tan\theta$ b. $y = \tan\left(\theta + \frac{\pi}{2}\right)$ c. $y = \tan\left(\theta - \frac{\pi}{4}\right)$

## The Tangent Function

analysis

14. What similarities and differences do you notice about the three functions? What in the algebraic function accounts for the similarities and differences?

15. Compare and contrast the sine, cosine, and tangent functions. How are they similar? How are they different?

## The Inverse Sine Function

input

In a previous course, you used inverse trigonometric functions to find the measure of an angle given the sides measures of a right triangle. Now, you will further investigate inverse trigonometric functions.

## The Inverse Sine Function

process: model

1. Below is a graph of the parent sine function. Sketch the inverse of the sine function on the grid.

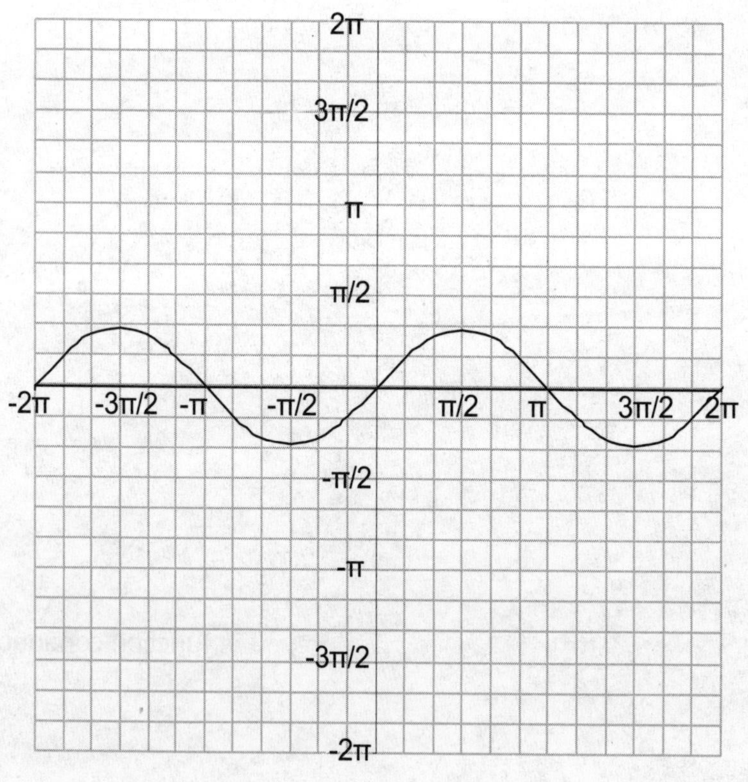

## The Inverse Sine Function

analysis

2. Is the inverse of the sine function also a function? Explain.

3. Is the sine function a one-to-one function?

## The Inverse Sine Function

input

To consider an inverse sine function, you need to restrict the domain of the sine function in a way that makes it a one-to-one function. There are many possible ways to restrict the domain of $y = \sin x$, and you need to choose one that is most appropriate. The following guidelines will help in determining an appropriate domain restriction.

- Since trigonometry is frequently used to find angles in a right triangle, the restricted domain should include all values for the acute angles of the right triangle. Therefore, the restricted domain should include angles between 0 and $\frac{\pi}{2}$.
- The restricted domain should represent all values of the original range. That is, the new range should still include all real numbers from -1 to 1.
- The restricted domain should keep the function continuous.

## The Inverse Sine Function

analysis

4. Each graph below shows a suggested restricted domain for the sine function to create an inverse sine function. Consider the guidelines from the previous page and decide whether each is an appropriate option. Explain why the option is appropriate or inappropriate.

a. Domain: $0 \le x \le \pi$

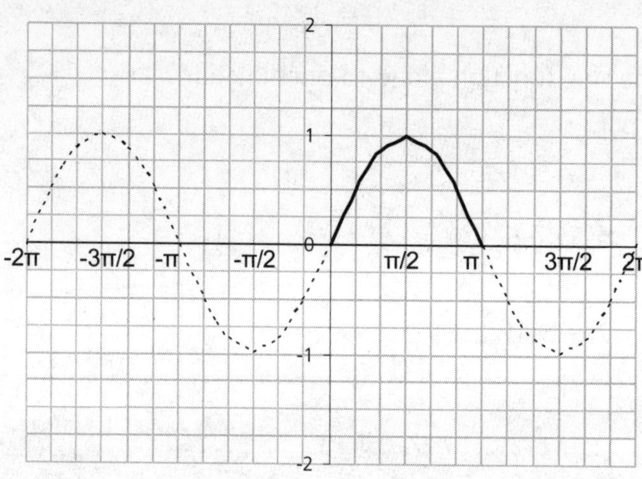


b. Domain: $\frac{\pi}{2} \le x \le \frac{3\pi}{2}$

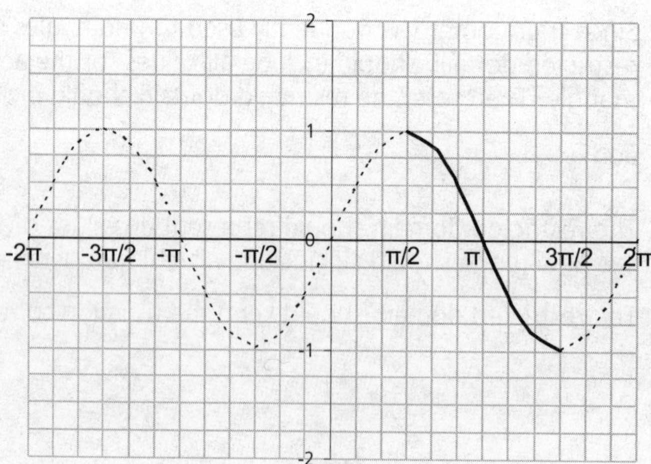

## The Inverse Sine Function

analysis

c. $-\frac{\pi}{2} \le x \le \frac{\pi}{2}$

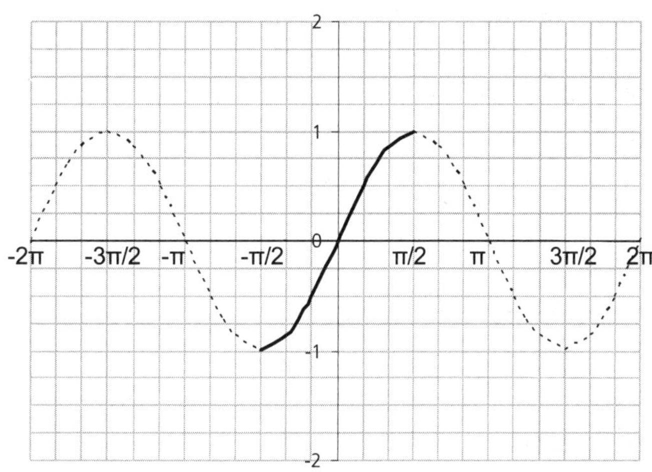


d. $0 \le x \le \frac{\pi}{2}$ and $\pi \le x \le \frac{3\pi}{2}$

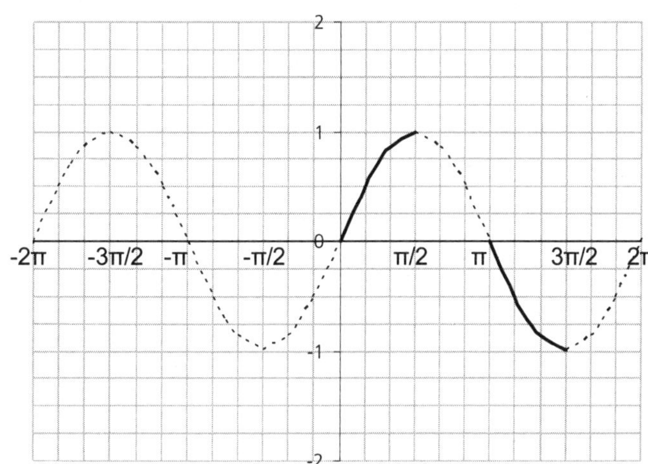

## The Inverse Sine Function

input

Restricting the domain of the sine function to $\left\{x \mid -\frac{\pi}{2} \le x \le \frac{\pi}{2}\right\}$ allows you to work with an **inverse sine function** – the inverse of the restricted sine function. The symbol for the inverse sine function is **sin$^{-1}$**, read "the inverse sine**.**" The notation, **arcsin**, is sometimes used instead of $\sin^{-1}$.

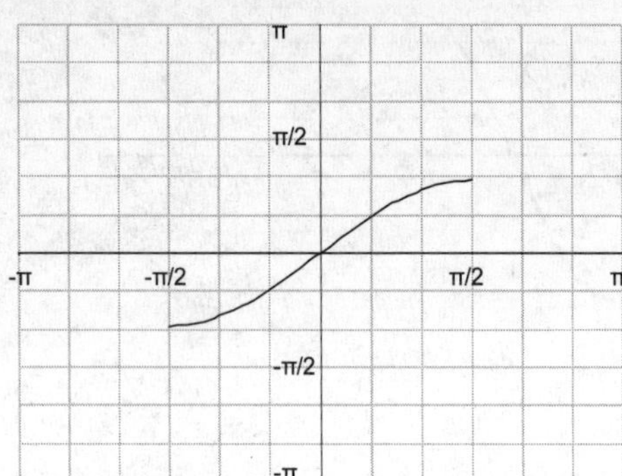


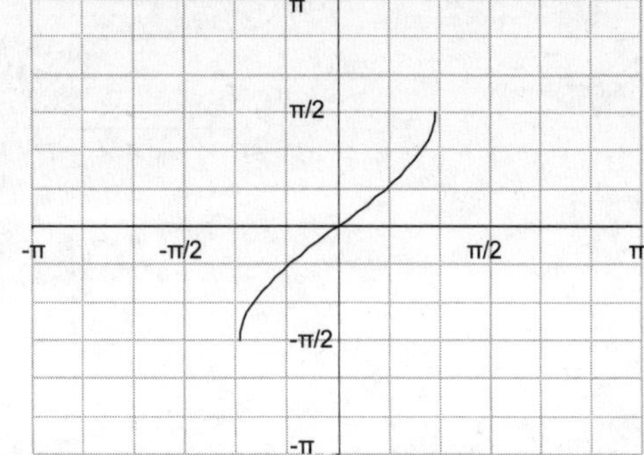


restricted sine function

domain: $\left\{x \mid -\frac{\pi}{2} \le x \le \frac{\pi}{2}\right\}$

range: $\{y \mid -1 \le y \le 1\}$

inverse sine function

domain: $\{x \mid -1 \le x \le 1\}$

range: $\left\{y \mid -\frac{\pi}{2} \le y \le \frac{\pi}{2}\right\}$

You can use the inverse sine function to find an angle of a right triangle when the lengths of the opposite side and hypotenuse are known. For example, if the hypotenuse of a right triangle measures 17 cm and the opposite side from angle $\theta$ measures 8 cm, what is the measure of $\theta$?

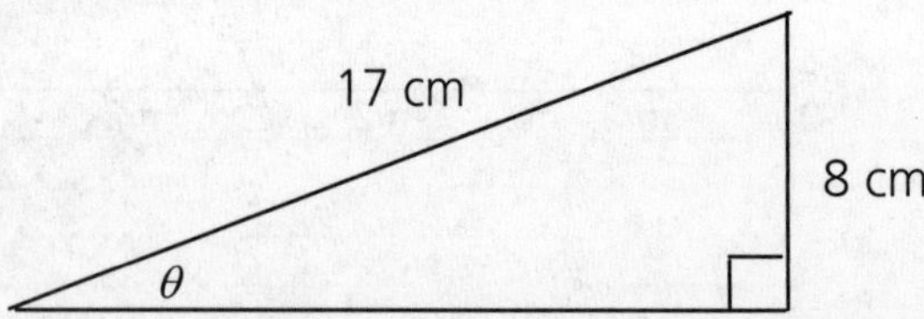


$$\sin\theta = \frac{opposite}{hypotenuse} \qquad \sin\theta = \frac{8}{17}$$

$$\theta = \sin^{-1}\left(\frac{opposite}{hypotenuse}\right) \qquad \theta = \sin^{-1}\left(\frac{8}{17}\right)$$

$$\theta \approx 0.49 \text{ radians or } 28.1 \text{ degrees}$$

## The Inverse Sine Function

process: practice

5. The law requires that wheelchair ramps have a ramp angle less than or equal to $8\frac{1}{3}°$. A ramp that is 20 feet in length is used for an 18 inch vertical rise. Is this ramp in compliance with the law? Hint: Drawing a diagram will help.

6. To exercise, you use a treadmill that simulates walking uphill at a constant slope. The readout on the treadmill indicates that you have traveled 10,000 feet and have reached an elevation of 1900 feet. At what simulated angle of elevation have you walked?

7. How could you determine the answers to questions 5 and 6 using the graph of the inverse sine function?

## The Inverse Cosine Function

input

To create an inverse sine function, you needed to restrict the domain of the sine function in a way that made it a one-to-one function. Similarly, the domain of the cosine function must be restricted to create an inverse cosine function. Recall the following guidelines for determining an appropriate domain restriction.

- Since trigonometry is frequently used to find angles in a right triangle, the restricted domain should include all values for the acute angles of the right triangle. Therefore, the restricted domain should include angles between 0 and $\frac{\pi}{2}$.
- The restricted domain should represent all values of the original range. That is, the new range should still include all real numbers from -1 to 1.
- The restricted domain should keep the function continuous.

## The Inverse Cosine Function

process: model

1. Below is a graph of the parent cosine function. Indicate an appropriate domain restriction to create a one-to-one function following the guidelines above.

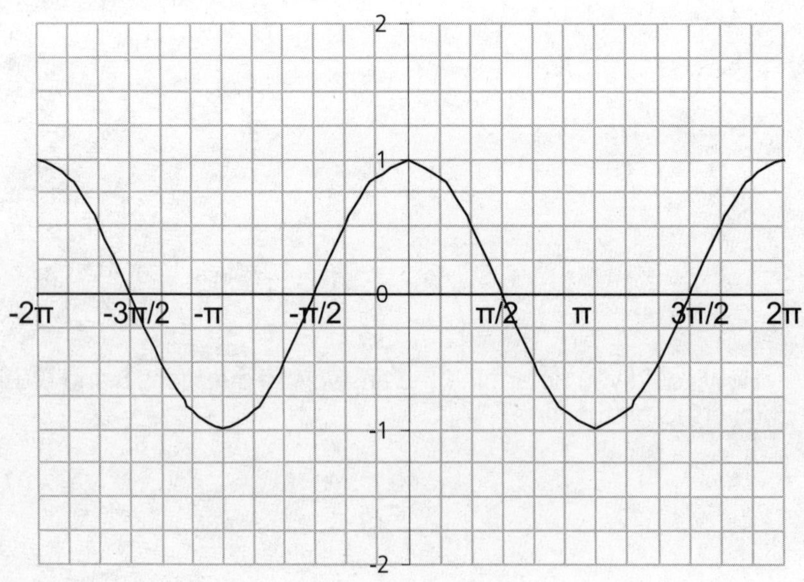

## The Inverse Cosine Function

input

Restricting the domain of the cosine function to $\{x \mid 0 \le x \le \pi\}$ allows you to work with an **inverse cosine function** – the inverse of the restricted cosine function. The symbol for the inverse cosine function is **cos$^{-1}$**, read "the inverse cosine." The notation, **arccos**, is sometimes used instead of $\cos^{-1}$.

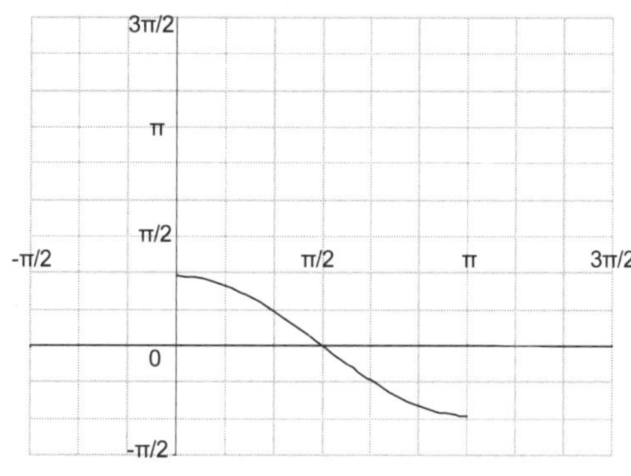


restricted cosine function

domain: $\{x \mid 0 \le x \le \pi\}$

range: $\{y \mid -1 \le y \le 1\}$

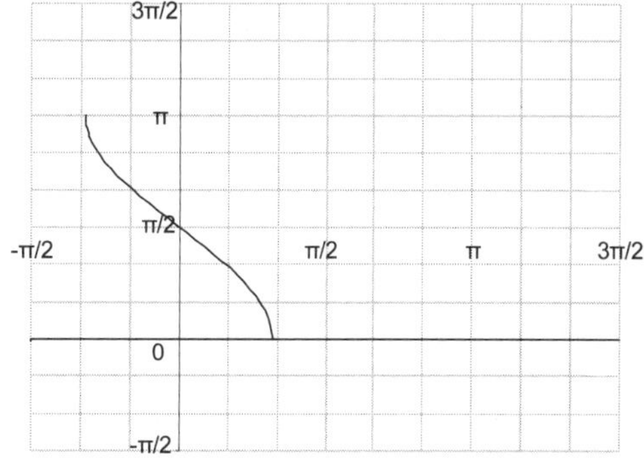


inverse cosine function

domain: $\{x \mid -1 \le x \le 1\}$

range: $\{y \mid 0 \le y \le \pi\}$

The inverse cosine function can be used to find an angle of a right triangle when the lengths of the adjacent side and hypotenuse are known. For example, if the hypotenuse of a right triangle measures 17 cm and the adjacent side to angle $\theta$ measures 8 cm, what is the measure of $\theta$?

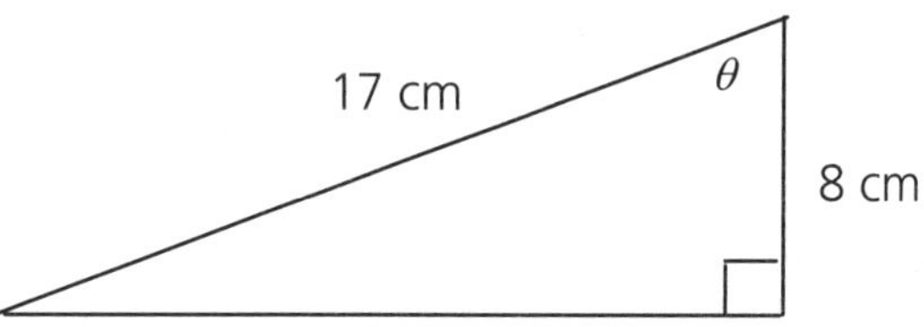


$$\cos\theta = \frac{adjacent}{hypotenuse} \qquad \cos\theta = \frac{8}{17}$$

$$\theta = \cos^{-1}\left(\frac{adjacent}{hypotenuse}\right) \qquad \theta = \cos^{-1}\left(\frac{8}{17}\right)$$

$$\theta \approx 1.08 \text{ radians or } 61.9 \text{ degrees}$$

process: practice

## The Inverse Cosine Function

2. The long water slide at the water park is a straight ramp that is 70 feet long. The top of the slide is on a tower that is 50 feet high. What is the angle the slide forms with the tower? Hint: Drawing a diagram will help.

3. The law requires that wheelchair ramps have a ramp angle less than or equal to $8\frac{1}{3}°$. You have a ramp that is 21 feet in length. Can it be used in a space with a horizontal distance of 20 feet and still be in compliance with the law?

4. How would shortening the vertical height of the ramp affect the angle?

5. How could you determine the answers to questions 2 and 3 using the graph of the inverse sine function?

## The Inverse Tangent Function

input

To create inverse sine and cosine functions, you needed to restrict the domain to create a one-to-one function. Similarly, the domain of the tangent function must be restricted to create an inverse tangent function. Recall the following guidelines for determining an appropriate domain restriction.

- Since trigonometry is frequently used to find angles in a right triangle, the restricted domain should include all values for the acute angles of the right triangle. Therefore, the restricted domain should include angles between 0 and $\frac{\pi}{2}$.
- The restricted domain should represent all values of the original range. That is, the new range should still include all real numbers.
- The restricted domain should keep the function continuous.

## The Inverse Tangent Function

process: model

1. Below is a graph of the parent tangent function. Indicate an appropriate domain restriction to create a one-to-one function following the guidelines above.

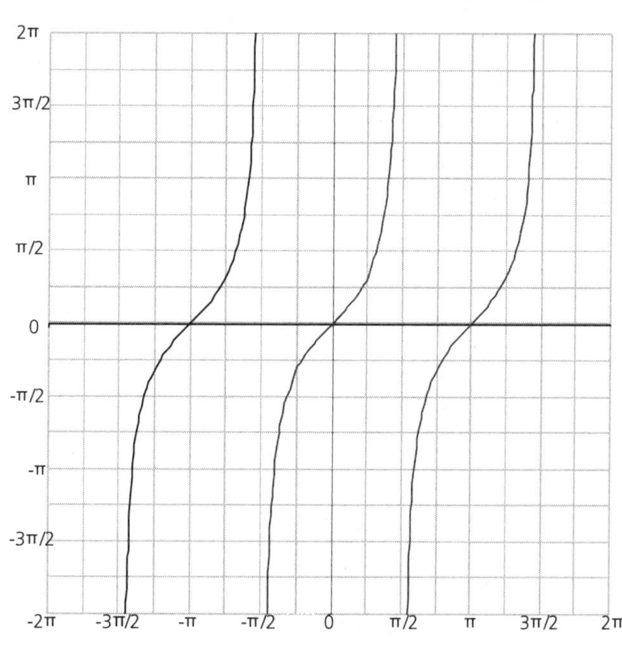

input

## The Inverse Tangent Function

Restricting the domain of the tangent function to $\left\{x \mid -\frac{\pi}{2} \le x \le \frac{\pi}{2}\right\}$ allows you to work with an **inverse tangent function** – the inverse of the restricted tangent function. The symbol for the inverse tangent function is **tan** $^{-1}$, read "the inverse tangent**.**" The notation, **arctan**, is sometimes used instead of tan $^{-1}$.

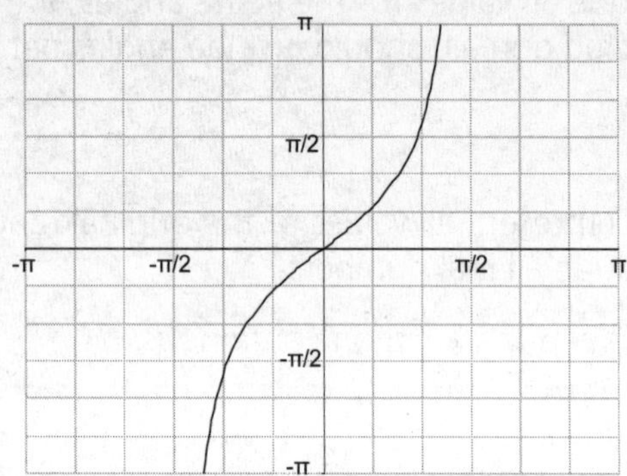


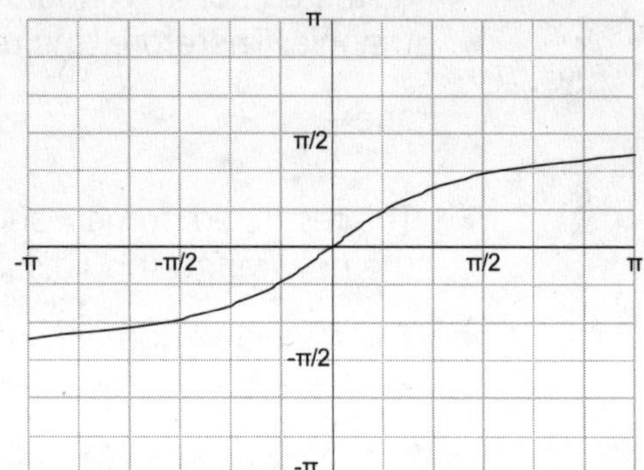


restricted tangent function

domain: $\left\{x \mid -\frac{\pi}{2} \le x \le \frac{\pi}{2}\right\}$

range: all real numbers

inverse tangent function

domain: all real numbers

range: $\left\{y \mid -\frac{\pi}{2} \le y \le \frac{\pi}{2}\right\}$

You can use the inverse tangent function to find an angle of a right triangle when the lengths of the adjacent side and opposite side are known. For example, if the adjacent side to angle $\theta$ measures 15 cm and the opposite side measures 8 cm, what is the measure of $\theta$?

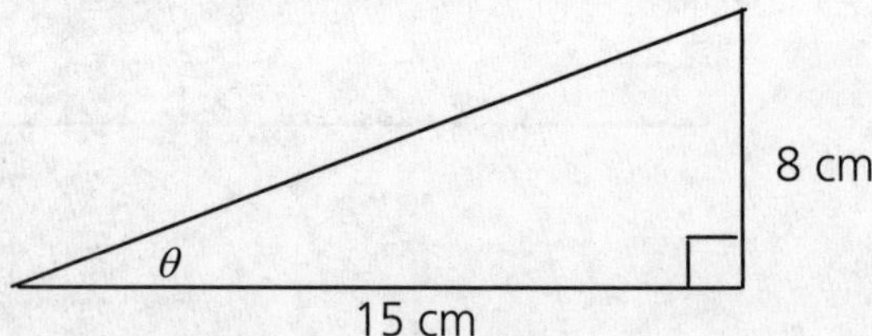


$$\tan\theta = \frac{opposite}{adjacent} \qquad \tan\theta = \frac{8}{15}$$

$$\theta = \tan^{-1}\left(\frac{opposite}{adjacent}\right) \qquad \theta = \tan^{-1}\left(\frac{8}{15}\right)$$

$$\theta \approx 0.49 \text{ radians or } 28.1 \text{ degrees}$$

## The Inverse Tangent Function

process: practice

2. A funicular is an inclined railway that uses a cable to pull a car on an inclined track where steepness prevents a traditional railway car from operating. There are two funiculars currently operating in Pittsburgh, Pennsylvania.

   a. The Monongahela Incline was built in 1870. It covers a horizontal distance of 516.5 feet and a vertical distance of 369.4 feet. What is the angle of incline?

   b. The Duquesne Incline was built in 1877. It covers a horizontal distance of 693 feet and a vertical distance of 400 feet. What is the angle of incline?

3. The horizontal part of a step is called the tread and the vertical part is called the riser. The ratio of the riser to the tread affects the safety of a staircase. Different builders may recommend different ratios.

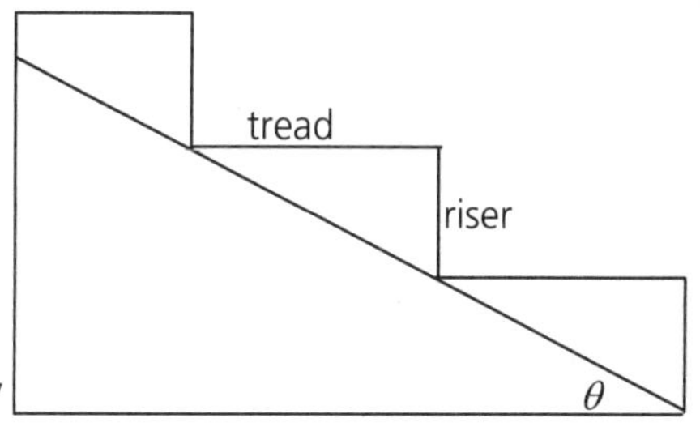


   a. One builder recommends a ratio of 7 inches to 11 inches. What is the angle of inclination, $\theta$, using this builder's recommendation?

   b. Another builder recommends a ratio of 7.5 inches to 10 inches. What is the angle of inclination using this recommendation?

## The Law of Sines

process: evaluate

1. Find the area of $\Delta ABC$ if AB = 120 cm, BC = 80 cm, and $m\angle B = 40°$.

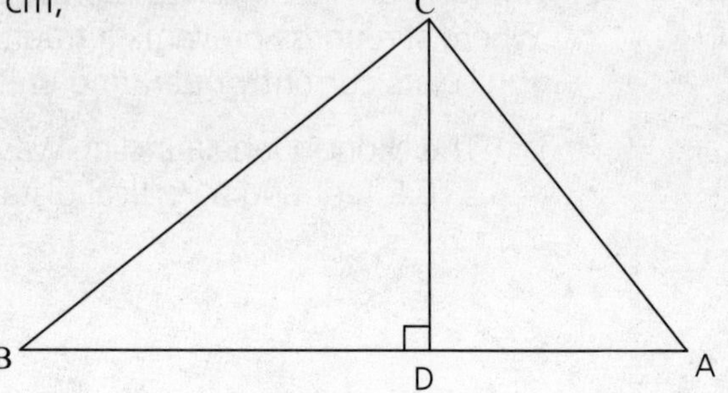


2. Find the area of $\Delta XYZ$ if XZ = 52 cm, XY = 37 cm, and $m\angle X = 74°$.

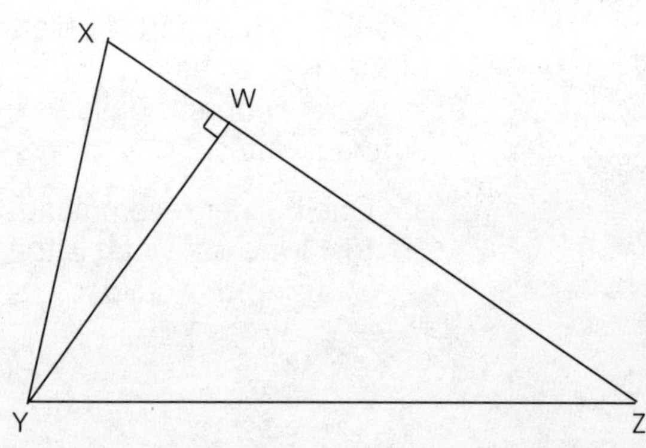

## The Law of Sines

process: model

3. Write an expression for $h$, the altitude, in terms of sin A and $b$.

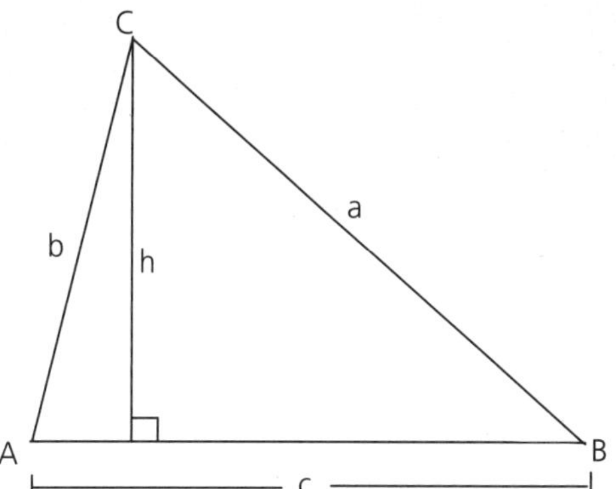


4. Write an expression for $h$, the altitude, in terms of sin B and $a$.

5. Show that $\frac{\sin A}{a} = \frac{\sin B}{b}$.

6. Write an expression for $k$, the altitude, in terms of sin B and $c$.

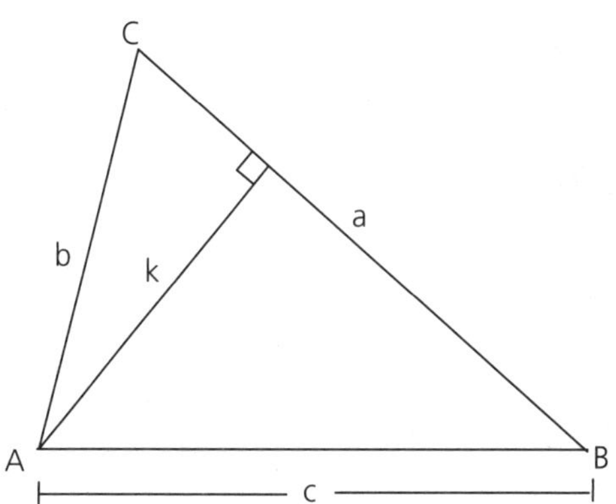


7. Write an expression for $k$, the altitude, in terms of sin C and $b$.

8. Show that $\frac{\sin A}{a} = \frac{\sin B}{b}$.

input

## The Law of Sines

The work you have done in this problem leads to a trigonometric property called **the law of sines**. The law of sines states that for a triangle with Angle measures of A, B, can C, and opposite side lengths of a, b, and c respectively,

$$\frac{\sin A}{a} = \frac{\sin B}{b} = \frac{\sin C}{c}.$$

Previously, you were able to use trigonometry to solve problems involving right triangles. The law of sines allows you to solve problems for any type of triangle if you know the measure of two angles and a side of the triangle.

process: model

## The Law of Sines

9. A forest ranger in an observation tower sights a fire 38° east of south. A ranger in a different tower located 10 miles due east of the first tower sights the fire at 43° west of south. How far is the fire from each tower?

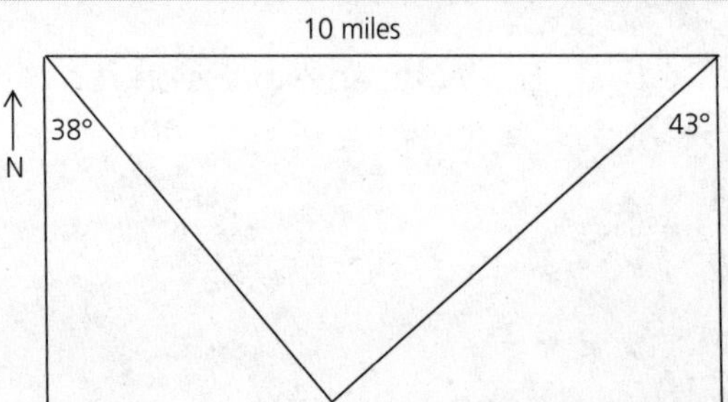

## The Law of Sines

process: model

10. To measure the height of a mountain, surveyors can measure the angle of elevation to the summit of the mountain from two different locations at the same elevation. In this case, the angle of elevation from point A measured 17° and the angle of elevation from point B measured 17°. Points A and B are 2200 feet apart. Determine the height of the mountain at its summit.

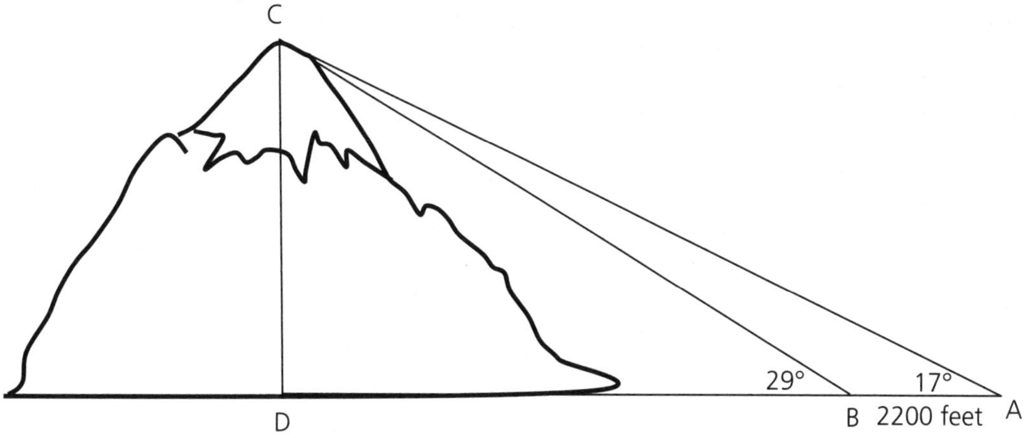

C

29° 17°

D B 2200 feet A

## The Law of Cosines

input

Trigonometry enables you to find the measures of the sides and angles of triangles. The sine, cosine, and tangent ratios by themselves allow you to work with right triangles. The law of sines allows you to work with other triangles given two angle measures and a side measure. The law of cosines will enable you to work with other triangles given the three side lengths, or two side lengths and an angle.

## The Law of Cosines

analysis

1. If two segments with lengths *a* and *b* meet at vertex C to form a right angle, the distance, *c*, between the other endpoints can be calculated using the Pythagorean Theorem.

$$c^2 = a^2 + b^2$$

   a. If the two segments with lengths *a* and *b* meet at vertex D to form an acute angle, what happens to the relationship between the square of *d* and the sum of the squares of *a* and *b*?

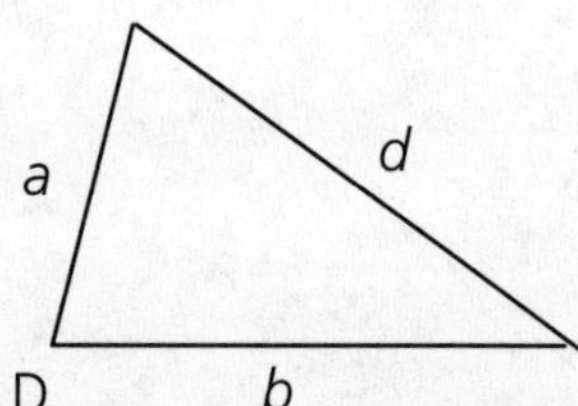


   b. If the two segments with lengths *a* and *b* meet at vertex E to form an obtuse angle, what happens to the relationship of the lengths of the sides of the triangle?

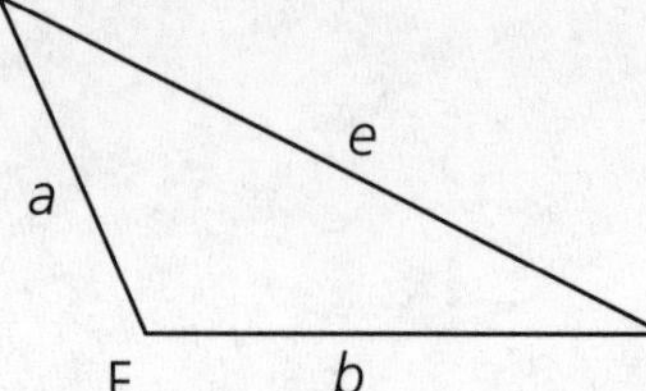

input

## The Law of Cosines

In the problem on the previous page, when C is an acute angle, $d^2 < a^2 + b^2$ (meaning $d^2 = a^2 + b^2 -$ something).

When C is an obtuse angle, $e^2 > a^2 + b^2$ (meaning $e^2 = a^2 + b^2 +$ something).

Mathematicians discovered the "something" is 2 *ab* cosC, which lead to a generalization of the Pythagorean Theorem known as **the law of cosines.**

For a triangle with side lengths *a*, *b*, and *c*, and angle C opposite the side with length c:

$$c^2 = a^2 + b^2 - 2ab \cos C$$

process: evaluate

## The Law of Cosines

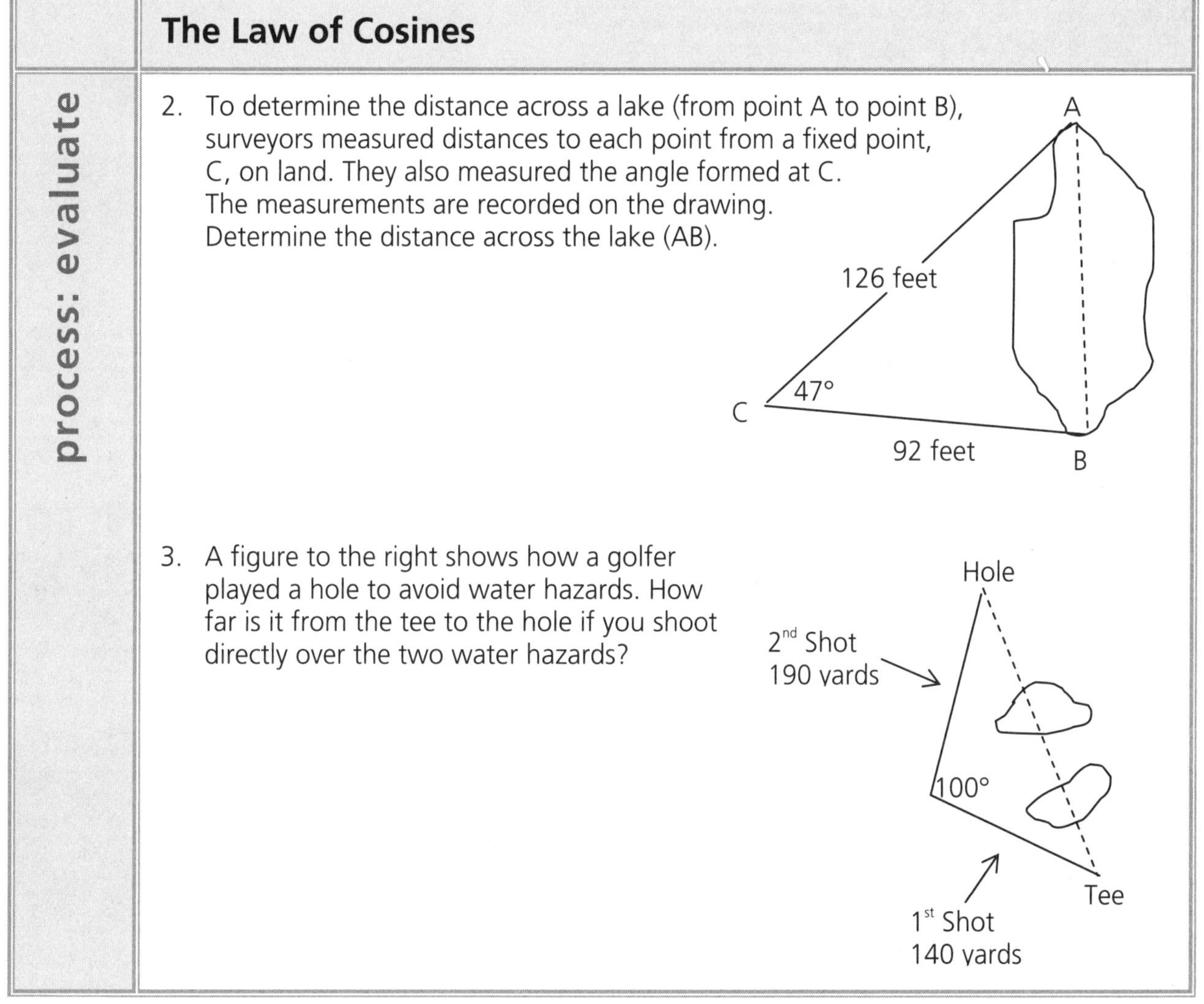


2. To determine the distance across a lake (from point A to point B), surveyors measured distances to each point from a fixed point, C, on land. They also measured the angle formed at C. The measurements are recorded on the drawing. Determine the distance across the lake (AB).

3. A figure to the right shows how a golfer played a hole to avoid water hazards. How far is it from the tee to the hole if you shoot directly over the two water hazards?

process: evaluate

## The Law of Cosines

4. You have been asked by a friend to make a frame for a triangular piece of art. She tells you that the sides of the artwork measure 26 inches, 32 inches, and 38 inches. Determine the angles of the triangle so that you can cut the wooden frame pieces to fit together properly.

analysis

## The Law of Cosines

5. The Pythagorean Theorem is a special case of the law of cosines. Show how this is true.

## Trigonometric Identities

input

In general, a trigonometric equation is any equation involving a trigonometric function. When a trigonometric equation is true for all values of the variable $\theta$, the equation is called a **trigonometric identity**.

In this section, you will explore some of the basic trigonometric identities involving the sine, cosine, and tangent functions.

## Trigonometric Identities

process: evaluate

1. Evaluate the following.

   a. $\sin\frac{\pi}{6}$

   b. $\sin\frac{13\pi}{6}$

   c. $\sin\frac{\pi}{3}$

   d. $\sin\frac{7\pi}{3}$

   e. $\sin\frac{\pi}{2}$

   f. $\sin\frac{5\pi}{3}$

   g. $\sin\pi$

   h. $\sin 3\pi$

## Trigonometric Identities

analysis

2. What do you notice about your answers in question 1?

   a. How do the answers in the left column compare to the answers in the right column?

   b. Why does this happen? (Hint: Consider the angle measures.)

**input**

## Trigonometric Identities

Since the period of the sine function is 2π, the values of the sine function repeat every 2π radians. This leads us to our first identity:

$$\sin(\theta + 2\pi) = \sin\theta$$

The identity is called a **periodicity identity** because it is based on the period of the function.

**analysis**

## Trigonometric Identities

3. Consider the period of the cosine function. Write a periodicity identity for the cosine function and explain why it makes sense.

4. Consider the period of the tangent function. Write a periodicity identity for the tangent function and explain why it makes sense.

**analysis**

## Trigonometric Identities

5. Consider the unit circle and the angles, $\theta$ and $-\theta$, which are reflection images over the x-axis.

   a. How are the x-coordinates of points A and $A'$ related?

   b. How are the y-coordinates of points A and $A'$ related?

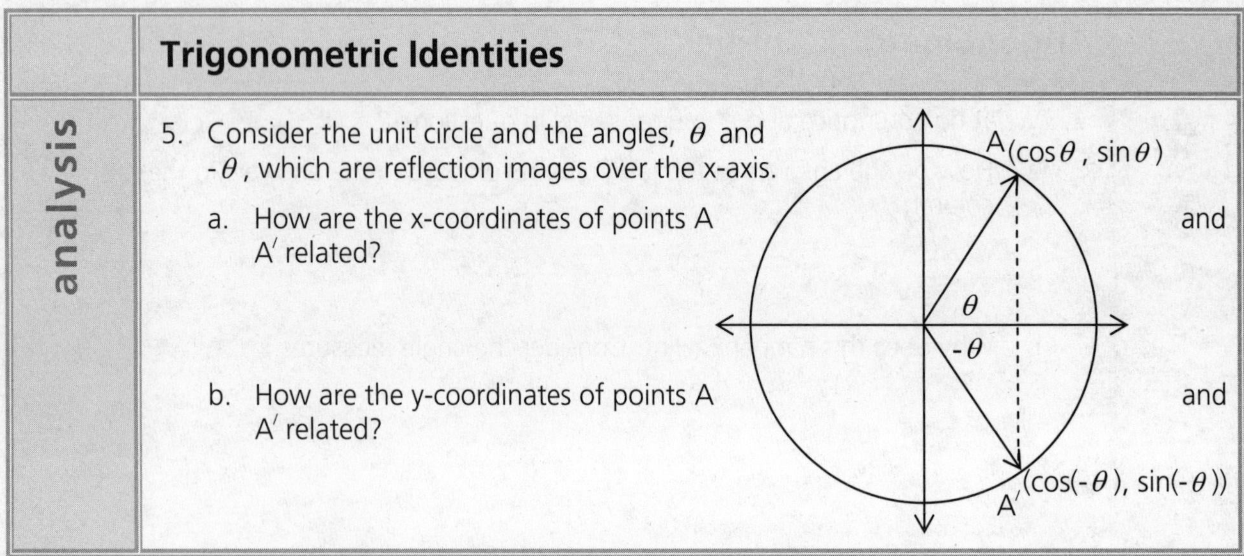

analysis

## Trigonometric Identities

6. How are trigonometric functions related to the coordinates?
   a. What trigonometric function is associated with the x-coordinate?

   b. What trigonometric function is associated with the y-coordinate?

7. Opposite angles generate a set of identities. Two of the identities are derived from your work in questions 5 and 6. Explain why the following identities make sense.

$$\sin(-\theta) = -\sin\theta$$

$$\cos(-\theta) = \cos\theta$$

analysis

## Trigonometric Identities

8. Consider the opposite angle identities above, the unit circle, and what you know about the tangent function.
   a. Complete the following identity statement: $\tan(-\theta)$ = ______________.

   b. Explain why your statement from part (a) is true.

## Trigonometric Identities

analysis

9. Consider the unit circle drawn to the right.

   a. Use the Pythagorean Theorem to write an equation showing the relationship between the sides of the right triangle.

   

   b. What trigonometric function is associated with *x*?

   c. What trigonometric function is associated with *y*?

   d. Rewrite the Pythagorean relationship using the trigonometric functions.

## Trigonometric Identities

analysis

The identity you found in question 8 is called the **Pythagorean Identity**. It is customary to write $(\cos\theta)^2$ as $\cos^2\theta$ and $(\sin\theta)^2$ as $\sin^2\theta$.

The Pythagorean Identity is usually written $\sin^2\theta + \cos^2\theta = 1$.

There are many more trigonometric identities than the few you have seen in this section. Trigonometric identities can help you solve equations or simplify expressions.

## Solving Trigonometric Equations

input

A **trigonometric equation** is an equation in which the variable to be found is associated with a trigonometric function. The number of solutions to a trigonometric equation can vary depending on the domain used for the variable.

Frequently used types of domains include:

1. The restricted domain used to generate the inverse function
2. An interval equal in size to the period of the function
3. The set of all real numbers for which the function is defined

For example, with the equation, $\sin\theta = 0.5$, you can find the solutions for:

a. $-\frac{\pi}{2} \le \theta \le \frac{\pi}{2}$

Using the inverse sine function gives you $\theta = \sin^{-1}(0.5)$. In radians, $\theta = \frac{\pi}{6}$.

b. $0 \le \theta \le 2\pi$

A graph of the sine function and a graph of the line $y = 0.5$ show two intersection points in the interval.

Consideration of the unit circle I indicates that the value of sine is 0.5 in the first and second quadrants. The supplement of $\frac{\pi}{6}$ is another solution.

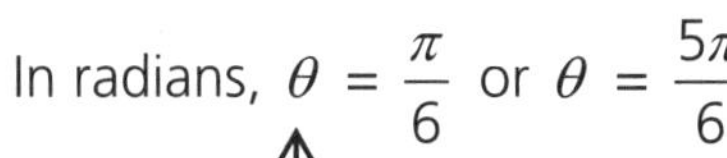

In radians, $\theta = \frac{\pi}{6}$ or $\theta = \frac{5\pi}{6}$

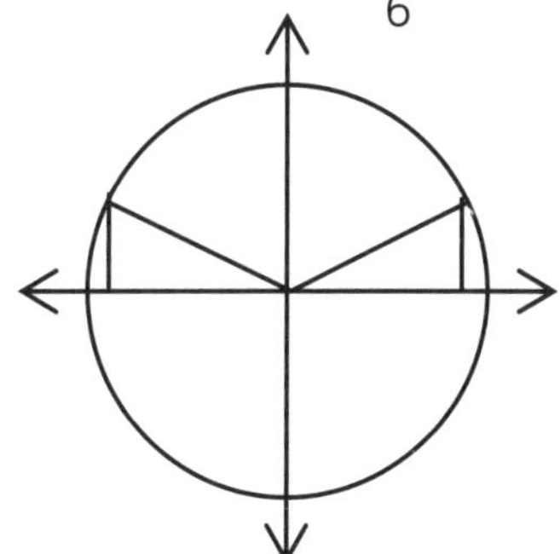

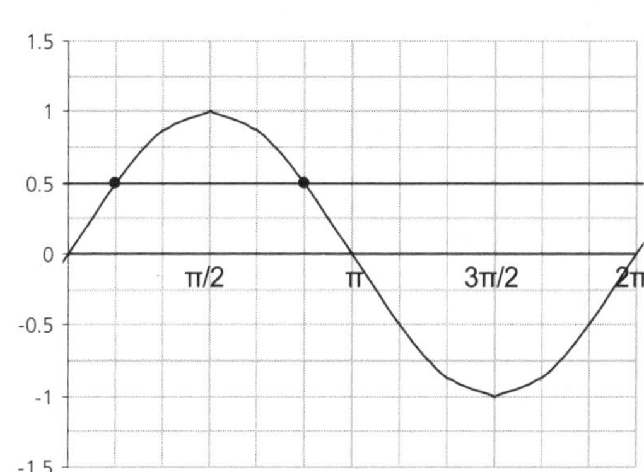


c. All real numbers (the general solution)

You can use the periodicity identity for sine to find the generalized solution from the solutions in part (b). Adding or subtracting multiples of 2π generates all solutions for $\sin\theta = 0.5$.

$\theta = \frac{\pi}{6} + 2\pi n$ or $\theta = \frac{5\pi}{6} + 2\pi n$ for all integers $n$

process: practice

## Solving Trigonometric Equations

1. Solve $\cos\theta = \frac{\sqrt{2}}{2}$ over the domain $0 \le \theta \le 2\pi$.

2. Solve $\tan\theta = \sqrt{3}$ over the domain of all real numbers.

3. Solve $\sin\theta = 0.6$ over the domain $-\frac{\pi}{2} \le \theta \le \frac{\pi}{2}$.

input

## Solving Trigonometric Equations

When a trigonometric equation involves transformation on the parent function, solving the equation requires the same techniques you used to solve other equations. For example,

| | | |
|---|---|---|
| Solve | $\sqrt{3} \bullet \tan\theta + 5 = 4$ | over the domain of all real numbers |
| | $\sqrt{3} \bullet \tan\theta = -1$ | subtract 5 from both sides |
| | $\tan\theta = \frac{-1}{\sqrt{3}}$ | divide both sides by $\sqrt{3}$ |
| | $\theta = \tan^{-1}\left(\frac{-1}{\sqrt{3}}\right) = -\frac{\pi}{6}$ | apply the inverse tangent function |
| | $\theta = -\frac{\pi}{6} + 2\pi n$ | for integer values of n (periodicity identity) |

process: practice

## Solving Trigonometric Equations

4. Solve $4\sin\theta + 3 = 6$ over the domain of all real numbers.

5. Solve $2\cos\theta = 3$ over the domain of all real numbers.

process: practice

## Solving Trigonometric Equations

6. Solve $2\tan\theta + 4 = 5$ over the domain of all real numbers.

extension

## Solving Trigonometric Equations

If an equation that can be written in the form, $ax^2 + bx + c = 0$, has $x$ replaced with a trigonometric function, the result is a trigonometric equation in quadratic form. These equations can be solved as you would solve other quadratic equations, by factoring or by using the quadratic formula. For example,

Solve $2\sin^2\theta + 5\sin\theta = 3$ over the domain of all real numbers

Start with a substitution to make the equation easier to solve. Since this equation involves the sine function, substitute $x$ in place of $\sin\theta$.

| | | | |
|---|---|---|---|
| $2x^2 + 5x = 3$ | | | |
| $2x^2 + 5x - 3 = 0$ | | | (get equation in standard form) |
| $(2x - 1)(x + 3) = 0$ | | | (factor the quadratic expression) |
| $2x - 1 = 0$ *or* $x + 3 = 0$ | | | (set each factor equal to zero) |
| $2x = 1$ | | | |
| $x = \frac{1}{2}$ | *or* | $x = -3$ | (solve each equation) |
| $\sin\theta = \frac{1}{2}$ | *or* | $\sin\theta = -3$ | (replace $x$ with the trig function) |
| $\theta = \frac{\pi}{6} + 2\pi n$ *or* $\theta = \frac{5\pi}{6} + 2\pi n$ | *or* | *no solutions* | (solve the trig equations using $\sin^{-1}$) |

Note that the quadratic equation in standard form could have been solved using the quadratic formula instead of factoring. Once solved for $x$, the last two steps would be the same.

## Solving Trigonometric Equations

extension

7. Solve $4\sin^2\theta - 1 = 0$ over the domain of all real numbers.

8. Solve $2\cos^2\theta - \cos\theta = 1$ over the domain of all real numbers.

extension

## Solving Trigonometric Equations

9. Solve $2\tan^2\theta + 3\tan\theta - 1 = 0$ over the domain of all real numbers.

10. Solve $6\sin^2\theta - 16\sin\theta - 33 = 0$ over the domain of all real numbers.

input

## Solving Trigonometric Equations

In this unit, you learned about special periodic functions called trigonometric functions. Trigonometric functions are useful for modeling real-world situations and for solving problems, particularly those involving measurements in triangles.

Trigonometric functions are based on the unit circle. In the following unit, *Conic Sections*, you will learn more about the unit circle.

# Unit 8: Conic Sections

## Contents

preview

## Making Planar Shapes

A cube with side length 10 cm will hold exactly one liter of liquid. Obtain a clear cube that will hold one liter of liquid. Fill the cube so that the water level is at 4 cm.

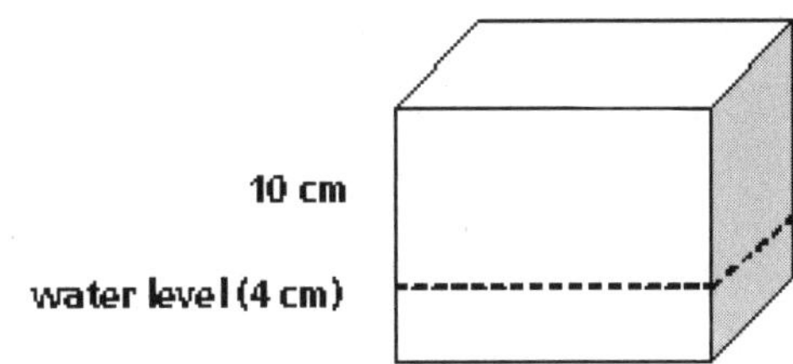


1. Place the cube on a flat surface. What shape does the top surface of the water form? Write a sentence explaining why.

2. Tilt the cube so that the top surface of the water creates a different shape. Try to create the following shapes from the surface of the water.

   For each shape, describe in as much detail as possible how you positioned the cube so that the top surface of the water formed the desired shape.

   a. Square

   b. Rectangle (that is not a square)

   c. Parallelogram (that is not a rectangle)

   d. Triangle

## Making Planar Shapes

preview

e. Isosceles Triangle

f. Equilateral Triangle

g. Trapezoid

h. Pentagon

i. Hexagon

3. Are there any additional shapes not previously mentioned that you were able to create? How did you position the cube to obtain these shapes?

## Two Cones, Please

input

When a three dimensional solid, such as a cube, is cut by a plane, such as the top surface of the water, the 2-dimensional figure that results is called a **plane section** or **cross section** of the solid. The shape of the cross section will depend on the positioning of the plane with respect to the solid.

Four special cross sections called **conic sections** are formed when a plane intersects a solid called a double napped cone. In addition, it is possible to form a point, a line, and intersecting lines. These are called **degenerate conics**.

Below is an example of a double napped cone. The upper and lower cones are referred to as nappes.

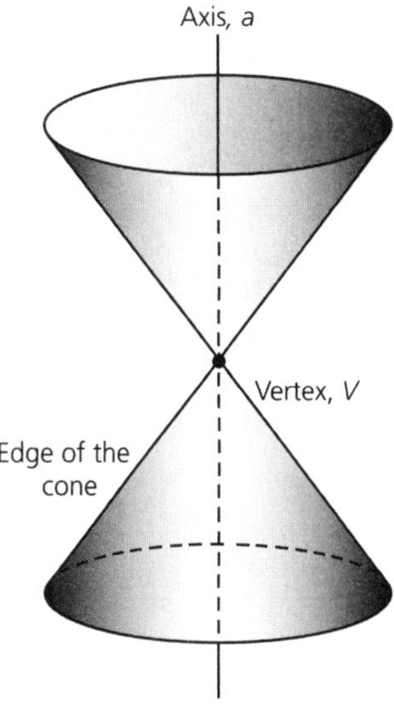


## Two Cones, Please

analysis

1. Describe how the intersection of the plane and the double napped cone could result in a point.

## Two Cones, Please

analysis

2. Describe how the intersection of the plane and the double napped cone could result in a line.

3. Describe how the intersection of the plane and the double napped cone could result in intersecting lines.

4. Describe the conic section that results when a plane intersects the double napped cone perpendicular to the axis.

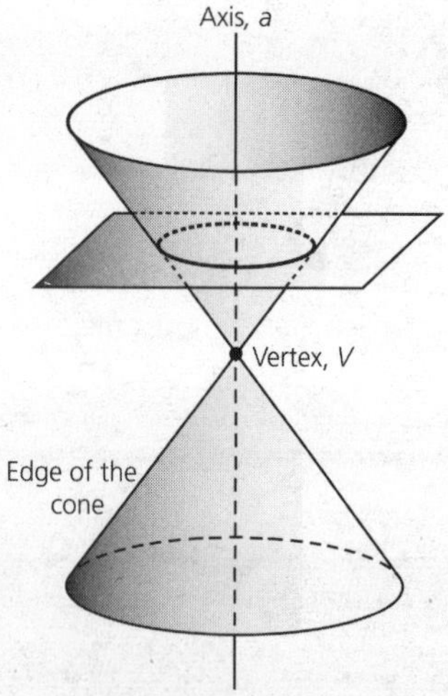

5. What mathematical term can be used to describe the cross section?

## Two Cones, Please

analysis

6. Describe the conic section that results when a plane at an angle between $0^{\circ}$ and $90^{\circ}$ to the axis intersects one nappe of the double napped cone.

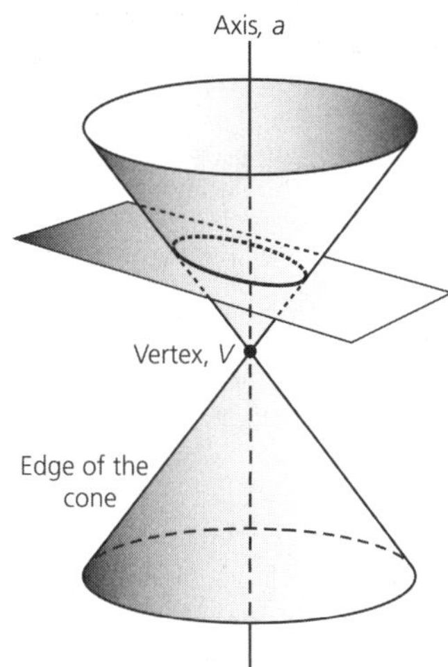


7. What mathematical term can be used to describe the cross section?

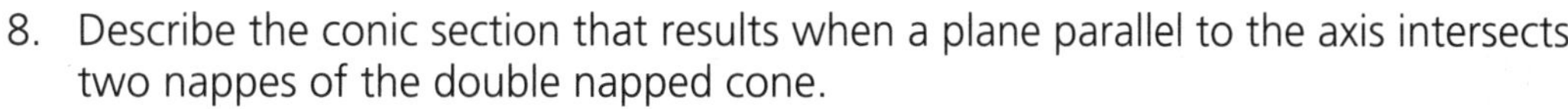

8. Describe the conic section that results when a plane parallel to the axis intersects two nappes of the double napped cone.

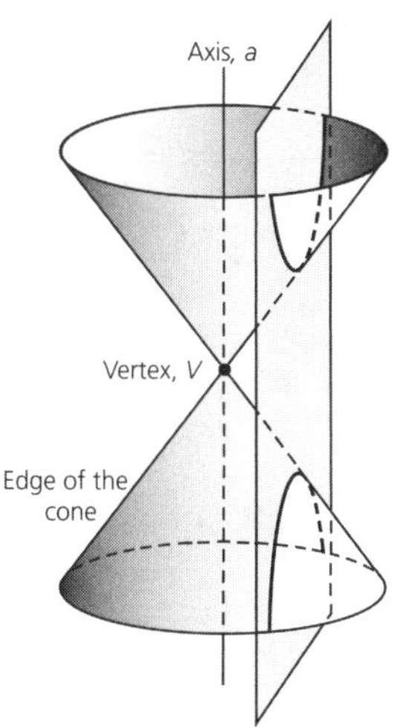


9. What mathematical term can be used to describe the cross section?

analysis

## Two Cones, Please

10. Describe the conic section that results when a plane parallel to the edge of the cone intersects one nappe of the double napped cone.

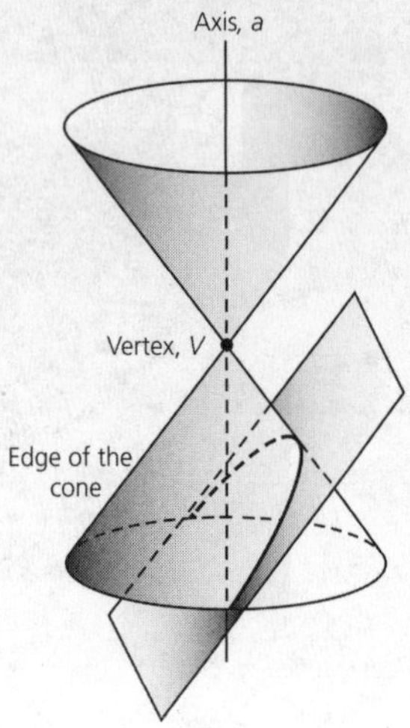

Axis, *a*

Vertex, *V*

Edge of the cone

11. What mathematical term can be used to describe the cross section?

input

## Two Cones, Please

When a plane intersects one nappe of the double-napped cone perpendicular to the axis of the cone, the curve that results is a **circle**.

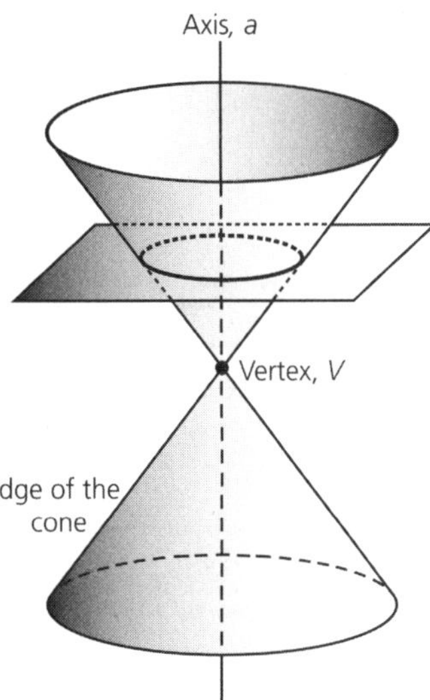


When a plane intersects one nappe of the cone at an angle between 0° and 90° to the axis of the cone, the curve that results is an **ellipse**.

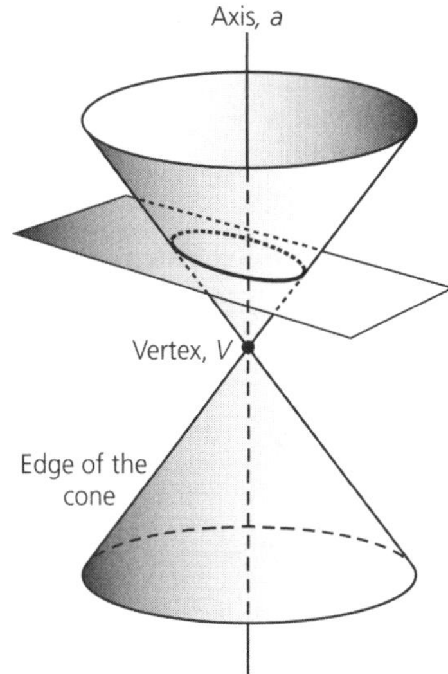


When a plane intersects both nappes of the cone, the curve that results is a **hyperbola**.

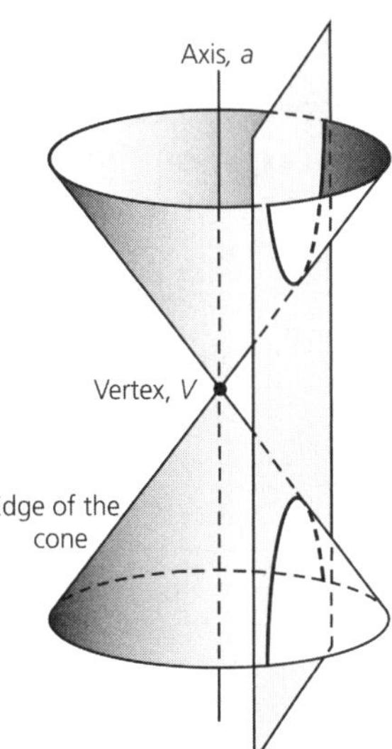


When a plane intersects one nappe of the double-napped cone parallel to the edge of the cone, the curve that results is a **parabola**.

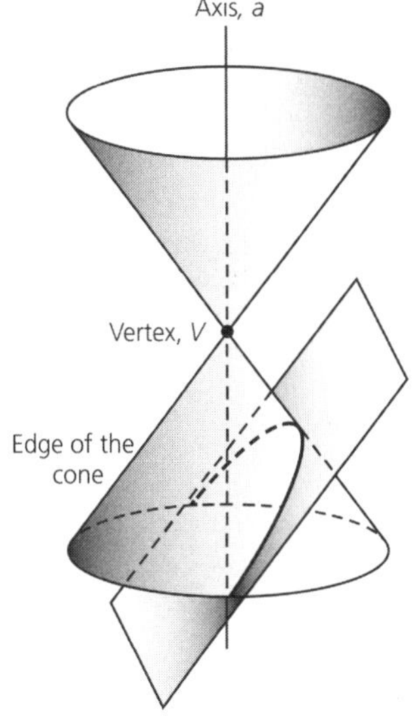

## Round and Round

process: summary

1. Based on your previous experience, draw a circle and label the parts of the circle in the diagram.

2. Write a definition for a circle. Reference your diagram in the definition.

The equation of a circle with a given center and radius can be derived using the definition of a circle and the distance formula.

## Round and Round

process: model

3. Use the distance formula to write an equation for a circle with the center at the origin and a radius of 5 units. Use the point (x, y) on the circle to write the equation. Simplify the equation completely and write the final equation without a square root.

## Round and Round

process: model

4. Create a graph of the equation below.

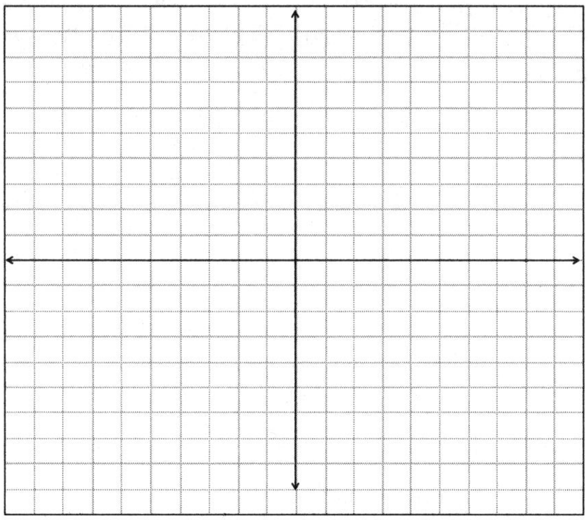

## Round and Round

analysis

5. Does the equation for a circle with the center at the origin and a radius of 5 units represent a function? Explain why or why not.

6. If the equation is not a function, how would you describe the equation?

7. What is the domain and range?

8. What are the x- and y-intercepts?

process: model

## Round and Round

9. Use the distance formula to write the general equation for a circle with the center at the origin and a radius of r units. Use the point (x, y) on the circle to write the equation. Simplify the equation completely and write the final equation without a square root.

analysis

## Round and Round

10. Does the general equation for a circle with the center at the origin and a radius of r units represent a function or a relation? Explain.

11. In general, what is the domain and range of a circle centered at the origin?

12. In general, what are the x- and y-intercepts of a circle centered at the origin?

input

## Round and Round

The general equation for a circle centered at the origin with a radius of r units is

process: model

## Round and Round

13. Use the distance formula to write an equation for a circle with the center at (1, 3) and a radius of 5 units. Use the point (x, y) on the circle to write the equation. Simplify the equation completely and write the final equation without a square root.

14. Create a graph of the equation below.

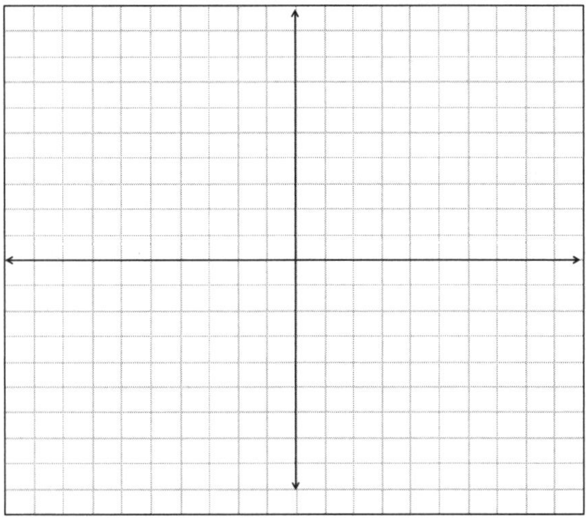

analysis

## Round and Round

15. Does the equation for a circle with the center at (1, 3) and a radius of 5 units represent a function? Explain why or why not.

16. If the equation is not a function, how would you describe the equation?

## Round and Round

analysis

17. Does the top half of the circle represent a function? The bottom half?

18. Write one equation to represent the top half of the circle and one equation to represent the bottom half of the circle.

19. What is the domain and range of the full circle?

20. Determine the x- and y-intercepts of the full circle algebraically.

## Round and Round

analysis

21. Describe the similarities and differences between the graph of $x^2 + y^2 = 25$ and the graph of $(x-1)^2 + (y-3)^2 = 25$.

22. Describe the transformations applied to the graph of $x^2 + y^2 = 25$ to get the graph of $(x-1)^2 + (y-3)^2 = 25$.

23. Write the equation for the circle that results when the graph of $x^2 + y^2 = 49$ is shifted to the

    a. Left 4 units

    b. Right six units

    c. Left 1 unit

24. In general, what is the equation for a circle that results when the graph of $x^2 + y^2 = r^2$ is shifted horizontally h units?

## Round and Round

analysis

25. Write the equation for the circle that results when the graph of $x^2 + y^2 = 16$ is shifted

a. Down 5 units

b. Up ten units

c. Down 2 units

26. In general, what is the equation for a circle that results when the graph of $x^2 + y^2 = r^2$ is shifted vertically k units?

27. Write the equation for the circle that results when the graph of $x^2 + y^2 = 81$ is shifted

a. Down 2 units and to the left 3 units

b. Up six units and to the left seven units

c. Down 7 units and to the right 1 unit

28. What is the general equation for a circle with the center at the point (h, k) and a radius r?

## Round and Round

process: practice

29 Identify the center and radius for each circle defined below.

a. $(x+2)^2+(y-8)^2=4$

b. $(x-3)^2+(y-1)^2=64$

c. $(x+11)^2+(y+5)^2=36$

d. $(x-4)^2+(y+7)^2=1$

30. Write the equation of a circle based on the given information.

a. Center (6, -3); radius = 3

b. Center (-9, 1); radius = 11

c. Center (-5, -6); radius = 12

d. Center (10, 7); radius = 7

process: practice

## Round and Round

31. Determine the domain, range, and x- and y-intercepts for each circle.

   a. $(x-3)^2+(y+7)^2=16$

process: practice

## Round and Round

b. $(x+3)^2+(y+1)^2=36$

## Halley's Comet

preview

Most planets and comets orbit around the sun in an elliptical path. The orbit of Halley's Comet, named after scientist Edmond Halley, can be modeled by the equation

$$\frac{(x-1{,}606{,}273{,}137)^2}{(1{,}673{,}201{,}184)^2}+\frac{y^2}{(468{,}496{,}330)^2}=1$$

In this model, the sun will appear at the origin of the graph.

Determine Halley's Comet's minimum and maximum distance from the sun.

## It's a Stretch

process: model

1. Graph the circle represented by the equation $x^2 + y^2 = 1$.

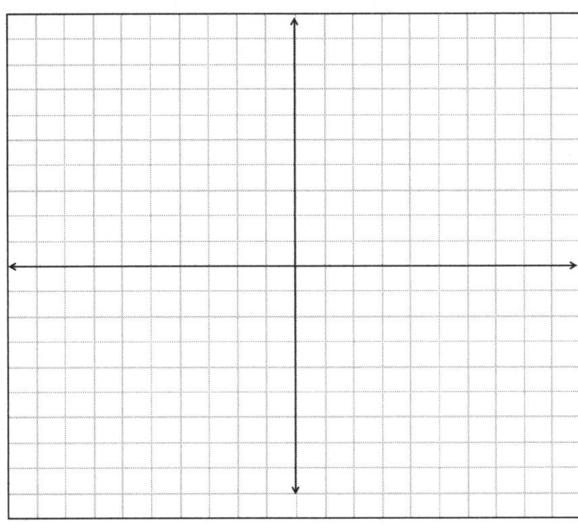

## It's a Stretch

analysis

2. Identify the radius of this circle represented by the equation $x^2 + y^2 = 1$.

3. The circle represented by the equation $x^2 + y^2 = 1$ is often referred to as the **Unit Circle**. Why do you think this name is used to describe this circle?

## It's a Stretch

process: model

4. Graph the circle represented by the equation $x^2 + y^2 = 4$.

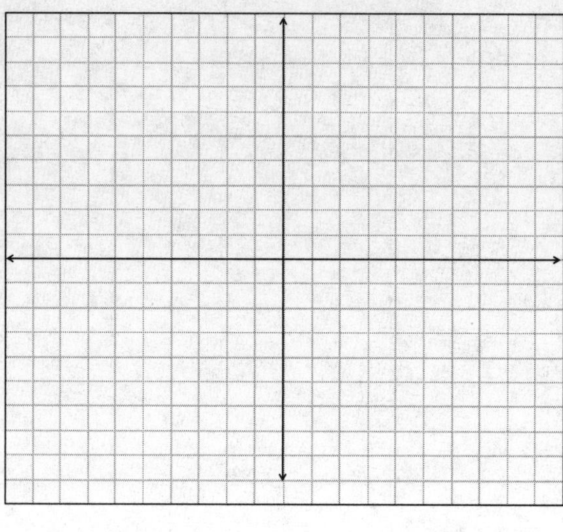

## It's a Stretch

analysis

5. Identify the radius of the circle represented by $x^2 + y^2 = 4$.

6. Describe the transformation applied to the unit circle to result in the graph of $x^2 + y^2 = 4$?

7. How is this transformation represented in the equation $x^2 + y^2 = 4$?

process: model

## It's a Stretch

8. Graph the circle represented by the equation $x^2 + y^2 = 9$.

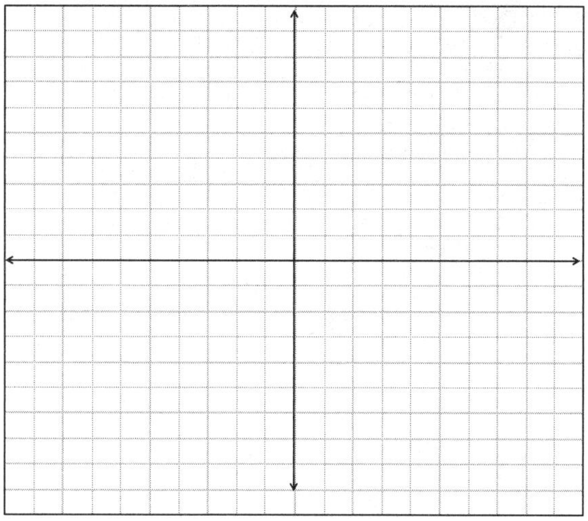

analysis

## It's a Stretch

9. Identify the radius of the circle represented by $x^2 + y^2 = 9$.

10. Describe the transformation applied to the unit circle to result in the graph of $x^2 + y^2 = 9$?

11. How is this transformation represented in the equation $x^2 + y^2 = 9$?

## It's a Stretch

analysis

12. Rewrite the equation $x^2 + y^2 = 4$ in the form $\frac{x^2}{r^2} + \frac{y^2}{r^2} = 1$.

13. Rewrite the equation $x^2 + y^2 = 9$ in the form $\frac{x^2}{r^2} + \frac{y^2}{r^2} = 1$.

14. What do you think the graphs would look like if the transformed equation was of the form $\frac{x^2}{r^2} + y^2 = 1$?

15. What do you think the graphs would look like if the transformed equation was of the form $x^2 + \frac{y^2}{r^2} = 1$?

process: model

## It's a Stretch

16. Graph the equations $\frac{x^2}{4} + y^2 = 1$ and $\frac{x^2}{9} + y^2 = 1$.

analysis

## It's a Stretch

17. Describe the shape of each graph. Do you know a mathematical term to describe each shape?

18. Do the graphs have a single radius? Explain why or why not.

## It's a Stretch

analysis

19. What is the distance from the center to the edge of each graph along the x-axis?

20. What is the distance from the center to the edge of each graph along the y-axis?

21. Describe the transformation applied to the unit circle to result in each graph.

22. How is each transformation represented in the equation?

process: practice

## It's a Stretch

23. Write an equation for the graph that results when the following transformations are applied to a unit circle.

    a. Dilated by a factor of 10 along the x-axis

    b. Dilated by a factor of six along the x-axis

    d. Dilated by a factor of a along the x-axis

process: model

## It's a Stretch

24. Graph the equations $x^2 + \frac{y^2}{4} = 1$ and $x^2 + \frac{y^2}{9} = 1$.

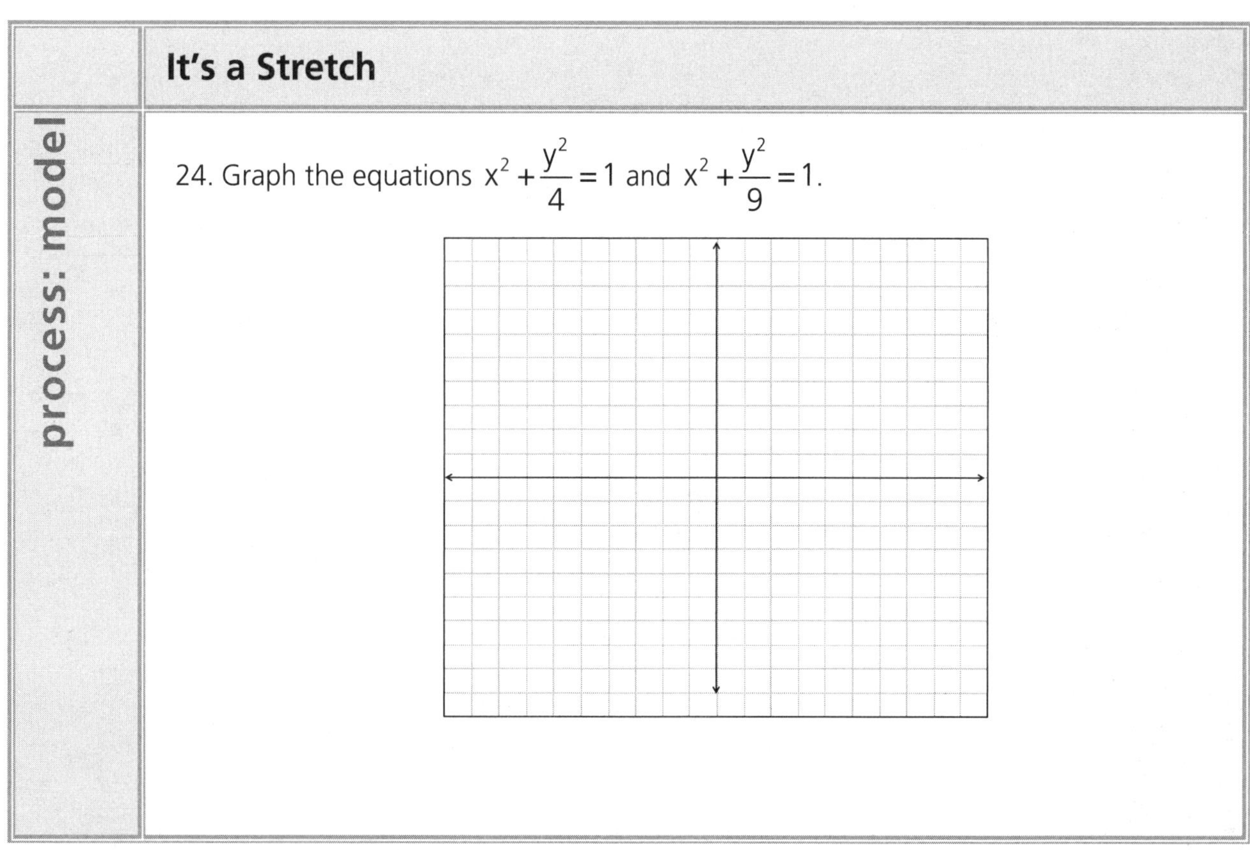

## It's a Stretch

analysis

25. What is the distance from the center to the edge of each graph along the x-axis?

26. What is the distance from the center to the edge of each graph along the y-axis?

27. Describe the transformation applied to the unit circle to result in each graph.

28. How is each transformation represented in the equation?

process: practice

## It's a Stretch

29. Write an equation for the graph that results when the following transformations are applied to a unit circle.

    a. Dilated by a factor of 7 along the y-axis

    b. Dilated by a factor of eleven along the y-axis

    d. Dilated by a factor of b along the y-axis

process: practice

## It's a Stretch

30. Sketch the graph that results if a unit circle is dilated along the x-axis by a factor of 2 and along the y-axis by a factor of 5. Write the equation representing the graph.

process: practice

## It's a Stretch

31. Sketch the graph that results if a unit circle is dilated along the x-axis by a factor of four and along the y-axis by a factor of two. Write the equation representing the graph.

32. Sketch the graph that results if a unit circle is dilated along the x-axis by a factor of 3 and along the y-axis by a factor of 4. Write the equation representing the graph.

## Paper Folding - Ellipses

input

The shapes generated by dilating a unit circle by different factors along the x- and y-axes are examples of **ellipses**.

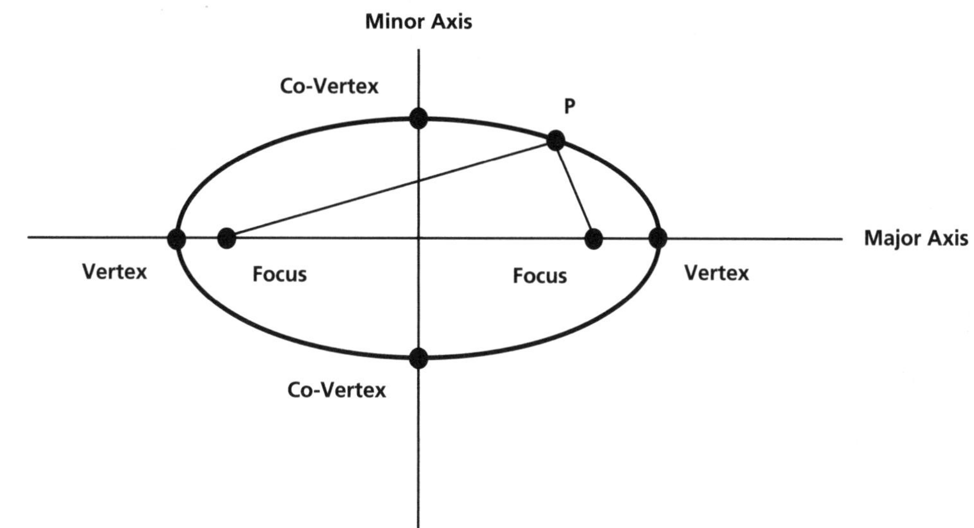


In an ellipse, the foci and vertices always lie on the major axis, while the co-vertices always lie on the minor axis. The major and minor axes may lie on either the x- or y-axes.

An ellipse is also symmetrical with respect to both the major and minor axes.

## Paper Folding - Ellipses

process: model

1. Take a piece of wax paper. Draw a large circle on the paper. Label the center of the circle point C. The center will represent one focus of the ellipse we will construct. Draw another point within the circle not at the center to represent the other focus. Label it point F.

2. Fold the paper so that point F and the edge of the circle meet. Repeat for various points along the circle.

## Paper Folding - Ellipses

analysis

3. What conic section is formed by the folds?

4. Draw any point on the circle and label it point D. Draw a line from point F to point D. Fold the paper so points F and D meet. What is the relationship between the line formed by the crease and $\overline{FD}$?

5. Draw a line from point C to point D. Label the point where the line intersects the crease as point P. Should point P be on the ellipse?

6. What is true about the distances from point P to point F and point D? How do you know this is true?

## Paper Folding - Ellipses

analysis

7. Write segment CD as the sum of two segments.

8. What segment is equal in length to segment PD? Substitute that segment into the equation above.

9. Will the length of segment CD change if the position of point D changes? Explain why or why not.

10. Repeat questions 4-9 for two other points on the circle. Do you see the same results?

11. Summarize what you discovered about the sum of the distances from any point on an ellipse to the foci.

## Paper Folding - Ellipses

analysis

12. Place point D on the ellipse along the major axis.
    a. Measure the distance from point D to each focus.

    b. What is the sum of the distances from point D to each focus?

    c. Measure the distance between the vertices.

    d. Is the sum of the distances from point D to the foci equal to any other measurements involving the ellipse? Explain why.

13. In general, how does the sum of the distances from any point on the ellipse to the foci relate to other distances within an ellipse?

process: model

## Try To Focus

1. Graph the ellipse defined by the function $\frac{x^2}{4}+\frac{y^2}{16}=1$.

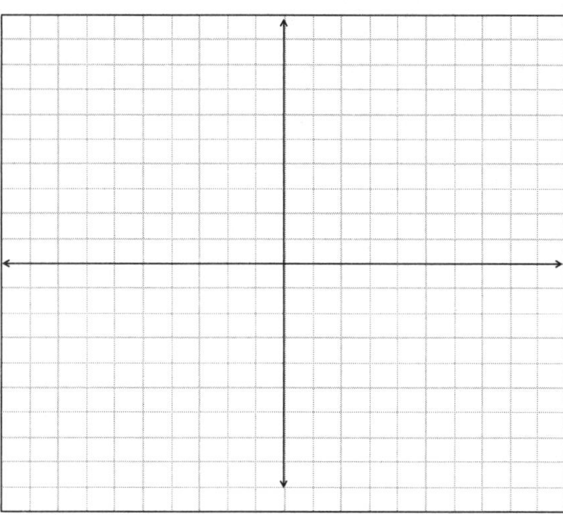

process: evaluate

## Try To Focus

2. Draw two points where you think the foci should be. Draw a triangle formed by the center, one focus, and one co-vertex. What type of triangle is formed?

3. What is the length of the side of the triangle connecting the center and one co-vertex?

4. What is the sum of the lengths from each focus to a co-vertex? Why?

5. What is the length of the side of the triangle connecting one focus and one co-vertex? How do you know?

process: evaluate

## Try To Focus

6. Compute the length of the side of the triangle connecting the center and one focus. Explain how you computed this value from the coordinates of the vertices and co-vertices.

7. What are the coordinates of the two foci?

process: model

## Try To Focus

8. Graph the ellipse defined by the function $\frac{x^2}{9} + \frac{y^2}{1} = 1$.

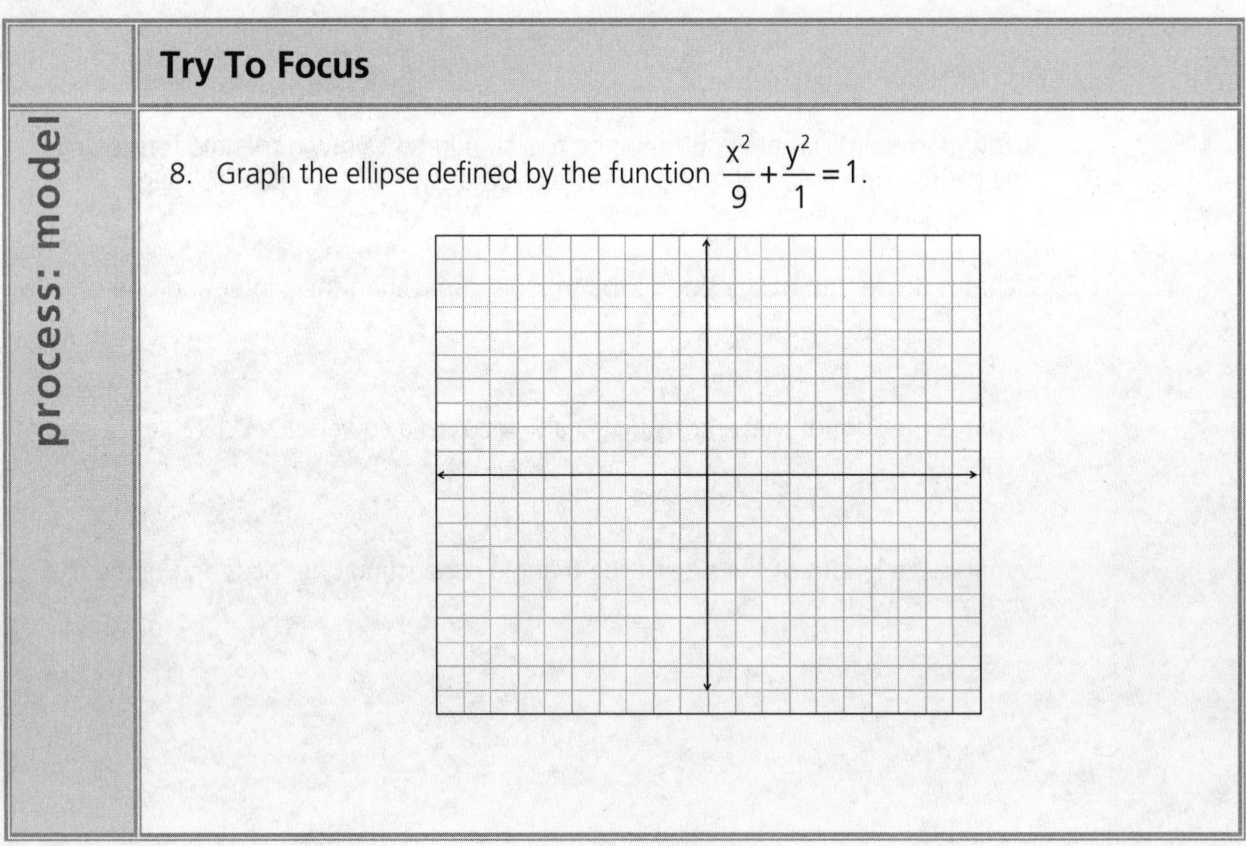

## Try To Focus

process: evaluate

9. Draw two points where you think the foci should be. Draw a triangle formed by the center, one focus, and one co-vertex. What type of triangle is formed?

10. What is the length of the side of the triangle connecting the center and one co-vertex?

11. What is the length of the side of the triangle connecting one focus and one co-vertex? How do you know?

12. Compute the length of the side of the triangle connecting the center and one focus. Explain how you computed this value from the coordinates of the vertices and co-vertices.

13. What are the coordinates of the two foci?

## Try To Focus

process: model

14. Graph the ellipse defined by the function $\frac{x^2}{36}+\frac{y^2}{49}=1$.

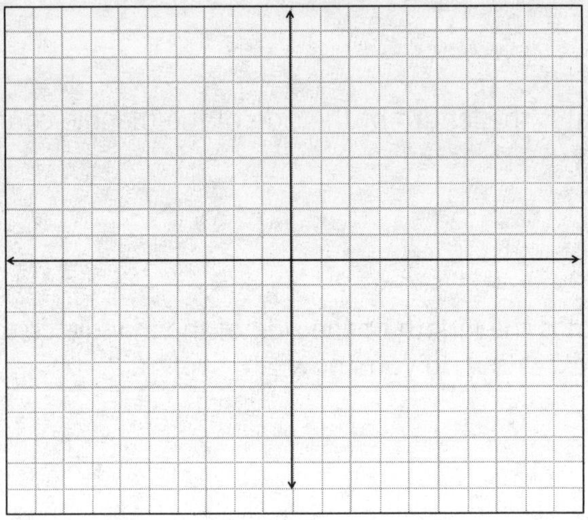

## Try To Focus

process: evaluate

15. Draw two points where you think the foci should be. Draw a triangle formed by the center, one focus, and one co-vertex. What type of triangle is formed?

16. What is the length of the side of the triangle connecting the center and one co-vertex?

17. What is the length of the side of the triangle connecting one focus and one co-vertex? How do you know?

## Try To Focus

process: evaluate

18. Compute the length of the side of the triangle connecting the center and one focus. Explain how you computed this value from the coordinates of the vertices and co-vertices.

19. What are the coordinates of the two foci?

## Try To Focus

process: summary

20. Given the equation for an ellipse, explain how to compute the coordinates of the foci from the vertices and co-vertices.

process: practice

## Try To Focus

21. Compute the coordinates of the foci for each ellipse.

a. $\frac{x^2}{81}+\frac{y^2}{16}=1$

b. $\frac{x^2}{4}+\frac{y^2}{64}=1$

c. $\frac{x^2}{100}+\frac{y^2}{36}=1$

input

## Whispering Galleries

A whispering gallery is a room with an elliptical-shaped ceiling. The shape of the ceiling allows two people standing at specific points to hear each other perfectly, even when whispering from a distance.

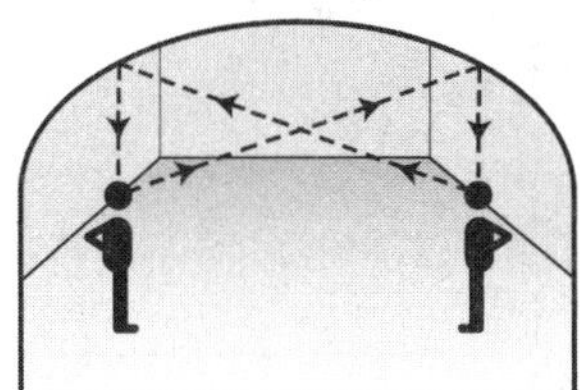

In the diagram above, assume the major axis is along the floor and the full top portion of the ellipse extends beyond the walls to the floor.

In terms of an ellipse, where do you think the two people need to stand in order to be able to hear each other whispering?

One example of a whispering gallery is the Mormon Tabernacle in Salt Lake City, Utah, whose inner chamber is 250 feet long and 80 feet high.

process: model

## Whispering Galleries

1. Write an equation to model the elliptical shape of the Mormon Tabernacle ceiling if the center of the room's floor is located at the origin. Explain how you determined the equation.

## Whispering Galleries

process: model

2. Create a graph of the whispering gallery.

## Whispering Galleries

process: evaluate

3. Determine how far two people would need to be from the center of the room to hear each other in the Mormon Tabernacle. Remember the two people must be standing on the foci.

4. How far apart will the two people be?

input

## Whispering Galleries

The Museum of Science and Industry in Chicago has had a whispering gallery on exhibit since 1937. The ellipse in Chicago is 47 feet, 4 inches in length and the foci are 40 feet, 7 inches apart.

process: model

## Whispering Galleries

5. Determine the height of the display.

process: model

## Whispering Galleries

6. Write an equation to model the elliptical shape of the exhibit if the center of the room's floor is located at the origin.

process: model

## Whispering Galleries

7. Create a graph of the whispering gallery.

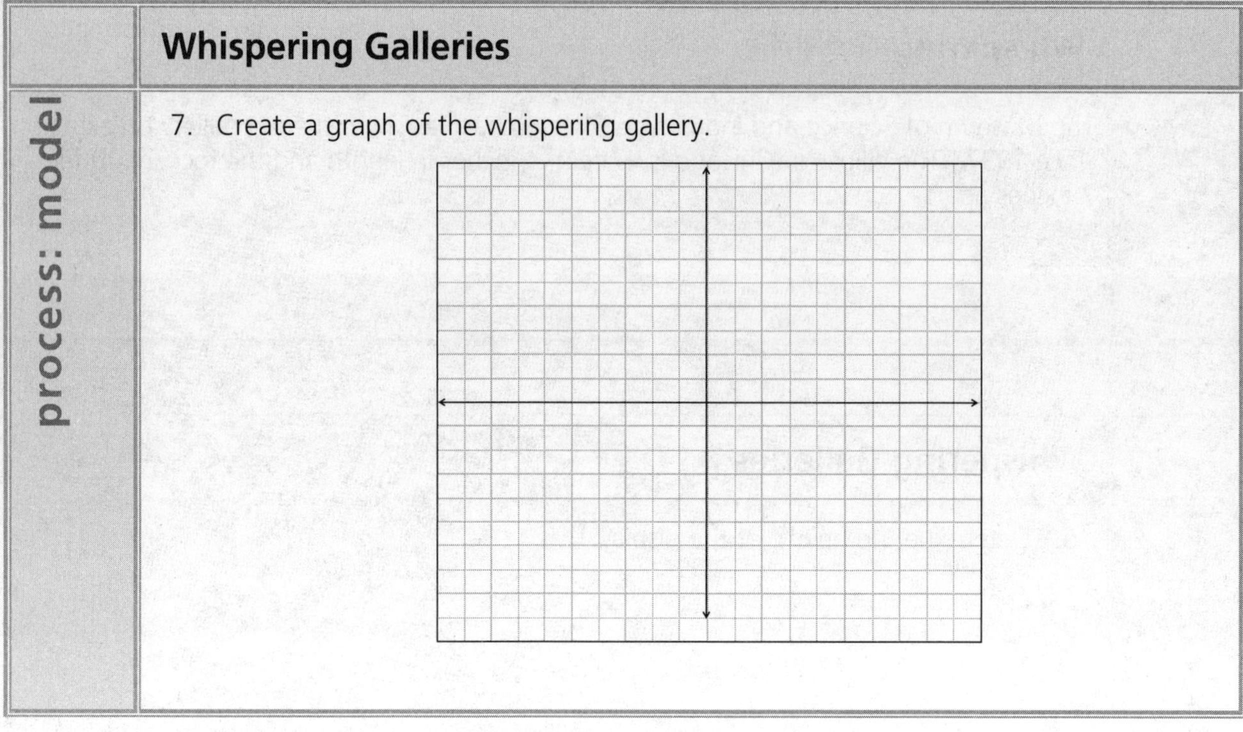

extension

## Whispering Galleries

8. After learning about whispering galleries, someone suggests building one in your classroom. Determine an equation to model the largest whispering gallery that could be built within your classroom.

## The Law of Orbits

input

Johannes Kepler published his three laws of planetary motion in 1609. The Law of Orbits states that all planets move in elliptical orbits, with the sun at one focus. Prior to Kepler's Laws, it was believed that planets moved in circular orbits, with the sun at the center.

Amazingly, Kepler's observations were made without the aid of a telescope and all calculations were performed by hand. Kepler's original calculations to compute the orbit of Mars took nearly 1000 pages!

For the ellipse modeling the orbit of Mercury, the distance between the vertices is approximately 0.774 AU (Astronomical units) and the distance between the foci is approximately 0.16 AU. One Astronomical unit is approximately 93,000,000 miles.

## The Law of Orbits

process: evaluate

1. For the ellipse modeling the orbit of Mercury, what are the coordinates of the vertices?

2. What are the coordinates of the foci?

3. What are the coordinates of the co-vertices?

4. What is the distance between the co-vertices?

process: model

## The Law of Orbits

5. Write the equation for an ellipse modeling the orbit of Mercury centered at the origin with the major axis along the x-axis and the sun to the left of the origin.

6. Graph the ellipse modeling the orbit of Mercury.

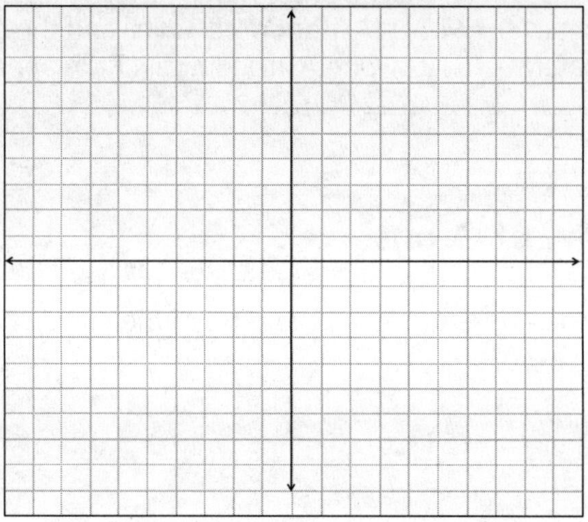

analysis

## The Law of Orbits

7. What are the coordinates of the sun?

8. Describe the transformation to the graph if you wanted the sun to appear at the origin.

process: model

## The Law of Orbits

9. Write the equation for an ellipse modeling the orbit of Mercury with the major axis along the x-axis and the sun at the origin.

10. Graph the ellipse modeling the orbit of Mercury.

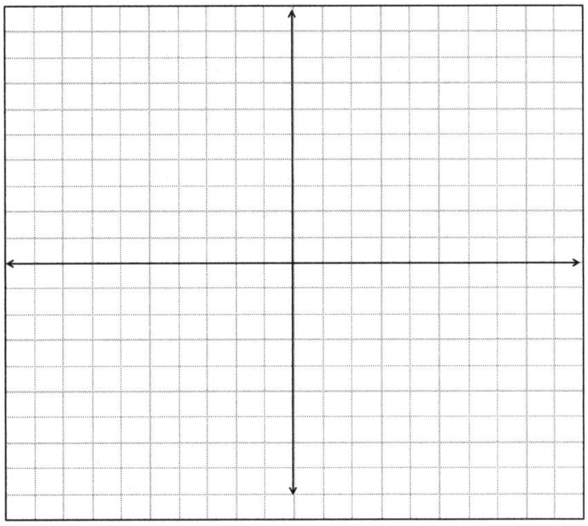

process: evaluate

## The Law of Orbits

11. If the orbit of Mercury is positioned with the major axis along the x-axis and the sun at the origin, identify the following:

 a. The coordinates of the center

 b. The coordinates of the vertices

process: evaluate

## The Law of Orbits

c. The coordinates of the co-vertices

d. The coordinates of the foci

e. The equation of the major axis

f. The equation of the minor axis

analysis

## The Law of Orbits

12. Describe the transformation to the original graph if you wanted the lower co-vertex to appear at the origin.

process: model

## The Law of Orbits

13. Write the equation for an ellipse modeling the orbit of Mercury with the low co-vertex at the origin.

14. Graph the ellipse modeling the orbit of Mercury.

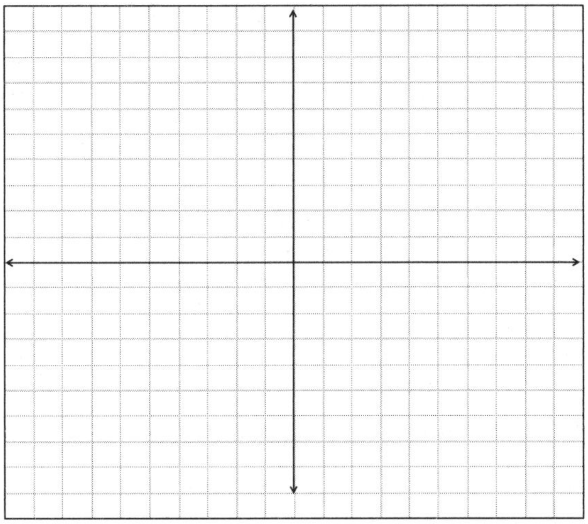

process: evaluate

## The Law of Orbits

15. If the orbit of Mercury is positioned with the lower co-vertex at the origin, identify the following:

    a. The coordinates of the center

    b. The coordinates of the vertices

process: evaluate

## The Law of Orbits

c. The coordinates of the co-vertices

d. The coordinates of the foci

e. The equation of the major axis

f. The equation of the minor axis

analysis

## The Law of Orbits

16. Describe the transformation to the original graph if you wanted the lower co-vertex to appear along the x-axis and the left vertex to appear along the y-axis.

process: model

## The Law of Orbits

17. Write the equation for an ellipse modeling the orbit of Mercury with the low co-vertex at the origin.

18. Graph the ellipse modeling the orbit of Mercury.

process: evaluate

## The Law of Orbits

19. If the orbit of Mercury is positioned with the lower co-vertex at the origin, identify the following:

    a. The coordinates of the center

    b. The coordinates of the vertices

## The Law of Orbits

process: evaluate

c. The coordinates of the co-vertices

d. The coordinates of the foci

e. The equation of the major axis

f. The equation of the minor axis

## The Law of Orbits

process: summary

20. Given an ellipse of the form $\frac{(x-h)^2}{a^2}+\frac{(y-k)^2}{b^2}=1$, explain how to compute the following:

a. The center

process: summary

## The Law of Orbits

b. The equation of the major axis

c. The equation of the minor axis

d. The coordinates of the vertices

e. The coordinates of the co-vertices

f. The coordinates of the foci

preview

## Ships at Sea

LORAN (Long Range Navigation System) is one method used to determine the location of ships at sea. LORAN consists of a network of land based transmitters. Using a ship's distance from three of these transmitters, the ship's exact location can be calculated.

Plotted on a coordinate plane with one unit equaling one mile, the coordinates of transmitter A are (0, 126), the coordinates of transmitter B are (0, 0) and the coordinates of transmitter C are (112, 0).

A ship is located 104 miles from transmitter A, 50 miles from transmitter B and 78 miles from transmitter C. Determine the coordinates of the ship.

## Paper Folding - Hyperbolas

input

Below is one example of a hyperbola. The solid curve is the actual hyperbola while the dotted lines indicate key components of a hyperbola.

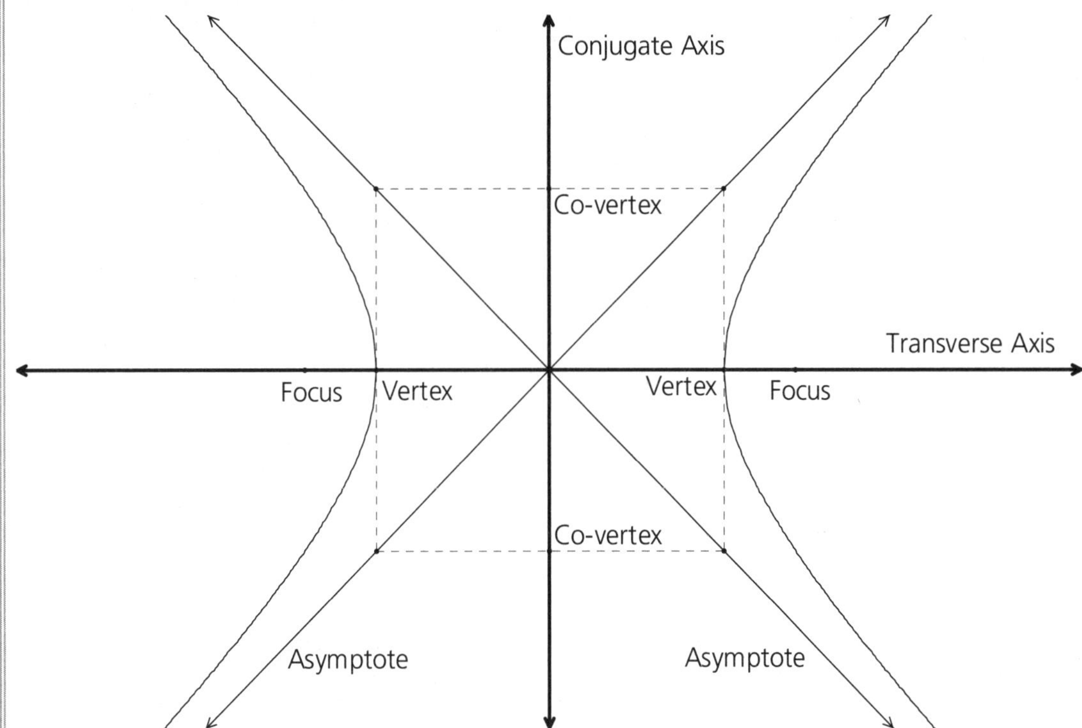


In a hyperbola, the foci and vertices always lie on the **transverse axis**, while the co-vertices always lie on the **conjugate axis**. A hyperbola is symmetrical with respect to both the transverse and conjugate axes.

## Paper Folding - Hyperbolas

process: model

1. Take a piece of wax paper. Draw a large circle on the paper. Label the center of the circle point C. The center will represent one focus. Draw another point outside the circle to represent the other focus. Label it point F

2. Fold the paper so that point F and the edge of the circle meet. Repeat for various points along the circle.

## Paper Folding - Hyperbolas

analysis

3. What shape is formed by the folds?

## Paper Folding - Hyperbolas

analysis

4. Draw any point on the circle and label it point D. Draw a line from point F to point D. Fold the paper so points F and D meet. What is the relationship between the line formed by the crease and $\overline{FD}$?

5. Draw a line containing points C and D. Label the point where the line intersects the crease as point P. Is point P on the hyperbola?

6. What is true about the distances from point P to point F and point D? How do you know this is true?

analysis

## Paper Folding - Hyperbolas

7. Write $\overline{CD}$ as the difference of two segments.

8. What segment is equal in length to $\overline{PD}$? Substitute that segment into the equation above.

9. Will the length of $\overline{CD}$ change if the position of point D changes? Explain.

10. Repeat questions 4-9 for two other points on the circle. Do you see the same results?

11. Summarize what you discovered about the difference of the distances from any point on a hyperbola to the foci.

## Paper Folding - Hyperbolas

analysis

12. Place point D along the transverse axis. Is the difference of the distances from point D to the foci equal to any other measurements involving the hyperbola? Explain why.

13. In general, how does the difference of the distances from any point on the hyperbola to the foci relate to other distances within a hyperbola?

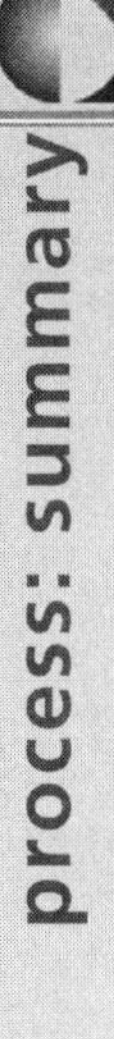


## Equations of Hyperbolas

1. Write the general equation of an ellipse centered at the origin. What operation is performed to the $x^2$ and $y^2$ terms?

2. For an ellipse, what operation makes the distances from any point on the ellipse to the foci constant?

3. Are the operations performed in questions 1 and 2 the same or inverse operations?

4. Examine the distance from the center to a vertex squared and the distance from the center to a co-vertex squared. What operation is performed to these quantities to compute the distance from the center to a focus squared?

5. Are the operations performed in questions 1 and 4 the same or inverse operations?

analysis

## Equations of Hyperbolas

6. For a hyperbola, what operation makes the distances from any point on the hyperbola to the foci constant?

7. The equation of a hyperbola is very similar to the equation of an ellipse. What operation do you think will be performed on the $x^2$ and $y^2$ terms in the equation of a hyperbola?

process: model

## Equations of Hyperbolas

8. Based on your hypothesis, write the equation of a hyperbola and create a graph. Does your equation generate a hyperbola?

## Equations of Hyperbolas

process: model

9. Write the equations of two hyperbolas that are identical except for the order of the x and y variables.

10. Create graphs for each equation.

## Equations of Hyperbolas

Analysis

11. In the equation of an ellipse, the order of the x and y terms can be reversed without affecting the shape of the ellipse because addition is commutative. Does the order of the x and y terms matter in the equation of a hyperbola? Explain why or why not.

## Equations of Hyperbolas

analysis

12. What do you notice about the graphs of the two equations you wrote? How are the graphs the same? How are they different?

13. How can you compute the distance from the center to a focus squared from the distances from the center to a vertex and to a co-vertex?

input

## Equations of Hyperbolas

For a hyperbola centered at the origin and intersecting the x-axis the general equation is

$$\frac{x^2}{a^2} - \frac{y^2}{b^2} = 1$$

where

The coordinates of the vertices are (-a, 0) and (a, 0)

The coordinates of the co-vertices are (0, -b) and (0, b)

The coordinates of the foci are (-c, 0) and (c, 0) where $c^2 = a^2 + b^2$

The equation of the transverse axis is $y = 0$

The equation of the conjugate axis is $x = 0$

The equations of the asymptotes are $y = \pm\frac{b}{a}x$

process: practice

## Equations of Hyperbolas

14. Graph each hyperbola and calculate the coordinates of the vertices, co-vertices and foci and the equations for the asymptotes, transverse and conjugate axes.

a. $\frac{x^2}{144} - \frac{y^2}{25} = 1$

process: practice

## Equations of Hyperbolas

b. $\frac{x^2}{16} - \frac{y^2}{9} = 1$

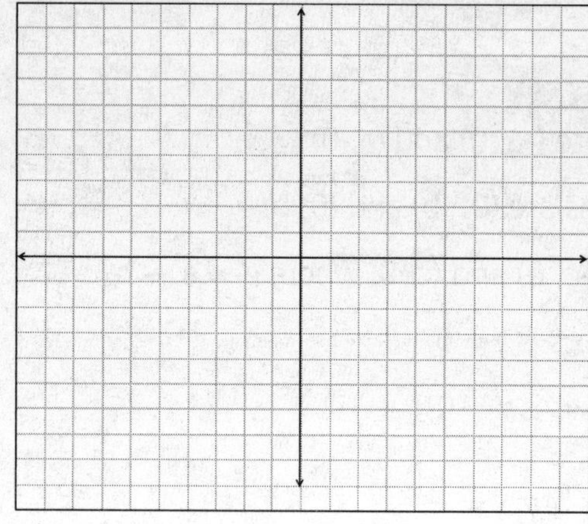

c. $\frac{x^2}{36} - \frac{y^2}{64} = 1$

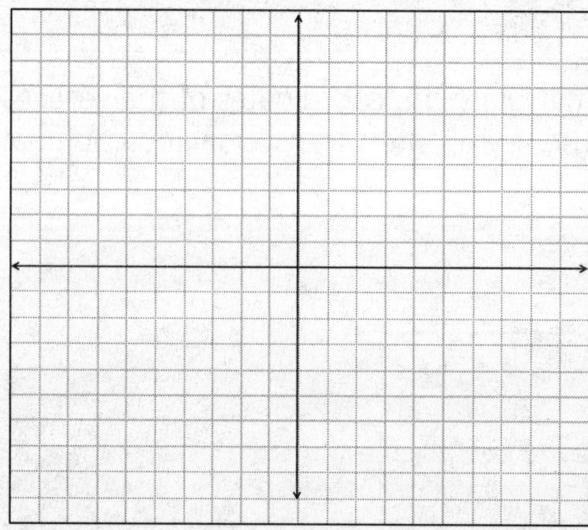

## Floodlight

input

The curve of the surface of a floodlight is designed in the shape of a hyperbola with the light source positioned at one focus. When light rays originating at one focus reflect off the surface of the hyperbola, the light travels in a path along the line from the other focus to the intersection point as shown below.

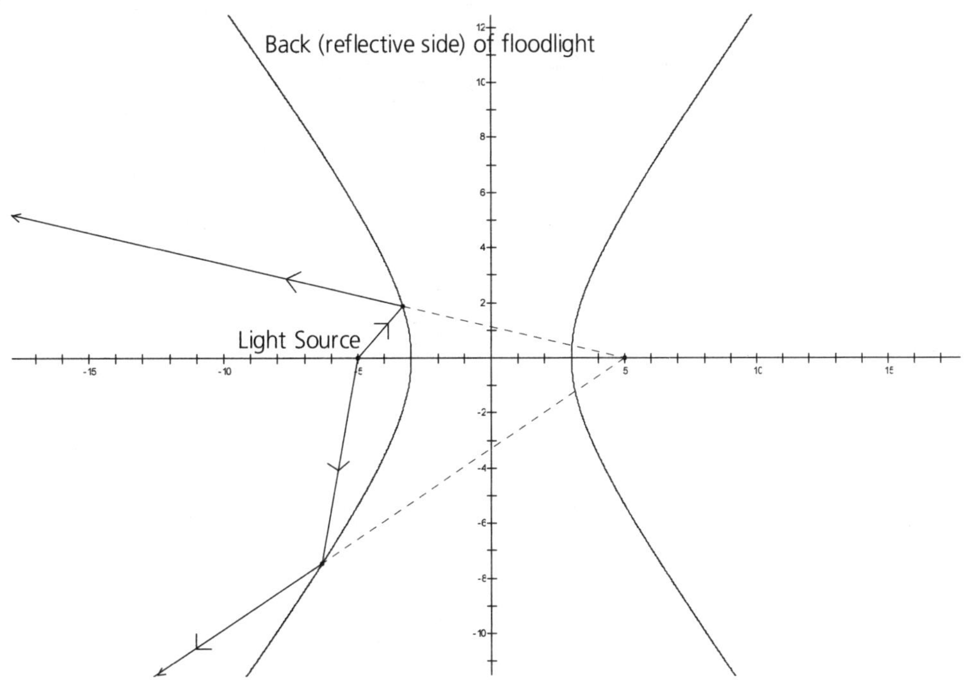


Illuminate brand floodlight offers one model whose surface can be represented by the equation $\frac{x^2}{9} - \frac{y^2}{16} = 1$

## Floodlight

process: model

1. Draw additional light rays in the diagram above to demonstrate how the floodlight spreads light.

process: evaluate

## Floodlight

2. Determine the coordinates of the vertices and co-vertices of the hyperbola.

3. Determine the coordinates of the foci.

4. How far should the light source be placed from the vertex of the hyperbolic surface to create a floodlight? Explain your answer.

## Change of Perspective - Hyperbolas

process: model

1. Graph the hyperbola $\frac{x^2}{16} - \frac{y^2}{9} = 1$ below.

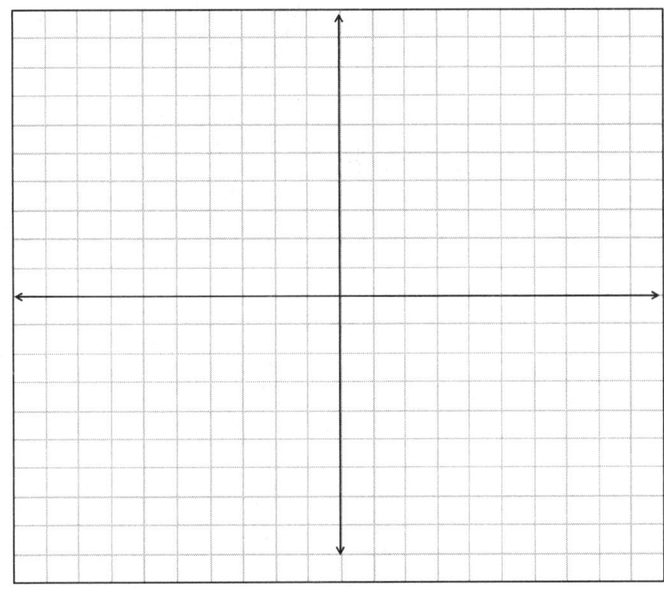

## Change of Perspective - Hyperbolas

process: evaluate

2. Determine the coordinates of the vertices and co-vertices.

3. Determine the coordinates of the foci.

4. Determine the equations for the asymptotes.

5. Determine the equations of the transverse and conjugate axes.

process: model

## Change of Perspective - Hyperbolas

6. Determine the inverse of $\frac{x^2}{16} - \frac{y^2}{9} = 1$.

7. Graph the inverse of the hyperbola $\frac{x^2}{16} - \frac{y^2}{9} = 1$ below.

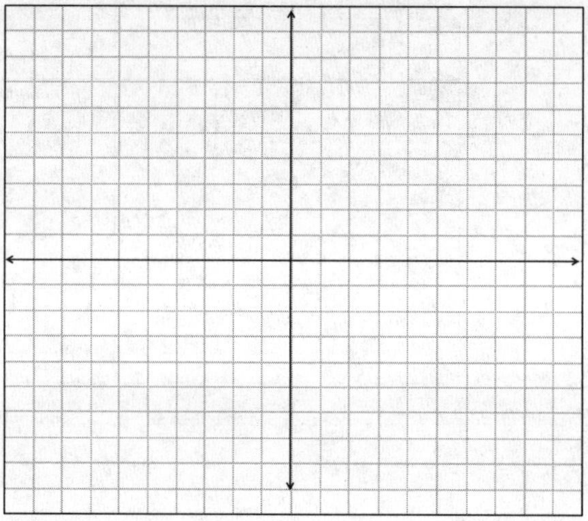

analysis

## Change of Perspective - Hyperbolas

8. Determine the coordinates of the vertices and co-vertices of the inverse.

analysis

## Change of Perspective - Hyperbolas

9. How can you compute the vertices and co-vertices of the inverse from the vertices and co-vertices of the original hyperbola? How can you compute the vertices and co-vertices of the inverse from the equation of the inverse?

10. Determine the coordinates of the foci of the inverse.

11. How can you compute the foci of the inverse from the foci of the original hyperbola? How can you compute the foci of the inverse from the equation of the inverse?

12. Determine the equation of the asymptotes.

analysis

## Change of Perspective - Hyperbolas

13. How can you compute the asymptotes of the inverse from the asymptotes of the original hyperbola? How can you compute the asymptotes of the inverse from the equation of the inverse?

14. Determine the equation of the transverse and conjugate axes.

15. How can you compute the transverse and conjugate axes of the inverse from the transverse and conjugate axes of the original hyperbola? How can you compute the transverse and conjugate axes of the inverse from the equation of the inverse?

process: practice

## Change of Perspective - Hyperbolas

16. Determine the coordinates of the vertices, co-vertices and foci and the equations for the asymptotes, transverse and conjugate axes for each hyperbola.

    a. $\frac{x^2}{25} - \frac{y^2}{144} = 1$

    b. $\frac{y^2}{81} - \frac{x^2}{4} = 1$

17. Determine the equation of the hyperbola defined by the given information.

    a. Center at (0, 0), one vertex at (0, 6) and one focus at (0, -10)

    b. Vertices at (-3, 0) and (3, 0). Co-vertices at (0, -1) and (0, 1)

## Change of Perspective - Hyperbolas

process: summary

18. Given a hyperbola in the form $\frac{x^2}{a^2} - \frac{y^2}{b^2} = 1$, explain how to compute the following:

a. The coordinates of the vertices and co-vertices

b. The coordinates of the foci

c. The equations for the asymptotes

d. The equation for the transverse and conjugate axes

19. Given a hyperbola in the form $\frac{y^2}{a^2} - \frac{x^2}{b^2} = 1$, explain how to compute the following:

a. The coordinates of the vertices and co-vertices

b. The coordinates of the foci

c. The equations for the asymptotes

d. The equation for the transverse and conjugate axes

process: evaluate

## Transforming Hyperbolas

1. Graph $\frac{x^2}{64} - \frac{y^2}{36} = 1$.

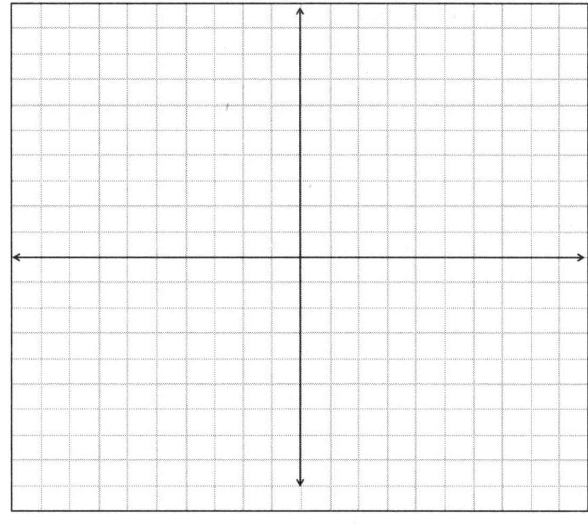

process: evaluate

## Transforming Hyperbolas

2. Determine the following:

   a. The coordinates of the center

   b. The coordinates of the vertices and co-vertices

   c. The coordinates of the foci

   d. The equations for the asymptotes

   e. The equation for the transverse and conjugate axis

process: model

## Transforming Hyperbolas

3. Write the resulting equation if $\frac{x^2}{64} - \frac{y^2}{36} = 1$ is shifted to the right 3 units.

4. Graph the transformed hyperbola on the same coordinate plane.

process: evaluate

## Transforming Hyperbolas

5. Determine the following for the transformed hyperbola based on the original calculations and the transformation applied:

   a. The coordinates of the center

   b. The coordinates of the vertices and co-vertices

   c. The coordinates of the foci

   d. The equations for the asymptotes

   e. The equation for the transverse and conjugate axis

6. Describe how you calculated each of the above.

## Transforming Hyperbolas

process: model

7. Graph $\frac{y^2}{144} - \frac{x^2}{25} = 1$.

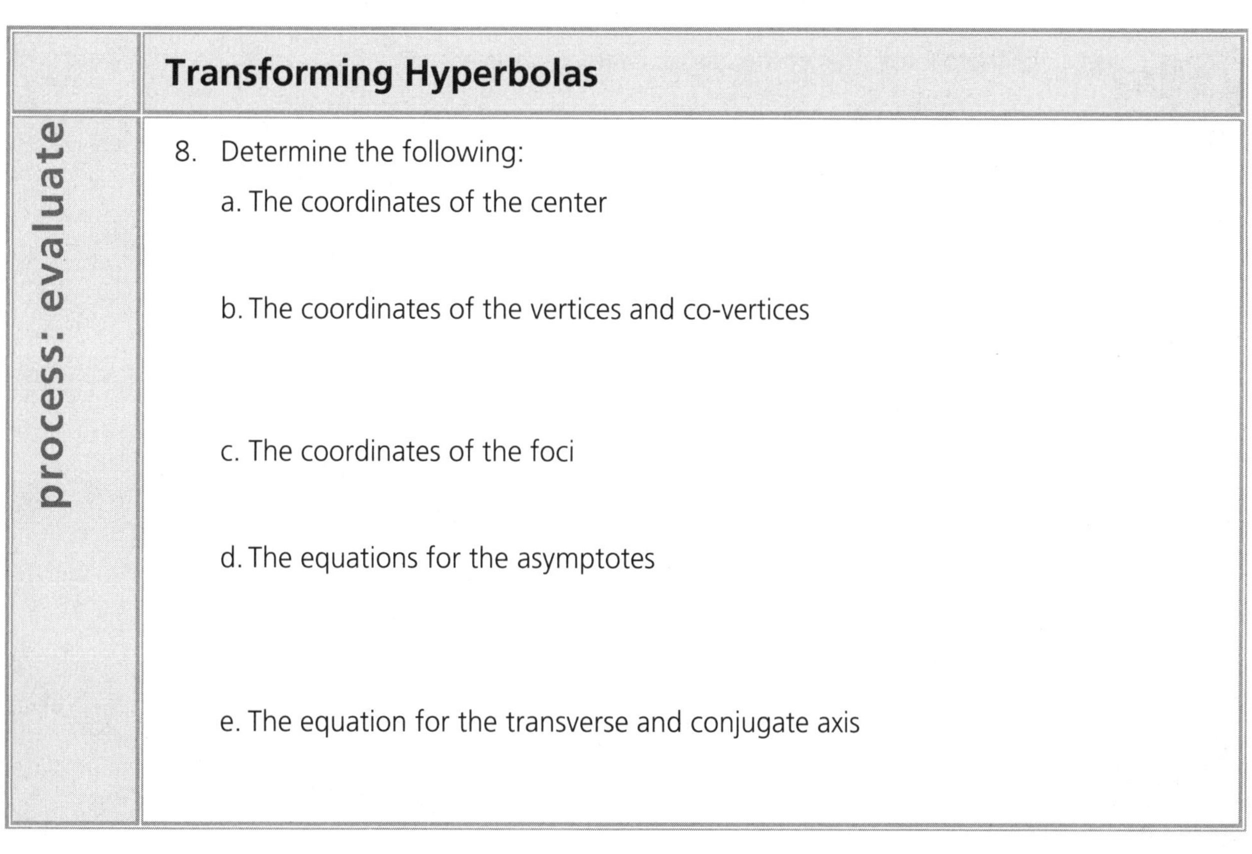

## Transforming Hyperbolas

process: evaluate

8. Determine the following:

   a. The coordinates of the center

   b. The coordinates of the vertices and co-vertices

   c. The coordinates of the foci

   d. The equations for the asymptotes

   e. The equation for the transverse and conjugate axis

## Transforming Hyperbolas

process: model

9. Write the resulting equation if $\frac{y^2}{144} - \frac{x^2}{25} = 1$ is shifted down 2 units.

10. Graph the transformed hyperbola on the same coordinate plane

## Transforming Hyperbolas

process: evaluate

11. Determine the following for the transformed hyperbola based on the original calculations and the transformation applied:

    a. The coordinates of the center

    b. The coordinates of the vertices and co-vertices

    c. The coordinates of the foci

    d. The equations for the asymptotes

    e. The equation for the transverse and conjugate axis

12. Describe how you calculated each of the above.

process: practice

## Transforming Hyperbolas

13. Graph each hyperbola and determine the coordinates of the center, vertices, co-vertices and foci and the equations for the asymptotes, transverse and conjugate axes for each hyperbola.

a. $\frac{(x-3)^2}{16} - \frac{(y+1)^2}{9} = 1$

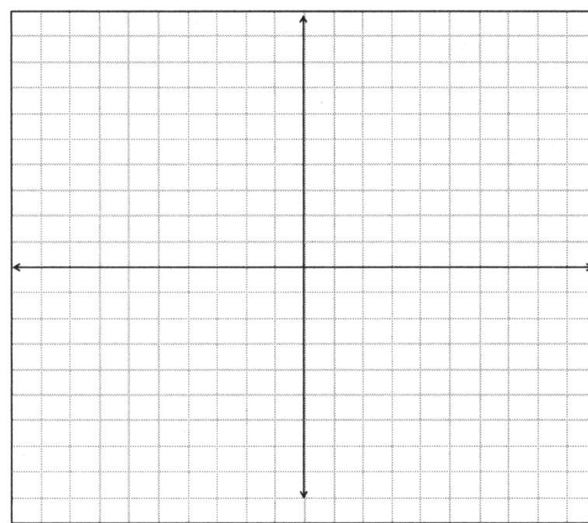

b. $\frac{(y+2)^2}{81} - \frac{(x+5)^2}{144} = 1$

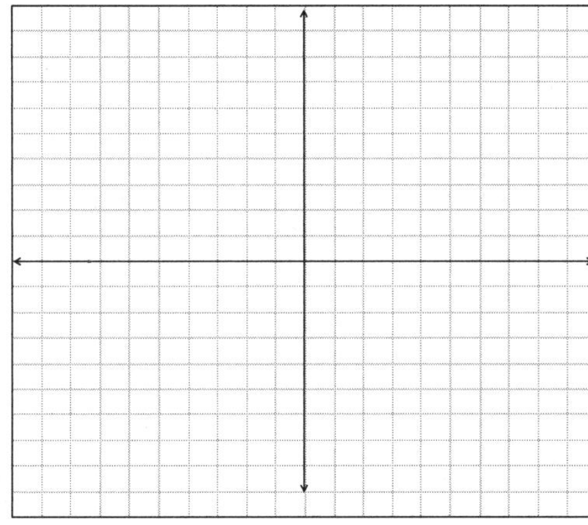

process: summary

## Transforming Hyperbolas

14. Given a hyperbola in the form $\frac{x^2}{a^2} - \frac{y^2}{b^2} = 1$, explain how to compute the following:

   a. The coordinates of the center

   b. The coordinates of the vertices and co-vertices

   c. The coordinates of the foci

   d. The equations for the asymptotes

   e. The equations for the transverse and conjugate axes

## Transforming Hyperbolas

analysis

15. Describe the transformation applied to $\frac{x^2}{a^2} - \frac{y^2}{b^2} = 1$ to result in $\frac{(x-h)^2}{a^2} - \frac{(y-k)^2}{b^2} = 1$

16. Given a hyperbola in the form $\frac{(x-h)^2}{a^2} - \frac{(y-k)^2}{b^2} = 1$, explain how to compute the following:

a. The coordinates of the center

b. The coordinates of the vertices and co-vertices

c. The coordinates of the foci

d. The equations for the asymptotes

e. The equations for the transverse and conjugate axes

process: summary

## Transforming Hyperbolas

17. Given a hyperbola in the form $\frac{y^2}{a^2} - \frac{x^2}{b^2} = 1$, explain how to compute the following:

    a. The coordinates of the center

    b. The coordinates of the vertices and co-vertices

    c. The coordinates of the foci

    d. The equations for the asymptotes

    e. The equations for the transverse and conjugate axes

analysis

## Transforming Hyperbolas

18. Describe the transformation applied to $\frac{y^2}{a^2} - \frac{x^2}{b^2} = 1$ to result in

$$\frac{(y-k)^2}{a^2} - \frac{(x-h)^2}{b^2} = 1$$

19. Given a hyperbola in the form $\frac{(y-k)^2}{a^2} - \frac{(x-h)^2}{b^2} = 1$, explain how to compute the following:

a. The coordinates of the center

b. The coordinates of the vertices and co-vertices

c. The coordinates of the foci

d. The equations for the asymptotes

e. The equations for the transverse and conjugate axes

## Ships at Sea Revisited

input

LORAN (Long Range Navigation System) is one method used to determine the location of ships at sea. LORAN consists of a network of land based transmitters. Using a ship's distance from three of these transmitters, the ship's exact location can be calculated.

Plotted on a coordinate plane with one unit equaling one mile, the coordinates of transmitter A are (0, 126), the coordinates of transmitter B are (0, 0) and the coordinates of transmitter C are (112, 0).

A ship is located 104 miles from transmitter A, 50 miles from transmitter B and 78 miles from transmitter C.

Specifically, LORAN determines the equation of a hyperbola with foci at transmitters A and B passing through the location of the ship and the equation of a hyperbola with foci at transmitters B and C passing through the location of the ship. The ships location is calculated by finding the intersection of the two hyperbolas.

## Ships at Sea Revisited

process: model

1. Plot points representing each transmitter on the grid below.

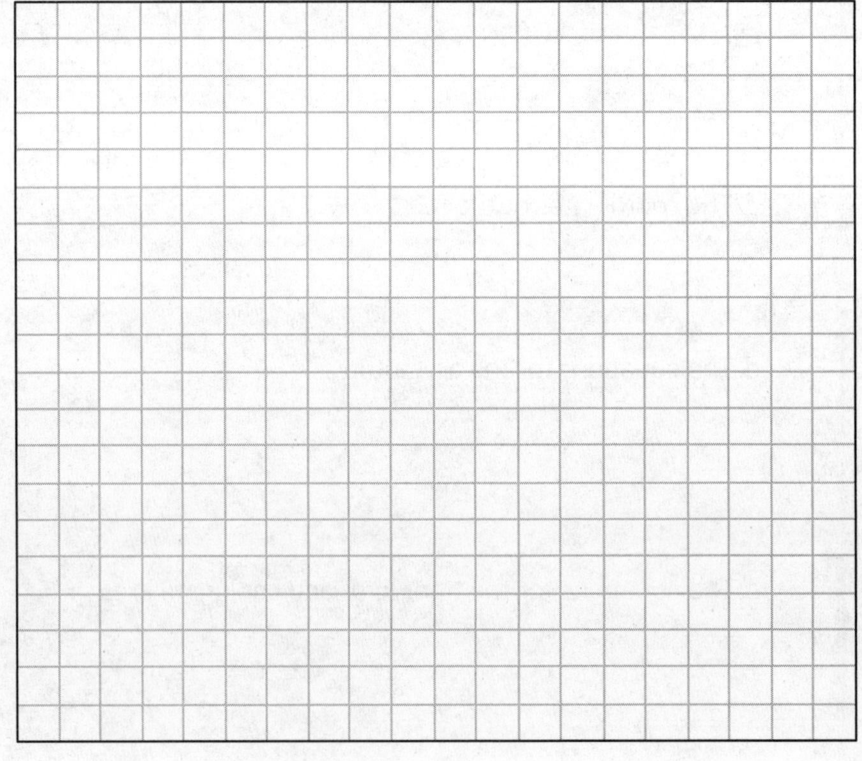

process: evaluate

## Ships at Sea Revisited

2. Determine the equation of a hyperbola with foci at transmitters A and B passing through the location of the ship by performing the following:

   a. Determine the center of the hyperbola.

   b. Determine the difference of the distances from the ship to the foci.

   c. What other distance of the hyperbola is this equal to?

   d. Determine the distance between the co-vertices.

process: model

## Ships at Sea Revisited

3. Write the equation of the hyperbola with foci at transmitters A and B passing through the location of the ship.

4. Sketch the graph of the hyperbola on the coordinate plane in question 1.

## Ships at Sea Revisited

process: evaluate

5. Determine the equation of a hyperbola with foci at transmitters B and C passing through the location of the ship by performing the following:

   a. Determine the center of the hyperbola.

   b. Determine the difference of the distances from the ship to the foci.

   c. What other distance of the hyperbola is this equal to?

   d. Determine the distance between the co-vertices.

## Ships at Sea Revisited

process: model

6. Write the equation of the hyperbola with foci at transmitters B and C passing through the location of the ship.

7. Sketch the graph of the hyperbola on the coordinate plane in question 1.

## Ships at Sea Revisited

process: evaluate

8. How many times do the two hyperbolas intersect?

9. Use a graphing calculator to determine the coordinates of each point of intersection.

## Ships at Sea Revisited

analysis

10. Which intersection point represents the actual position of the ship? Explain how you determined your answer.

## Satellite TV

preview

Many houses are now equipped with Satellite Television. In order to receive the television signals transmitted from a satellite, each subscriber must install a small satellite dish on their house.

Satellite dishes are designed in the shape of a parabola because parabolas have the property that waves traveling parallel to the axis of symmetry of the parabola are reflected from the surface to a fixed point called the focus. A receiver is placed at the focus to collect the incoming television signals.

One popular brand of satellite dish has a diameter of 12 inches and a depth of 2.52 inches.

Write an equation to model the shape of the satellite dish and determine the position of the receiver.

## Paper Folding - Parabolas

input

Below is one example of a parabola.

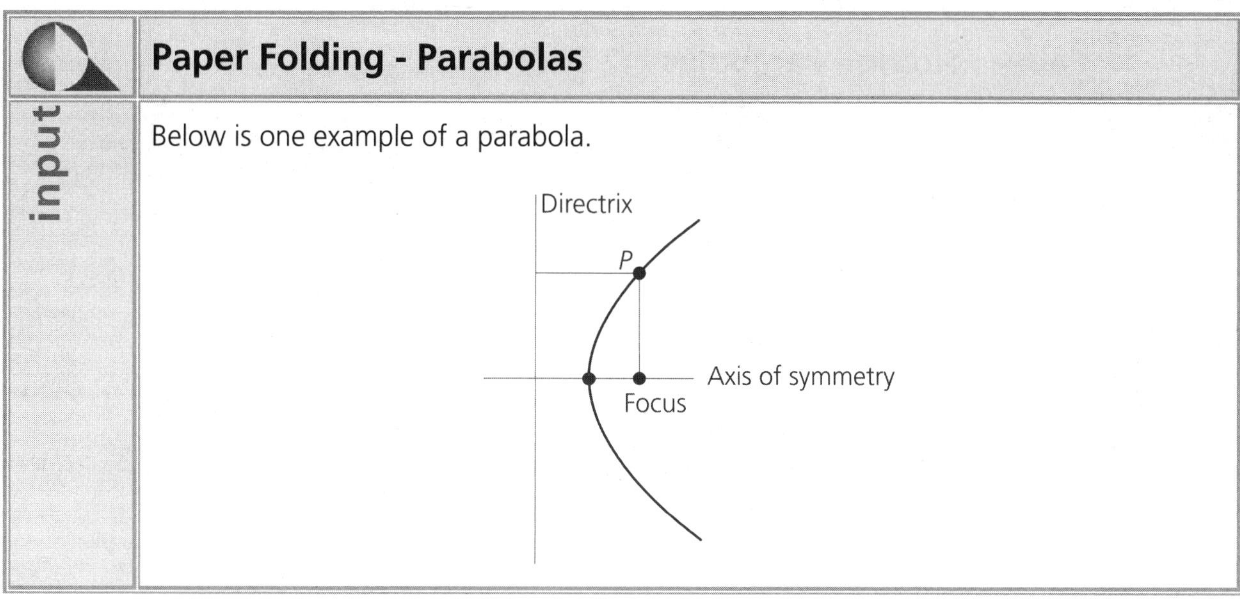


## Paper Folding - Parabolas

process: model

1. Take a piece of wax paper. Near the bottom of the paper, draw a line. Draw a point above the line and label it point F.

2. Fold the paper so the point and the line meet. Repeat for various points along the line. What shape is formed by the folds?

## Paper Folding - Parabolas

input

The **focus** of a parabola always lies along the axis of symmetry. The **directrix** is a line passing through the focus perpendicular to the axis of symmetry.

Label the focus and directrix on your wax paper.

analysis

## Paper Folding - Parabolas

4. Draw any point on the directix and label it point D. Draw a line from the focus to point D. Fold the paper so the focus and point D meet. What is the relationship between the line formed by the crease and $\overline{FD}$ ?

5. Draw a line perpendicular to the directix through point D. Label the point where the perpendicular line intersects the crease as point P. Is point P on the parabola?

6. What is true about the distances from point P to point F and point D? How do you know this is true?

7. Repeat questions 4-6 for other points on the directrix. Do you see the same results?

8. Summarize what you discovered about the distance from any point on a parabola to the focus and the directrix.

## Flashlight

input

The cross section of the curve of the reflective surface of a flashlight is designed in the shape of a parabola with the light source positioned at the focus. When light rays originating at the focus reflect off the surface of the parabola, the light travels parallel to the axis of symmetry.

*Illuminate* brand flashlight offers one model whose surface can be represented by the equation $y = \frac{1}{16}x^2$.

## Flashlight

process: model

1. Draw the graph of the equation representing the surface of the flashlight.

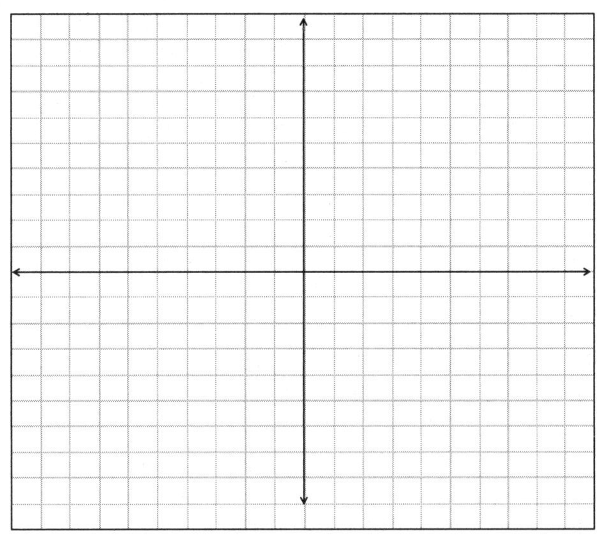

2. Define a variable to represent the distance from the vertex to the focus. Use this variable to write the general coordinates of the focus.

3. Based on the general coordinates of the focus; determine the general equation for the directrix.

process: model

## Flashlight

4. Locate a point on the parabola with whole number coordinates. Label this point P and list the coordinates of point P.

5. Draw a line from point P perpendicular to the directrix. Determine the coordinates of the point where the perpendicular line intersects the directrix.

process: evaluate

## Flashlight

6. Use the distance formula / Pythagorean Theorem to compute the distance from point P to the focus. Simplify the expression representing the distance completely.

7. Compute the distance from point P to the directrix. Simplify the expression representing the distance completely.

process: evaluate

## Flashlight

8. Write an equation to represent the relationship between the distance from point P to the focus and the distance from point P to the directrix. Solve the resulting equation.

9. What are the coordinates of the focus?

10. What is the equation representing the directrix?

analysis

## Flashlight

11. Look at the coefficient in the equation and the coordinates of the focus. How could you compute the focus directly from the coefficient?

input

## Flashlight

*Illuminate* offers another model of flashlight whose surface can be represented by the equation $y = \frac{1}{32}x^2$.

process: model

## Flashlight

12. Draw the graph of the equation representing the surface of the flashlight.

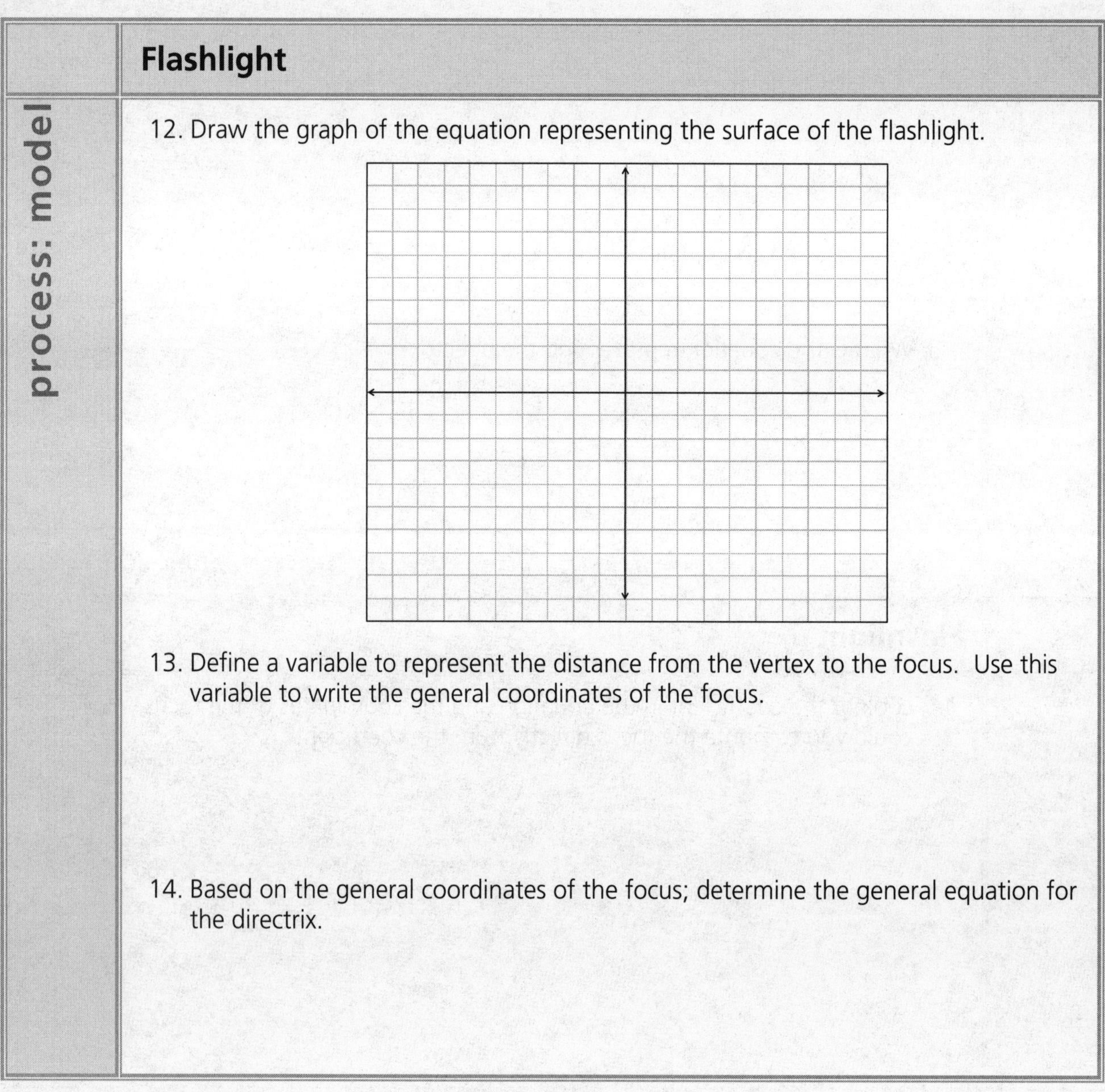

13. Define a variable to represent the distance from the vertex to the focus. Use this variable to write the general coordinates of the focus.

14. Based on the general coordinates of the focus; determine the general equation for the directrix.

## Flashlight

process: model

15. Locate a point on the parabola. Label this point P and list the coordinates of point P.

16. Draw a line from point P perpendicular to the directrix. Determine the coordinates of the point where the perpendicular line intersects the directrix.

## Flashlight

process: evaluate

17. Use the distance formula / Pythagorean Theorem to compute the distance from point P to the focus. Simplify the expression representing the distance completely.

18. Compute the distance from point P to the directrix. Simplify the expression representing the distance completely.

process: evaluate

## Flashlight

19. Write an equation to represent the relationship between the distance from point P to the focus and the distance from point P to the directrix. Solve the resulting equation.

20. What are the coordinates of the focus?

21. What is the equation representing the directrix?

analysis

## Flashlight

22. Look at the coefficient in the equation and the coordinates of the focus. How could you compute the focus directly from the coefficient?

23. Is the method you just described the same as the method described in question 11?

process: practice

## Flashlight

24. Determine the coordinates of the focus and the equation for the directrix for each equation.

a. $y = \frac{1}{4}x^2$

b. $y = \frac{1}{10}x^2$

c. $y = 4x^2$

d. $y = 2x^2$

process: summary

## Flashlight

25. Summarize how to compute the coordinates of the focus and the equation of the directrix from the equation of a parabola.

## Radio Waves

input

Radio stations use parabolic dishes to transmit radio waves.

Station WGEO uses a dish with the focus 2 feet from the vertex.

Assume that the dish is positioned with the axis of symmetry along the y-axis and the vertex at the origin. Determine an equation to model the parabolic dish.

## Radio Waves

process: evaluate

1. What are the coordinates of the focus?

2. What is the equation for the directrix?

## Radio Waves

process: model

3. Sketch a diagram of the radio dish that includes the focus and directrix. Pick one point P on the parabola and label it (x, y).

## Radio Waves

process: evaluate

4. Draw a line from point P perpendicular to the directrix. Determine the general coordinates of the point where the perpendicular line intersects the directrix. Label this point in your diagram.

5. Use the distance formula / Pythagorean Theorem to compute the distance from point P to the focus.

6. Compute the distance from point P to the directrix.

7. Write an equation to represent the relationship between the two distances. Solve the resulting equation for y.

## Radio Waves

analysis

8. How could you write the equation directly from the focus?

| | **Radio Waves** |
|---|---|
| input | Station WALG uses a dish with the focus 3 feet from the vertex.<br><br>Assume that the dish is positioned with the axes of symmetry along the y-axis and the vertex at the origin. Determine an equation to model the parabolic dish. |

| | **Radio Waves** |
|---|---|
| process: evaluate | 9. What are the coordinates of the focus?<br><br>10. What is the equation for the directrix? |

| | **Radio Waves** |
|---|---|
| process: model | 11. Sketch a diagram of the radio dish that includes the focus and directrix. Pick one point P on the parabola and label it (x, y). |

process: evaluate

## Radio Waves

12. Draw a line from point P perpendicular to the directrix. Determine the general coordinates of the point where the perpendicular line intersects the directrix. Label this point in your diagram.

13. Use the distance formula / Pythagorean Theorem to compute the distance from point P to the focus.

14. Compute the distance from point P to the directrix.

15. Write an equation to represent the relationship between the two distances. Solve the resulting equation for y.

analysis

## Radio Waves

16. Based on the previous two examples, how could you write the equation directly from the focus?

process: practice

## Radio Waves

17. Determine the equation of the parabola with the axis of symmetry along the y-axis and the vertex at the origin.

    a. Focus at (0, 0.75)

    b. Directrix at $y = -0.1$

    c. Focus at (0, 0.2)

    d. Directrix at $y = -4$

process: summary

## Radio Waves

18. Summarize how to write the equation of a parabola with the axis of symmetry along the y-axis and the vertex at the origin when given either the coordinates of the focus or the equation of the directrix.

| | **Radio Waves** |
|---|---|
| input | Consider a parabola with the focus p units from the vertex with the axes of symmetry along the y-axis and the vertex at the origin. Determine a general equation to model the parabola. |

| | **Radio Waves** |
|---|---|
| process: evaluate | 19. What are the coordinates of the focus?<br><br>20. What is the equation for the directrix? |

| | **Radio Waves** |
|---|---|
| process: model | 21. Sketch a diagram of the radio dish that includes the focus and directrix. Pick one point P on the parabola and label it (x, y). |

process: evaluate

## Radio Waves

22. Draw a line from point P perpendicular to the directrix. Determine the coordinates of the point where the perpendicular line intersects the directrix. Label this point in your diagram.

23. Use the distance formula to compute the distance from point P to the focus.

24. Use the distance formula to compute the distance from point P to the directrix.

25. Write an equation to represent the relationship between the two distances. Solve the resulting equation for y.

analysis

## Radio Waves

26. Could you have written the equation for the parabola after finding the coordinates of the focus?

## Change of Perspective - Parabolas

process: model

1. Graph the parabola $y = \frac{1}{4}x^2$ below.

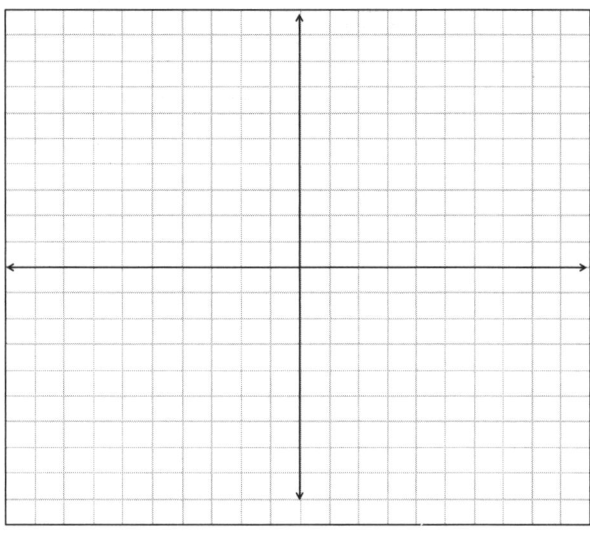

## Change of Perspective - Parabolas

process: evaluate

2. Determine the coordinates of the focus. Draw the focus in your graph.

3. Determine the equation of the directrix. Draw the directrix in your graph.

4. Determine the inverse of $y = \frac{1}{4}x^2$. Write the equation in the form x equals.

## Change of Perspective - Parabolas

process: model

5. Graph the inverse of the parabola $y = \frac{1}{4}x^2$ below.

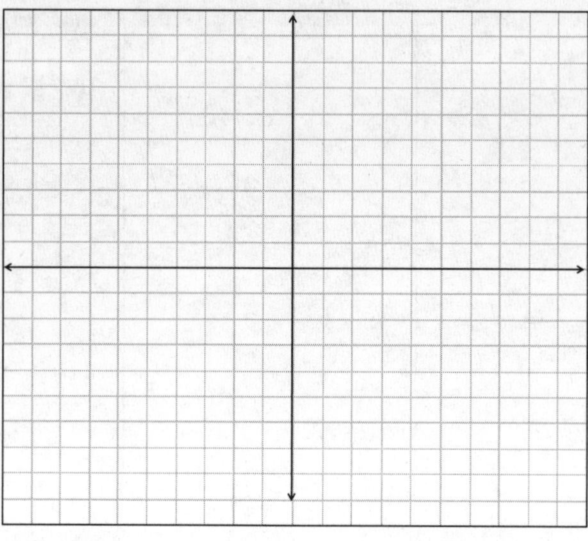

## Change of Perspective - Parabolas

analysis

6. Determine the coordinates of the focus of the inverse. Draw the focus in your graph.

7. How did you compute the focus of the inverse from the focus of the original parabola? How could you compute the focus of the inverse from the equation of the inverse?

## Change of Perspective - Parabolas

analysis

8. Determine the equation of the directrix of the inverse. Draw the directrix in your graph.

9. How did you compute the equation of the directrix of the inverse from the directrix of the original parabola? How could you compute the directrix of the inverse from the equation of the inverse?

## Change of Perspective - Parabolas

process: model

10. Graph the parabola $y = \frac{1}{8}x^2$ below.

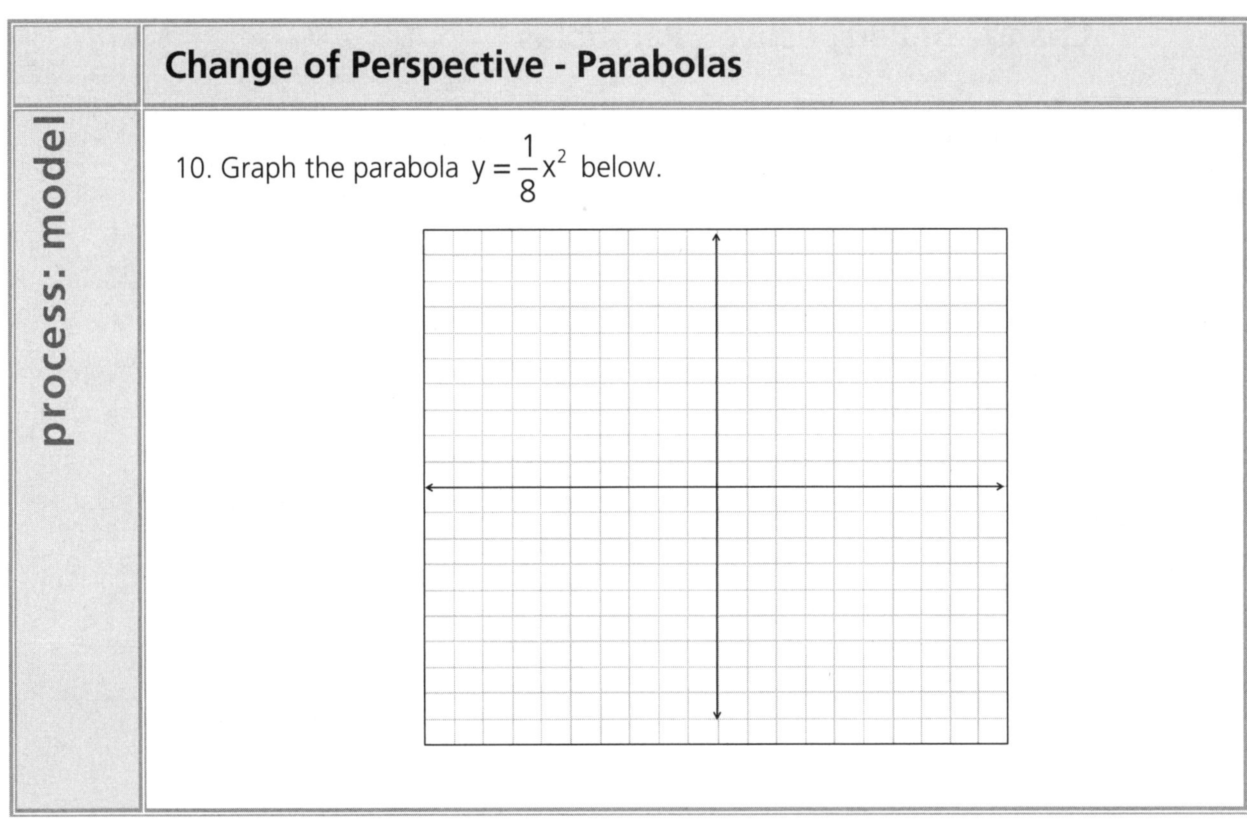

## Change of Perspective - Parabolas

process: evaluate

11. Determine the coordinates of the focus. Draw the focus in your graph.

12. Determine the equation of the directrix. Draw the directrix in your graph.

13. Determine the inverse of $y = \frac{1}{8}x^2$. Write the equation in the form x equals.

## Change of Perspective - Parabolas

process: model

14. Graph the inverse of the parabola $y = \frac{1}{8}x^2$ below.

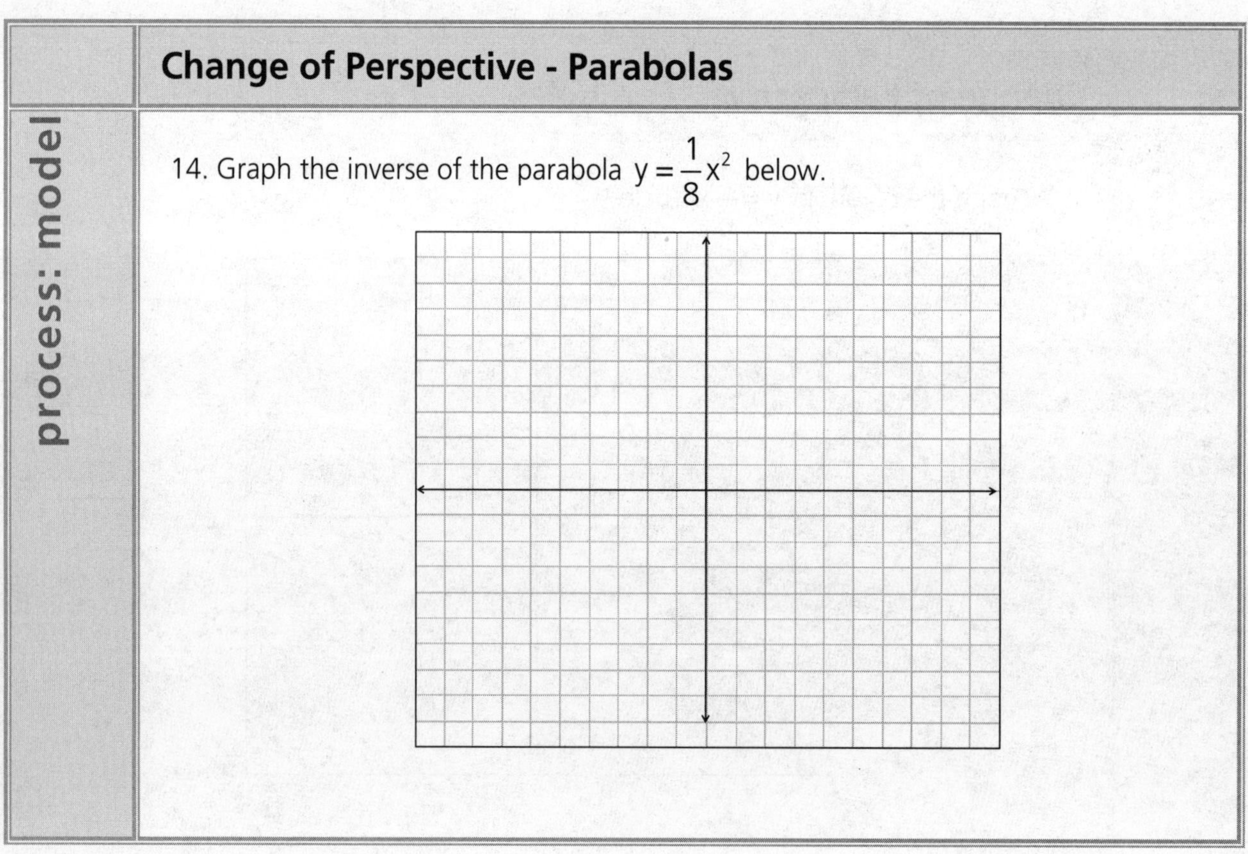

## Change of Perspective - Parabolas

analysis

15. Determine the coordinates of the focus of the inverse. Draw the focus in your graph.

16. How did you compute the focus of the inverse from the focus of the original parabola? How could you compute the focus of the inverse from the equation of the inverse?

17. Determine the equation of the directrix of the inverse. Draw the directrix in your graph.

18. How did you compute the equation of the directrix of the inverse from the directrix of the original parabola? How could you compute the directrix of the inverse from the equation of the inverse?

## Change of Perspective - Parabolas

process: summary

19. Given a parabola in the form $y = \frac{1}{4p}x^2$, compute the following:

   a. The coordinates of the vertex

   b. The equation for the line of symmetry

   c. The coordinates of the focus

   d. The equation for the directrix

20. Given a parabola in the form $x = \frac{1}{4p}y^2$, explain how to compute the following:

   a. The coordinates of the vertex

   b. The equation for the line of symmetry

   c. The coordinates of the focus

   d. The equation for the directrix

process: practice

## Change of Perspective - Parabolas

21. Determine the coordinates of the focus and the equation of the directrix for each parabola.

    a. $x = 2y^2$

    b. $x = \frac{1}{2}y^2$

    c. $y = \frac{1}{10}x^2$

    d. $x = \frac{1}{12}y^2$

22. Determine the equation of the parabola defined by the given focus or directrix with the vertex at the origin.

    a. Focus at (0, 4)

    b. Directrix at x = -0.25

    c. Focus at (5, 0)

    d. Directrix at y = -0.2

## Transforming Parabolas

process: model

1. Graph the parent parabola $y = \frac{1}{4}x^2$

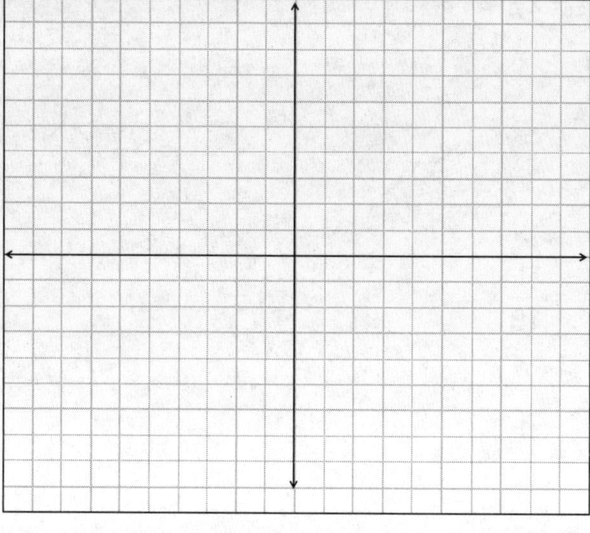

## Transforming Parabolas

process: evaluate

2. Compute the following for $y = \frac{1}{4}x^2$:

a. The coordinates of the vertex

b. The equation for the line of symmetry

c. The coordinates of the focus

d. The equation for the directrix

process: model

## Transforming Parabolas

3. Graph the transformed parabola $y = \frac{1}{4}(x+2)^2$

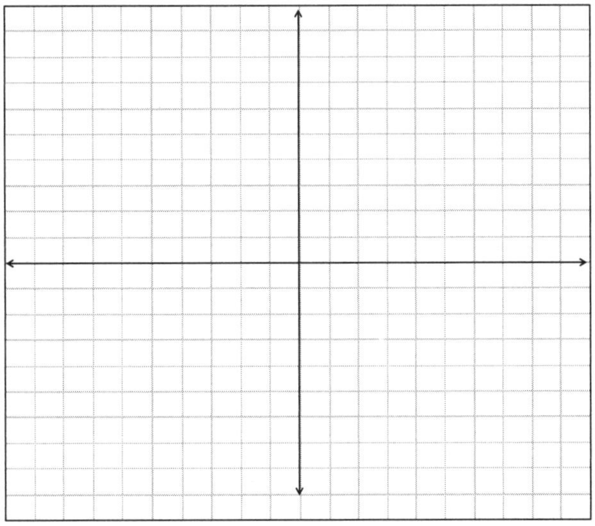

process: evaluate

## Transforming Parabolas

4. What transformation is applied to the parent function to get $y = \frac{1}{4}(x+2)^2$

5. Compute the following for $y = \frac{1}{4}(x+2)^2$:

   a. The coordinates of the vertex

   b. The equation for the line of symmetry

   c. The coordinates of the focus

   d. The equation for the directrix

## Transforming Parabolas

process: model

6. Graph the transformed parabola $y = \frac{1}{4}(x - 3)^2$

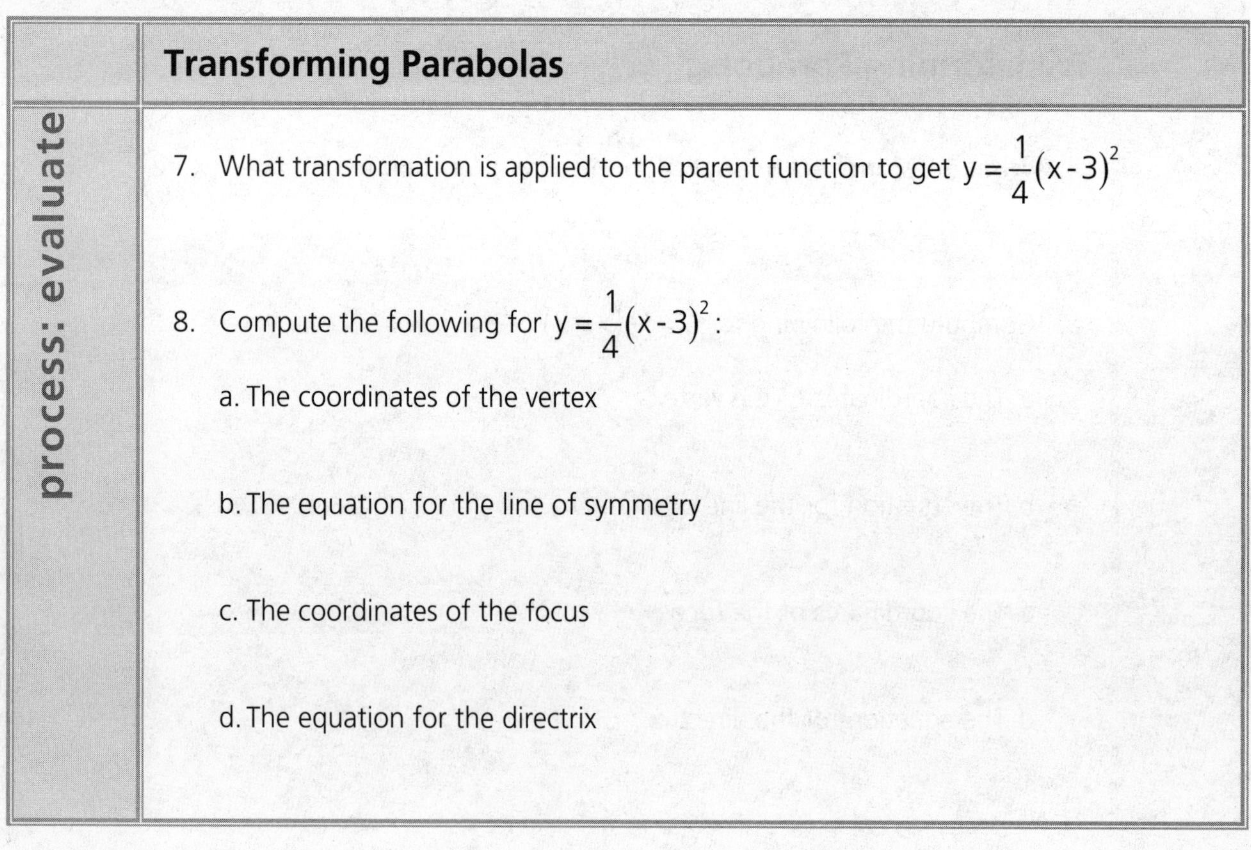

## Transforming Parabolas

process: evaluate

7. What transformation is applied to the parent function to get $y = \frac{1}{4}(x - 3)^2$

8. Compute the following for $y = \frac{1}{4}(x - 3)^2$:

   a. The coordinates of the vertex

   b. The equation for the line of symmetry

   c. The coordinates of the focus

   d. The equation for the directrix

process: model

## Transforming Parabolas

9. Graph the transformed parabola $(y-1)=\frac{1}{4}x^2$

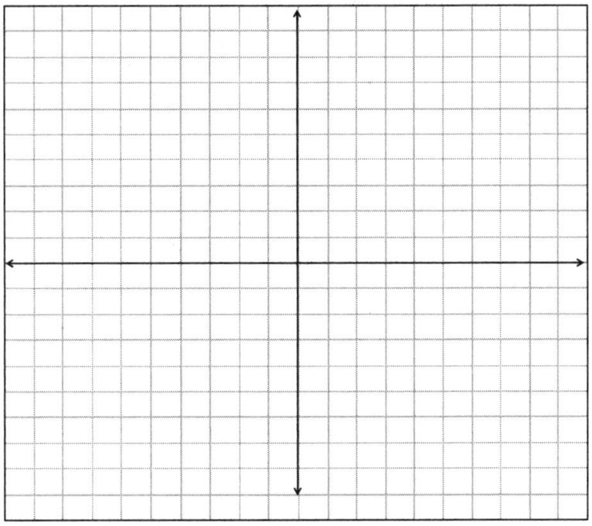

process: evaluate

## Transforming Parabolas

10. Solve the equation for y.

11. What transformation is applied to the parent function to get $(y-1)=\frac{1}{4}x^2$

12. Compute the following::

   a. The coordinates of the vertex and focus

   b. The equations for the line of symmetry and directrix

process: model

## Transforming Parabolas

13. Graph the transformed parabola $(y+4)=\frac{1}{4}x^2$

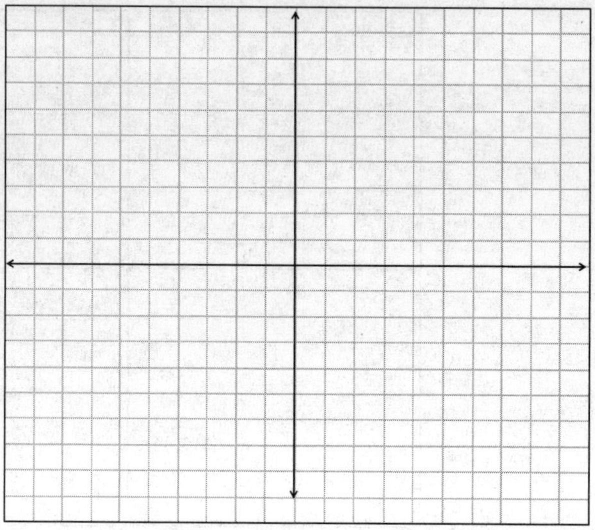

process: evaluate

## Transforming Parabolas

14. Solve the equation for y.

15. What transformation is applied to the parent function to get $(y+4)=\frac{1}{4}x^2$

16. Compute the following:

a. The coordinates of the vertex and focus

b. The equations for the line of symmetry and directrix

process: model

## Transforming Parabolas

17. Graph the transformed parabola $(y+2)=\frac{1}{4}(x-1)^2$

process: evaluate

## Transforming Parabolas

18. Solve the equation for y.

19. What transformations are applied to the parent function to get $(y+2)=\frac{1}{4}(x-1)^2$

20. Compute the following:

    a. The coordinates of the vertex and focus

    b. The equations for the line of symmetry and directrix

## Transforming Parabolas

process: summary

21. Given a parabola in the form $y = \frac{1}{4p}x^2$, explain how to compute the following:

a. The coordinates of the vertex

b. The equation for the line of symmetry

c. The coordinates of the focus

d. The equation for the directrix

22. What transformations are applied to $y = \frac{1}{4p}x^2$ to get $(y-k) = \frac{1}{4p}(x-h)^2$

23. Compute the following for $(y-k) = \frac{1}{4p}(x-h)^2$:

a. The coordinates of the vertex

b. The equation for the line of symmetry

c. The coordinates of the focus

d. The equation for the directrix

process: summary

## Transforming Parabolas

24. Given a parabola in the form $x = \frac{1}{4p}y^2$, explain how to compute the following:

a. The coordinates of the vertex

b. The equation for the line of symmetry

c. The coordinates of the focus

d. The equation for the directrix

25. What transformations are applied to $x = \frac{1}{4p}y^2$ to get $(x-h) = \frac{1}{4p}(y-k)^2$

26. Compute the following for $(x-h) = \frac{1}{4p}(y-k)^2$:

a. The coordinates of the vertex

b. The equation for the line of symmetry

c. The coordinates of the focus

d. The equation for the directrix

## Transforming Parabolas

process: practice

27. Graph each parabola and compute the coordinates of the vertex, the equation for the axis of symmetry, the coordinates of the focus and the equation for the directrix.

a. $(x-1)=\frac{1}{8}(y+3)^2$

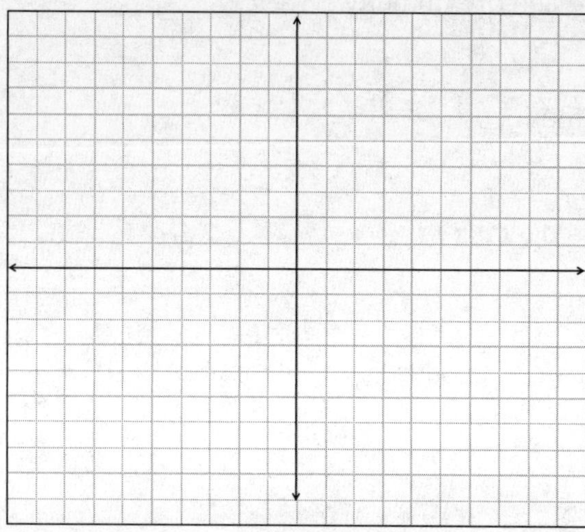

b. $(y+4)=(x+1)^2$

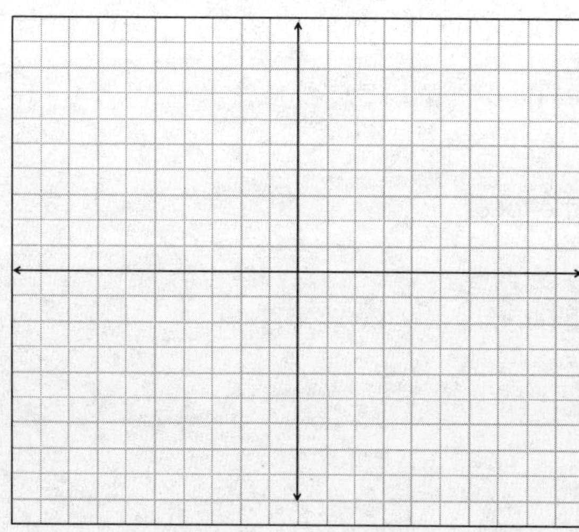

## Transforming Parabolas

process: practice

c. $(x-5)=\frac{1}{12}(y-4)^2$

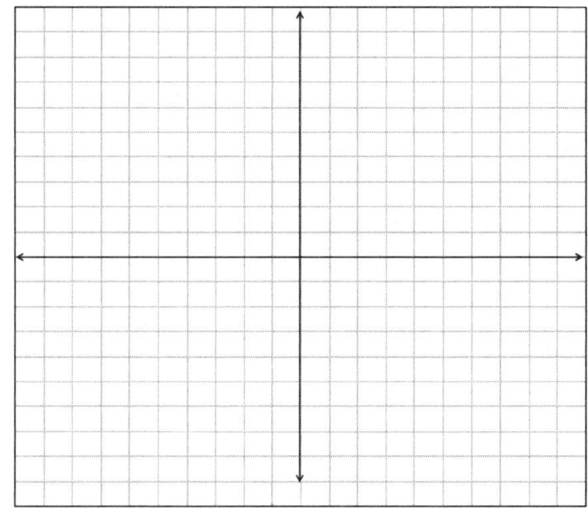

d. $(y-1.5)=\frac{1}{32}(x+3)^2$

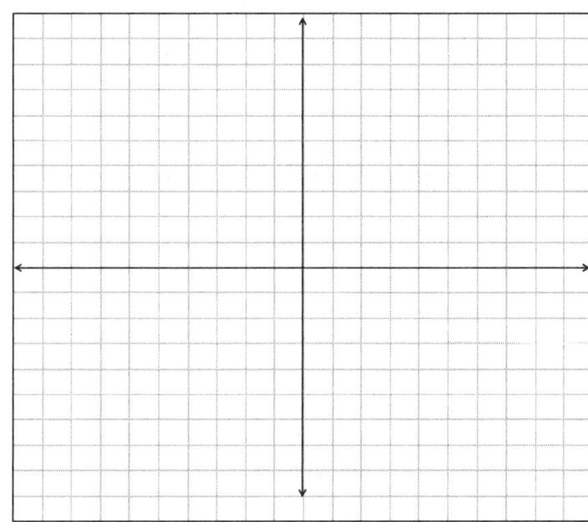

## Satellite TV Revisited

input

Your neighbor buys a satellite dish with a diameter of 16 inches and a depth of 3.2 inches.

## Satellite TV Revisited

process: model

1. Write three points on the parabola representing the cross section of the satellite dish if the vertex is at the origin and the axis of symmetry is along the y-axis.

2. Use a graphing calculator to determine the equation of the parabola.

3. Draw a graph of the equation representing the satellite dish.

process: evaluate

## Satellite TV Revisited

4. Determine the coordinates representing the location of the receiver.

input

## Satellite TV Revisited

Your neighbor installs the satellite dish on a 4 foot pole located 3 feet from the house.

If plotted on a coordinate plane, the corner of the house would be located at the origin.

process: model

## Satellite TV Revisited

5. Describe the transformations that must be applied to the original equation to model the position of the satellite dish on the pole.

6. Write an equation to model the satellite dish if the edge of the house is at the origin.

process: model

## Satellite TV Revisited

7. Draw a graph of the equation representing the satellite dish.

process: evaluate

## Satellite TV Revisited

8. Determine the coordinates representing the location of the receiver with respect to the edge of the house.

# Unit 9: Three-Dimensional Geometry

## Contents

## Michael and Freddie

preview

You and your friend are both real horror movie buffs. For your birthday a few months ago, she bought an action figure of Freddie Kruger from the movie *Nightmare on Elm Street*. Now, it's her birthday. You don't want to be outdone, so you decide to purchase a replica of Jason Vorhees, of *Friday the Thirteenth* fame, which is three times as big as the replica she bought you.

1. If you saved the package from your Freddie Kruger model, how can you determine how big the package has to be for your friend's Jason model?

2. Compared to the amount of wrapping paper your friend used to wrap your birthday present, how much more wrapping paper will you need to use?

3. Compared to the amount of ribbon your friend used to wrap your present, how much more ribbon will you need to use? Assume you want to make your bow three times as big as hers as well.

## Get to the Hearth of the Matter

input

Last summer, you worked part-time for a building contractor. You worked on an interesting project for a wealthy client for most of the summer. First, the client asked that a patio be built in his back yard. Then, the patio was covered with 12″ square marble tiles. Next, the client asked for a hearth to be built on one particular rectangular section of the patio. The bricks used to construct the hearth were specially designed. The bottom of each brick fit precisely on one of the marble tiles. The bricks were all 12″ high.

Below is a top view of the patio that was built. The area that was covered by the hearth is shaded dark grey. The height of the hearth is precisely 4 feet.

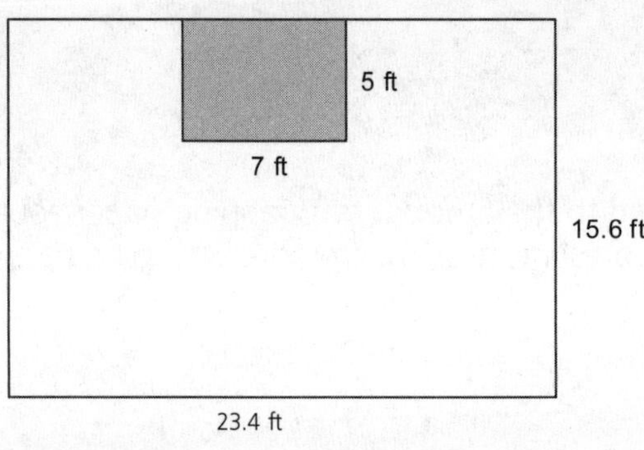


## Get to the Hearth of the Matter

process: evaluate

1. Find the area of the entire patio. Show all your work.

When the building contractor orders tiles for a job like this, any tiles that will be broken by the builders because a full tile won't fit in a particular place, like at the edge of the patio or in a corner, must be ordered. Additionally, an extra 10% is included for breakages in shipping or mistakes made while attempting to lay the tile.

2. Using the guidelines above, how many tiles had to be used to completely cover the patio?

## Get to the Hearth of the Matter

process: evaluate

3. Find the area of the portion of the patio covered by the hearth. Show your work.

4. How many measurements do you need in order to determine the area of the patio covered by the hearth? Which measurements do you need?

5. Which units are best to use to measure the area of the patio? Why?

6. How many dimensions does a length have? How many dimensions does a width have? How many dimensions does an area have? How do you know?

7. How many measurements are involved when building the hearth atop the patio? Which measurements are needed?

## Get to the Hearth of the Matter

process: evaluate

8. How many dimensions does a volume have? How do you know?

9. Based upon the units used for measuring length and area, which units seem best for measuring the volume of the hearth?

10. Find the volume occupied by the hearth.

11. Complete the following analogy. *Area is to square centimeter as volume is to* ___. Explain your response.

12. When working with an area in two dimensions, the region is enclosed by *sides*, which are one-dimensional segments. What kinds of sides enclose the hearth?

13. How many dimensions do the sides of the hearth have?

## Moving from Two Dimensions to Three

input

You previously studied the geometry of two-dimensional figures. You calculated areas and perimeters of two-dimensional objects like parallelograms, triangles, hexagons, and circles. (What is the specific name given to the "perimeter" of a circle?) You found unknown lengths using the Pythagorean Theorem, the trigonometric ratios, and the properties of similarity and congruency.

Now, you will study the geometry of three-dimensional figures. You will calculate measurements analogous to area and perimeter, and you will find other unknown measurements as well.

Recall when you learned about area, one of the first activities you performed was counting squares enclosed within a region. Area was defined as the amount of the surface enclosed within the boundary of the two-dimensional region.

When these two-dimensional regions are "built-up" into three-dimensions, you can then start to find the amount of *space* enclosed within the boundary instead of simply the amount of surface. The amount of space enclosed within the boundary of a three-dimensional object is called the **volume** of the object.

## Moving from Two Dimensions to Three

process: model

Think about a small rectangle of cardboard. You can measure certain dimensions of the cardboard and calculate its perimeter and area.

1. Create a model of a small rectangular piece of cardboard in the space below.

2. What dimensions should you measure to calculate its perimeter and area? Make sure to answer in a full sentence.

process: evaluate

## Moving from Two Dimensions to Three

3. Measure the dimensions you cited in question 2 in centimeters and record them on the figure.
4. Calculate the perimeter of the rectangular piece of cardboard. Show your work, and include the units in your final answer.
5. Calculate the area of the rectangular piece of cardboard. Show your work, and include the units in your final answer.

Imagine covering your rectangle with sugar cubes, which are approximately a centimeter on each edge.

6. How many sugar cubes does it take to cover your rectangular piece of cardboard? Explain your response.

Now, imagine creating another layer of sugar cubes on top of the first layer. In fact, create 4 layers of sugar cubes atop one another.

7. Calculate the number of sugar cubes it takes to create this structure. Show all your work.

input

## Moving from Two Dimensions to Three

You just calculated the volume of the structure. You took a two-dimensional figure that was a rectangle and created a three-dimensional figure by stacking the sugar cubes on top of one another to form layers.

What did you use as units to build the sugar cube structure?

A two-dimensional shape, like a rectangle or a pentagon can become a three-dimensional shape when layers are stacked on top of one another. A shape that is created in this manner is called a **prism**. The prism you created is called a **rectangular prism.** When a pentagon is "built up" in the same way, the three-dimensional shape formed is called a **pentagonal prism.**

process: model

## Moving from Two Dimensions to Three

8. A **prism** is a three-dimensional figure with two parallel surfaces called **bases** that are congruent to one another. What shape are the bases of this prism? (Note that the dotted edges are not visible parts of the three-dimensional shape.)

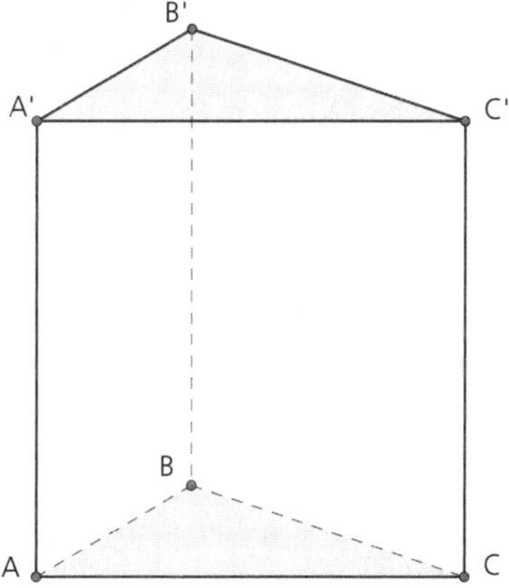

## Moving from Two Dimensions to Three

process: model

9. Using the shape of the bases, what should be the name of this prism?

10. The surfaces of the prism that are not bases are called **lateral faces**. In general, all of the surfaces of the prism are called **faces**. A prism has polygons for all of its faces. What shape are the lateral faces of this prism? Be as specific as you can in your answer.

11. When two of the faces intersect in a prism, what geometric object is created?

12. These line segments are called **edges** of the prism. The intersections of the lateral faces are specifically called **lateral edges.** How many lateral edges are there in this prism? Name the lateral edges.

13. Will the lateral edges ever intersect? Why or why not?

14. How many edges are there in this prism (including the lateral edges)? Name the edges that are not lateral edges.

15. Points where edges intersect are called **vertices.** How many vertices are there in the triangular prism? Name these vertices

process: evaluate

## Moving from Two Dimensions to Three

16. If you need 10 sugar cubes to cover the bottom of the triangular prism pictured here and it takes 6 layers of the sugar cubes to fill the prism, calculate the volume of the triangular prism. Show your work and include the units used for volume in your answer.

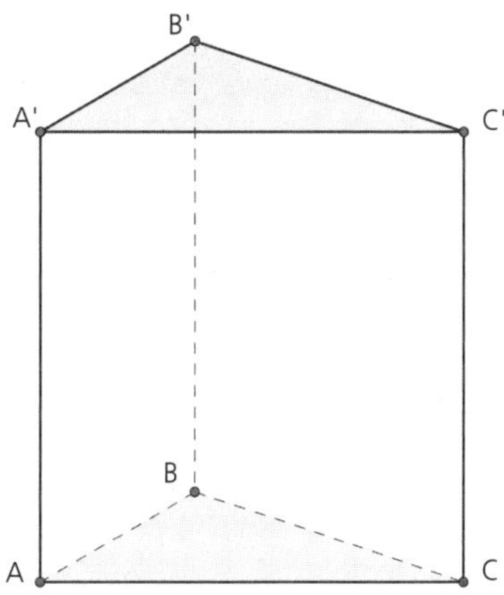


process: model

## Moving from Two Dimensions to Three

17. Now, you will draw your own prism.

    a. In the space below, use a straightedge to create a pentagon. This pentagon will be the bottom base of your prism, so draw it close to the bottom of the page. Label the vertices of your pentagon LMNOP.

process: model

## Moving from Two Dimensions to Three

b. Create another pentagon that is congruent to pentagon LMNOP. Draw this one directly above the first pentagon. If the new pentagon overlaps the first pentagon, that is not a problem. This pentagon will serve as the top base of the prism. Label the vertices of the second pentagon as L'M'N'O'P'.

c. Connect the matching or corresponding vertices of the top and bottom pentagons. Once you have done this, your prism is completed. However, if you want to give your prism a more three-dimensional feel, then you should dot the line segments that would not be seen. These segments are at the back (or top) of the bottom base and in the rear of the prism itself.

d. Using the shape of the bases, what should be the name of this prism?

e. What shape are the lateral faces of this prism? Be as specific as you can in your answer.

process: evaluate

## Moving from Two Dimensions to Three

18. The following regular octagon has an apothem (distance from the center to any side) that measures 2 cm. Each side of this octagon measures 1.7 cm. Calculate the area of the octagon. Show your work and include the units in your answer.

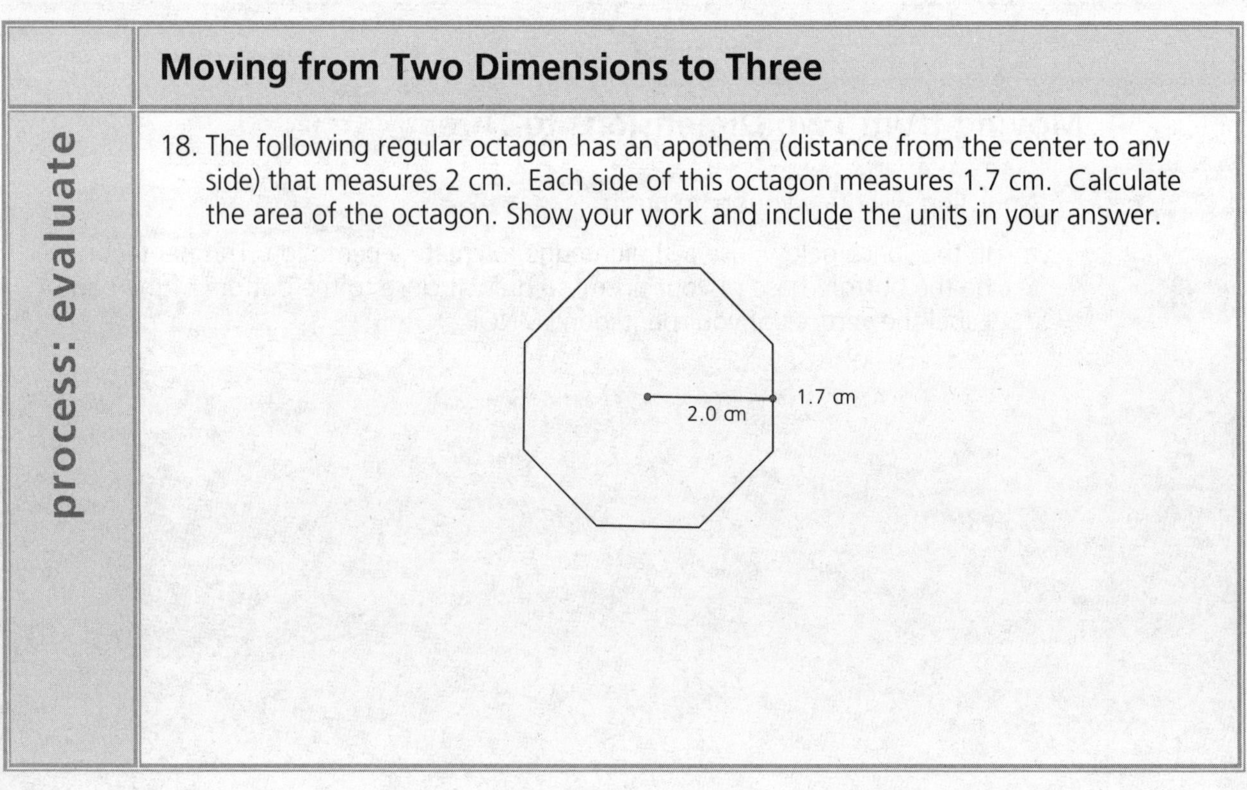

process: evaluate

## Moving from Two Dimensions to Three

19. Now build this octagon up into three dimensions to a height of 1 cm. This means there is only one layer of sugar cubes. What is the volume of this one-layer high prism?

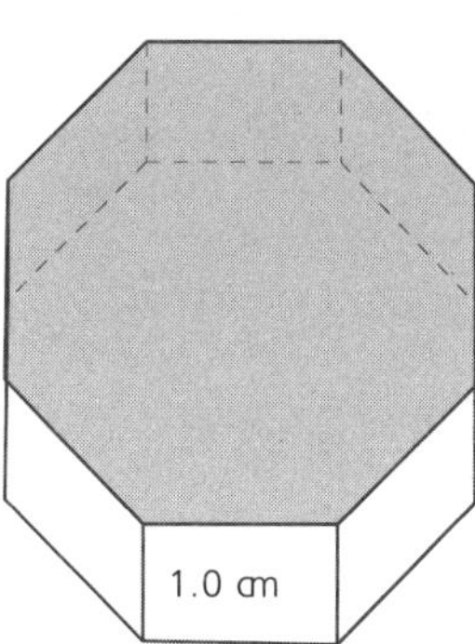


analysis

## Moving from Two Dimensions to Three

20. How does the area of the octagon compare to the volume of the one-layer high prism?

21. Square units are always used to measure area. Why do we use this type of unit?

22. What three-dimensional shape is analogous to the two-dimensional square?

## Moving from Two Dimensions to Three

analysis

23. Using this analogy, what type of unit should be used to measure volume?

24. Why should this type of unit be used to *fill* three-dimensional shapes?

## Moving from Two Dimensions to Three

process: model

25. Return to the octagon, with base area equal to 10.2 square cm. Build it to a height of 3.5 cm. This would use three layers of sugar cubes and a layer of sugar cubes cut in half. Calculate its volume and use the appropriate units in the answer.

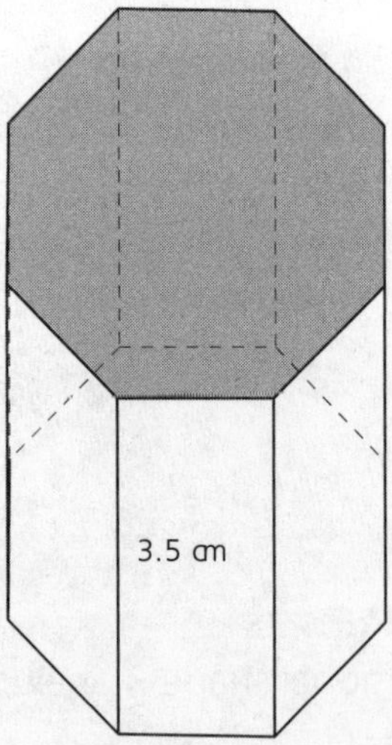

3.5 cm

process: model

## Moving from Two Dimensions to Three

26. Based on the prisms you encountered in this section, how can you use the area of the base and the height to calculate the volume of the prism?

27. What is a general formula for calculating the volume of a prism?

input

## Moving from Two Dimensions to Three

You were just working with an octagonal prism. This octagonal prism, the pentagonal prism, and the triangular prism before it are all called **right prisms**. In **right prisms,** the angle formed between the base and any lateral face is a right angle. Not all prisms are right prisms. The angles formed between the bases and the lateral faces do not have to be right angles. If the angles are not right angles, then the prism is called an **oblique prism.** Below is one example of an oblique triangular prism.

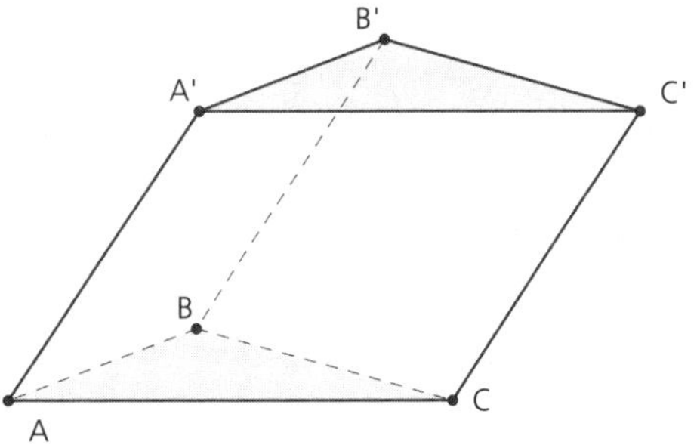

process: model

## Moving from Two Dimensions to Three

28. Create a model of an oblique hexagonal prism below. When you create the two base hexagons do not create them directly above one another. Move one of the hexagons to the left or right of the other.

29. What shape are the lateral faces of the oblique triangular prism on the previous page? Be as specific as you can in your answer.

30. What shape are the lateral faces of the oblique hexagonal prism you just drew? Be as specific as you can in your answer.

analysis

## Moving from Two Dimensions to Three

31. Note that in the right prisms on the previous pages, the height of the prism is the same as the length of any of the lateral edges. Is that true with the oblique prisms Why or why not?

analysis

## Moving from Two Dimensions to Three

32. In an oblique prism the length of the lateral edges is known as the **slant height** instead of the height. Should you use the height or the slant height to calculate the volume of an oblique prism? Why?

process: evaluate

## Moving from Two Dimensions to Three

33. Calculate the volume of the oblique triangular prism pictured below.

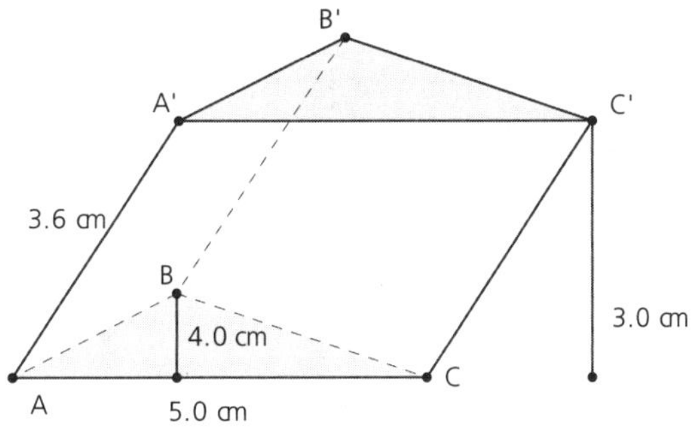

## Moving from Two Dimensions to Three

process: summary

34. Decide if the following statements are true or false. If the statement is true, explain why it is true. If the statement is false, rewrite the statement so it becomes a true statement.

    a. The bases of prisms are always congruent to one another.

    b. The lateral faces of a prism are always congruent to one another.

    c. The lateral faces of prisms are always rectangles.

    d. The lateral faces of prisms are always parallelograms.

    e. The lateral edges in any prism are parallel to one another.

    f. The volume of a prism is calculated by multiplying the area of the base and the height.

input

## Nets Aren't Just for Basketball Anymore

When you go to the grocery store to buy food or household goods, the person bagging your groceries always asks the same question. "Do you want paper or plastic?" If you answer, "paper" you get a sack similar to the one pictured below.

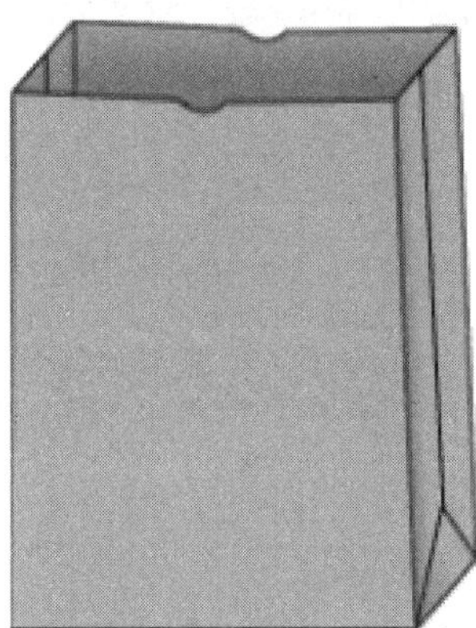

What type of three-dimensional shape is represented by the paper grocery sack in the picture above?

Leaving the base of the sack in tact, imagine cutting down each of the lateral edges to form four rectangles from the lateral faces. The lateral faces remain connected to the base, but not to each other. The resulting shape will look something like this.

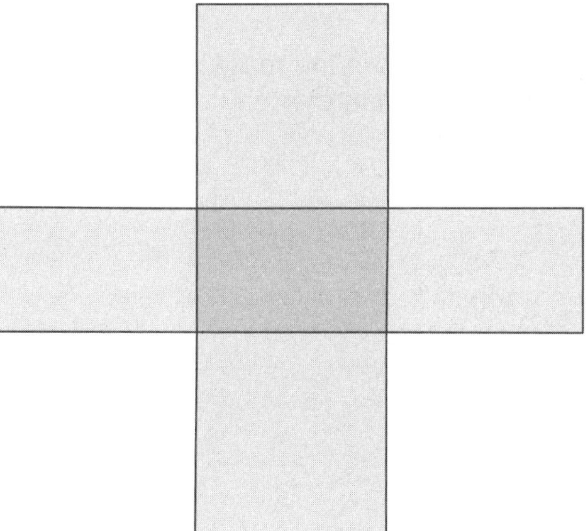

The shape above is called a **net** of the paper bag. A geometric **net** is a two-dimensional model showing the faces of the figure. It is formed by "flattening" a three-dimensional shape. When the net is folded up again, the original three-dimensional shape is formed. Nets can help you analyze a three-dimensional shape by simplifying how it is viewed.

process: model

## Nets Aren't Just for Basketball Anymore

1. Examine the following nets. Determine which nets can be folded to form completed cubes. You can fold along any of the solid segments, but you cannot cut or tear the net in order to form the cube. Also, none of the faces can overlap another existing face. For each of the nets that cannot be folded to form a cube, explain why a cube cannot be formed.

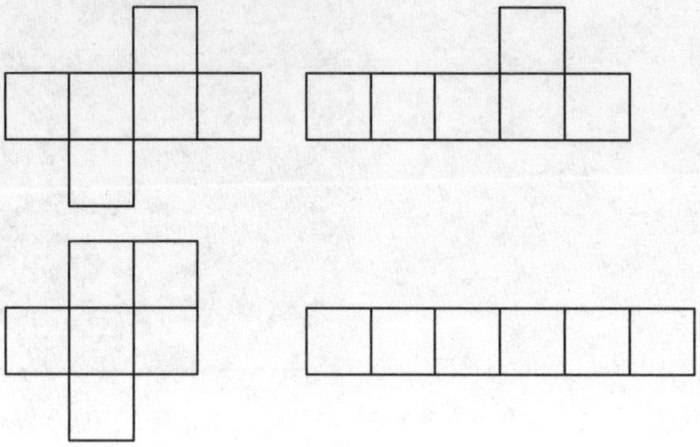

2. Create two more nets in the space below that can be folded to form cubes. To get started, you may wish to create a cube and then decompose it into a net.

input

## Nets Aren't Just for Basketball Anymore

Nets can be useful when trying to find the area of the surfaces of a three-dimensional figure. The combined area of all the faces or surfaces of a three-dimensional figure, like a prism, is called the **surface area**.

Use your grocery sack net to find the surface area and the lateral area of the sack.

process: evaluate

## Nets Aren't Just for Basketball Anymore

3. Measure the length and width of the base of your grocery sack. If you do not have an actual sack to measure, use the model provided here. Record your measurement on the sack with the appropriate units.

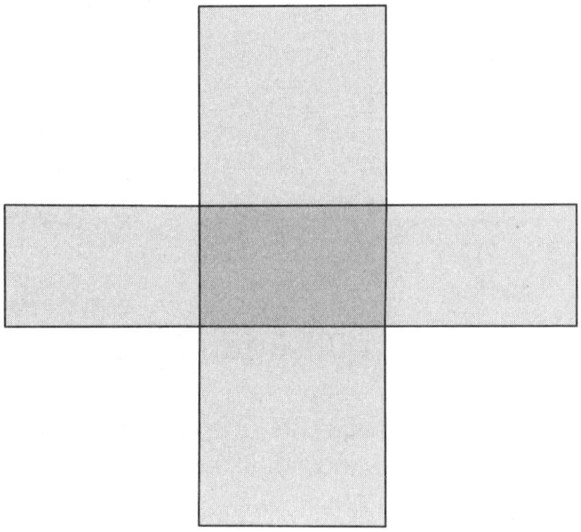

4. Find the area of the base of the sack and record it on the sack with the units.

5. Measure the length and width of the back face of the sack. In the picture above this is the top rectangle. Record your measurements on the sack.

6. Find the area of this face and record it on the sack with the appropriate units.

7. Is there another face with the same measurements as the back face? If so, which face(s)?

8. Find the area of the remaining faces of the sack. Record the measurements of the length and width as well as the area of each face.

process: evaluate

## Nets Aren't Just for Basketball Anymore

9. Find the total surface area (sometimes abbreviated at SA) by adding the areas of the 5 faces together. Show your work and make sure to include the appropriate units in your answer.

10. The area of the lateral surfaces of a three-dimensional figure (the area of the surfaces that are not bases) is called the **lateral area**. Find the lateral surface area (sometimes abbreviated LA) by adding the areas of the lateral faces together. Do not include the area of the base in your calculation of the lateral area. Show your work and make sure to include the appropriate units in your answer.

analysis

## Nets Aren't Just for Basketball Anymore

11. What "shortcuts" did you use (or could you use) to find the surface area so you did not have to add all five individual areas? What "shortcuts" did you use (or could you use) to find the lateral area, so you did not have to add all four individual areas?

process: model

## Nets Aren't Just for Basketball Anymore

12. Create a net for a right rectangular prism with the same measurements as the grocery sack. Make sure to include the top of the prism.

13. How is the net for the right rectangular prism similar to the net for a cube? How is the net for the right rectangular prism different from the net for a cube?

process: evaluate

## Nets Aren't Just for Basketball Anymore

14. Calculate the surface area of the rectangular prism above.

process: evaluate

## Nets Aren't Just for Basketball Anymore

15. Calculate the lateral area of the rectangular prism above.

analysis

## Nets Aren't Just for Basketball Anymore

16. What "shortcuts" did you use (or could you use) to find the surface area so you did not have to add all six individual areas? What "shortcuts" did you use (or could you use) to find the lateral area, so you did not have to add all four individual areas?

17. Return to the nets you created for a cube in question 2. Are the "shortcuts" for finding the surface area of a cube the same as or different than the shortcuts for finding the surface area of the grocery sack? Explain your response in full sentences.

## Nets Aren't Just for Basketball Anymore

process: evaluate

18. Return to the section on Moving from Two Dimensions to Three. Use the right triangular prism. Measure the height of the prism as well as the length of each side of the base. Sketch the net for this prism below.

19. Calculate the surface area of the right triangular prism. Show all your work and include the units in your answer.

20. Return to the section on Moving from Two Dimensions to Three. Use the right pentagonal prism you created. Measure the height of the prism as well as the length of each side of the base. Sketch the net for this prism below.

21. Calculate the surface area of the right pentagonal prism. Show all your work and include the units in your answer.

process: practice

## Nets Aren't Just for Basketball Anymore

22. Find the surface area and lateral area of the prisms below.

 a. The following prism models a new packaging design for a water cooler. The hexagons are regular hexagons.

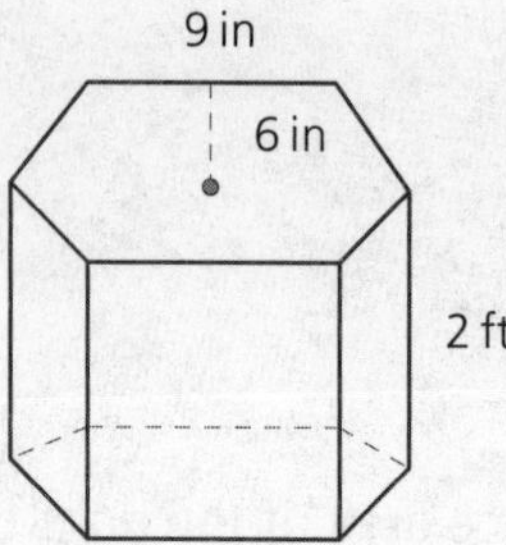


 b. The following prism models the packaging design for the cups alongside the water cooler.

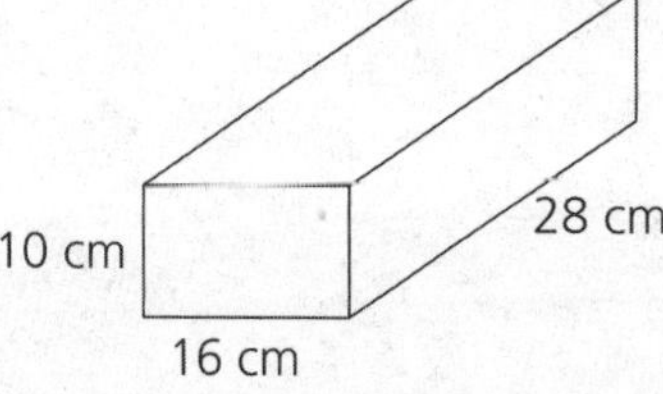

process: summary

## Nets Aren't Just for Basketball Anymore

23. You need to wrap a birthday present. The picture of the box is shown below. The dashed lines in the second view represent the edges of the hidden sides of the box. The box is 1.5 feet wide by 2 feet long by 3 inches high. If the wrapping paper is 2 feet wide, find the minimum length of paper you need to wrap the box. Also find the length of ribbon you need to go around the box as shown in the figure. Don't worry about the bow. Show your work and include the correct units in your answer.

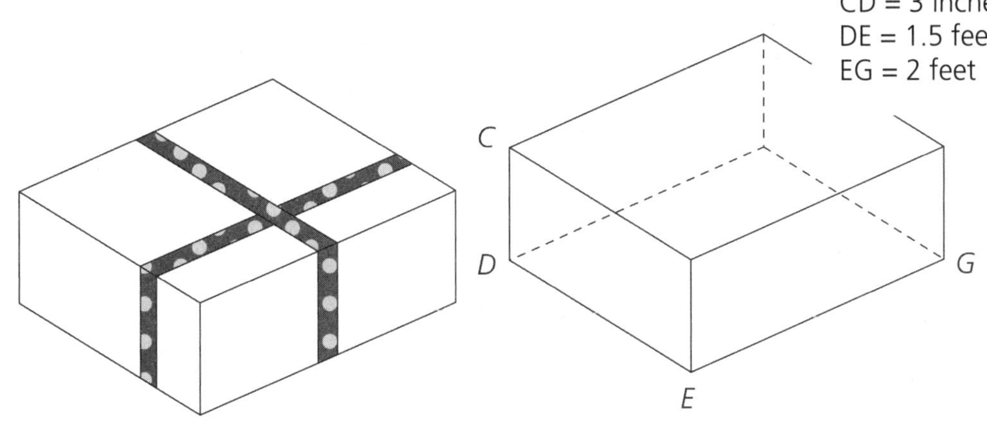

process: summary

## Nets Aren't Just for Basketball Anymore

24. What types of units do you use when finding surface area?

25. In the past few problems, you have found the surface area for three-dimensional objects. Are the units you use to measure surface area three-dimensional? Explain why or why not.

26. How are nets helpful in calculating surface area?

## Polyhedra

input

In the last section, you created nets for prisms. Nets can be created for other three-dimensional figures besides prisms. Many of the three-dimensional shapes you will study in this unit are called **polyhedra** (plural for polyhedron). Prisms are only one type of polyhedron.

What does the prefix *poly-* mean in words like *polyhedron* or *polygon*?

A **polyhedron** is a solid that has polygons for all of its faces and encloses a single region of space. Polyhedrons are named according to the number of faces they have. Similar to how polygons are named; the specific name of a polyhedron is formed by combining a prefix that indicates the number of faces of the shape followed by the root *–hedron*. For example, a pentagon has five sides, and a pentahedron has five faces. The following is an example of a pentahedron. (You know this as a square pyramid.)

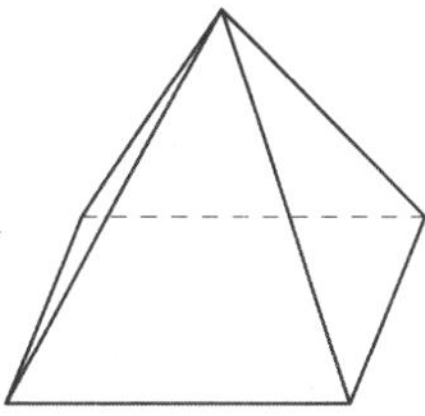

One exception to the similar naming convention between two-dimensional and three-dimensional figures is the name for a figure with four sides. While a quadrilateral names a polygon with four sides, the name for a four-sided polyhedron is a tetrahedron.

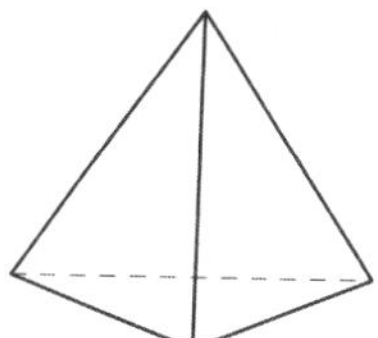

A tetrahedron is shown here. This tetrahedron is a regular tetrahedron. In general, a **regular polyhedron** is analogous to a regular polygon. In fact, a **regular polyhedron** is made up of regular polygons. While a regular polygon must have all sides equal and all angles equal, a **regular polyhedron** has all faces congruent and all faces meeting in the same manner.

What is the measure of the angle formed between the faces of a regular tetrahedron?

Create another regular polyhedron. (Hint: The angles between faces can be right angles.)

## Polyhedra

input

Now, you have sketches of two of the regular polyhedra. In total, there are only five regular polyhedra. The regular polyhedra are called the **Platonic solids**. Plato (429 – 347 B.C.) used the regular polyhedra as representations of the elements of the universe. According to Plato, all things were composed of fire, water, air, earth, or the cosmos. Plato let the regular tetrahedron shown on the previous page represent fire.

Water was represented by a regular isocahedron, which has 20 triangular faces.

Air was represented by a regular octahedron. The faces of an octahedron are also triangular.

How many faces should an octahedron have? Sketch an octahedron in the space provided.

You may have already drawn a sketch of the regular hexahedron used by Plato to represent Earth.

What is another more common name for a regular hexahedron?

The last of the Platonic solids is a regular dodecahedron. According to Plato, this solid represented the Cosmos. It has 12 pentagonal faces.

analysis

## Polyhedra

1. In previous sections, you drew prisms and found the surface and volume of some prisms. Are prisms polyhedra? Explain your response.

2. Are prisms regular polyhedra? Why or why not?

input

## Polyhedra

You have viewed **pyramids** before. **Pyramids** are defined as three-dimensional figures with one and only one polygonal base and lateral faces in the shape of triangles. The lateral faces all meet at one vertex called the **vertex of the pyramid**.

The pyramids with which you are most familiar are probably the Great Pyramids of Giza in Egypt. These pyramids are **square pyramids** because the base of each of these pyramids is a square. Pyramids are always named according to the shape of their base.

Below is a picture of a hexagonal pyramid.

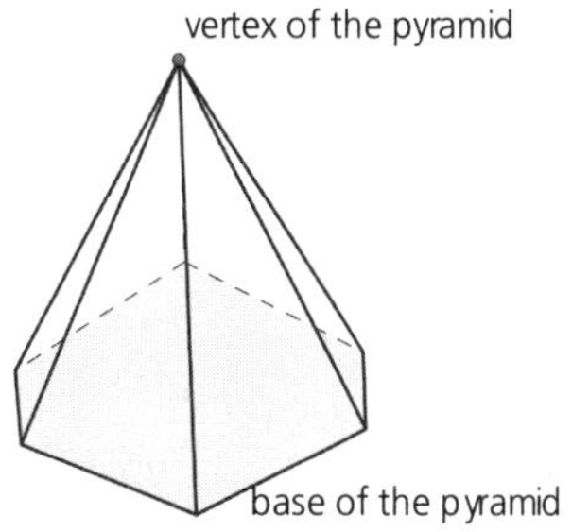

vertex of the pyramid

base of the pyramid

process: model

## Polyhedra

3. Name at least two other places (besides math texts or classrooms), where you can find pyramids in the world.

4. Use construction paper or oak tag paper provided by your teacher to create the net of a pyramid. Sketch that net in the space provided here.

process: evaluate

## Polyhedra

5. Use your ruler to measure the lengths needed to calculate the surface area and the lateral area of your pyramid. Record these measurements on your net. Show your work below, and calculate the surface area. Include the appropriate units with your answers.

## Polyhedra

analysis

6. Are pyramids polyhedra? Explain your response.

7. Are pyramids regular polyhedra? Why or why not?

8. Are there any pyramids that are also prisms? If so, sketch an example in the space provided. If not, explain why not.

9. Are there other polyhedra aside from prisms and pyramids? If so, sketch an example in the space provided. If not, explain why not.

process: practice

## Polyhedra

10. Classify each of the following polyhedra in as many ways as possible. For example, a cube can also be called a square prism, or a hexahedron. Circle the most specific name for each polyhedron. Then, calculate the surface area of the polyhedra for which length measurements are provided.

a.

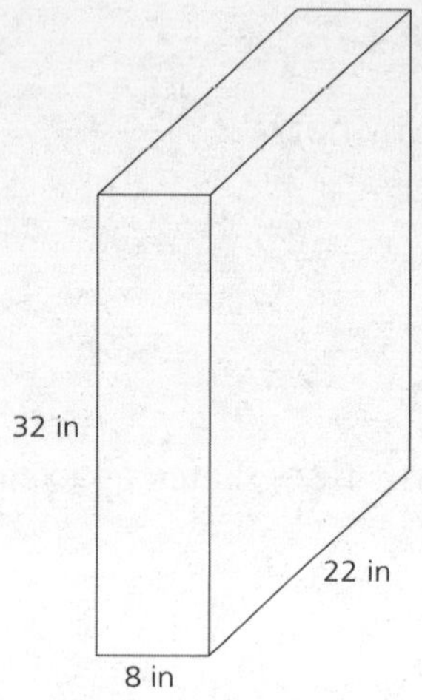

32 in

22 in

8 in

b. Assume the base is a square.

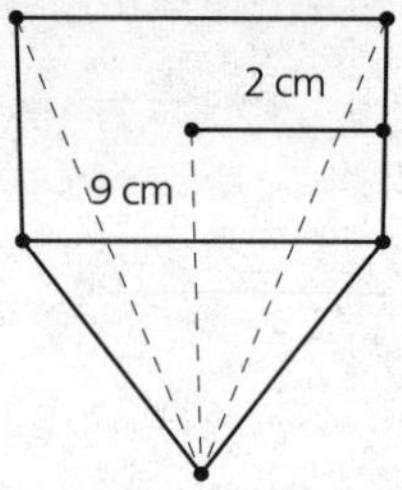

2 cm

9 cm

c.

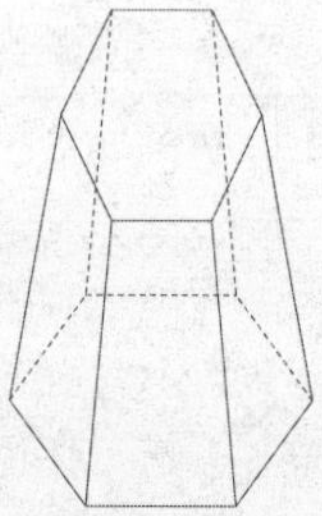

d.

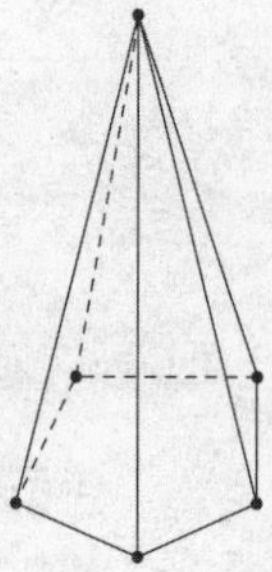

## Euler's Formula

input

Every polyhedron has vertices, edges, and faces. Recall that the faces of a polyhedron are the plane surfaces that enclose it. The edges of a polyhedron are the segments that form the boundaries of the faces. The vertices are the points where two or more edges intersect. For example, this square pyramid has five faces (4 lateral faces and one base). It also has eight edges (4 surrounding the base and 4 separating the lateral faces), and five vertices (4 in the base and the vertex of the pyramid).

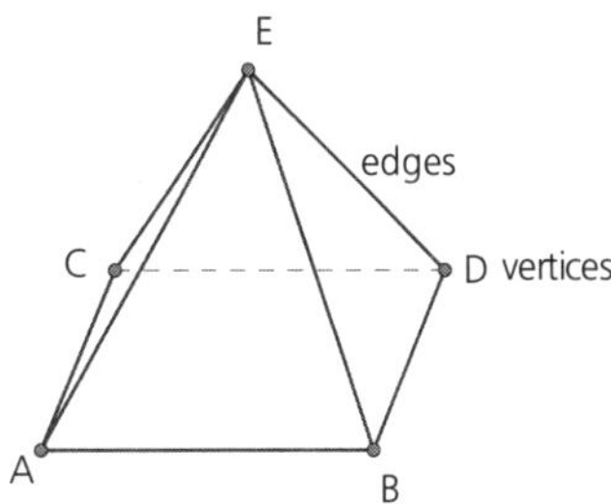


Is there a relationship among the number of vertices, edges, and faces of a polyhedron? In this activity, you will attempt to find a relationship.

## Euler's Formula

process: model

1. For the polyhedra you examined in the previous two sections of this text, record the name of the polyhedron in the table. Also record the number of faces, the number of edges, and the number of vertices. Start with the square pyramid pictured above.

| Name of Polyhedron | Number of Edges (E) | Number of Vertices (V) | Number of Faces (F) |
|---|---|---|---|
| Square Pyramid | 8 | 5 | 5 |
| | | | |
| | | | |
| | | | |
| | | | |
| | | | |
| | | | |
| | | | |
| | | | |
| | | | |

analysis

## Euler's Formula

2. What patterns do you notice for the pyramids listed in your table?

3. What patterns do you notice for the prisms in your table?

4. What patterns hold for all the polyhedra listed in your table? (Hint: What do you notice about the sum of the number of vertices and the number of faces compared to the number of edges?)

process: model

## Euler's Formula

5. Create an equation or formula to model the relationship among the number of edges (E), the number of vertices (V), the number of and faces (F) in a polyhedron.

extension

## Euler's Formula

The formula you found in this activity was first found by a Swiss mathematician named Leonhard Euler who lived in the 18th century (1707 – 1783). For that reason, the formula is named after Euler (pronounced like the word Oiler). Euler's work formed the start of a new branch of mathematics called **topology**. **Topology** is related to geometry. The upcoming section on Networks also deals with topology.

Euler extended his theory beyond solid three-dimensional shapes to three-dimensional shapes with one hole or shapes with two holes, like those pictured below.

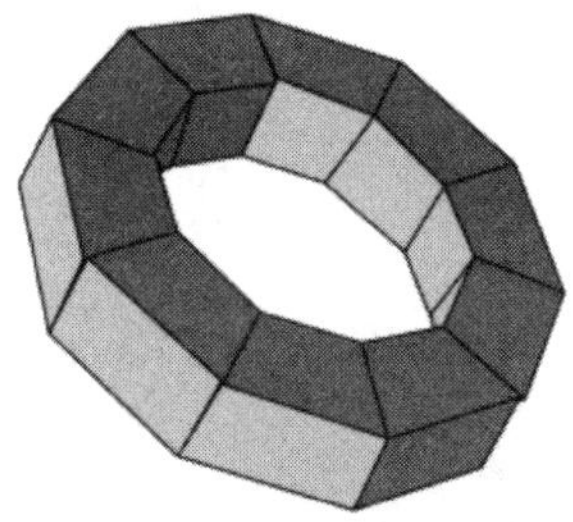

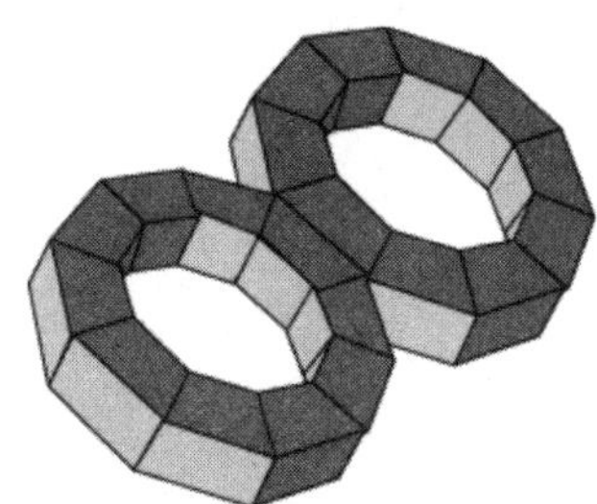

extension

## Euler's Formula

6. Build a few of these shapes of your own. Use them to help you complete the table. It is similar to the one you completed for polyhedra without holes.

| | E | V | F |
|---|---|---|---|
| | | | |
| | | | |
| | | | |
| | | | |
| | | | |
| | | | |
| | | | |
| | | | |

7. Create an equation or formula to model the relationship among the number of edges (E), the number of vertices (V), the number of and faces (F) in a polyhedron with one hole, with two holes, and for all polyhedra (if possible).

## Some Networks Aren't on Television

extension

In the previous activity, you found a pattern among the number of vertices, edges, and faces in polyhedra. This formula (V + F = E + 2) was found by Leonhard Euler. Euler also helped solve a number of other interesting and perplexing problems. Among these problems is the Koenigsburg Bridge Problem also known as the Seven Bridges of Koenigsburg.

Koenigsburg was a town is Russia. Today it is called Kalingrad. The river Pregal runs through the town, and there are two islands in the river. The banks are connected to the islands and the islands are connected to one another by a series of seven bridges. A sketch of the town is included below.

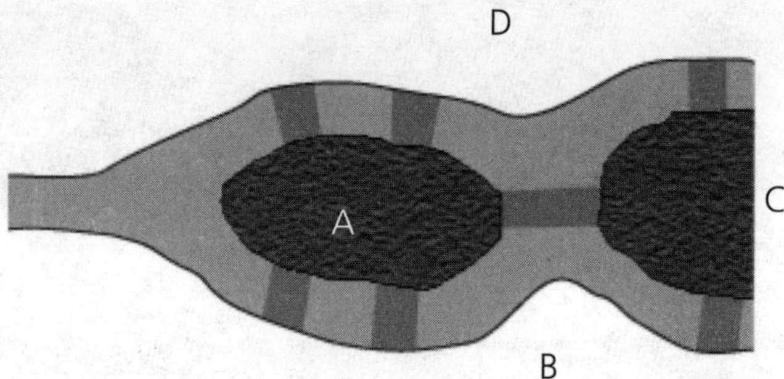


Citizens of the town would routinely stroll throughout the city, crossing from the banks to the islands and back again. Is it possible for the citizens of Koenigsburg to leave their homes, cross all seven bridges once and only once, and return to the starting place?

Euler was able to solve the problem by thinking of the banks, islands, and bridges as a network of paths. Each bank or island in the sketch below is represented by a point or **node**. Each bridge is represented by a **path** connecting the nodes like the bridges connect the regions of land. A and C represent the islands, and B and D represent the banks.

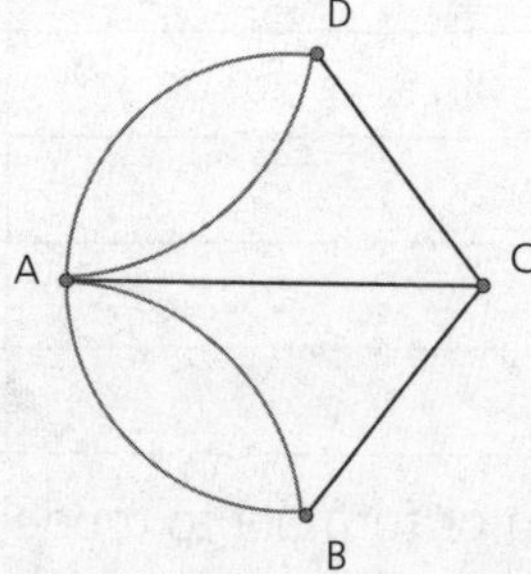


Now the problem is simplified to the task of tracing a path from any point over each of the paths and returning to the original point without retracing any path. (You can touch each node more than once.)

extension

## Some Networks Aren't on Television

This problem is similar to challenges you may have been given in the past in which you are asked to trace a figure without lifting your pencil from the page. The following is a typical problem like that.

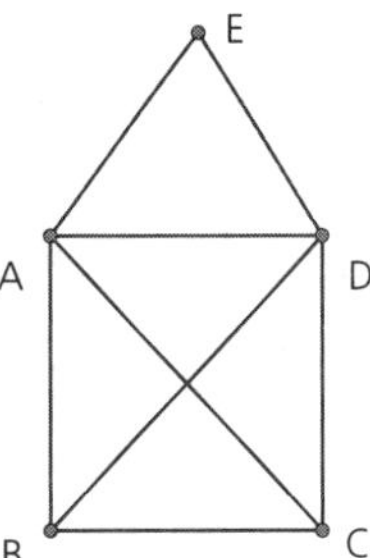


1. Pick a node and attempt to trace every path only once before returning to your starting point. Can you do it? If not, try again. Pick a different starting point or go in a different direction.

Notice that each node in each network has a number of paths connected to it. If a node has an even number of paths connected to it, it is called an **even node**.

2. If the node has an odd number of paths connected to it, what do you think it is called?

3. For the network above and for each of the networks below record the number of even nodes, the number of odd nodes, and whether or not the network can be traveled without lifting your pencil from the page.

| | | A | B | C | D | E | F | G | H | I | J |
|---|---|---|---|---|---|---|---|---|---|---|---|
| **Odd** | 2 | | | | | | | | | | |
| **Even** | 3 | | | | | | | | | | |
| **Y/N** | No | | | | | | | | | | |

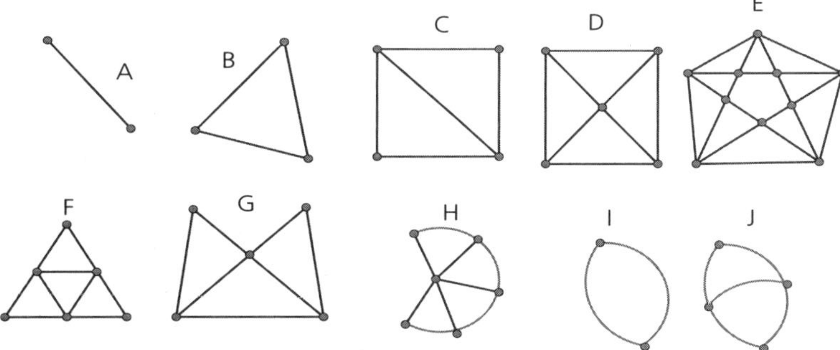

## Some Networks Aren't on Television

extension

4. Using the data in your table, what generalizations can you make about when you can and cannot trace a network and return to your starting point?

If you change the stipulation about returning to the starting point, perhaps that will change the results. Return to the Koenigsburg Bridge and the networks on the previous page. Attempt to retrace them without lifting your pencil, but this time you can start and end on different nodes. In mathematical terms, a network is **traversable** if it can be traced without lifting your pencil and without retracing any of the individual paths.

5. Add a row to your table to record whether each network is traversable.
6. How did changing the rule about returning to the starting point change the results? Were any additional networks traversable?

7. Using the data in your table, what generalizations can you make about when a network is traversable?

8. Why are certain networks traversable while others are not? Hint: Focus on the characteristics of an even versus an odd nodes in this explanation.

## Some Networks Aren't on Television

extension

9. Is the town of Koenigsburg traversable? Explain your response in terms of even and odd nodes.

A traversable network in which you end at the same node where you began is called an **Euler Circuit.** Euler Circuits are used in logistical applications in many businesses. For example, Euler Circuits can be used to map the route of a snow removal vehicle. Euler Circuits might also be used to map flight routes for airlines so the crew can return to their home city at the end of a day of work.

10. Think of at least one other area where Euler Circuits are useful.

11. Why are Euler Circuits useful for things like snow removal?

## Volume of Prisms and Pyramids

input

In the section on Moving from Two Dimensions to Three, you found the volume of prisms by building or envisioning layers of sugar cubes atop one another. Later in that section, you found a formula to calculate the volume of any prism by multiplying the area of the base by the height (the number of layers).

Volume = Area of the Base * Height

Finally, you used that formula to calculate the volume of a few prisms.

## Volume of Prisms and Pyramids

process: model

1. Draw a model for a rectangular prism with a square base that is 3 inches by 3 inches and a height of 6 inches.

process: evaluate

## Volume of Prisms and Pyramids

2. Find the volume of the rectangular prism your just drew.

process: model

## Volume of Prisms and Pyramids

3. Draw a model for a square pyramid with the same base and height as the prism you just created.

## Volume of Prisms and Pyramids

process: evaluate

4. Estimate the number of square pyramids needed to fill the volume of the rectangular prism.

5. Based on your teacher's demonstration, how many pyramids does it actually take to fill the volume of the rectangular prism?

6. Find the volume of the square pyramid you just drew, and explain what you did to calculate this volume.

## Volume of Prisms and Pyramids

process: model

7. Based on your explanation above and on the formula for the volume of a prism, create a formula for the volume of a pyramid.

process: practice

## Volume of Prisms and Pyramids

8. Find the volume of the following prisms and pyramids. Assume the polygonal bases are regular polygons.

a.

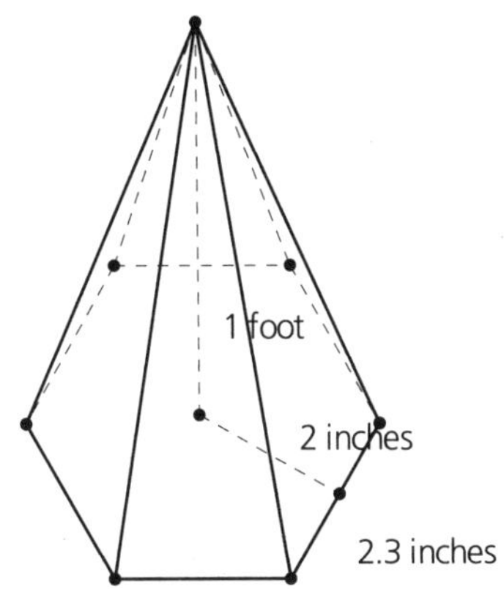


b.

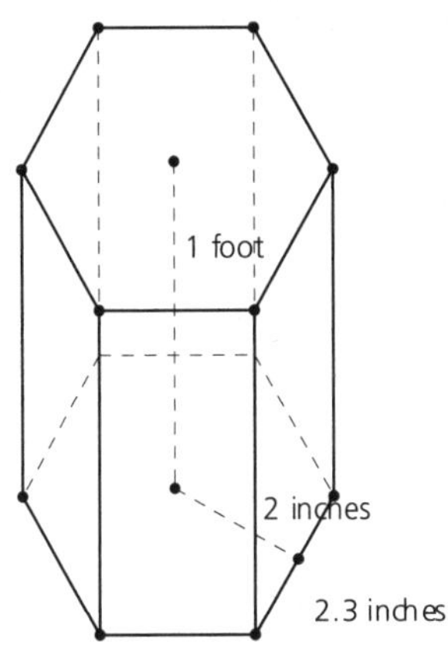


c.

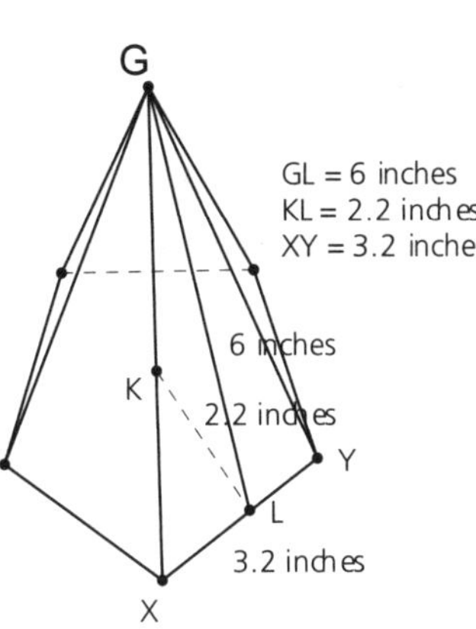


d.

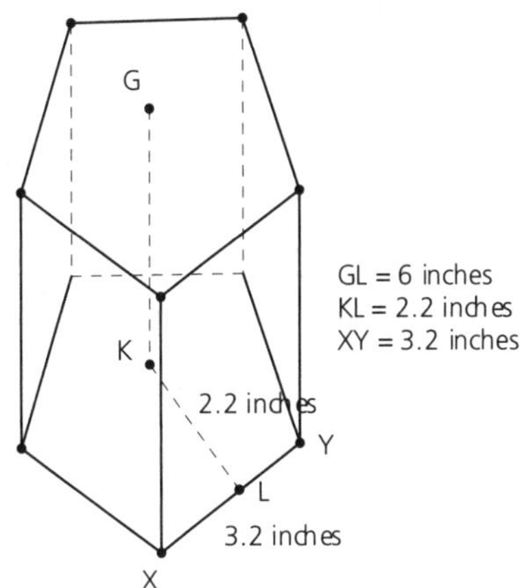


9. You may have studied the Great Pyramids of Giza, in Egypt. There is a pyramid in North America that rivals the Great Pyramids in size. The Pyramid of the Sun at Teotihuacan, Mexico is 738 feet on each side of its square base and 213 feet tall. Find the volume of this Mexican pyramid, showing all your work.

## Volume of Prisms and Pyramids

process: practice

10. The Muttart Conservatory in Edmonton, Alberta consists of 4 square pyramids. There are two sizes of pyramids in the Conservatory. The Temperate and Tropical pyramid dimensions are approximately 25.7 meters long and 24 meters high at the apex. The Show Pyramid and Arid Pyramid dimensions are approximately 19.5 meters in length and 18 meters in height at the apex. Find the volume consumed by each of the pyramid that makes up this conservatory in Canada.

## Volume of Prisms and Pyramids

process: summary

11. Cite at least two similarities and at least two differences between a prism and a pyramid.

input

## More Three-Dimensional Figures

So far, the three dimensional figures you have seen in this unit have been polyhedra.

Are all three-dimensional figures polyhedra? Can you name at least one three-dimensional figure that is not a polyhedron?

Some three-dimensional figures have curved surfaces instead of planar surfaces. Those are the figures you will study in this section.

process: model

## More Three-Dimensional Figures

1. What is the name of a prism with six lateral faces? ...with eight lateral faces?

2. Is it possible for a prism to have 100 lateral faces? ...one-thousand lateral faces?

3. Imagine what prisms with thousands or millions of lateral faces look like. The lateral edges separating one lateral face from another are almost imperceptible. The vertices separating one edge of the base from another are also imperceptible. What shape does the base of a prism with one million lateral faces approximate?

4. What is the name of a pyramid with seven lateral faces? ...with ten lateral faces?

process: model

## More Three-Dimensional Figures

5. Is it possible for a pyramid to have 200 lateral faces? ...one-million lateral faces?

6. Imagine what pyramids with thousands or millions of lateral faces look like. The lateral edges separating one lateral face from another are almost imperceptible. The vertices separating one edge of the base from another are also imperceptible. What shape does the base of a pyramid with one billion lateral faces approximate?

input

## More Three-Dimensional Figures

The figures you have been envisioning are called cylinders and cones. Below are depictions of cylinders and cones.

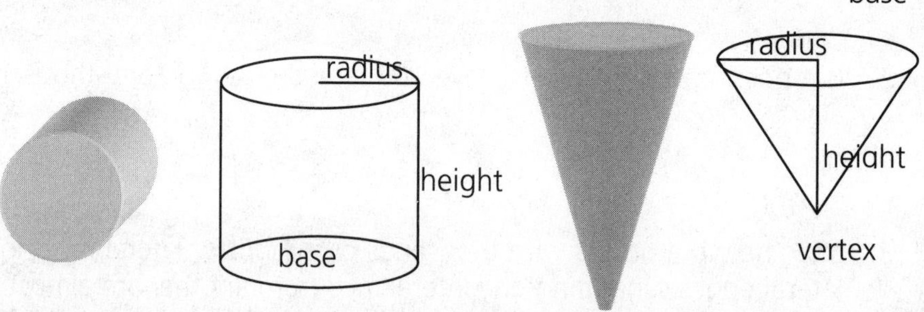


The bases of cylinders and cones are circles. Polygons with many sides approximate this shape. Like prisms, cylinders have two parallel bases and a height that is perpendicular to the planes of the bases.

Like pyramids, cones have a base and a height that are perpendicular. The **vertex of a cone** is a point that is in a different plane than its base. For a right cone, the vertex will be in the same vertical line as the center of the base. Is that true for an oblique cone?

What are some everyday objects that are cylinders or cones?

process: model

## More Three-Dimensional Figures

7. When you drew prisms in a previous section, you sketched two bases that were congruent to one another but in different planes. To sketch cylinders, you do the same thing, but the circular bases may need to be drawn more like ovals or ellipses to have a three-dimensional look. Sketch a cylinder in the space provided.

8. The pyramids in a previous section had one polygonal base. The vertices of the base were all joined to the vertex of the pyramid via line segments. There are no vertices on the circular base of a cone, and your sketch cannot show the entire circumference of the circular base connected to the vertex. To represent the connection of the base to the vertex of the cone, show only two of the points along the base connected to the vertex of the cone. Sketch a cone in the space below.

9. Just as there are right or oblique prisms and pyramids, there are also right or oblique cylinders and cones. Create an oblique cylinder in the space provided. If you think of music disks that have been stacked and then bumped so they are tilted, it is easier to envision an oblique cylinder. Create an oblique cone beside the oblique cylinder. To make this task easier, do not place the vertex of the cone directly above the center of the base of the cone.

process: model

## More Three-Dimensional Figures

10. Recall that a net of a three-dimensional figure is a "flattening" or an "unfolding" of the figure. A net of a cylinder can be pictured if you envision a tin can. Picture the top and bottom attached to the unrolled label of the can. Complete the partial net of a cylinder pictured below.

analysis

## More Three-Dimensional Figures

11. Refer to the net you completed above. The radius of the top base of the cylinder is 1.0 cm. Use that measurement to calculate the length of the lateral surface of the cylinder (the unrolled label). Record this measure on the net. Explain the reasoning you used to deduce the length.

process: evaluate

## More Three-Dimensional Figures

12. Measure the width of the lateral surface of the cylinder's net you just completed. Record that measurement on the net.
13. Calculate the lateral area of the cylinder. Show your work. Record the result on the net.
14. Calculate the area of each of the bases of the cylinder. Show your work and record the results on the net.
15. Calculate the surface area of the cylinder.

process: model

## More Three-Dimensional Figures

16. To find the surface area of any cylinder, which individual areas do you need to calculate and add together.
17. What is the formula you know for the area of any circle?
18. What is the formula you know for the area of a rectangle?

process: model

## More Three-Dimensional Figures

19. As you have seen, the lateral surface of the cylinder becomes a rectangle when it is "unrolled." The length of this rectangle is the same as what measurement of the circular base of the cylinder?

20. Change the formula for the area of the rectangle to show that the length is the same as the circumference of the circular base. Use the formula for the circumference of a circle instead of the length.

21. Use the formula you developed in the previous question and the formula for the area of a circle to write a formula for the surface area of any cylinder.

process: practice

## More Three-Dimensional Figures

22. Find the surface area of the following cylinders. Show your work and your units in the result.

| a. | The radius of the base is 5 feet and the height is 30 feet. | b. | The height of the straw is 15 cm and its radius is ½ cm. |
|---|---|---|---|

## More Three-Dimensional Figures

process: model

23. Just as your completed the net of a cylinder a few pages ago, complete the net of a cone in the space below. When you create the base as a part of your completed net, it should have a radius of 1.5 cm.

## More Three-Dimensional Figures

process: evaluate

24. Calculate the area of the base of this net of a cone. Show your work and include the units in your answer.

## More Three-Dimensional Figures

process: evaluate

25. Calculate the circumference of the base circle. Where is this distance represented on the lateral surface of the net of the cone?

26. The slant height of this cone is 4 cm. Trace the segment that represents the slant height in the net.

27. What fraction of a circle is represented by the lateral surface of this cone?

28. The lateral surface of the cone is represented by a part of a circle called a **sector**. Find the area of the full circle of which the sector is a part.

29. Find the area of the sector of the circle that represents the lateral surface of the cone. Recall that the fraction of the circle that is the lateral surface was calculated above. This is the lateral area of the cone.

30. Finally, use the area of the base and the lateral area to find the complete surface area of the cone.

## More Three-Dimensional Figures

input

That method for finding the lateral area and surface area of a cone works, but as you experienced, it is somewhat complicated and tedious. There is a different way of thinking about the lateral area. This method may be better for you.

## More Three-Dimensional Figures

input

Think about the lateral area for a pyramid with many lateral faces.

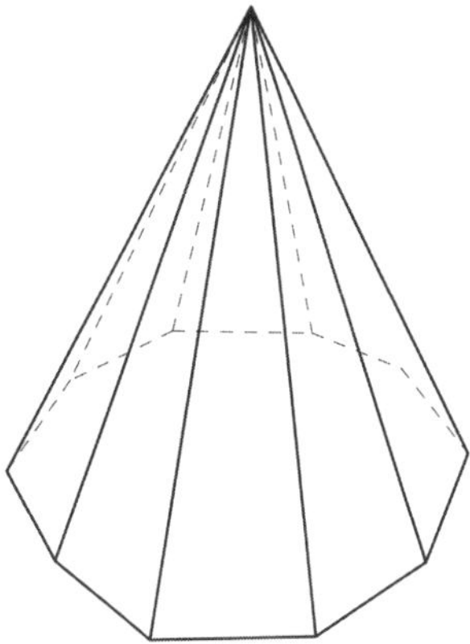

You can find the lateral area by finding the area of each triangular lateral face and multiplying by the number of triangular faces (10 in this case).

LA = number of triangle or faces * ( ½ * length of one edge of the base * slant height)

LA = 10 ( ½ 1.7 cm * 6.1 cm)

Since the number of triangles or faces multiplied by the length of one edge of the base is the perimeter of the base, you can find the lateral area by multiplying the perimeter of the base by one-half of the slant height.

number of faces * length of one edge = perimeter of base

LA = perimeter of the base * ½ * slant height

This formula can be extended to pyramids with any number of sides. If you think of a cone as something similar to a pyramid with many, many faces, you can extend this formula to a cone as well.

What is the "perimeter" of a cone's base?

What does the formula for lateral area become for a cone?

process: practice

## More Three-Dimensional Figures

31. Find the surface area of the cones. Show your work and the units of your answer.

a. $C = 12\pi$ cm

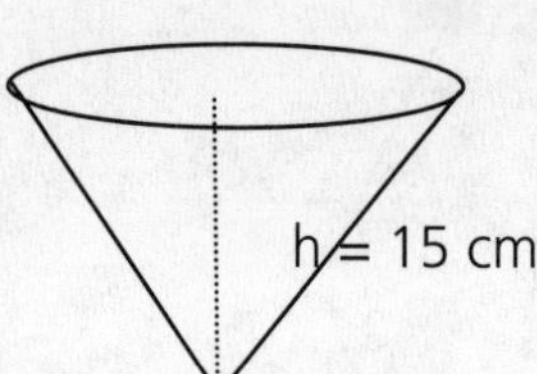


b.

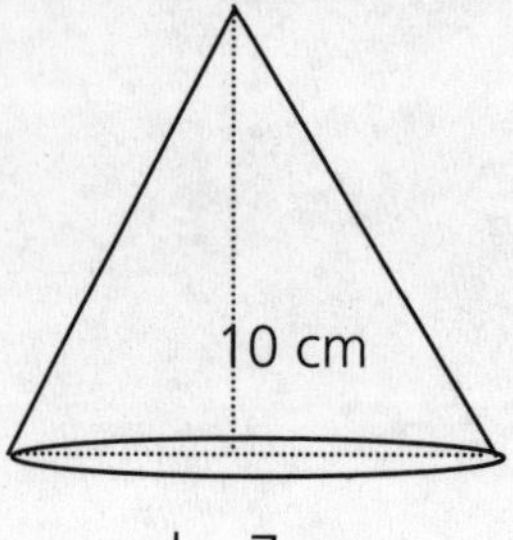


32. Kip's ice cream shop sells waffle cones as well as regular sugar cones. The manager is trying to determine which type of cone gives the store more of a profit per unit of cone. The slant height of the sugar cone is about 4.25 inches, and its diameter is about 1.75 inches. The waffle cone has a slant height and diameter twice that of the sugar cone. Find the area of each cone.

The store currently charges twice as much for a waffle cone as it does for a sugar cone. Considering *only* the areas of the cones, which cone gives the store a greater percentage profit?

## More Three-Dimensional Figures

process: summary

33. Compare and contrast the similarities and differences between prisms and pyramids to the similarities and differences between cylinders and cones.

    a. What are the similarities between a prism and a cylinder?

    b. What are the differences between a prism and a cylinder?

    c. What are the similarities between a pyramid and a cone?

    d. What are the differences between a pyramid and a cone?

## Volume of Cylinders and Cones

input

The last question in the previous section asked you to compare and contrast cylinders and prisms. The volume of prisms and cylinders is another point for comparison. Just as the volume of a prism can be found by imagining layers, so can the volume of a cylinder.

What is the difference between the layers in a prism and the layers in a cylinder?

## Volume of Cylinders and Cones

analysis

1. Pictured below is a cylinder with the one layer shown. How can you find the number of cubes in one layer of a cylinder?

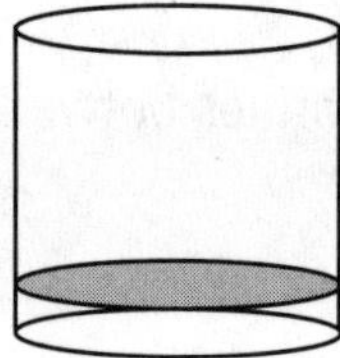

2. How can you find the area of the base of a cylinder?

3. How can you determine the number of layers in a cylinder?

process: evaluate

## Volume of Cylinders and Cones

4. The cylinder below has a diameter of 12 units and a height of 15 units. Find the area of the base of this cylinder.

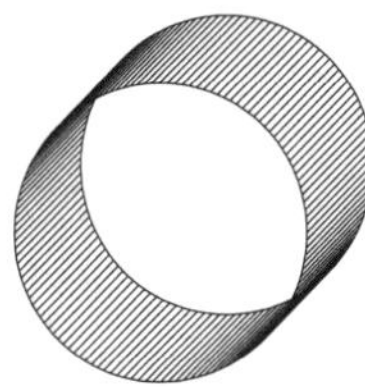

5. How many cubic units are in the first layer of the cylinder?

6. How many layers in are in this cylinder?

7. Find the volume of this cylinder. Show your work and include the units in your answer.

process: model

## Volume of Cylinders and Cones

8. The cylinder below has a radius of r units and a height of h units. Find the area of the base of this cylinder.

9. How many cubic units are in the first layer of the cylinder?

10. How many layers in are in this cylinder?

11. Find the volume of this cylinder.

12. Based on your answer to the previous question, what is the formula for the volume of a cylinder?

analysis

## Volume of Cylinders and Cones

13. How is finding the volume of a cylinder the same as finding the volume of a prism?

process: practice

## Volume of Cylinders and Cones

14. Find the volume of these cylinders.

   a. The radius of the base is 5 feet and the height is 30 feet.

   c. The radius of the coffee cup is 1 ½ inches and the height is 4 inches.

## Volume of Cylinders and Cones

process: model

16. Create a model of a cylinder with the radius of the base measuring 3 cm and the height measuring 6 cm.

17. Find the volume of this cylinder.

## Volume of Cylinders and Cones

process: model

18. Create a model of a cone with the radius of the base measuring 3 cm and the height measuring 6 cm.

## Volume of Cylinders and Cones

analysis

19. Based on the similarities and differences you have cited for three-dimensional shapes, how many cones will it take to fill the volume of a cylinder with the same base and height?

20. Based on the relationship you saw between prisms and cylinders and the relationship you saw between pyramids and cones, how do you think the volume of a cone will compare to the volume of a cylinder?

## Volume of Cylinders and Cones

process: evaluate

21. Find the volume of the cone you created, and explain what you did to calculate this volume.

## Volume of Cylinders and Cones

process: model

22. Generate a formula to find the volume of a cone, based on the formula for the volume of a cylinder.

process: practice

## Volume of Cylinders and Cones

23. Find the volume of the cones below.

a. $C = 12\pi$ cm

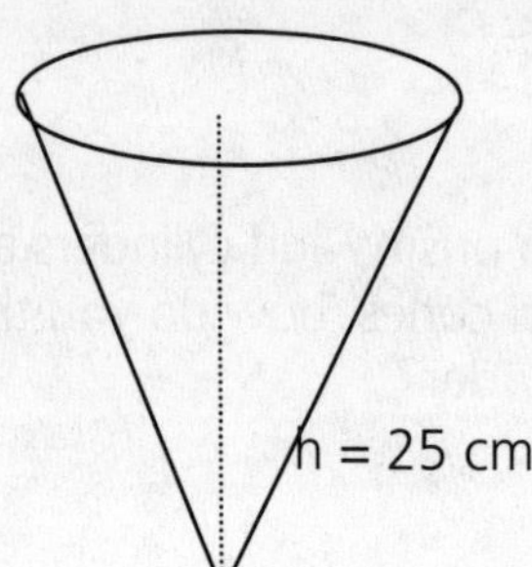


b.

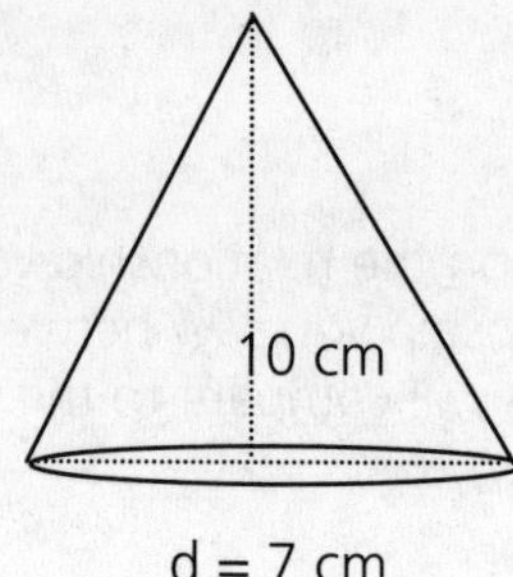


24. Kip's ice cream shop sells waffle cones as well as regular sugar cones. The manager is trying to determine which type of cone gives the store more of a profit based upon the volume of the cones. The slant height of the sugar cone is about 4.25 inches, and its diameter is about 1.75 inches. The waffle cone has a slant height and diameter twice that of the sugar cone. Find the volume of each cone.

The current charge for a waffle cone is twice that of a sugar cone. Considering *only* the volumes, how much more should the store charge for a waffle cone?

## Spheres

input

**Spheres** are the last of the three-dimensional figures you will study in detail in this unit. Spheres are unique among the figures you are studying because they have no flat surfaces.

What are some items that approximate the shape of a sphere?

It is useful to think of a sphere as a three-dimensional circle. The next several questions will help you do just that.

## Spheres

process: model

1. Pick a point in the space provided. Label the point A. Place a point 3 cm from A. Label it B. Place another point 3 cm from A. Label it C. Continue to place several more points 3 cm from A. What do you see forming?

2. Imagine lifting your original point up from the plane of the paper. You can lift your paper up to better imagine this. Instead of limiting the points you place 3 cm from A to the surface of the paper, now you can imagine placing more points 3 cm from A. However, these new points will not all be in the same plane as the points you placed in question number 1. If you continue to place points 3 cm from point A, what shape will eventually form?

input

## Spheres

Based on what you have just imagined, a **sphere** is the set of all points that are a certain distance away from a given point. The given point is the **center** of the sphere. The **radius** of a sphere is the "certain distance" mentioned in the definition for a sphere.

One part of a sphere that is unique is called a **great circle**. A great circle of a sphere is any circle along the sphere's surface that has the same center as the sphere. Any circle along the sphere's surface that does not have the same center as the sphere is called a **small circle**.

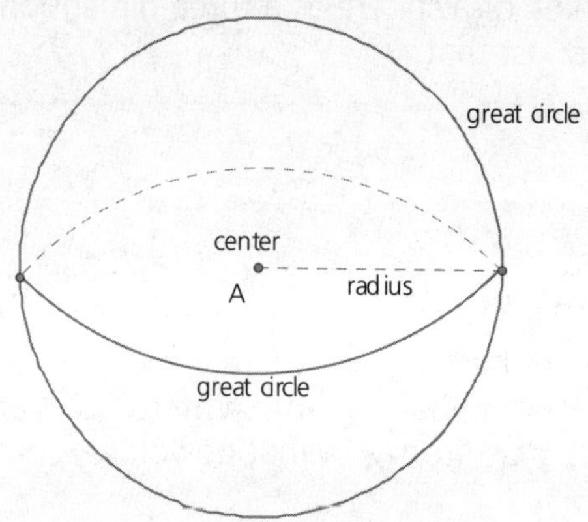


Just as a semi-circle is half of a circle, a **hemisphere** is half of a sphere.

analysis

## Spheres

3. How is the center of a sphere the same as the center of a circle?

4. How is the center of a sphere different from the center of a circle?

| analysis | **Spheres** |
|---|---|
| | 5. How is the radius of a sphere the same as the radius of a circle?<br><br>6. How is the radius of a sphere different from the radius of a circle? |

| process: model | **Spheres** |
|---|---|
| | 7. If you consider the Earth a sphere, what are some examples of great circles?<br><br>8. If you consider the Earth a sphere, what are some examples of small circles? |

| input | **Spheres** |
|---|---|
| | For the other three-dimensional shapes you studied, you found the volume and you created nets. You will do the same thing for spheres. However, with other three-dimensional figures, the nets helped you find the surface area. With a sphere, the net is a bit more difficult to create. |

input

## Spheres

Mapmakers have long struggled with the idea of representing a nearly spherical Earth on a flat surface. When you attempt to "flatten" or "unfold" a sphere, there are points that just can't stay connected to one another. For that reason, a two-dimensional map of the Earth misrepresents the size of land masses. As you move away from the Equator towards the poles, the exaggeration of the size of the land masses worsens. For example, Greenland looks very much larger than it actually is relative to the United States.

To make the calculation of surface area of spheres simpler, you will use a smaller sphere (or at least an item that comes close to a sphere). You need an orange and a partner for the following activity, as well as some paper towels to clean up after yourself.

process: model

## Spheres

9. Start by placing your orange on plain white paper. Hold your pencil against the edge of the orange and perpendicular to your paper. Trace the circumference of the orange by having your partner hold the orange in place. Project a great circle of the orange onto your paper. If you make a mistake, turn the paper over and try again to get the circumference of the orange (a great circle) traced on the paper.
10. Peel your orange and gather all of the pieces on your desk. (Set the orange aside for later consumption.) If you have small to medium-sized pieces of peel, this activity will work better. Small to medium sized pieces of peel will lie relatively flat on the desktop. If you have any large pieces of peel, rip them into smaller pieces.
11. Now comes the most important part of the activity. Take the peel pieces and fill the circle. Make sure you fill the circle completely, but do not overlap the orange pieces or make them protrude from the circle. For this reason, your pieces of peel should lie relatively flat. After you fill the circle once, discard the pieces you have already used so as not to reuse them. Fill the circle with orange pieces once again. Discard those pieces.
12. Keep going until you have used all your peel pieces. Remember, do not overlap and do not go outside the circle, but make sure to fill the circle completely with relatively flat pieces of peel each time.
13. How many times did you manage to fill the great circle of the sphere with the peel from the sphere?

process: evaluate

## Spheres

14. Gather the information from your class. Make a data table by listing the number of times each of your classmates could fill the great circle of the orange with its peel.

| Name | # of times the circle was filled | Name | # of times the circle was filled |
|---|---|---|---|
| | | | |
| | | | |
| | | | |
| | | | |
| | | | |
| | | | |
| | | | |
| | | | |
| | | | |
| | | | |
| | | | |
| | | | |
| | | | |
| | | | |
| | | | |

15. Find the mean, median, and mode of the numbers in your table, so you can decide on the best number to represent the class data as a whole.

16. On average, how many great circles will cover the surface of the orange?

## Spheres

process: evaluate

17. Measure your circle to find the diameter and consequently, the radius. Record that measurement. What part of the sphere is the radius of your great circle?

18. Use the measure of the radius to find the area of your great circle. Show your work and the units on your answer.

19. Using the number of great circles needed to cover the surface of the sphere, calculate the surface area of your orange.

## Spheres

process: model

20. Based on the activity you just completed, describe an easy way to find the amount of peel on an orange or the surface area of any sphere?

21. Use your description above to create a formula for the surface area of a sphere.

input

## Spheres

Now, you know how to find the surface area of any sphere based on it radius. Next, you will find a way to calculate the volume of any sphere, also based on it radius. First you need to return to the section on Volume of Cylinders and Cones. In that activity, you sketched a cylinder with a radius of 3 cm and a height of 6 cm. You also sketched a cone with the same dimensions.

How does the volume of the cone compare to the volume of the cylinder with the same base and height?

In the following experiment, you will again use water or rice to demonstrate a volume relationship just as you did when comparing the volume of pyramids with the volume of prisms. In this experiment you will use the volume of a hemisphere compared to the volume of a cylinder with the same radius. The height of the cylinder must be the same as its diameter (twice the radius) just as it was in the cylinder you sketched in Volume of Cylinders and Cones. As a result, the diameter of the hemisphere must be the same as that of the cylinder.

process: model

## Spheres

22. If the radius of the cylinder is *r*, write a formula for the volume of the cylinder based only on its radius. Simplify this formula.

23. What is the formula for the volume of a cone with the same base and height as the cylinder in the previous question?

process: model

## Spheres

24. As your teacher fills the cylinder with the material from the hemisphere, estimate how many hemispheres it will take to fill the cylinder.

25. Using the formulas you developed in questions 21 and 22, write a formula for the volume of the hemisphere.

26. Knowing that a sphere has twice the volume of a hemisphere, create a formula for the volume of a sphere.

process: evaluate

## Spheres

27. Return to the cylinder you sketched in Volume of Cylinders and Cones that has a radius of 3 cm and a height of 6 cm. What is its volume?

28. Calculate the volume of a hemisphere with the same radius.

29. Calculate the volume of a sphere with the same radius.

process: practice

## Spheres

30. Find the volume and surface area of the following spheres or hemispheres.

a.

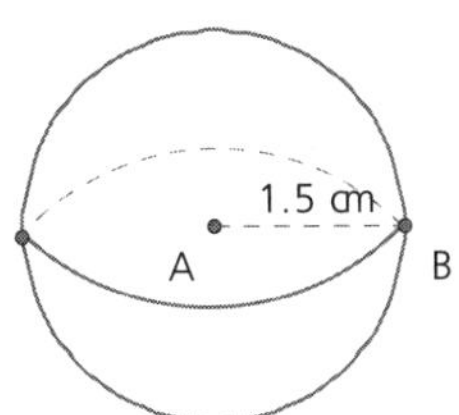


b. The diameter of the sphere shown here is 24 inches.

31. The volume of the planet Mercury is estimated at about 65 billion cubic kilometers. Calculate the radius of the planet.

## Applications of Volume

process: practice

1. Silo Problem

   A farmer's silo is used to store grain that the cows eat during the winter months. The silo is presently empty. If the truck hauling grain to the silo has a bed that is 8 feet long, 5 feet wide, and 4 feet high, find the number of truckloads of grain required to fill the silo.

   The top of the silo is a hemisphere with radius of 8 feet. The body of the silo is 40 feet in height. Show your work. Note that in reality the top of the silo does not get filled with grain.

Picture is not drawn to scale.

process: practice

## Applications of Volume

2. Ice Cream Problem

   You are going to a birthday party and it's your responsibility to bring the ice cream. Each person at the party will eat one ice cream cone. The closest grocery store sells ice cream in rectangular half-gallon containers. The dimensions of the container are 6.75 inches by 5 inches by 3.5 inches. It takes two scoops of ice cream to fill each cone. Each scoop is a sphere with a radius of 1 inch. Show your work for each part of this problem.

   a. What is the volume of one scoop?

   b. What is the volume of two scoops?

   c. What is the volume of ten scoops?

   d. What is the volume of the half-gallon container?

   e. If 30 people attend the party, how many half-gallons of ice cream should you buy?

   f. If you can afford to buy only 4 half-gallons of ice cream, how many people will get a cone?

process: practice

## Applications of Volume

3. Storage Problem

You need some cash, and because you are a good craftsman, you figure out a way to make coffee tables. Each table must be packed in a cardboard box measuring 4 feet by 4 feet by 2 feet. You need a place to store all these boxes, so you consider renting storage space. When you call to inquire how much it costs, the storage agent tells you a shed measuring 8 feet by 12 feet by 8 feet rents for $25 per month. The big question is how much space do you need, and how much will it cost to store all your boxes?

a. What is the volume of one box?

b. What is the volume of two boxes?

c. What is the volume of twelve boxes?

d. How many boxes will one storage shed hold?

e. If you have 35 boxes, how many sheds will you need to rent?

f. If you can afford to rent only 3 sheds for one month, how many tables can you store?

process: practice

## Applications of Volume

4. Driveway Problem

   Premixed cement is sold by the cubic yard. Find the number of cubic yards it takes to build a driveway that is 55 feet long, 6 inches thick, and 15 feet wide. Show your work.

5. Parking Lot Problem

   To convert a football field into a parking lot, you must pave the field. The field is 65 yards wide and 120 yards long. The cement must be 4 inches thick. How many cubic yards of cement will it take to complete the job?

## Applications of Volume

process: practice

6. Firewood for the Winter

   A cord of firewood measures 4 feet by 4 feet by 8 feet when stacked. If the average height of a tree is 15 feet, and the average diameter of a tree trunk is 1 foot, find the number of trees required to make 3 cords of firewood. Show your work.

7. Basketball Packaging

   The diameter of a basketball is about 9.5 inches. If a basketball is packaged in a cubic box measuring 9.5 inches on each edge, find the volume of empty space in the box. Show your work.

process: practice

## Applications of Volume

8. Will the Frozen Yogurt Cone Overflow?

The frozen yogurt cone is 12 cm deep, and its diameter is 6 cm. A scoop of frozen yogurt sits on the top of the cone. The scoop is a sphere with a diameter of 6 cm.

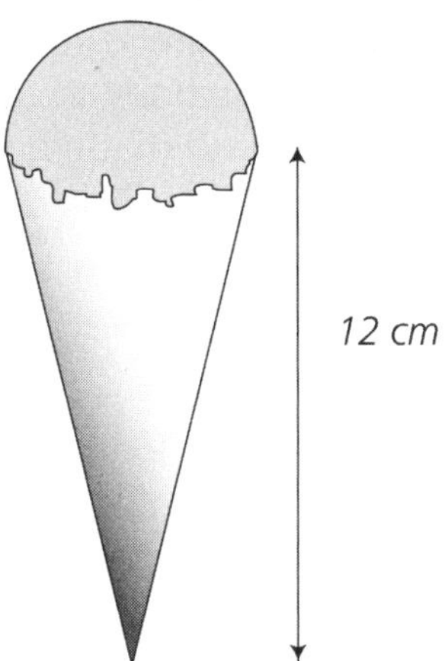


a. If the entire scoop of frozen yogurt melts into the cone, will the cone overflow? Show your work and explain how you arrived at your answer.

b. If everything else in this problem situation remains the same, but the cone is now only 8 cm deep, will the cone overflow? Show your work and explain how you arrived at your answer.

## Cross-Sections

input

A **cross-section** of an object can be viewed by imagining the object being cut or sectioned, so you can see what is usually on the inside. You have seen cross sections of certain objects before. For example, a slice of bread shows a cross-section of the loaf. Many biology texts show cross-sections of cells or dissected animals to give the reader a better picture of what is present inside the cell or animal. Magnetic Resonance Imaging (MRI) gives doctors and scientists a cross-sectional view of the body so they can examine muscles and tissues. This allows problems to be found that are not identifiable with X-rays.

Think of at least two other cross-sections you have seen before.

## Cross-Sections

process: model

Think of the Earth. You have likely seen a cross-section of the Earth in geography textbooks or in encyclopedias. A cross-sectional view of the Earth shows the molten core, the mantel, and the Earth's crust.

1. What shape is the cross-sectional view of the Earth you have seen?

2. Sketch a cross-sectional view of the Earth in the space below.

## Cross-Sections

analysis

3. When you created the cross-sectional view of the Earth in the previous question, you may have thought about cutting the Earth along the Equator, or cutting the Earth along the Prime Meridian. Using the terminology from the section on Spheres, what term describes a cross-section of the Earth created by slicing along the Equator? If your cross-section in question 2 was not cut along the Equator, sketch the cross-section created by cutting the Earth at the Equator. (You do not need to include the layers of the Earth in the sketch.)

4. What term describes the cross-section created by cutting along the Prime Meridian? If your cross-section in question 2 was not created by cutting along the Prime Meridian, sketch that cross-section here. (There is no need to include the layers of the Earth.)

5. Think about slicing the Earth along any other longitude lines, except for the Prime Meridian. (Lines of longitude run north and south.) Will you still create a great circle with the cross-section from any or all of these slices? Explain your response.

6. Think about slicing the Earth along any other latitude lines, except for the Equator. (Lines of latitude run east and west.) Will you still create a great circle with the cross-section from any or all of these slices? Explain your response. Sketch the cross-section in the space provided.

## Cross-Sections

analysis

7. Do not limit your thinking to the approximate sphere of the Earth for the next few questions. Think about any and all spheres in order to respond.

   a. How many cross-sections can be created from a sphere?

   b. How many cross-sections can be created from great circular regions? Sketch at least one of these cross-sections, if possible.

   c. How many cross-sections can be created from small circular regions? Sketch at least one of these cross-sections, if possible.

8. How many cross-sections of a sphere can be created that are not circular regions? Sketch at least one of these cross-sections, if possible.

9. Jose said, "Creating a cross-section of a sphere gives a two-dimensional version of a sphere." What does he mean by this statement, and what is a two-dimensional version of a sphere.

input

## Cross-Sections

You witnessed that any cross-section of a sphere creates a circle. The cross-sections of the sphere seem to collapse the sphere from three dimensions to two. Next, you will explore cross-sections of other three-dimensional shapes, to answer, among other questions, whether this "collapsing" happens in the same way with other figures.

In other words, will cutting a cube, a triangular prism, or a tetrahedron always give a two-dimensional version of a cube, a triangular prism, or a tetrahedron? Find out.

process: model

## Cross-Sections

10. Use the cube in the space below to visualize the answers to the series of questions below.

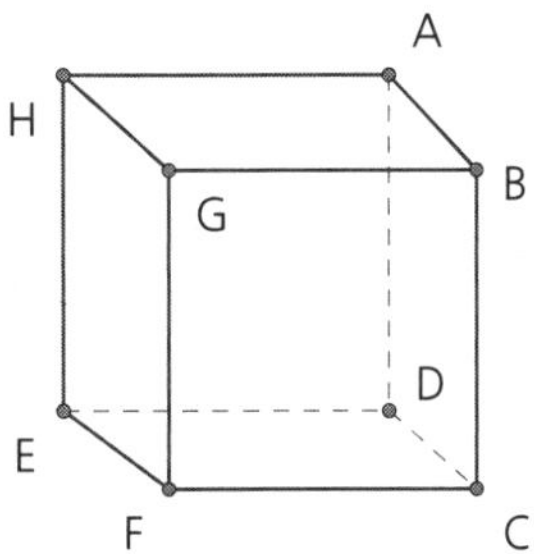


a. Describe the cross-section formed when a cube is cut with a plane parallel to any face of the cube. Be as specific as you can when describing the shape formed.

b. Describe the cross-section formed when a cube is cut with a plane that intersects the cube at vertices A, B, E, and F.. Be as specific as you can.

process: model

## Cross-Sections

c. Visualize tilting the plane that cuts the cube so it intersects four different faces of the cube. Tilt the plane in any different direction, but maintain the intersection of four faces. Describe the shape(s) that are created. Again, be as specific as you can.

d. Change the way the plane intersects the cube so the plane only intersects three faces. It is simpler to visualize this if you imagine the plane at a 45° to a vertex of the cube. Then move the plane into the space of the cube. Now, the plane intersects three faces. What types of shapes are created when a plane intersects three faces of a cube? Be specific.

e. Is it possible to intersect more than four faces of the cube with one plane? If so, how many faces can one plane intersect? What do(es) the cross-section(s) form? If one plane cannot intersect more than four faces of the cube, explain why not.

analysis

## Cross-Sections

11. Does cutting a cube with a plane always produce a square as the cross-section? Explain your response.

process: model

## Cross-Sections

12. Use the regular tetrahedron below to visually explore the possible cross-sections of any regular tetrahedron.

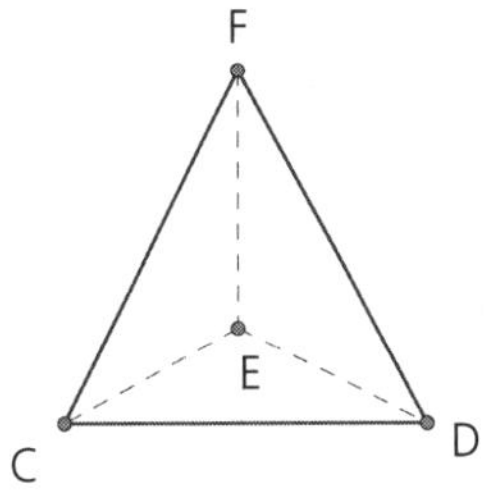

F

E

C D

a. What two-dimensional cross-sections of the regular tetrahedron are formed by a plane parallel to a face of the tetrahedron?

b. Are other cross-sections possible when a regular tetrahedron is intersected with a plane that is not parallel to a face of the tetrahedron? If so, what cross-sections are formed?

analysis

## Cross-Sections

13. Does cutting a regular tetrahedron with a plane always produce an equilateral triangle as the cross-section? Explain your response.

14. Make a generalization about the maximum number of sides a planar cross-section of a polyhedron can have.

## Cross-Sections

extension

During the 1880's, a little book called *Flatland* was written by Edwin A. Abbott. Within it, the author takes the role of a Square who lives in a land he calls Flatland, because it is strictly two-dimensional. A passage from the book follows. In this passage, a sphere has entered the square's world (Flatland) and is attempting to explain the world of three dimensions.

"I am in indeed in a certain sense a circle," replied the voice," and a more perfect circle than any in Flatland; but to speak more accurately, I am many circles in one."...I began to approach the stranger with the intention of taking a nearer view and of bidding him be seated: but his appearance struck me dumb and motionless with astonishment.

*I*: [B]efore your lordship enters into any further communications, would he deign to satisfy the curiosity of one who would gladly know from whence his visitor came?

*Stranger*: From space, from space, Sir. Whence else?

*I*: Pardon me, my Lord, but is not your Lordship already in space...?

*Stranger*: Pooh! What do you know of space? Define space.

*I*: Space, is height and breadth indefinitely prolonged.

*Stranger*: You think of [space] as two dimensions only; but I have come to announce to you a third – height, breadth, and length.

*I*: We also speak of length and height, or breadth and thickness, thus denoting two dimensions by four names.

*Stranger*: I mean not only three names, but three dimensions.

*I*: [E]xplain to me in what direction is this third dimension.

*Stranger*: I came from it. It is up above and down below.

*I*: Northward and Southward.

*Stranger*: I mean a direction in which you cannot look, because you have no eye in your side...In order to see into space you ought to have an eye, not on your perimeter, but on your side, that is, on what you probably call your inside; but we in Spaceland call it your side....by height, I mean a dimension like your length.

You are living on a plane. What you style Flatland is the vast level surface of what I may call a fluid on, or in, the top of which you and your countrymen move about, without rising above it or falling below it. I am not a plane figure, but a solid. You call me a circle; but in reality I am not a circle, but an infinite number of circles, of size varying from a point to a circle of thirteen inches in diameter, one placed on top of the other. When I cut through your plane as I am doing now, I make in your plane a section which you, very rightly, call a circle. For even a sphere – which is my proper name in my own country – if he manifests himself at all to an inhabitant of Flatland – must manifest himself as a circle.

[D]o you not remember, I say, how, when you entered the realm of Lineland, you were compelled to manifest yourself to the King, not as a square, but as a line, because that linear realm had not dimension enough to represent the whole of you, but only a slice or section of you? It is precisely the same way your country has two dimensions but is not spacious enough to represent me, a being of three, but can only exhibit a slice or section of me, which is what you call a circle.

Tell me, Mr. Mathematician; if a point moves northward and leaves a luminous wake, what name would you give to the wake?

I: A straight line.

Stranger: And a straight line has how many extremities?

I: Two

extension

## Cross-Sections

Stranger: Now conceive the northward straight line moving parallel to itself, east and west, so that every point in it leaves behind it the wake of a straight line. What name will you give to the figure therby formed? We will suppose it moves through a distance equal to the original straight line.

I: A square.

Stranger: And how many sides has a square? How many angles?

I: Four sides and four angles.

Stranger: Now stretch your imagination a little, and conceive a square in Flatland, moving parallel to itself upward...out of Flatland altogether....One point produces a line with two terminal points. One line produces a square with four terminal points. Now, 1, 2, 4, are evidently in geometric progression. What is the next number?

I: Eight.

Stranger: Exactly. The one square produces a *something-which-you-do-not-yet-know-a-name-for-but-which-we-call-a-cube*.

Abbot, Edwin A.; *Flatland*; 1884; excerpt from pp. 78 - 87

Write a paragraph or two explaining Spaceland (our world of three-dimensions) in terms an outsider, who comes from a world of four dimensions, would understand. Use the passages above to guide your explanation.

## Try a Different Point of View

input

Designers, architects, engineers, and craftsmen must all be skilled in creating and interpreting three-dimensional drawings. However, anyone who has assembled a piece of furniture, a bike, or even a gas grill has interpreted three-dimensional drawings. In this section, you will create and interpret three dimensional drawings. Most of the sketches in this section are created using a special type of graph paper called **isometric dot paper**. Isometric dot paper makes the creation and interpretation of three-dimensional pictures easier. The dots on the paper are at 30° angles, so most shapes look as if they have a depth associated with them. A sample of a cube drawn on isometric dot paper is shown below. Note that the dots form diagonal rows.

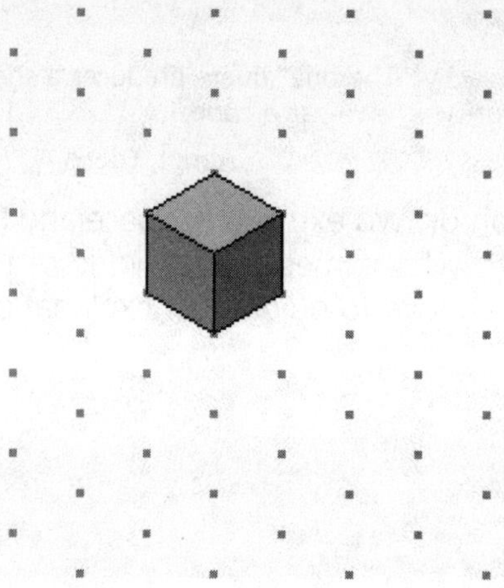

The cube shown here displays the top face with the lightest gray. You can pick either the middle or darkest shade of gray as the front face of this cube. Let the left face be the middle shade of gray. Then the front visible face is the darkest shade.

If you view this cube strictly from the top, you will see only the top face, a square as is shown below.

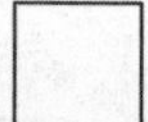

If you view this cube from the front or from the left, you will also see only one face. That faces is a square congruent to the square shown above. Most objects have different views when shown from the top, left, and front. These views are called **elevations.** The hidden faces of three-dimensional shapes are usually not shown with isometric dot paper.

Some of the activities in this section will provide front, top, and left or right views for you. These activities will ask you to construct the drawing or the physical object based on the three views given to you. The next page of this text is a sample of isometric dot paper. Use it for the explorations you will perform in this section. Your teacher will provide you with additional sheets of isometric dot paper as you need it.

process: model

## Try a Different Point of View

1. For this first set of questions, you are given a drawing on isometric dot paper. Choose or construct the different views or elevations. Use the darkest gray color as the front of the object.

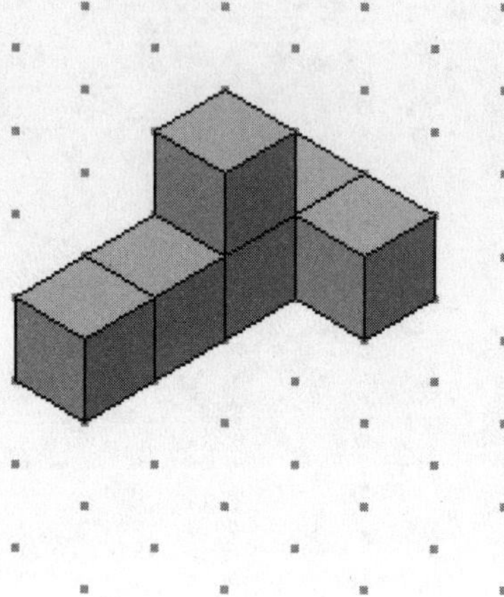

a. Which choice is the front view of the object above? Justify your response.

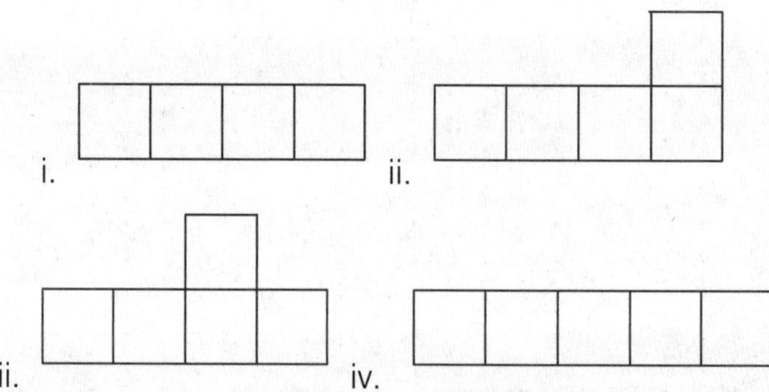


b. Label each of the blocks in the correct front view with the depth of that section of the structure. For example, if there are three block behind one another, the depth is three, as is shown for picture below. The top view shows the height.

| 2 | 1 |
|---|---|
| 3 | 3 |

Front View

| 2 | 1 | |
|---|---|---|
| 2 | 2 | 2 |

Left View

| 2 | 1 | 1 |
|---|---|---|
| 2 | 2 | 1 |

Top View

process: model

## Try a Different Point of View

c. How many total blocks are used to construct the structure?

d. Which is the top view for the structure from question 1? Justify your choice.

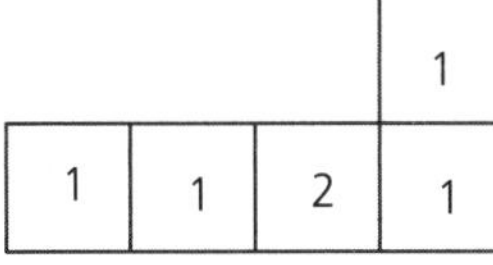

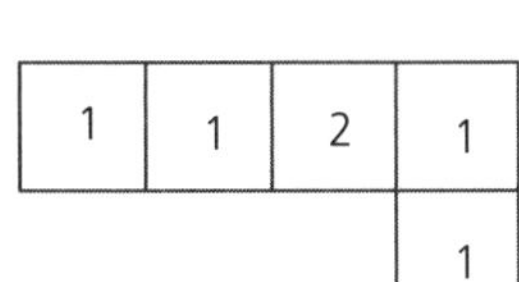

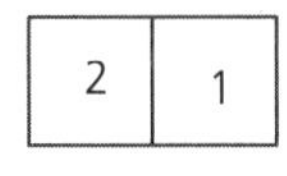

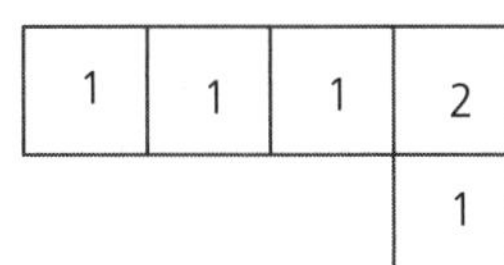


e. Choose the left view for the structure. Label the correct elevation with the depths. Justify your choice.

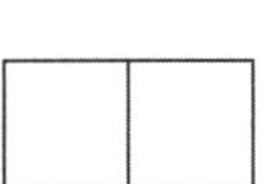
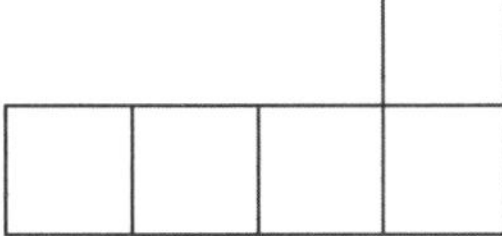

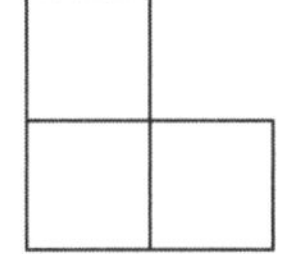

2. Given the structure below, create the top, left, and front views. Show the height or depth of each view on the elevations you create.

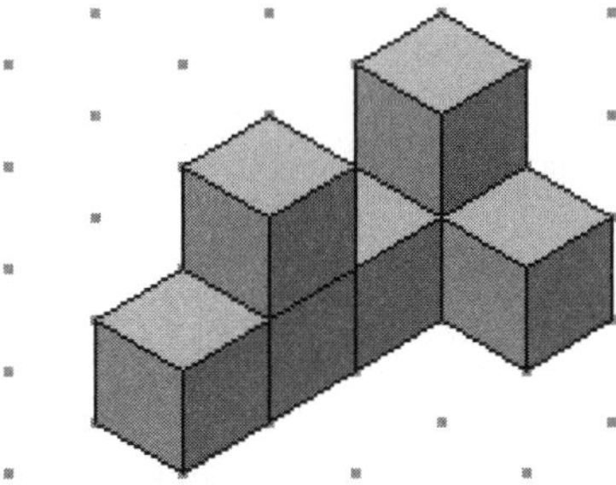

## Try a Different Point of View

process: model

3. You are given elevations for a structure below. Use your isometric dot paper to create a three-dimensional drawing of the structure.

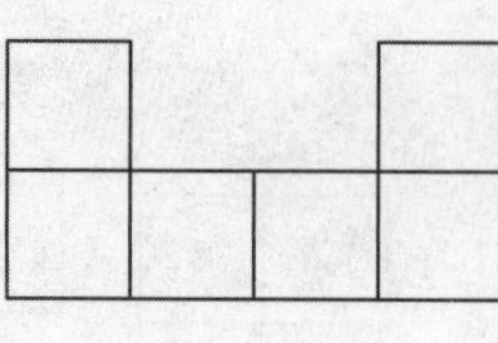

Front View

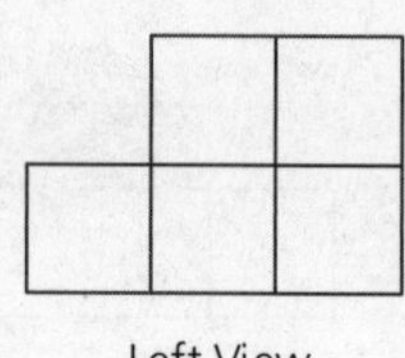

Left View

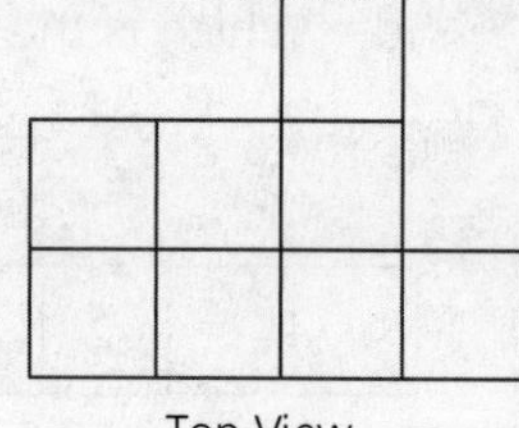

Top View

a. How many stories tall is the structure? Which view(s) gives this information?

b. How deep is the structure from front to back? Which view(s) give this information?

c. How wide is the structure? Which view(s) give this information?

d. Where are the tallest parts of the building? How do you know?

4. You are given the top view of a structure below. The elevation is labeled with heights. Use your isometric dot paper to create at least one possible configuration for the structure.

| | | |
|---|---|---|
| 4 | 2 | 3 |
| 3 | | 1 |

## Try a Different Point of View

process: summary

5. What was difficult for you in this activity?

6. What was least difficult for you in this activity?

7. You have been working on developing your spatial reasoning skills in the last two sections of text. Of what use are good spatial reasoning skills?

## Attack of the Sixty Foot Teenager

input

In a recent episode of *Sweetie, I Enlarged the Kids* on the children's television network Walt's Worldwide Network, the family's teenage son, Bill was enlarged. Originally, 6 feet tall, the dysfunctional time machine enlarged his height to 10 times normal. As you might imagine, there are some potential problems associated with being 10 times your normal height. Aside from not being able to fit through the front door, you would need different clothes, and much more food.

## Attack of the Sixty Foot Teenager

process; evaluate

1 Calculate Bill's height after being enlarged. Show your work.

2. Bill's torso measured 23 inches from the base of his neck to his bellybutton and almost 20 inches across. Calculate the area of Bill's chest before and after the enlargement. Show your work.

3. Bill's friend, Fausto, bought him a baseball hat for his birthday. The circumference of Bill's head before the enlargement was approximately 26 ½ inches. The hat Fausto bought was in the shape of a hemisphere with the same radius as Bill's head. Calculate the amount of material used for the hat.

analysis

## Attack of the Sixty Foot Teenager

4. Fausto reasons that the new circumference of Bill's head is 265 inches…so he needs a new hat. There are no hat shops that make hats with 265-inch circumferences. Fortunately, Fausto's grandmother can sew a new hat for Bill. Since Bill is 10 times his original height, Fausto reasons that his grandma should buy 2650 square inches of material. Where is the flaw in Fausto's reasoning?

process: evaluate

## Attack of the Sixty Foot Teenager

5. The depth of Bill's torso is about 10 inches. Calculate the original volume of Bill's torso and the new volume of Bill's torso.

6. Since the new volume of Bill's torso is so large, it is not practical to measure it in cubic inches. There are approximately 231 cubic inches per gallon. What is the volume of Bill's large torso in gallons?

## Attack of the Sixty Foot Teenager

analysis

7. The average human body contains about 5 liters of blood. Based on the enlarged volume of Bill's body, how much blood should circulate through his body?

8. Is it reasonable for Bill's heart to pump this volume of blood through his body? Explain your response.

9. Bill's foot is a size 12. It measures 12″ by 4″ before his size change. What are the dimensions and area of Bill's newly enlarged foot?

10. Bill weighed 200 pounds before the machine enlarged him. If you assume weight depends on the three dimensions of your body, how much will Bill weigh after the machine enlarges him?

11. How much pressure per square inch did Bill's feet absorb before he "grew?"

## Attack of the Sixty Foot Teenager

analysis

12. How much pressure per square inch do Bill's enlarged feet absorb with respect to his increased weight?

13. Is it reasonable to expect Bill's feet to absorb such pressure?

14. In reality it is physically impossible for someone to be 60 feet tall. Explain why. Consider bodily functions as well as forces outside the body.

# Unit 10: Compositions of Transformations

## Contents

## Using Matrices with Transformations

process: practice

In Integrated Math II, you used a variety of different methods and notations to represent transformations. In this section, you will expand your ability to represent transformations by using matrices.

1. Before moving on, review the work you have done on transformations. The simplest representation involves using a prime or double prime with each of the vertices of a transformed figure. Use that notation to describe the transformation.

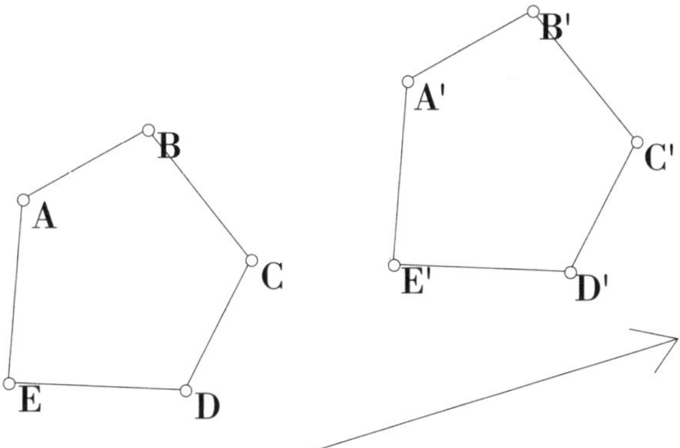


2. You also learned a shorthand way of writing about the transformation types. This notation used lowercase or capital letters, like **r**, or **T,** or **S** to represent the transformation. Use that notation to describe each of the following four transformations.

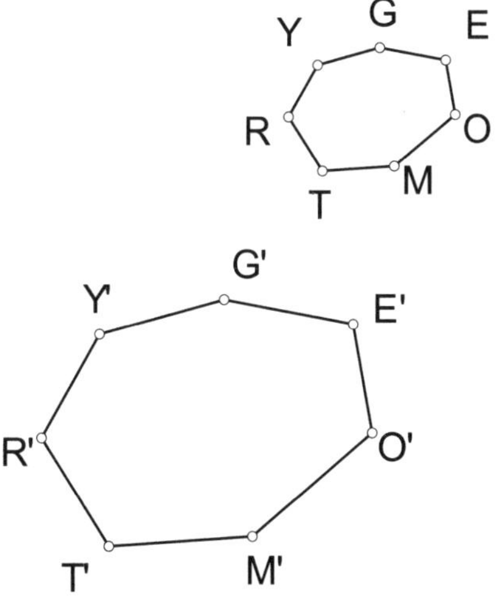


a.

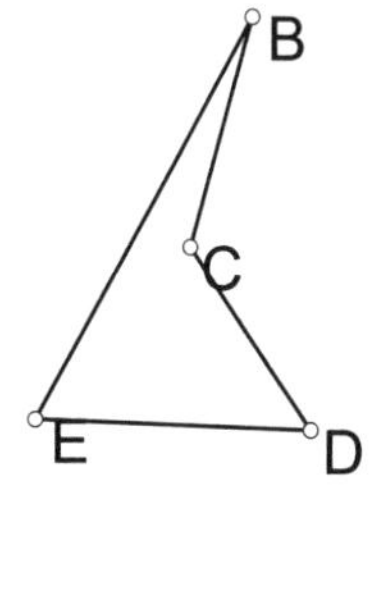


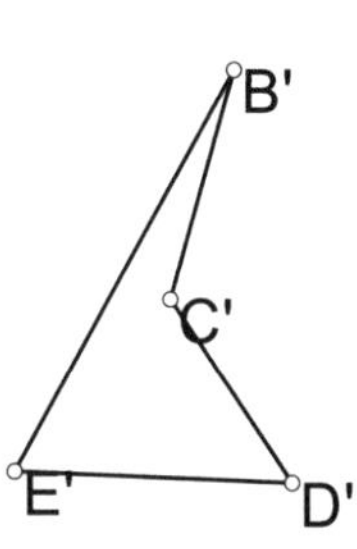


b.

process: practice

## Using Matrices with Transformations

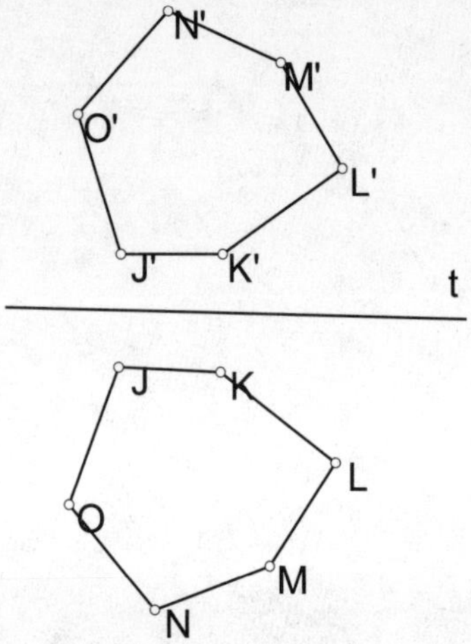

c.

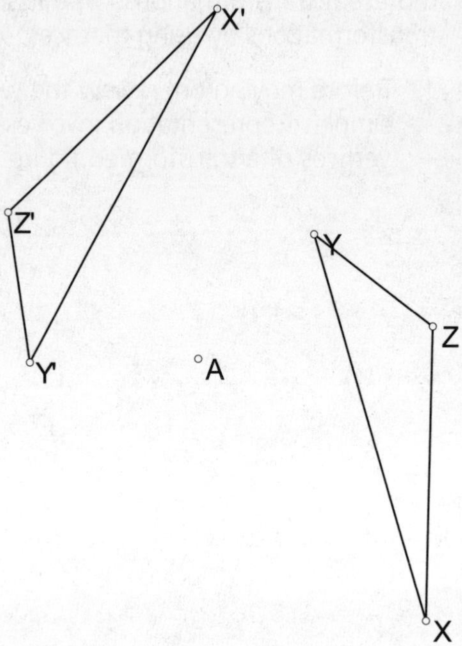

d.

3. When transformations take place in the coordinate plane, you can use the mapping notation for each vertex. Sometimes you can generalize the mapping notation so that it applies to all points. Then, you can use (x, y) to represent any point or all points of the figure. Use this generalized mapping notation to describe each of the transformations below. If you have a hard time generalizing, write the mapping for each of the vertices and generalize from these specific points.

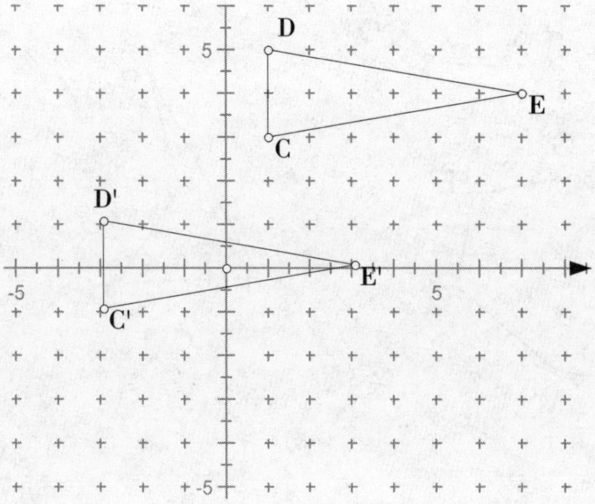

a.

process: practice

## Using Matrices with Transformations

b.

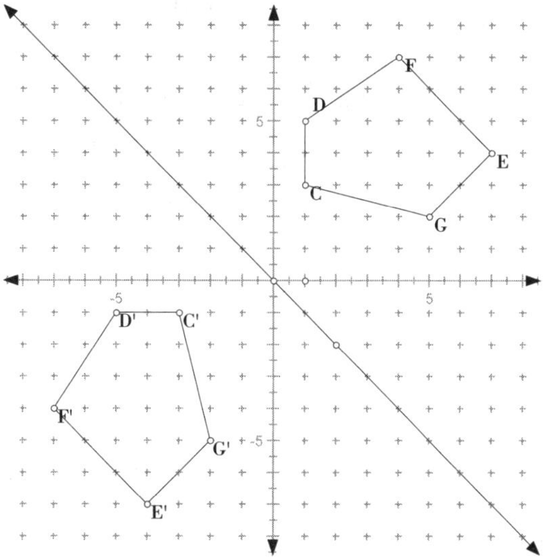

c.

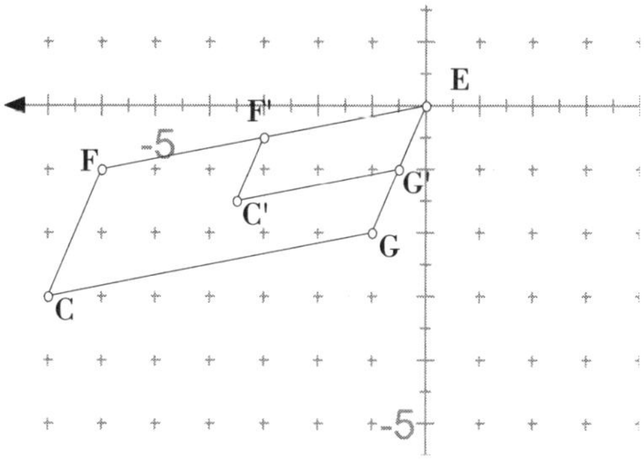

d.

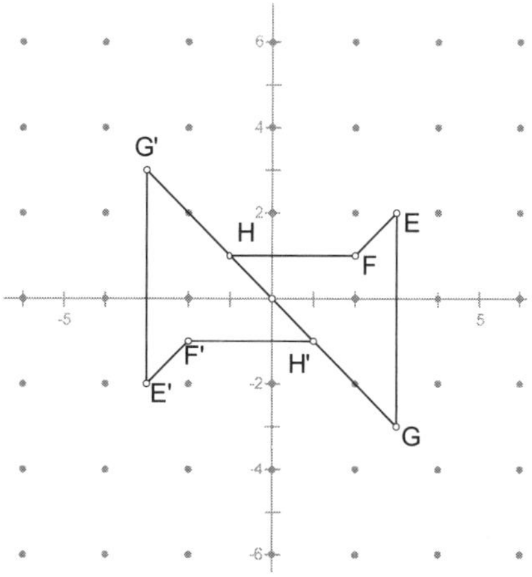

input

## Using Matrices with Transformations

A matrix is another representation that can be used when transformations take place in the coordinate plane. A **matrix** is simply a rectangular arrangement of numbers, ordered in rows and columns.

When using a matrix to represent a point, like (9, 13), write the point so that the x-coordinate is in row 1, column 1, and the y-coordinate is in row 2, column 1. So, the matrix becomes a 2 row by 1 column (2 x 1 – say "2 by 1") matrix. This matrix is also called a "column matrix" because it only has one column.

The column matrix $\begin{bmatrix} 9 \\ 13 \end{bmatrix}$ would represent the point (9, 13).

process: model

## Using Matrices with Transformations

4. Write a matrix to represent the point (-5, 7).

5. Write a matrix to represent the origin.

6. Write a matrix to represent any general point.

7. Create a triangle in the coordinate plane so that no vertices are on the x- or y-axes. Use a matrix to represent the coordinates of the vertices in the triangle?

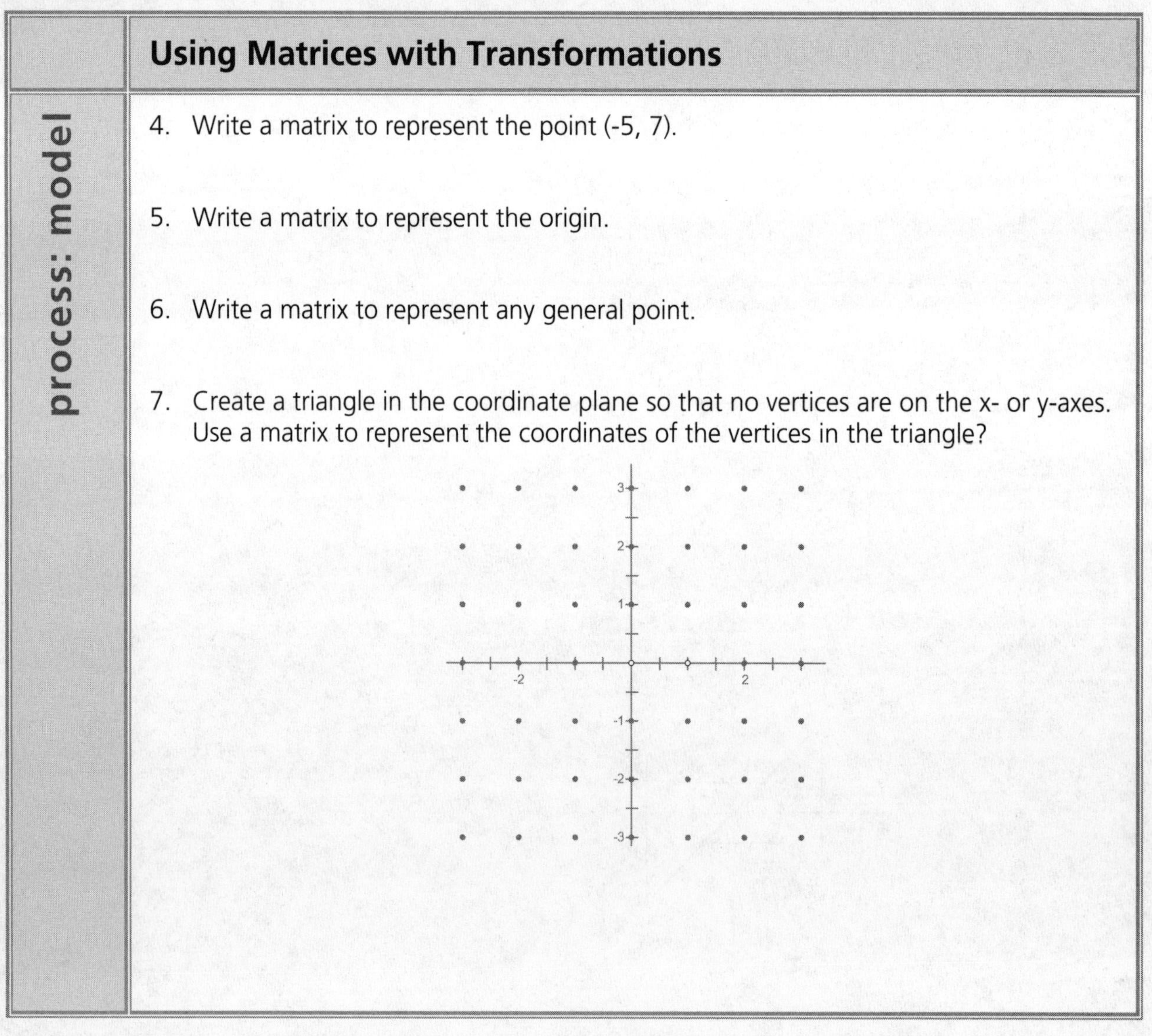

input

## Using Matrices with Transformations

When you were asked to use the mapping notation with (x, y) to represent transformations, you examined a few points before making your conjecture about the generalization. Use the same method with matrices and transformations. If you know what happens to two points in the coordinate plane, then you can make a generalization about all the points involved in the transformation. You can use *any* two points, but certain points are better to use than others.

process: model

## Using Matrices with Transformations

8. Theorize which points are the "key" points in the coordinate plane? In order to know what happens to (x, y), you must know what happens to x and you must know what happens to y. Aside from the origin, what is the simplest point along the x-axis? What about along the y-axis?

9. Represent these two points as column matrices.

10. Use these two column matrices to form one 2 x 2 matrix, writing the point along the x-axis as the first column.

process: evaluate

## Using Matrices with Transformations

This new matrix is called the **identity matrix**.

11. Try multiplying this matrix by any point of your choosing. What is the result?

12. Multiply this matrix by the 2 x 3 matrix you created to represent the triangle in the coordinate plane on page 10-6.

process: evaluate

## Using Matrices with Transformations

13. Where do all of the x-coordinates appear in the matrix for the triangle?

14. Where do all the y-coordinates appear in the matrix for the triangle?

15. Next, multiply this matrix by the general point (x, y).

$$\begin{bmatrix} 1 & 0 \\ 0 & 1 \end{bmatrix} * \begin{bmatrix} x \\ y \end{bmatrix} =$$

16. In each case, did you get what you expected?

input

## Using Matrices with Transformations

**Transformation matrices** rely on 2 x 2 matrices like the identity matrix.
A transformation matrix can be formed by discovering where the key points, (1, 0) and (0, 1), map under your transformation and then creating a new 2 x 2 matrix from the results.

process: model

## Using Matrices with Transformations

Use a *reflection over the x-axis* as an example. (If you need to, plot these points on a coordinate grid and reflect them to answer the following.)

17. $\mathbf{r}_{x\text{-axis}}(1, 0) = ($ $\quad\quad)$ Write this result as the first column in the transformation matrix.

18. $\mathbf{r}_{x\text{-axis}}(0, 1) = ($ $\quad\quad)$Write this as the second column in the transformation matrix.

$$\begin{bmatrix} \quad & \quad \\ \quad & \quad \end{bmatrix}$$

process: model

## Using Matrices with Transformations

19. Multiply your transformation matrix by any point that you wish to reflect over the x-axis. Keep in mind what the result should be. (Do not pick a point on either of the axes. If you aren't sure which point should be the result, perform the reflection of your point in the coordinate plane.)

$$\begin{bmatrix} \quad & \quad \\ \quad & \quad \end{bmatrix} * \begin{bmatrix} \quad \\ \quad \end{bmatrix} =$$

The column matrix you got as the result represents the reflection of your chosen point over the x-axis.

20. Use the same transformation matrix to reflect the triangle on 10-6 over the x-axis.

$$\begin{bmatrix} \quad & \quad \\ \quad & \quad \end{bmatrix} * \begin{bmatrix} \quad & \quad & \quad \\ \quad & \quad & \quad \end{bmatrix} =$$

Your result should be the matrix to represent the vertices of the triangle reflected over the x-axis.

analysis

## Using Matrices with Transformations

21. What happened to each of the entries in the top row of your resultant matrix?

22. What does this show you about the vertices of the image triangle?

23. What happened to each of the entries in the bottom row of your resultant matrix?

24. What does this show about the vertices of the image triangle?

process: model

## Using Matrices with Transformations

25. Now use the same transformation matrix to calculate what the general result of reflecting over the x-axis should give.

$$\begin{bmatrix} & \\ & \end{bmatrix} * \begin{bmatrix} x \\ y \end{bmatrix} =$$

This result should agree with the mapping notation you used to represent the reflection over the x-axis. (x, y) → (x, –y)

process: evaluate

## Using Matrices with Transformations

Try finding the transformation matrix for a reflection over the line **y = x**.

26. Where does (1, 0) map when it is reflected over the line y = x?

27. Where does (0, 1) map when it is reflected over the line y = x?

28. Create your 2 x 2 transformation matrix from the results you just got for the two previous questions.

29. Test that you have the correct matrix by multiplying this matrix by any chosen point. How does multiplication show you whether your transformation matrix is correct or not?

process: evaluate

## Using Matrices with Transformations

30. Confirm that you have the correct matrix by multiplying this matrix by your triangle (page 10-6) matrix. How does multiplication confirm whether your transformation matrix is correct?

31. Now use the same transformation matrix to calculate the general result of reflecting over the line $y = x$. Make sure your results match with the points you get when actually performing the reflection in the coordinate plane. Does this result agree with the result you obtained using the mapping notation? How do you know?

input

## Using Matrices with Transformations

Like reflections, transformation matrices can also represent **rotations**. However, rotations of certain degrees are more easily represented than others.

Which rotations do you think we will represent with transformation matrices?

process: evaluate

## Using Matrices with Transformations

Hopefully, one of your responses above was a half-turn or 180° rotation. To find the transformation matrix for a 180° rotation, use the same points as with reflections.

32. Where does (1, 0) map under a half-turn?

33. What about (0, 1)?

34. Show your transformation matrix for a half-turn below.

35. Use the triangle you created on page 10-6 to confirm that your result is correct by multiplying your transformation matrix by this triangle's matrix.

$$\begin{bmatrix} \quad & \quad \\ \quad & \quad \end{bmatrix} * \begin{bmatrix} \quad & \quad & \quad \\ \quad & \quad & \quad \end{bmatrix} =$$

36. Use the same transformation matrix to calculate the general result of rotating by 180°, by multiplying by the matrix for the point (x, y). Does your result here agree with the result from using the mapping notation in the section on rotations? Explain how you know.

process: evaluate

## Using Matrices with Transformations

Find the transformation matrix for a rotation of 90° around the origin.

37. Where does (1, 0) map under a quarter-turn?

38. Where does (0, 1) map under a quarter-turn?

39. Show your transformation matrix for a 90° turn below.

process: evaluate

## Using Matrices with Transformations

40. Use a triangle or any polygon you have graphed whose vertices are not on either axis to confirm that your result is correct. Do this by multiplying your transformation matrix by the polygon's matrix.

$$\begin{bmatrix} & \\ & \end{bmatrix} * \begin{bmatrix} & & \\ & & \end{bmatrix} =$$

41. Use the same transformation matrix to calculate the general result of rotating by 90°, by multiplying by the matrix $\begin{bmatrix} x \\ y \end{bmatrix}$. Does your result here agree with the result from using the mapping notation? Explain how you know.

input

## Using Matrices with Transformations

Back at the CLI Company, where you are still working as the webmaster, more requests keep rolling in. For your next project, a gif must appear to blink on the computer screen, going from small to large to small again each second.

What transformation represents this change?

In the process of completing your next project, you start dabbling with a few different computer software programs. One of these uses matrices as an input to determine the location and size of the image file. Since you already know about reflections and rotations using matrices, you decide to learn about size changes too.

How are dilations different from reflections or rotations?

The transformation matrices for dilations about the origin are also a bit different. However, you will still use the idea of key points for finding all you need to represent dilations with matrices.

process: model

## Using Matrices with Transformations

42. Form the quadrilateral A(0, 0); B(1, 0); C(0, 1); D(2, 2). Dilate it by a factor of 2. Create the dilated quadrilateral and provide the coordinates of each of the vertices.

process: evaluate

## Using Matrices with Transformations

43. Under this dilation, (1, 0) → (        ), and (0, 1) → (        ).

44. So the transformation matrix for a dilation of scale factor 2 (with respect to the origin) is: $\begin{bmatrix} & \\ & \end{bmatrix}$.

45. Dilate the original ΔABC by a factor of ½ with respect to the origin on the coordinate grid. Provide the coordinates of each of the new vertices.

46. What is the transformation matrix for this dilation? $\begin{bmatrix} & \\ & \end{bmatrix}$

process: summary

## Using Matrices with Transformations

47. Describe what happens to (1, 0) under these dilations.

48. Describe what happens to (0, 1) under these dilations.

49. If you dilated by a scale factor "a", what would the transformation matrix become for this dilation?

analysis

## Using Matrices with Transformations

For the project you have as Webmaster, you decide to take the original figure and enlarge it to five times its original size. Then, you will make the figure half of its original size, before bringing it back to its original size again and repeating the enlargement. This will create the blinking effect you are looking for. What transformation matrices should you use to complete this process?

## Using Matrices with Transformations

input

For each of the previous transformation matrices, you used matrix multiplication to transform the original point or figure. With translations, you do not need to use matrix multiplication, nor do you need to use a 2 x 2 transformation matrix.

## Using Matrices with Transformations

process: model

50. Create a new triangle in the coordinate plane.

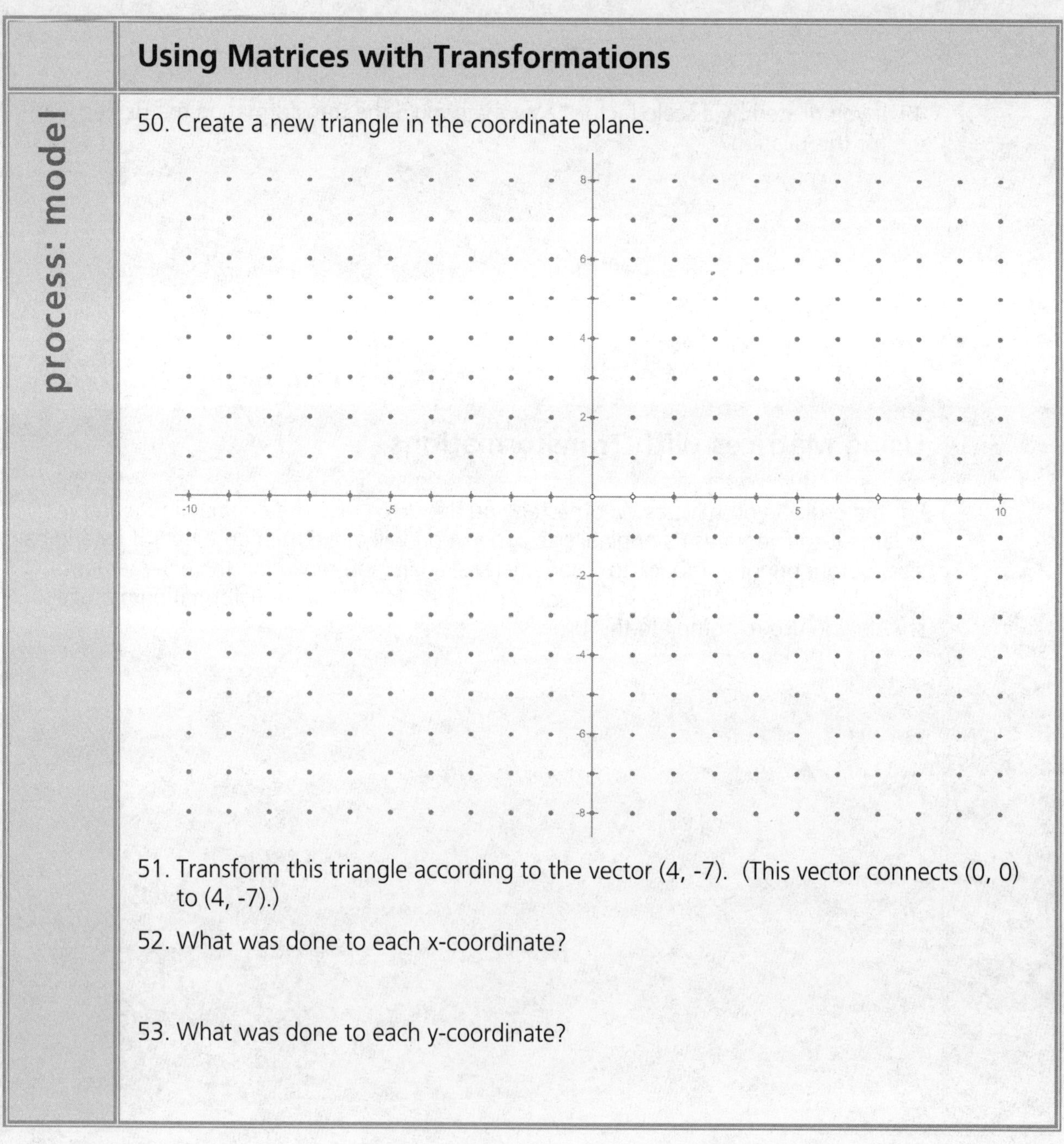


51. Transform this triangle according to the vector (4, -7). (This vector connects (0, 0) to (4, -7).)

52. What was done to each x-coordinate?

53. What was done to each y-coordinate?

input

## Using Matrices with Transformations

The most straightforward technique for representing the translation of a point is: add the x-coordinate of the point to the horizontal component of the vector, and add the y-coordinate of the point to the vertical component of the vector. Therefore, matrices for translations are column matrices that come from the vectors. This column matrix can be added to the matrix for your point or figure.

process: model

## Using Matrices with Transformations

54. Do this with your triangle.

$$\begin{bmatrix} \quad & \quad & \quad \\ \quad & \quad & \quad \end{bmatrix} + \begin{bmatrix} 4 \\ -7 \end{bmatrix} =$$

55. Does your matrix agree with the result you got when you translated the triangle in the coordinate plane? Explain why.

process: practice

## Using Matrices with Transformations

56. Use the mapping (x, y)→(x-3, y+6) to transform your triangle on the coordinate grid.
57. Represent this translation with a column matrix.
58. Perform the matrix addition with the coordinates of the vertices of your triangle to confirm that you plotted the correct points.

## Using Matrices with Transformations

process: summary

59. Cite at least two similarities between using matrices to represent transformations and using mappings to represent transformations.

60. Cite at least two differences between using matrices to represent transformations and using mappings to represent transformations.

## Using Matrices with Transformations

analysis

61. What is an advantage to using matrix representations for transformations?

62. What is a disadvantage to using matrix representations for transformations?

## Glide Reflections

input

You decided to enter the miniature golf competition in your town. (You never dreamed you would get this far, but there you were, putting for the championship.) You were on the 18th hole at the tee. If you could somehow manage to get a hole-in-one, you would beat the reigning champ by one stroke, unseating him from a title he had held for three years running. The trophy (and free miniature golf for a year) would be yours for the taking!

Winning was going to require some really good strategy. You knew the layout of the hole well. It is pictured below.

**Hole Number 18**

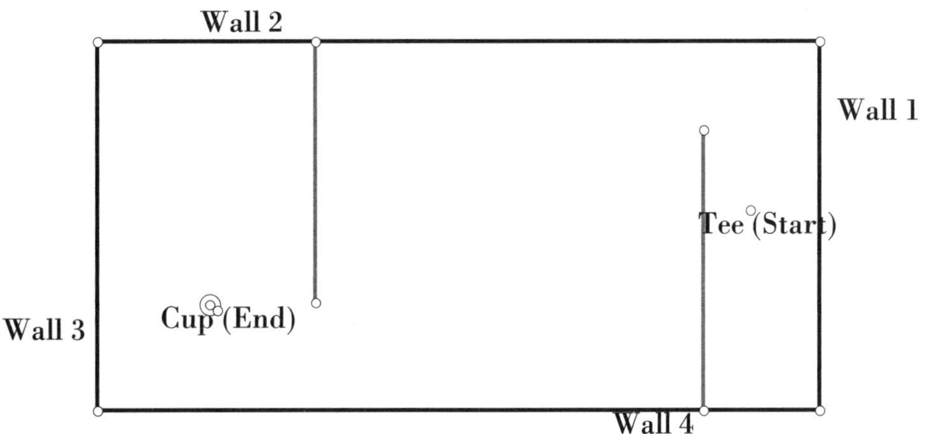


*Part 1:*

You had played this hole many times before and knew you had to bank the ball off some walls to get the hole-in-one, but you had never managed to do it before. You took a deep breath, stepped back, and decided to let geometry be your guide.

1. Obviously, a direct shot is out of the question. Why?

## Glide Reflections

input

Keeping in mind that the shortest distance would be a direct shot, your next options involve banking the ball off one of the walls. If you envision reflecting the cup over one of the walls, you can shoot for the reflected cup and perhaps you can make a hole-in-one. This strategy had just worked for you on hole number 17, which is pictured below.

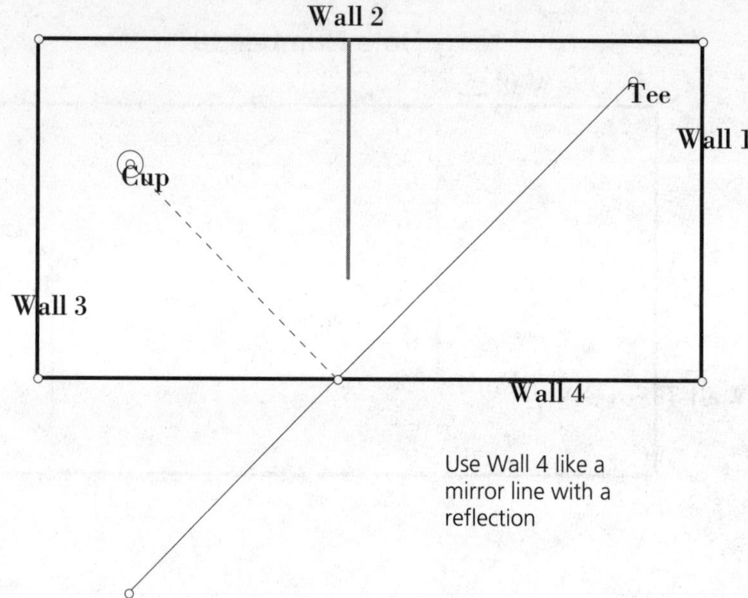


On #17 you banked the shot off wall 4 as you aimed for the reflection of the cup over that wall. The ball was heading for the reflection of the cup until it hit the wall.

## Glide Reflections

analysis

2. If you think about *the golf shot*, what is the dotted segment representing?

3. If you think about *geometry*, what is the dotted segment representing?

4. What happened after the ball hit the wall?

5. Measure the angle formed between your shot and wall 4 as your ball hits the wall, called the **angle of incidence**. Record this measure on the diagram.

6. Measure the angle formed between your shot and wall 4 as your ball leaves the wall, called the **angle of reflection**. Record this measurement.

7. Why did this shot work to get a hole-in-one on #17?

8. Pick a wall on the 18th hole (pictured two pages ago) and try this strategy.

9. Will banking the shot off one wall on the 18th hole work to get you a hole-in-one? Explain, making sure to use complete sentences.

## Glide Reflections

input

*Part 2*

A strategy that is a little more complicated, but might work on hole number 18, is to bank the shot off two walls. You used this strategy back on the first hole of the course. That was a good start in the contest.

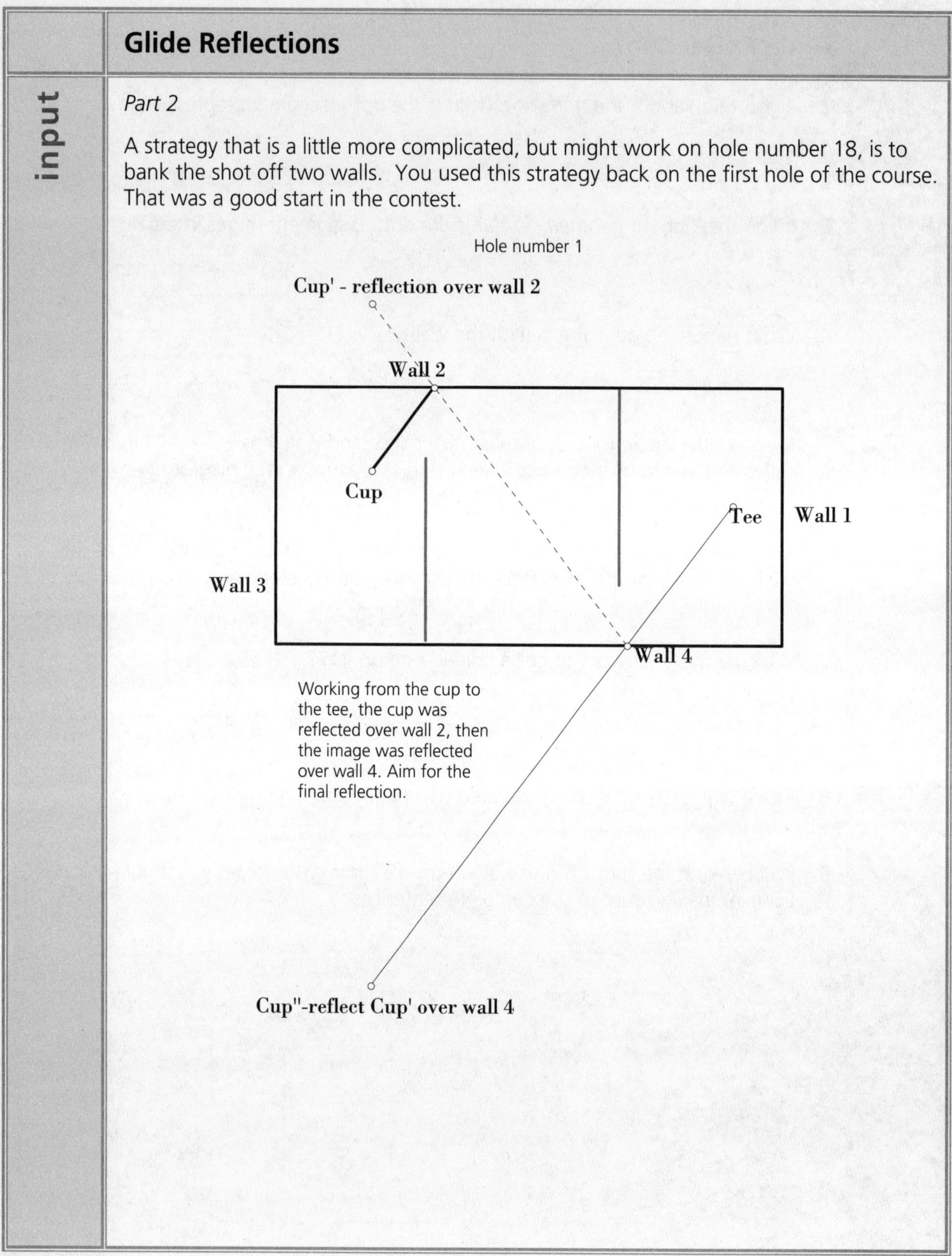

## Glide Reflections

analysis

10. Thinking *geometrically*, what is Cup′ with regard to Cup?

11. What is Cup′′ with regard to Cup′?

12. What is Cup′′ with regard to Cup? (Hint: What is a double reflection over parallel lines?)

13. Still thinking *geometrically*, what is the dotted segment representing?

14. What is the bold segment representing?

15. Measure the two angles formed by the path of your ball and wall 4. Record them on the diagram of hole number 1.

16. Measure the two angles formed by the path of your ball and wall 2. Record them on the diagram.

17. Explain *mathematically* why this shot worked to get you a hole-in-one on the 1st hole.

18. Give this strategy a try on hole #18, which is copied for you below. You may want to try a couple times with different pairs of walls. You do *not* necessarily have to use parallel walls. (If you use intersecting walls, will you still have a translation? If not, what transformation will you have performed?)

**Hole Number 18**

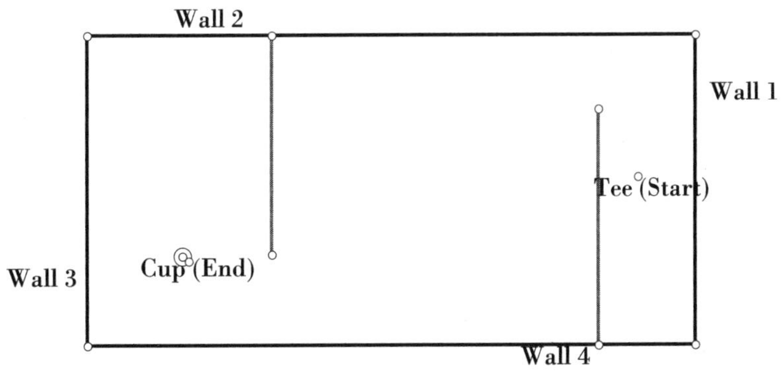

analysis

## Glide Reflections

19. Did any of the pairs of walls give you a hole-in-one? If so, explain mathematically why the walls you picked worked. If not, which of the pairs of walls comes the closest to giving you a hole-in-one?

input

## Glide Reflections

*Part 3*

The greatest number of walls you have ever used to get a hole-in-one is three. That was on a very different course, but by now, you know that using three reflections is your only option on hole number 18. So, the question becomes, which walls should you use and in which order?

In thinking back to your hole-in-one at Lukewarm Breezes, you remember banking off the two side walls first and then the back wall.

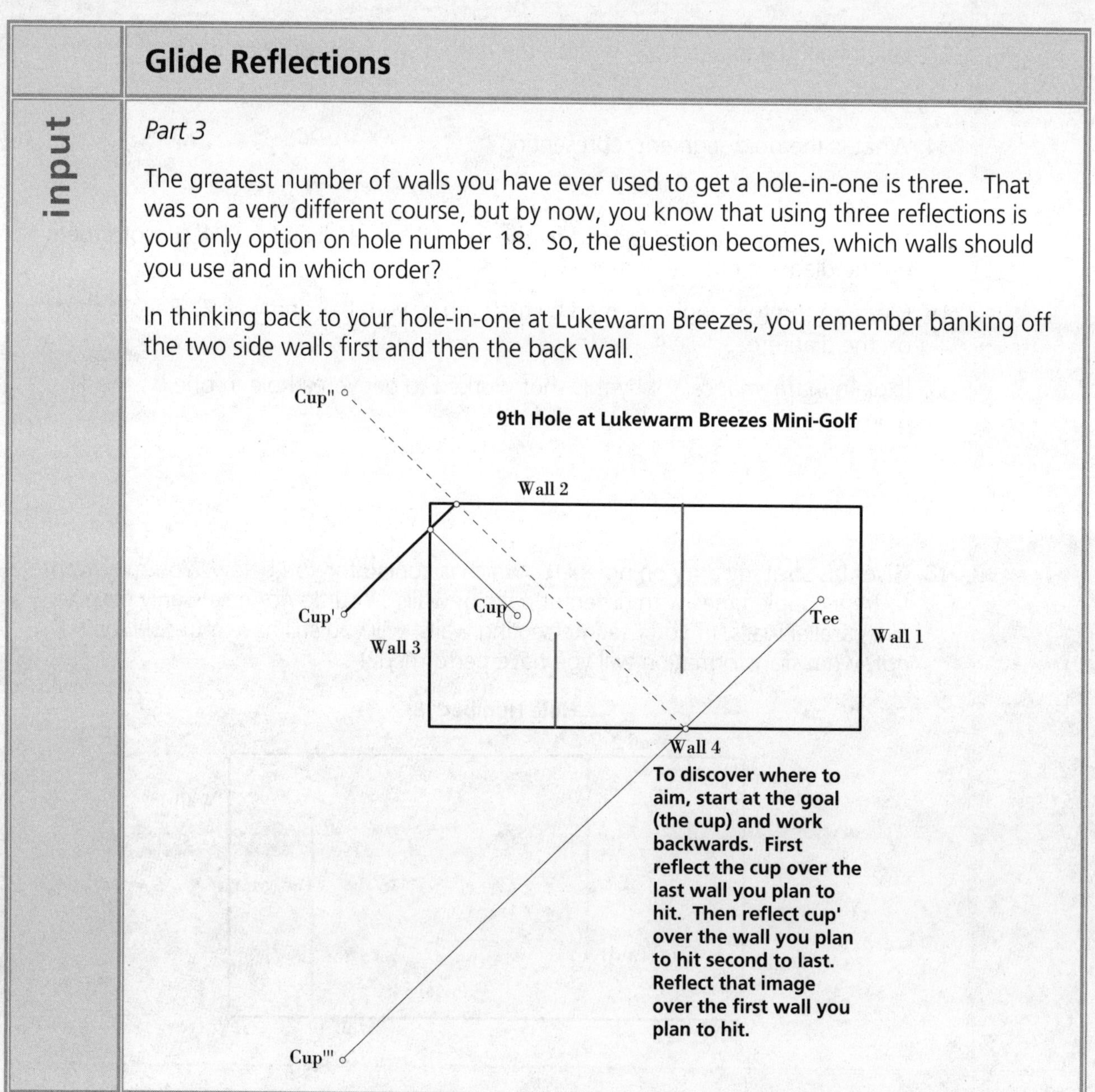

## Glide Reflections

analysis

20. What *geometric* relationship does the back wall have to the two side walls?

21. Thinking *geometrically*, what is Cup′ with regard to Cup?

22. What is Cup′′ with regard to Cup′?

23. What is Cup′′′ with regard to Cup′′?

24. What is Cup′′′ with regard to Cup′? (Hint: What is a double reflection over parallel lines?)

25. The relationship between Cup and Cup′′′ is the combination of three reflections *or* which two transformations?

When you combine a translation (double reflection over two parallel lines) with another reflection over a line that is *perpendicular* to the other two, as is done here, the transformation can be called a **glide reflection.**

26. Still thinking *geometrically*, what is the dotted segment representing?

27. What is the bold segment representing?

28. What should be true about the angles formed where your shot intersects wall 3? … wall 2? … wall 4? Measure these angles.

## Glide Reflections

analysis

29. Explain *mathematically* why this shot worked to get a hole-in-one at Lukewarm Breezes.

30. This strategy should work on hole #18, which is copied for you below. You may need to try a couple times with different combinations of walls. You should try to use the side walls and one of the walls perpendicular to them. Remember that this shot is for all of the marbles!

**Hole Number 18**

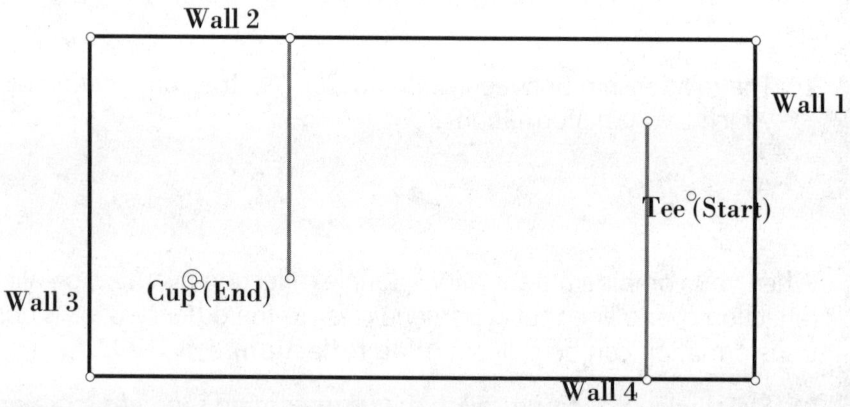

Wall 2
Wall 1
Tee (Start)
Wall 3
Cup (End)
Wall 4

## Glide Reflections

extension

31. After you make this shot, the owner of the course approaches you to congratulate you and give you the trophy. He poses a challenge question to you. "Is there more than one way to use a glide reflection to get a hole-in-one on this hole?"

    First you think, "I didn't know Mr. Lopez knew about geometry." Then, you try to answer his question.

process: summary

## Glide Reflections

32. Can you think of any other games where this strategy of using reflections or combinations of reflections might come in handy? What are they?

33. Give one example and illustrate how you could use this strategy in your example.

34. In the miniature golf scenario, you learned about one more type of transformation that is a combination of reflections. What is this new type of transformation?

35. Is this transformation an isometry? Explain. If so, does it maintain orientation? Why or why not?

36. Use a glide reflection to transform the polygon pictured below.

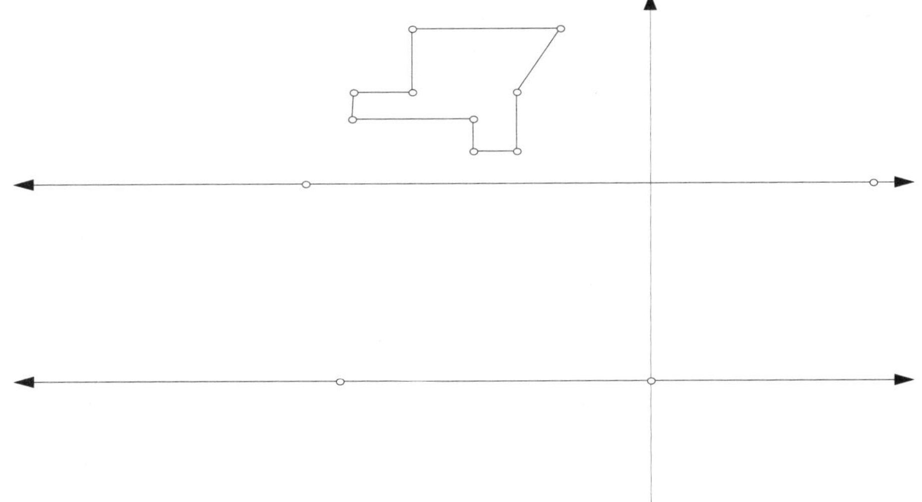

## Tessellations

input

Artists like Mondrian, Picasso, Vasarely, and M.C Escher used geometric patterns in their art. Escher, in particular, used geometry that is directly linked to what you are studying right now, transformations. Escher used transformations of shapes that would create **tessellations**. Tessellations or tilings use shapes to cover a plane without any overlapping and without leaving any gaps between the shapes. For example, the illustration below is an M.C. Escher design called *Reptiles*. In this picture, you see that the reptile design could be continued without end and without space between each subsequent reptile.

In this activity, you will discover which polygons will tessellate and which will not. You will also create your own tessellation, similar to Escher.

process: model

## Tessellations

Creating a tessellation or tiling pattern can involve reflections, translations, rotations, or glide reflections. Start with the simplest polygon, a triangle, and attempt to tile the plane with it by following these directions.

1. On a separate piece of paper, use your straightedge to create a scalene triangle of your choosing.
2. Measure each of the angles in the triangle and record the measures on the triangle.
3. Cut out the triangle and use it as a template to create about 10 more copies.
4. Color one side (the same side) of each triangle so that one side of the paper is blank and one side is colored.
5. Use only rotations and translations to tile a part of your desktop using only your own triangle copies.
6. Create a scaled-down sketch of this tessellation below.

7. Pick one vertex and measure the angles in your tiling that surround that vertex. Record the measures on your sketch.
8. Pick another vertex and do the same.

process: model

## Tessellations

9. Create a different tiling of the plane, but use reflections as a part of this design.
10. Create a sketch of this tiling below.

11. Again, pick a vertex and measure the angles in your tiling that surround that vertex. Record the measures on your sketch.

analysis

## Tessellations

12. What do you notice about the angles surrounding any vertex point in your tiling?

13. Do you think that other scalene triangles besides your own will tessellate the plane? Why?

14. Do you think that isosceles or equilateral triangles will tessellate the plane? Why?

process: model

## Tessellations

You have likely seen tessellation patterns that involve squares or rectangles on kitchen or bathroom floors, but what about other quadrilaterals? Will they tile the plane?

Try to tile the plane with a quadrilateral. (Do not choose a parallelogram or an isosceles trapezoid.)

15. As you did with the triangle in the last task, use your straightedge to create your quadrilateral on a separate piece of paper.
16. Measure each of the angles in it and record the measures on the quadrilateral.
17. Cut out the quadrilateral and use it as a template to create about 10 more copies.
18. Mark or color one side so that you can tell the difference between the two sides of the paper.
19. Try to tessellate a part of your desktop using only your own copies.
20. Create a sketch below.

21. Pick one vertex and measure the angles in your tiling that surround that vertex. Record the measures on your sketch.
22. Pick another vertex and do the same.

analysis

## Tessellations

23. What do you notice about the angles surrounding any vertex point in your tiling?

process: model

## Tessellations

Continue your investigation of which polygons will tessellate, by considering regular pentagons. (All the angles are equal and all the sides are equal.)

24. Why do you think you are being asked to use **regular** pentagons?

25. Attempt to create a tessellation using **regular** pentagons. Make a sketch below.

26. Measure the angles in the pentagons around any vertex in your tiling attempt.

analysis

## Tessellations

27. Is the angle sum the same as it was with the triangle and the quadrilateral? Why or why not?

28. Can you change your **regular** pentagon tiling to make the sum of the angles at each vertex the same as it was with the triangle or quadrilateral tiling? If so, sketch your tiling below. Explain what you changed to make the tessellation possible. If not, explain why not.

process: model

## Tessellations

29. Before attempting each of the next couple of tessellations with **regular** polygons, (hexagon, heptagon, octagon), conjecture about whether each will tile the plane and explain your reasoning. Then, attempt a tessellation for each of the polygons you think will tile the plane. Show any attempts you make by making a sketch below.

   a. Hexagon:

      Conjecture and Reasoning:

      Tessellation Attempt:

   b. Heptagon:

      Conjecture and Reasoning:

      Tessellation Attempt:

## Tessellations

process: model

c. Octagon:

Conjecture and Reasoning:

Tessellation Attempt:

d. Nonagon:

Conjecture and Reasoning:

Tessellation Attempt:

e. Decagon:

Conjecture and Reasoning:

Tessellation Attempt:

input

## Putting It All Together

Below is a partial tiling of the plane with a non-isosceles trapezoid.

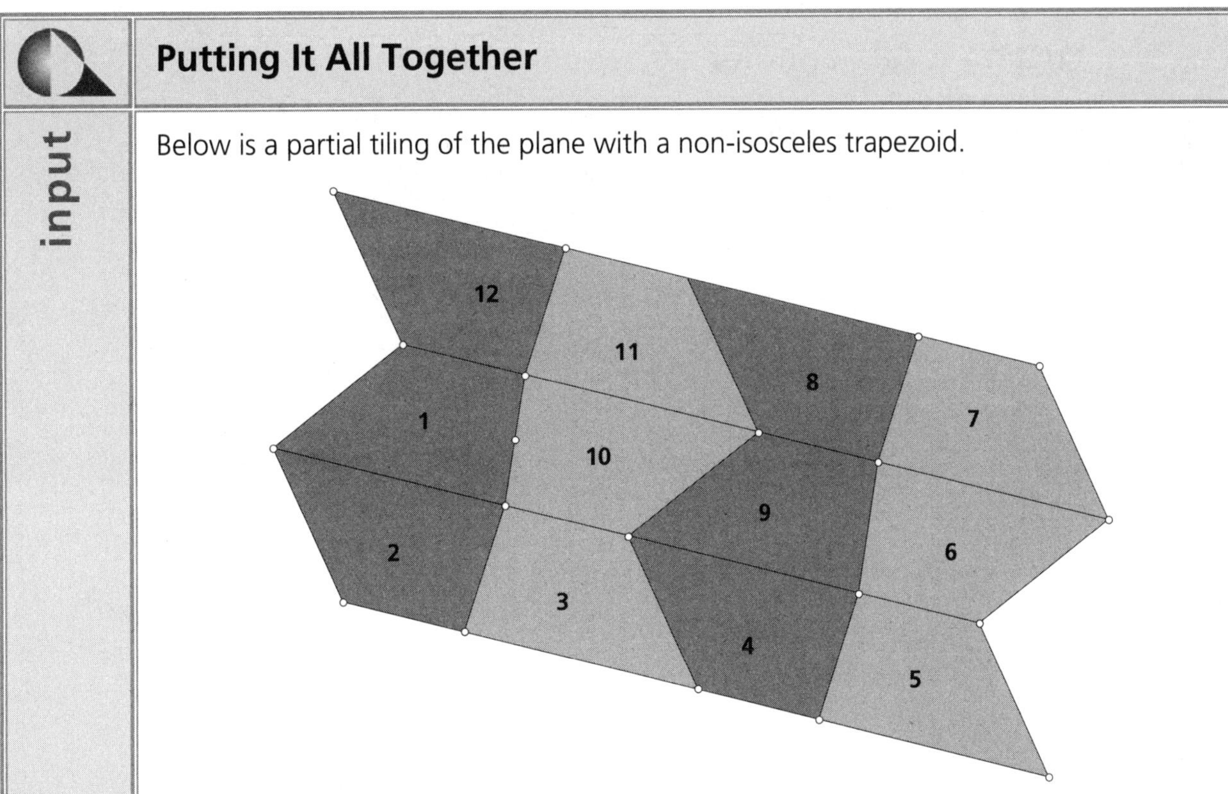


process: practice

## Putting It All Together

1. Which transformation or combination of transformations can be used to map figure 1 onto figure 9?

2. Which transformation or combination of transformations can be used to map figure 1 onto figure 10?

3. Which transformation or combination of transformations can be used to map figure 1 onto figure 12?

4. Which transformation or combination of transformations can be used to map figure 1 onto figure 4?

## Putting It All Together

analysis

5. Only figure 1 and figure 8 are shown here. Use your geometry tools (ruler, protractor, compass, Mira®, etc...) to show each of the transformations figure 1 went through to become figure 8.

1

8

   a. How many transformations did you need to perform to map figure 1 onto figure 8?

   b. Could you have performed this mapping using fewer transformations? If so, which transformation(s) could you have performed? If not, explain why not.

   c. Is this transformation an isometry? Why or why not?

## Putting It All Together

analysis

6. Perform the same transformation of figure 1 to figure 8, but this time ONLY use reflections to get to figure 8 from figure 1.

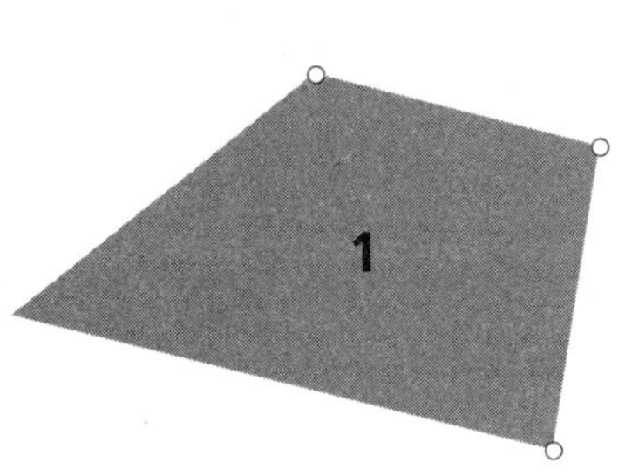


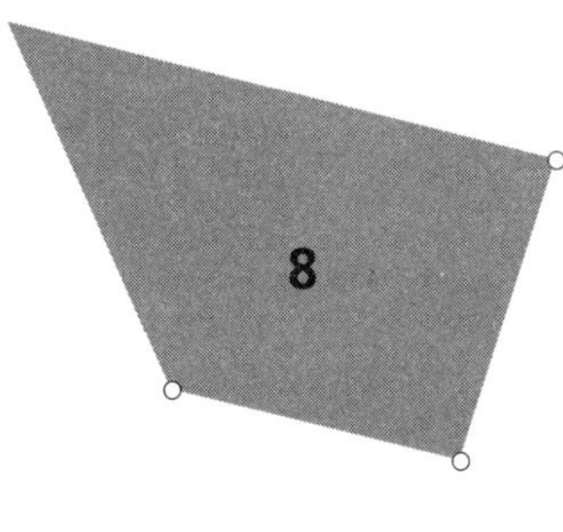


   a. How many reflections did you need to use? Could you have used fewer reflections to map figure 1 onto figure 8? Why or why not?

Create two copies of a figure of your choosing below.

7. Can you map one figure onto the other? If so, perform this transformation. If not, explain why this transformation cannot occur.

8. How many transformations did you need to perform to do this mapping?

9. Could you have performed this mapping using fewer transformations? If so, show this. If not, explain why not.

## Putting It All Together

analysis

10. Create two new copies of the same figure below. Can you map one figure onto the other using ONLY reflections? If so, perform this transformation. If not, explain why this combination of reflections cannot occur.

11. Is this transformation an isometry?

12. Do you think that any two identical figures can be mapped onto one another using isometries? Explain your answer in complete sentences.

13. What is the minimum number of transformations that are needed to map one figure onto a copy of that figure? Explain and illustrate your response.

14. What is the maximum number of transformations that are needed to map one identical figure onto another? Explain and illustrate your response.

process: summary

## Putting It All Together

As you have seen, transformations and tessellations have some applications in art, but are they useful in other areas? Provide two examples of where and when you have seen transformations used in the world.

# ASSIGNMENTS

# Assignment for It's a Function of....

**Name:** ______________________________

1. Without graphing the function $f(x) = 2x^2 + 4x - 6$
   a. Describe its domain and range

   b. Identify the possible number of x and y intercepts

   c. Describe positive and negative rate of change

   d. Indicate the number of relative extrema

   e. Describe the end behavior

2. Graph the function in question 1 and determine it has is one real root, two real roots or no real roots? If there are real roots, find the zeros.

   a. Write the function in factored form.

   b. What is the relationship between the zeros and factored form?

3. At what point does the graph reach its relative maximum or minimum?

4. Using the list of key characteristics from question 1 above, which characteristic(s) best distinguish the difference between a linear and quadratic function.

5. Generating revenue is always the bottom line for a manufacturer of goods. A company that sells audio/video equipment is still trying to determine what price to charge for their newest line of high tech sound equipment. To determine the pricing on the new model, the company will use the following revenue function

$$f(x) = 2.5x^2 + 2825x - 115812$$

   a. Based on what you know about polynomial functions describe the graph of this function.

   b. Based on the graph of the function and what you know about generating revenue, do you think this is a realistic model?

# Assignment for Getting the Third Degree

**Name:** ______________________________

1. Using key characteristics describe the difference between the quadratic and cubic equations?

2. You were told total attendance at American League baseball games per year can be modeled by the polynomial function

$$f(x) = 8{,}350x^3 - 386{,}040x^2 + 2{,}423{,}160x + 24{,}414{,}310$$

and total attendance per year at National League baseball games can be modeled by the polynomial function

$$f(x) = 19{,}320x^3 - 266{,}310x^2 + 1{,}195{,}580x + 22{,}655{,}750$$

Based on what you know the behaviors of cubic functions, can these models actually represent this situation?

# Assignment for Fourth Degree and Beyond

**Name:** ______________________________

1. Review the activities It's a Function of…, Getting the Third Degree and Fourth Degree and Beyond. Based on the work you have completed, summarize the behaviors and characteristics of the family of polynomial functions.

| **Behavior** | **Polynomial Functions of Odd Degree** | **Polynomial Functions of Even Degree** |
|---|---|---|
| Domain and Range | | |
| Intercepts | | |
| Rate of Change | | |
| Extrema | | |
| End Behavior | | |

2 Construct the following functions:

a. an even degree function with one real root and one maximum value

b. an odd degree polynomial function that approach negative infinity as x approaches positive infinity with no local maxima.

c. a function whose end behavior goes to positive infinity as x goes to positive or negative infinity.

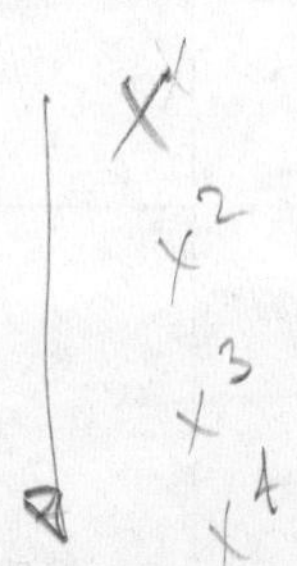

# Assignment for Down on Main Street

**Name:** ___________________________

1. Evaluate each of the following absolute values.

a. $|-4.2|$

b. $|6|$

c. $|342.9|$

d. $|-0.45|$

e. $|23 - 56|$

f. $|-3 + 8|$

g. $|2^3 - 5^2|$

h. $|30 - 4(5.5)|$

i. $\left|-\frac{52}{4} + 13\right|$

j. $\left|\frac{-4^2}{6^2}\right|$

2. Graph the function, $f(x) = |x| - 2$. Hint: A table of values may help you.

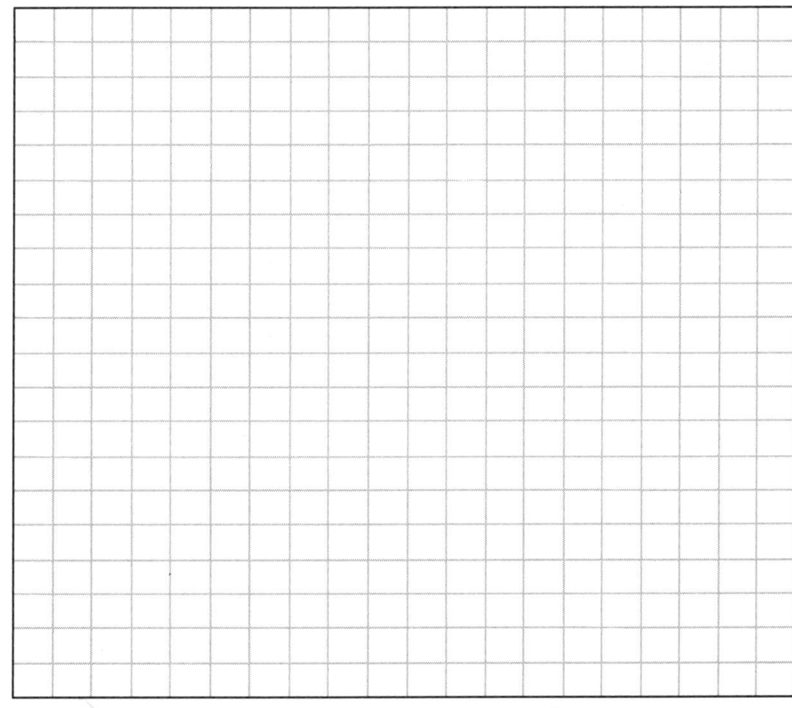

# Assignment for Absolutely Functional

**Name:** ______________________________

1. For the function, $y = -|x| + 3$, write a description of the transformations from the parent function, $y = |x|$, and sketch the graph.

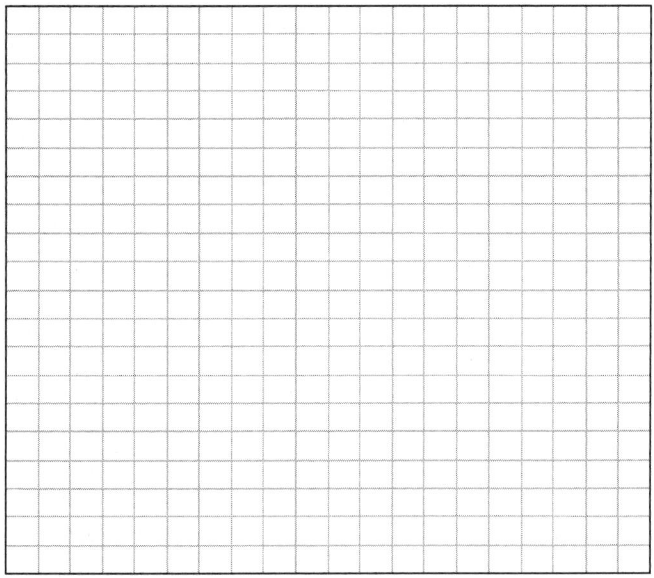

2. The parent function, $y = |x|$ is transformed with a dilation by a factor of 0.5, and a horizontal translation 2 units to the right. Write the new function and sketch its graph.

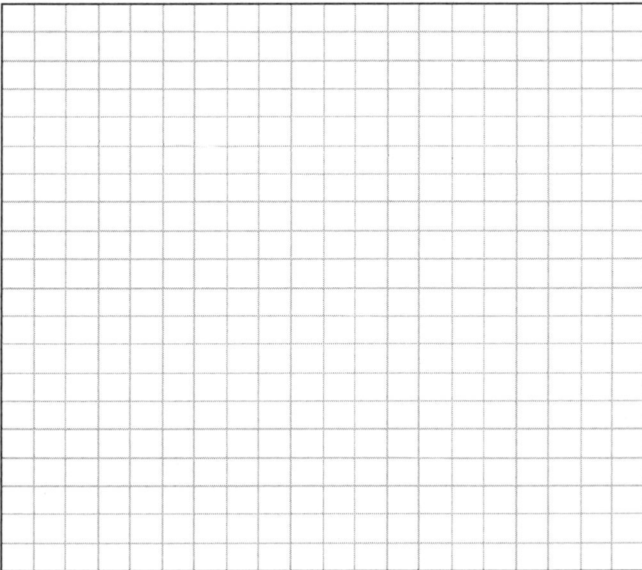

3. Write the algebraic function for the graph provided.

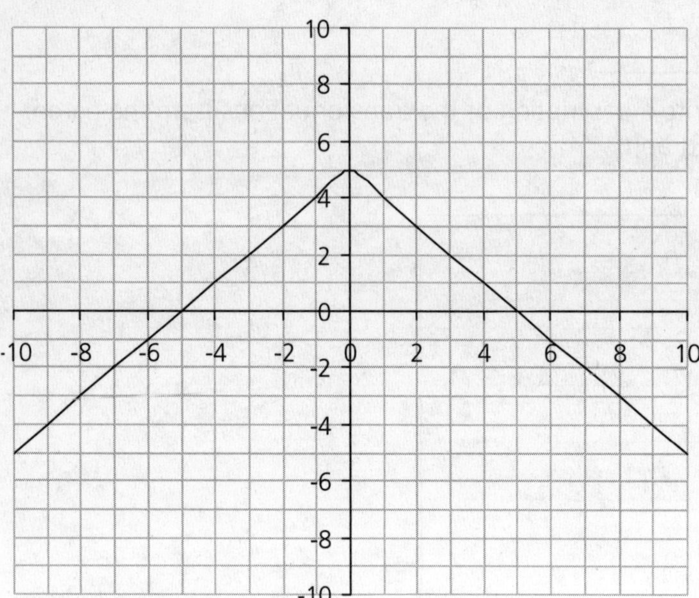

4. Write the algebraic function for the graph provided.

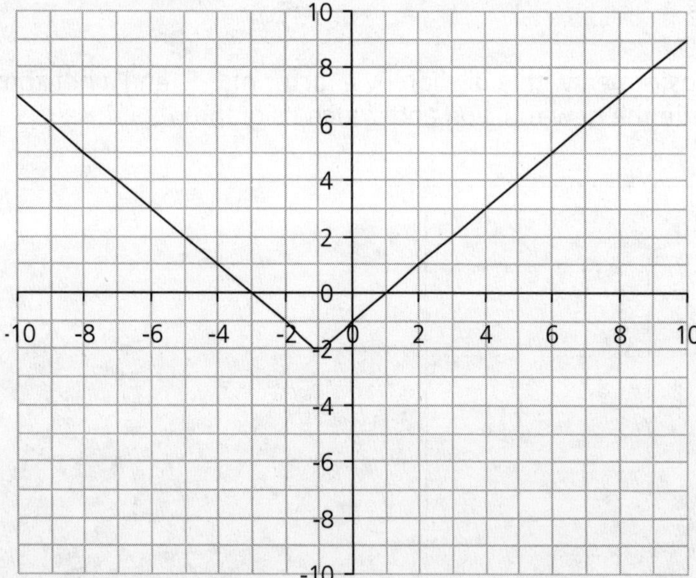

# Assignment for A Moving Experience

**Name:** ______________________________

Write a story that could be illustrated by the graph below. Scale and label the graph appropriately. Write an algebraic piecewise linear function for the graph.

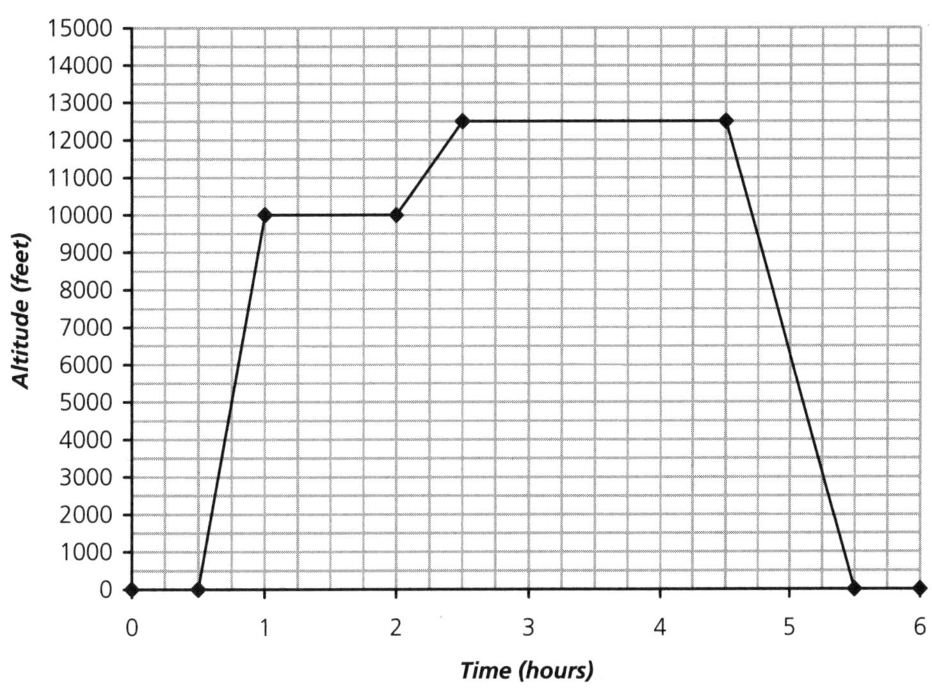

# Assignment for Comparing Long Distance Rates

**Name:** ______________________

Your friend, Andresa, decided to use U-Dial for her long-distance calls. Her phone has a timer that shows the duration of dialed calls. She used the timer to keep track of her call time for the month of September. The timer showed a total call time of 229 minutes, 13 seconds. When her phone bill arrived, it listed the following calls.

| Time of Call | To: | Duration | Amount Billed |
|---|---|---|---|
| Sept. 2, 8:06 p.m. | Los Angeles, CA | 38 | $2.28 |
| Sept. 5, 6:15 p.m. | San Jose, CA | 12 | $0.72 |
| Sept. 9, 11:59 a.m. | St. Louis, MO | 37 | $2.22 |
| Sept. 11, 7:45 p.m. | Los Angeles, CA | 1 | $0.06 |
| Sept. 13, 8:13 p.m. | Los Angeles, CA | 1 | $0.06 |
| Sept. 13, 8:47 p.m. | Los Angeles, CA | 26 | $1.56 |
| Sept. 19, 2:14 p.m. | Philadelphia, PA | 8 | $0.48 |
| Sept. 19, 2: 23 p.m. | Chicago, IL | 35 | $2.10 |
| Sept. 22, 8:02 p.m. | Los Angeles, CA | 17 | $1.02 |
| Sept. 25, 9:15 p.m. | San Jose, CA | 10 | $0.60 |
| Sept. 26, 11:10 a.m. | Chicago, IL | 9 | $0.54 |
| Sept. 28, 8:42 p.m. | Los Angeles, CA | 42 | $2.52 |
| **Totals** | | **236** | **$14.16** |

1. The total time billed by U-Dial was 236 minutes. Andresa only expected to be billed for 229 minutes, 13 seconds. How do you account for the difference?

2. Andresa wants to know if she chose the right company for her phone calls. What do you tell her? Should she stay with U-Dial or switch to another company?

# Assignment for Media Mail

**Name:** ____________________________

Most long distance and cell phone companies charge for each minute or part of a minute when calculating phone charges. Assume the phone companies in the Comparing Long Distance Rates problem charge this way.

1. Recall that AmeriCall charges $1.00 for the first 30 minutes and 5¢ for each additional minute. If they charge for each minute or part of a minute, how can you write the function algebraically?

2. How is the graph of the function different when the customer is charged for each minute or part of a minute instead of the exact time? You may want to sketch a graph or make a table on the back of this page to help you answer the question.

3. In your opinion, what is the most useful representation (words, graph, table, or algebraic equation) of a problem like this for a company trying to get customers to sign up? Why?

4. In your opinion, what is the most useful representation (words, graph, table, or algebraic equation) of a problem like this for an employee of the company working in the billing department?

# Assignment for The Bike Rental Shop

**Name:** ______________________________

Ms. Owens is reconsidering her grading policy. She calculates students' grades by dividing each student's earned points by the total possible points in the class. She multiplies that number by 100 to get a "percent grade." Currently, Ms. Owens rounds grades to the nearest percent using the traditional rule. That is, any value from 0 to 4 in the tens place indicates "rounding down" or leaving the integer part alone. Any value from 5 to 9 in the tens place indicates "rounding up" or increasing the integer part by one.

Ms. Owens is thinking about changing to either a ceiling function, $grade = \left\lceil 100 \cdot \frac{\text{earnedpts.}}{\text{total pts.}} \right\rceil$ or a floor function, $grade = \left\lfloor 100 \cdot \frac{earned\ pts.}{total\ pts.} \right\rfloor$. With the floor function, each value would be rounded to the largest integer less than or equal to the determined value.

1. Should she change her grading policy? If so, how should she calculate grades? Explain your answer.

2. In your opinion, which method is the fairest way to calculate grades? Why?

# Assignment for The Bike Rental Customer

**Name:** ______________________________

1. Evaluate each of the following.

   a. $\lfloor 12.25 \rfloor$

   b. $\lfloor -4.56 \rfloor$

   c. $\lceil -1.001 \rceil$

   d. $\lceil 24 \rceil$

   e. $\lfloor 2.4 + 1.3 \rfloor$

   f. $[-12.2 + 1.5 \bullet 3]$

   g. $\left\lfloor \frac{65}{12} \right\rfloor$

   h. $\left\lceil \frac{18}{24} + 1 \right\rceil$

   i. $\left\lfloor \frac{3^2 - |1.5|}{3} \right\rfloor$

   j. $\left| \left\lceil \frac{-8}{3} \right\rceil \right|$

2. Use what you know about the parent greatest integer function and transformations to sketch a graph of $y = 2\lfloor x \rfloor + 1$.

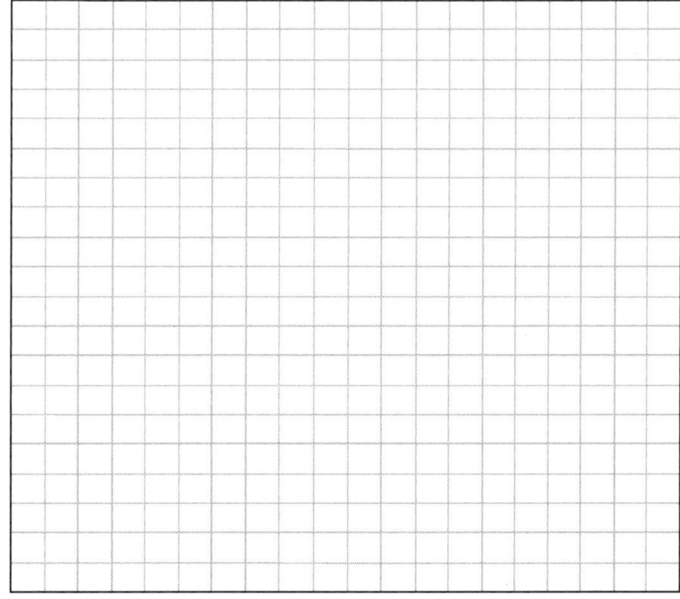

3. Compare and contrast the graph of $y = 2\lfloor x \rfloor + 1$ with the graph of $y = 2\lceil x \rceil + 1$. It may help to add the graph of $y = 2\lceil x \rceil + 1$ to the grid from the previous question.

## Assignment for Working at the Mall

**Name:** ______________________

Adult humans usually have a heart rate of about 60 to 90 beats per minute. Domestic house cats typically have a heart rate between 160 and 220 beats per minute.

Assume that Jessie has an average heart rate of 75 beats per minute and her cat, Quark, has an average heart rate of 190 beats per minute.

1. How many times does Jessie's heart beat in one day? How many times does Quark's heart beat in one day?

2. Write a function, j, for the number of beats of Jessie's heart in $x$ days, and a function, q, for the number of beats of Quark's heart in $x$ days.

3. What will the graphs of the functions look like? How will they be similar to each other? How will they be different?

4. An average human in the United States lives about 76 years. An average indoor house cat in the United States lives about 17 years. If Jessie and Quark are average, how many times will their hearts beat in their lifetimes?

5. Do you think there is a relationship between an animal's heart rate and its lifespan? If you can, find some evidence to support your claim.

# Assignment for Saving for a Car

**Name:** ______________________

1. Write an algebraic function for each of the following situations and state whether the relationship is a direct variation or an inverse variation.

   a. You earn $8.50 per hour.

   b. You are calculating your share of a $10,000,000 lottery prize based on the number of people who contribute to buy tickets.

   c. Your school has $2000 in award money to give away in equal amounts. You need to calculate the value of each award depending on how many students are granted an award.

   d. For every second delay from seeing lightning to hearing thunder, you are about 0.2 miles from the lightning.

2. You are joining a volunteer group that is building a community playground. The expectation is that the playground construction requires 180 person-days to complete.

   a. Write an algebraic equation for the situation and define your variables.

b. Create a table of values for the situation.

| Volunteers | Completion Time |
|---|---|
| People | Days |
| 1 | |
| 2 | |
| 3 | |
| 4 | |
| | 36 |
| | 30 |
| 8 | |
| | 20 |
| 10 | |

c. Make a graph of the situation.

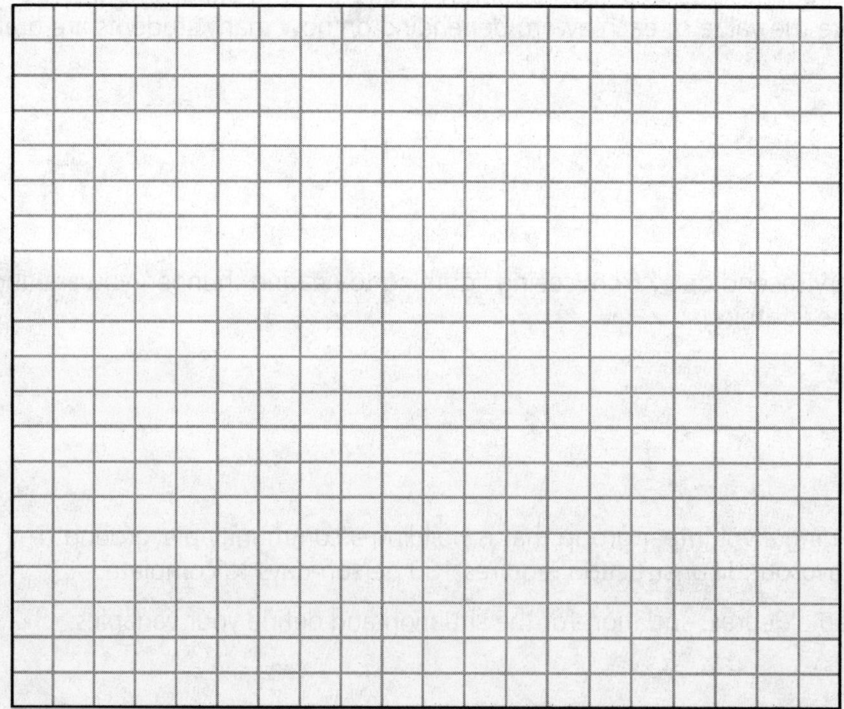

# Assignment for An Age Old Problem

**Name:** ______________________________

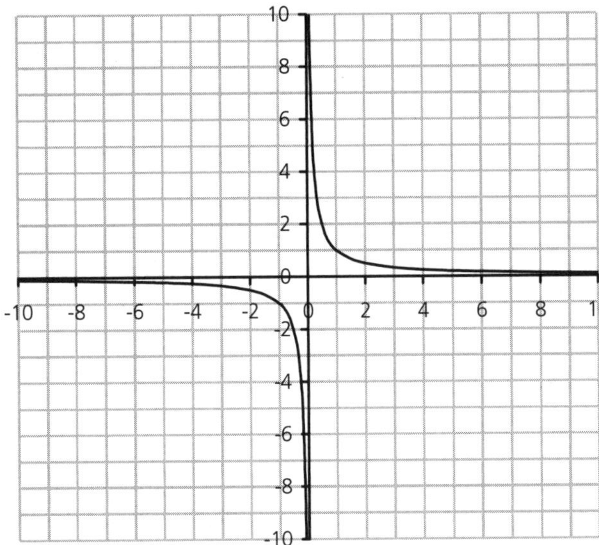

The parent function for inverse variation is $y = \frac{1}{x}$. Its graph is shown to the left.

Use your knowledge of transformations on functions to answer the following questions.

1. For the function, $y = \frac{1}{x} + 2$, write a description of the transformation(s) from the parent function, $y = \frac{1}{x}$, and sketch the graph.

2. The parent function, $y = \frac{1}{x}$, is transformed with a reflection over the x-axis and a horizontal translation 2 units to the left. Write the new function and sketch its graph.

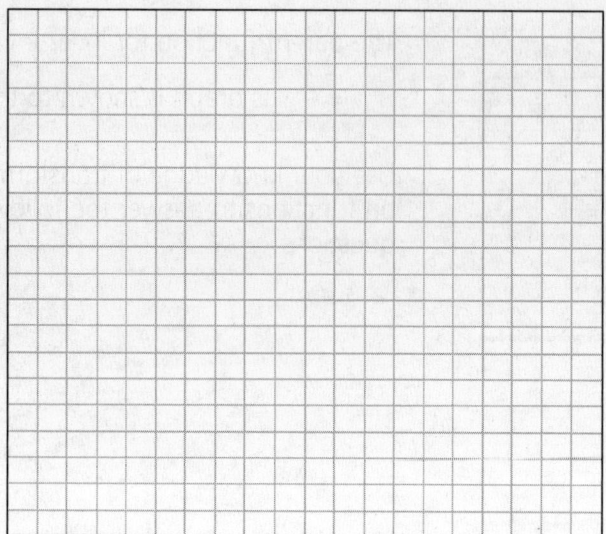

3. Write the algebraic function for the graph provided.

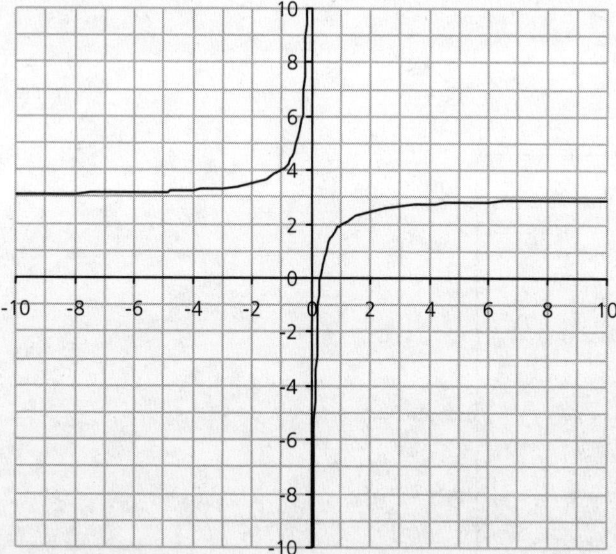

# Assignment for Rational Functions: Discontinuity

**Name:** ______________________________

For each rational function below, determine the domain of the function, state whether the function is continuous or discontinuous, identify any locations of discontinuity, and classify them as essential or removable.

1. $f(x) = \dfrac{1}{x^4}$

2. $g(x) = \dfrac{2}{(x+1)(2x-3)}$

3. $h(x) = \dfrac{x^2 + x}{(x+1)(x+2)}$

4. $j(x) = \dfrac{3x+6}{x^2 - 3x - 10}$

5. $k(x) = \dfrac{x^2}{4}$

# Assignment for Rational Functions: Asymptotes

**Name:** ____________________________

For each rational function below, provide the following information.

a. Determine the domain of the function

b. State whether the function is continuous or discontinuous,

c. Identify any locations of discontinuity and classify them as essential or removable

d. Identify any asymptotes

1. $f(x) = \dfrac{3x^2}{x^2 - 3x - 4}$

2. $g(x) = \dfrac{x^3 + 7x^2}{x + 7}$

3. $h(x) = \dfrac{10x}{x^2 + 2}$

4. $j(x)=\dfrac{x+3}{x-4}$

5. $k(x)=\dfrac{9x^2+1}{6x}$

# Assignment for Rational Functions: Graphing

**Name:** ______________________________

1. Sketch the graph for the rational function, $f(x) = \dfrac{2x+5}{x-1}$ by hand as instructed in the following steps:

   a. Find the x- and y-intercept and plot it on the coordinate grid below.

   b. Determine any locations of discontinuity. If there is removable discontinuity, plot the point with an open circle. If there is essential discontinuity, sketch the vertical asymptote with a dashed vertical line.

   c. Determine the horizontal asymptote and sketch it with a dashed horizontal line.

   d. Find and plot at least one point between and one point beyond each x-intercept and vertical asymptote.

   e. Complete the graph by drawing smooth curves.

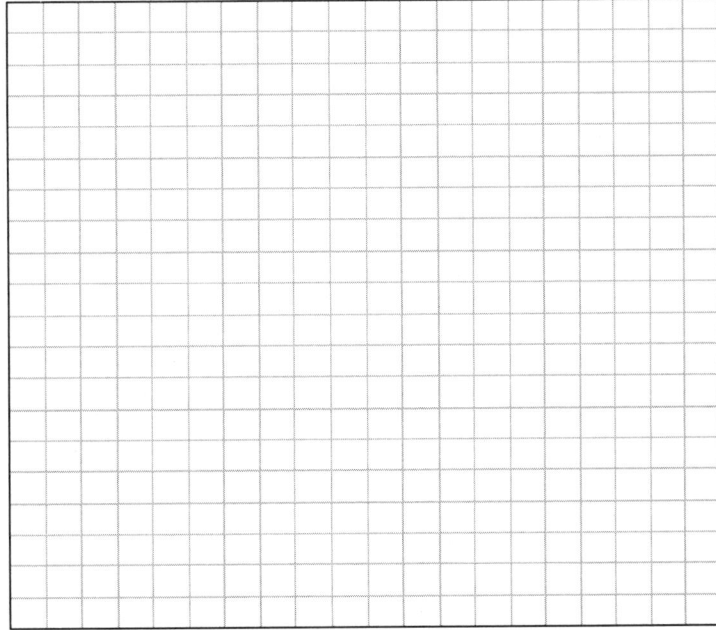

Use what you know about intercepts, continuity, asymptotes, and locating points to sketch a graph for each of the following rational functions. Check your work using a graphing calculator.

2. $g(x) = \dfrac{2}{x+3}$

3. $h(x) = \dfrac{3x}{x+1}$

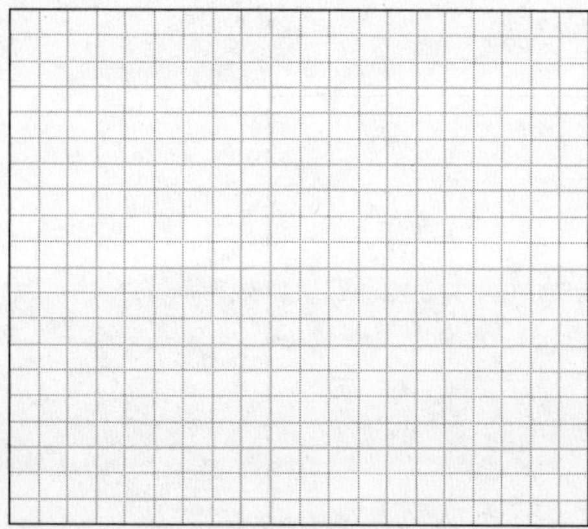

# Assignment for Light Intensity

**Name:** ______________________________

Annual fuel costs for your car can be calculated using the following formula:

$$\text{Fuel cost for one year} = \frac{\text{Miles driven} \cdot \text{Price per gallon}}{\text{Fuel efficiency rate}}.$$

You estimate that you drive about 12,000 miles each year and that gasoline costs on average about \$1.50 per gallon.

1. You are considering buying a new car and want to know the annual fuel costs. Write a rational function that will determine the annual fuel cost based on the fuel efficiency of the vehicle.

2. Use your function to determine the annual fuel costs if you purchased:
   a. a gas/electric hybrid car with a fuel efficiency of 60 miles per gallon
   b. a compact car with a fuel efficiency of 36 miles per gallon
   c. a midsize car with a fuel efficiency of 28 miles per gallon
   d. a full size car with a fuel efficiency of 22 miles per gallon
   e. a sports utility vehicle with a fuel efficiency of 18 miles per gallon

3. Complete the following table of values for your function.

| Fuel Efficiency | Annual Cost |
|---|---|
| miles per gallon | $ |
| 10 | |
| 20 | |
| 30 | |
| 40 | |
| 50 | |
| 60 | |

4. Create a graph of the function.

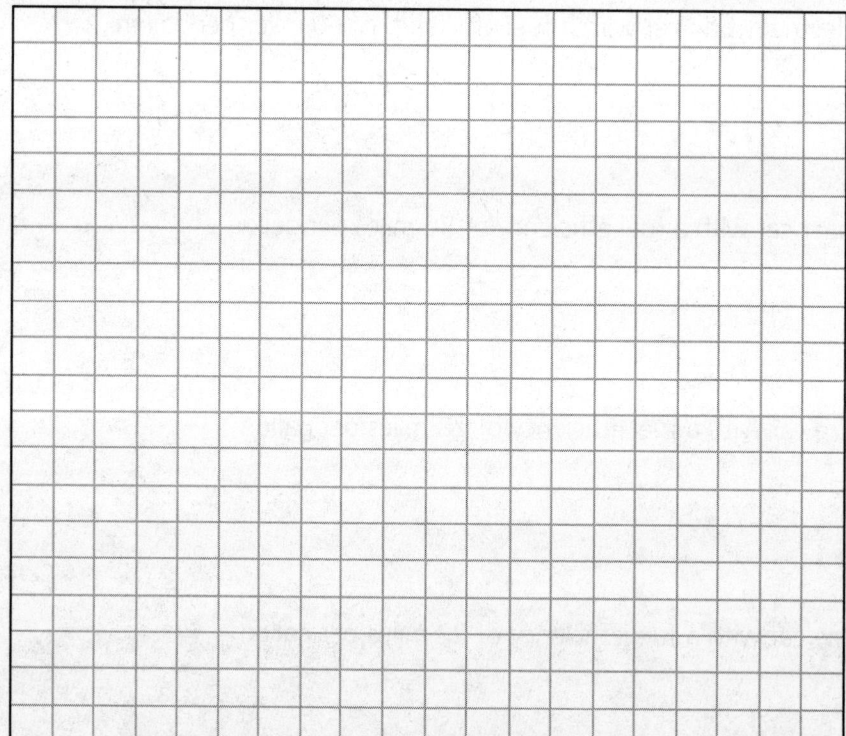

# Assignment for Change for a Dollar

**Name:** ______________________________

The local Gas 'n' Go station is currently charging $1.639 per gallon of gasoline.

1. Complete the tables of values for determining the cost of gasoline based on the amount purchased, and for determining how much gasoline can be purchased with a given amount of money.

| Gasoline | Cost |
|---|---|
| gallons | dollars |
| 0 | |
| 2 | |
| 4 | |
| 6 | |
| 8 | |
| 10 | |
| 12 | |
| 14 | |
| 16 | |

| Cost | Gasoline |
|---|---|
| dollars | gallons |
| 0 | |
| 2.50 | |
| 5.00 | |
| 7.50 | |
| 10.00 | |
| 12.50 | |
| 15.00 | |
| 17.50 | |
| 20.00 | |

2. Write an algebraic function for determining the cost of gasoline based on the amount purchased.

3. Write an algebraic function for determining how much gasoline can be purchased with a given amount of money.

4. Create a graph for each of the situations on the grid below. Also graph the function $y = x$ using a dotted line. Use the same intervals on both the x- and y-axes.

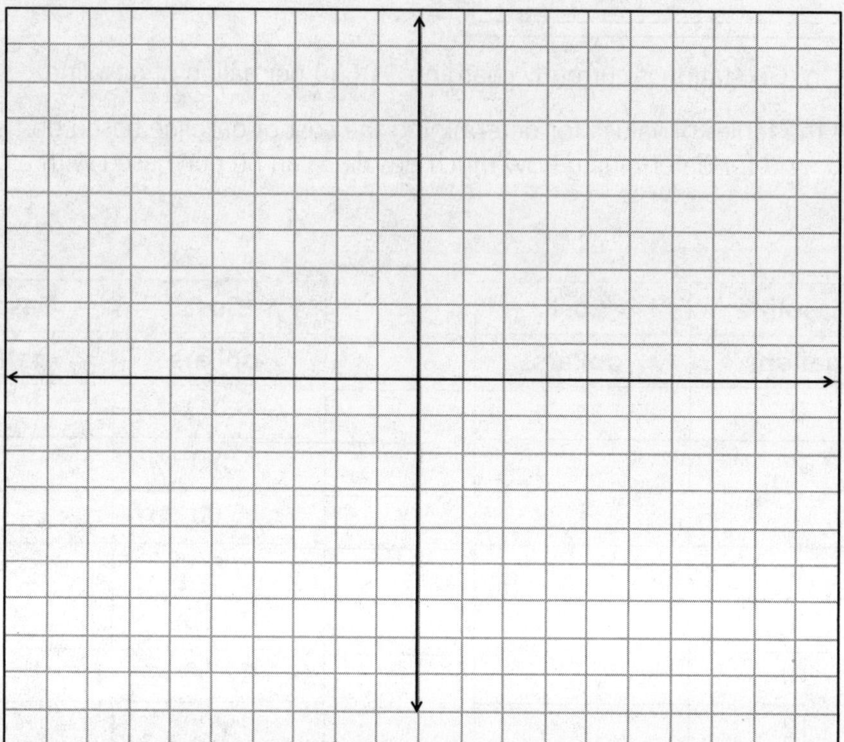

5. Are determining the cost of gasoline based on the amount purchased and determining how much gasoline can be purchased with a given amount of money inverses of each other? Explain your answer using evidence from the problem situation, tables, equations, and graph.

# Assignment for A Square Deal

**Name:** ______________________________

You decide to participate in a bicycle ride for charity. For long bike rides, you estimate that you can average 15 miles per hour. On the day of the ride, you wake up late and start the ride a half hour after the start of the event.

1. Create a table of values showing the distance you ride based on the time from the start of the event.

| Time from start of event | Distance |
|---|---|
| **Hours** | **miles** |
| 1 | |
| 2 | |
| 3 | |
| 4 | |
| | |
| | |

2. Write an algebraic function for finding the distance based on the time from the start of the event

3. How long after the start of the event will you have ridden:

   a. 30 miles?

   b. 60 miles?

   c. 100 miles?

4. Write an algebraic function for finding the time from the start of the event based on the distance.

5. Do the two parts of this problem represent inverse functions? Explain your answer.

# Assignment for Motions of the Square

**Name:** ____________________________

1. The marching commands given to soldiers can be used as elements with the operation "followed by." Using the following commands, complete the Cayley table below.

| | |
|---|---|
| **A**ttention: | do not move. |
| **R**ight Face: | turn 90° to the right |
| **L**eft Face: | turn 90° to the left |
| A**B**out Face | turn 180° |

| * | A | R | L | B |
|---|---|---|---|---|
| **A** | | | | |
| **R** | | | | |
| **L** | | | | |
| **B** | | | | |

2. Does the operation "followed by" demonstrate closure for the marching commands? That is, will the result always be one of the elements (motions of a square)?

3. What is the identity element for the operation "followed by" for the marching commands? How do you know it is an identity?

4. What are the inverse elements for the marching commands under the operation of "followed by?" How do you know what elements are inverses?

5. Does the associative property apply to the operation "followed by" for the marching commands? Explain.

6. Does the commutative property apply to the operation "followed by" for the marching commands? Explain.

# Assignment for Picture Perfect

**Name:** ______________________________

A company allows you to order coffee via the Internet. They will charge \$10 shipping and handling on any order. Their premium brand of coffee is priced at \$12 per pound.

1. Determine the algebraic function to calculate the cost of an order based on the number of pounds of coffee ordered.

2. Describe the function.

   a. What are the domain and range?

   b. What are the intercepts?

   c. What is the slope?

3. Determine the inverse of your function algebraically.

4. What does the inverse function represent in terms of the problem situation?

5. Describe the inverse function.
   a. What are the domain and range?

   b. What are the intercepts?

   c. What is the slope?

6. How are the descriptions of the function and its inverse related?

# Assignment for Inverses of Linear Functions

**Name:** ______________________________

1. To determine the temperature on the Fahrenheit scale if you know the temperature on the Celsius scale, you can use the function $f(x) = \frac{9}{5}x + 32$. Determine the inverse of the function and explain what it represents.

2. Find the inverse of $g(x) = 1.2x - 4.2$.

3. Find the inverse of $h(x) = 5 + \frac{1}{3}x$.

4. Are the functions $f(x) = 2x + 5$ and $g(x) = 5x + 2$ inverse functions? How do you know?

5. Are the functions $h(x) = 0.5x - 1$ and $j(x) = 2x + 2$ inverse functions? How do you know?

# Assignment for Classified Ad

**Name:** ____________________________

A local pizza shop charges $5.00 for a large pizza plus $1.25 per topping.

1. Write a sentence to describe the inverse of the problem situation.

2. Represent the situation and its inverse with algebraic functions.

3. Complete the table of values for the function and the inverse function.

| **Toppings** | **Charge** |
|---|---|
| **toppings** | **dollars** |
| | |
| | |
| | |
| | |
| | |

| **Charge** | **Toppings** |
|---|---|
| **dollars** | **toppings** |
| | |
| | |
| | |
| | |
| | |

4. Create a graph of the function and its inverse on the grid provided. Use the same intervals on both the x- and y-axes.

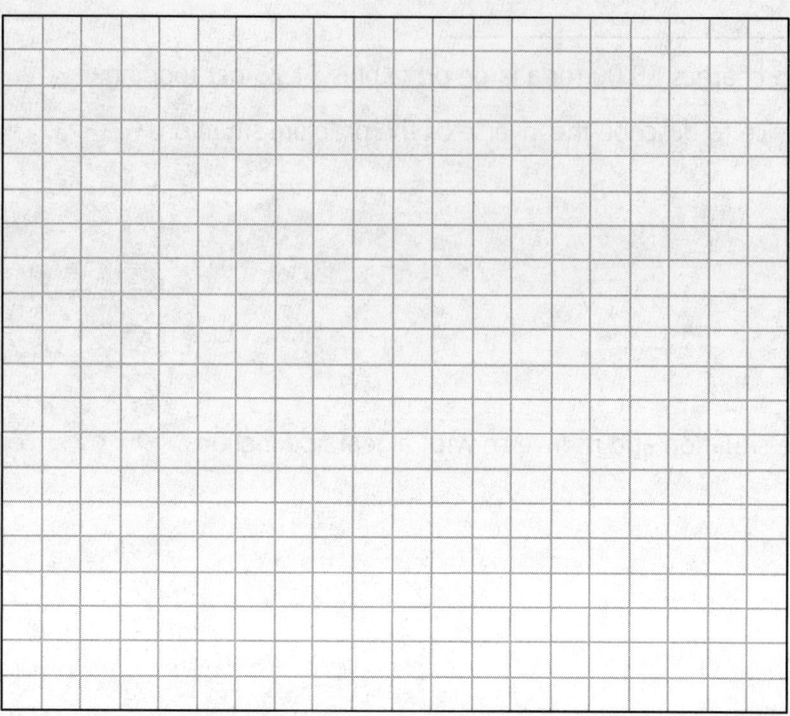

# Assignment for Selling Appliances

**Name:** ______________________________

You got a job at with the local dry cleaner and laundry. Since so many people bring in shirts to be cleaned, the owner has posted a table showing the cost based on the number of shirts.

| Number of Shirts | Cost ($) |
|---|---|
| 1 | 1.75 |
| 2 | 3.50 |
| 3 | 5.25 |
| 4 | 7.00 |
| 5 | 8.75 |
| 6 | 10.50 |
| 7 | 12.25 |
| 8 | 14.00 |
| 9 | 15.75 |
| 10 | 17.50 |

1. Create a table showing the inverse of the situation.

| | |
|---|---|
| | |
| | |
| | |
| | |
| | |
| | |
| | |
| | |
| | |
| | |

2. Create a graph of the function and its inverse on the grid below. Use the same intervals on both the x- and y-axes.

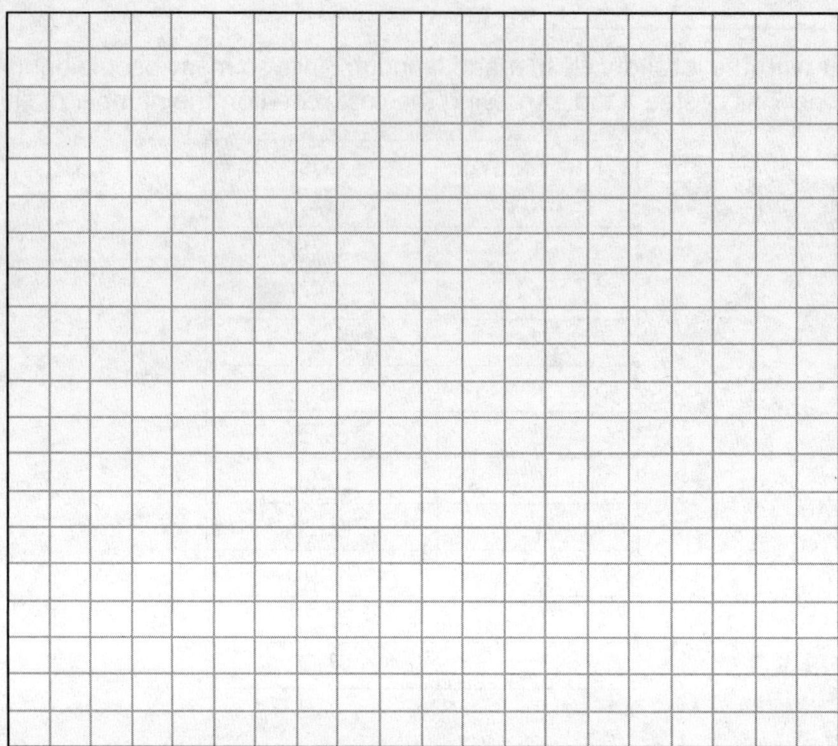

3. Write an algebra function for the original situation and the inverse situation.

4. Describe the original function and its inverse in sentences.

# Assignment for Burning Calories

**Name:** ______________________________

A magazine article about exercise shows the following graph as a way to calculate your target heart rate when exercising.

**Target Heart Rate By Age**

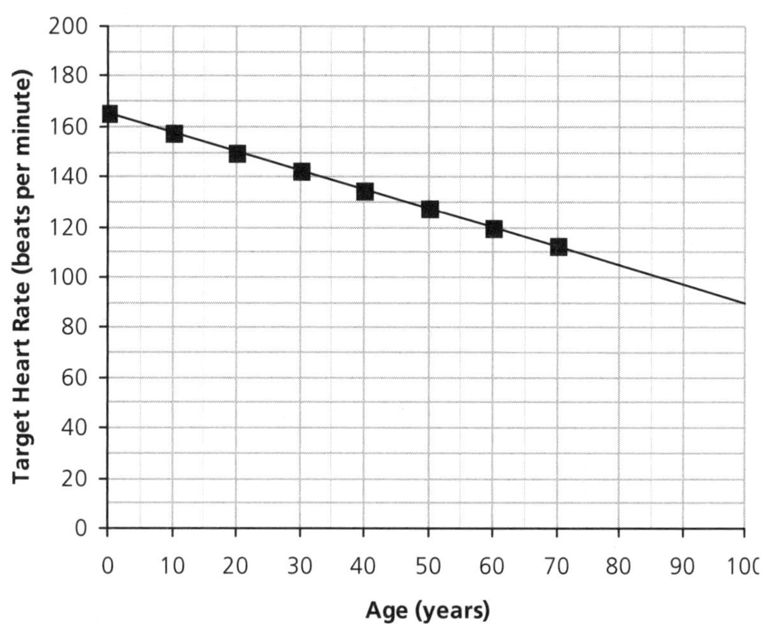


1. Create a graph showing the inverse of the graph from the newspaper.

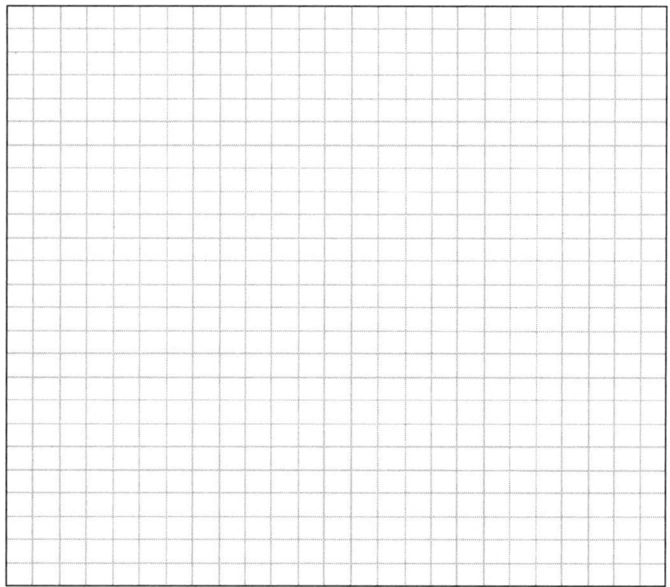

2. Complete the tables of values for the function and its inverse function.

| Age | Target Heart Rate |
|---|---|
| years | bpm |
| | |
| | |
| | |
| | |
| | |
| | |
| | |

| Target Heart Rate | Age |
|---|---|
| bpm | years |
| | |
| | |
| | |
| | |
| | |
| | |
| | |

3. Describe the original function and its inverse in sentences.

4. Represent the situation and its inverse with algebraic functions.

# Assignment for Inverses of Power Functions

**Name:** ____________________________

1. A function has a domain of every student enrolled in courses at Carnegie Community College, and a range of their student identification numbers. Is the function a one-to-one function? Explain.

2. A caller ID function takes the phone numbers of callers as the domain and the names of the callers as the range. Is the function a one-to-one function? Explain.

3. A function for calculating taxes for a local municipality takes income amounts as the domain and 2% of the income amounts as the range. Is the function a one-to-one function? Explain.

For problems 4 through 6, determine the inverse of the function algebraically.

4. $f(x) = 3x^2 + 2$

5. $g(x) = x^3 + 5$

6. $h(x) = x^2 + 6x$

# Assignment for Designing Cologne Bottles

**Name:** ____________________________

The aspect ratio of a television screen is the ratio of its length to its width. A standard television screen has an aspect ratio of 4:3. A high definition television (HDTV) screen has an aspect ratio of 16: 9. The areas of a standard television screen and an HDTV screen of the same given screen size will be different because of the different aspect ratios.

Television screen sizes are usually given as the length of the diagonal of the screen. A function to determine the area of a standard TV screen given its diagonal measurement is $f(x)=\frac{12x^2}{25}$. To determine the area of an HDTV screen, the function is $g(x)=\frac{144x^2}{337}$.

1. How large is the screen area of a standard television with a diagonal measurement of:
   a. 27 inches?

   b. 36 inches?

   c. 52 inches?

2. How large is the screen area of an HDTV with a diagonal measurement of:
   a. 27 inches?

   b. 36 inches?

c. 52 inches?

3. Find the inverse of the function for the standard television screen area algebraically.

4. Find the inverse of the function for the HDTV screen area algebraically.

5. If you wanted a television with a screen area of 500 square inches, what size screen do you need (measured by its diagonal) for a standard television? For an HDTV?

6. If you wanted a television with a screen area of 1000 square inches, what size screen do you need (measured by its diagonal) for a standard television? For an HDTV?

# Assignment for The Square Root Function

**Name:** ______________________________

1. What is the inverse of the function $f(x) = \sqrt{x-4}$ ? Remember to consider an appropriate domain.

2. Verify that f($x$) and $f^{-1}(x)$ are inverses by evaluating:

   a. $f(f^{-1}(x))$

   b. $f^{1}(f(x))$

3. What is the inverse of the function $g(x) = \sqrt{x} + 3$? Remember to consider an appropriate domain.

4. Verify that $g(x)$ and $g^{-1}(x)$ are inverses by evaluating:

   a. $g(g^{-1}(x))$

   b. $g^{-1}(g(x))$

# Assignment for The Rotor

**Name:** ______________________________

A hurricane is a severe tropical storm that can produce violent winds, incredible waves, torrential rains, and floods. When hurricanes move onto land, they often cause heavy damage to buildings, trees and cars.

The wind velocity of a hurricane can be estimated based on the air pressure inside the hurricane using the function, $v(p) = 14.1\sqrt{1013 - p}$, where p is the pressure measured in millibars (mb) and v is the wind velocity measured in miles per hour (mph).

1. The air pressure measurements in the table below were charted for a storm. Use the function above to predict the associated wind velocities.

| Air Pressure | Wind Velocity |
|---|---|
| **millibars (mb)** | **miles per hour (mph)** |
| 994 | |
| 982 | |
| 970 | |
| 958 | |
| 949 | |
| 942 | |
| 934 | |

2. Determine the inverse of the function, *v*(*p*), algebraically.

3. A storm reaches hurricane strength when it has sustained winds of at least 74 mph. What air pressure would indicate a storm of that strength?

4. A category two hurricane (moderate strength) has sustained winds of at least 96 mph. What air pressure would indicate a storm of that strength?

5. A category four hurricane (very strong) has sustained winds of at least 131 mph. What air pressure would indicate a storm of that strength?

# Assignment for Becoming More Profitable

**Name:** ______________________

Your friend owes you a dollar. Instead of paying off the debt, he suggests that you wager him on a card game, double or nothing. Since you do not seem interested, he tries to entice you with a bet of triple or nothing. You think this might not be a bad idea.

1. Create a table showing how much your friend would owe if he continued to lose a triple or nothing bet over several games.

| Number of Games | Amount Owed ($) |
|---|---|
| 0 | |
| 1 | |
| 2 | |
| 3 | |
| 4 | |
| 5 | |

2. Write an algebraic function to model the situation.

3. Find an algebraic function to model the inverse of this situation.

4. Explain what the inverse function means in terms of the problem.

5. If your luck holds out, how many games would you need to win for your friend to owe you $729?

6. If your luck holds out, how many games would you need to win for your friend to owe you $5,000?

# Assignment for E. Coli

**Name:** ____________________________

Under ideal conditions for growth, baker's yeast (Saccharomyces cerevisiae) can double every two hours.

1. A culture starts with just one yeast organism. Complete the chart to show how the yeast population grows over time.

| Time | Yeast |
|---|---|
| hours | organisms |
| 0 | 1 |
| 2 | |
| 4 | |
| 6 | |
| 8 | |
| 10 | |

2. Write an algebraic function to model the yeast growth.

3. How many yeast organisms will be present after 15 hours?

4. Find an algebraic function that models the inverse of the situation.

5. How long would it take before there were 128 yeast organisms?

6. How long would it take before there were 1000 yeast organisms.

# Assignment for Inverses of Rational Functions

**Name:** ______________________________

Determine the inverse of each of the following rational functions algebraically.

1. $f(x) = \dfrac{3}{x+2}$

2. $g(x) = \dfrac{4}{3x}$

3. $h(x) = \dfrac{x}{2x+3}$

4. $j(x) = \dfrac{3x+1}{x-3}$

# Assignment for Adding and Subtracting Polynomials

**Name:** ______________________________

Simplify each expression by finding the sum or difference.

1. $(2x^3 - 5x + 1) + (x^3 - 3x^2 + 3x)$

2. $(r^3 - r^2 + 2r - 5) - (2r^3 + 3r^2 - 6r + 1)$

3. $(3a - ab + 4b) - (5a - 2ab - b)$

4. $(x^2 + 8x^2y^2 - 2x^2y + 6xy^2) + (3y^2 + x^2y - xy^2 - 7x^2y^2)$

5. $(2g^3 - 3g) + (6g^2 + 3g) - (4g^3 - g^2 + 5g)$

6. $(3t^6 - t^4 + 7t^2) - (t^5 + 4t^4 + 2t^3) - (2t^6 - t^5 + 9t^2)$

7. $(r^2 - s^2) - (4r^2s - rs^2 + r^2) + (s^2 - rs^2)$

8. $\left(-w^3+3w\right)-\left(2w^3-4w^2+7\right)-\left(4w^4+9w^2\right)-\left(5w^4+4w^2-7w\right)$

# Assignment for Multiplying Polynomials

**Name:** ____________________________

Compute each product and simplify.

1. $2r^3s(r^2 - 6rs^2 + 3r^2s^2 - 4s^2 + 2)$

2. $(2h - 3)(h^2 - h + 3)$

3. $(3x^2 - 2x - 1)(3x^2 + 2x + 1)$

4. $(x^2 + x - 2)^2$

5. $(x - 2)^3$

6. $(2d^2 - 5d + 2)(d^3 - 2d^2 + d - 3)$

7. $(2x - 5)^2(-x - 3)$

# Assignment for Dividing Polynomials

**Name:** ______________________________

Use long division to compute each quotient. Verify your answer.

1. $(6x^2 + x - 12) \div (2x + 3)$

2. $(2x^3 - 6x^2 - 7x - 4) \div (x - 4)$

3. $\left(16x^2 - 3\right) \div \left(8x - 4\right)$

4. $\left(3x^4 - x^3 + 5x^2 + 6\right) \div \left(x^2 + x + 1\right)$

# Assignment for U.S. Shirts Revisited

**Name:** ______________________________

Another local T-shirt shop, Hot Shirts, charges $5.50 per shirt plus a one-time charge of $49.95 to set up the design.

1. Write an algebraic equation to calculate the total cost of the shirts.

2. Write an algebraic equation to calculate the average cost of a shirt.

3. Create a graph for the total cost of the shirts.

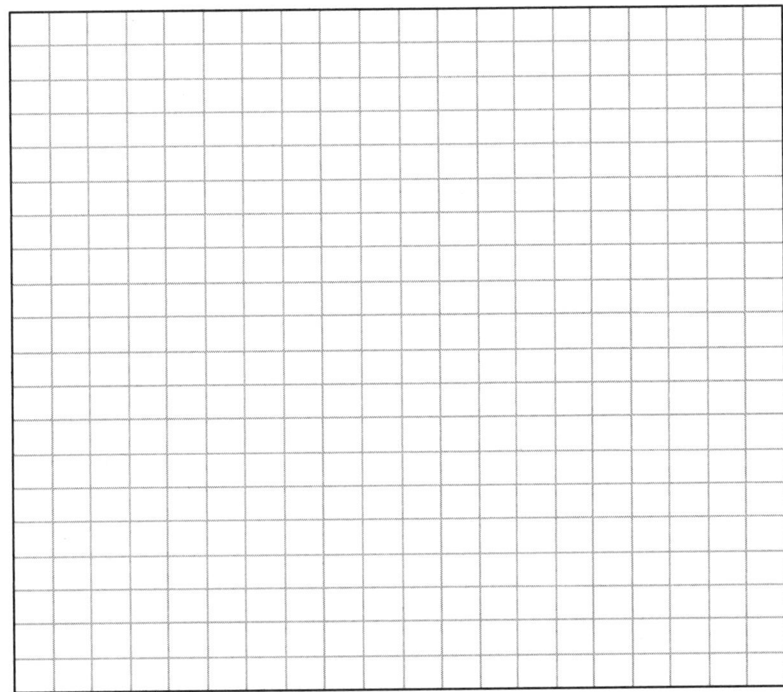

4. Create a graph for the average cost of a shirt.

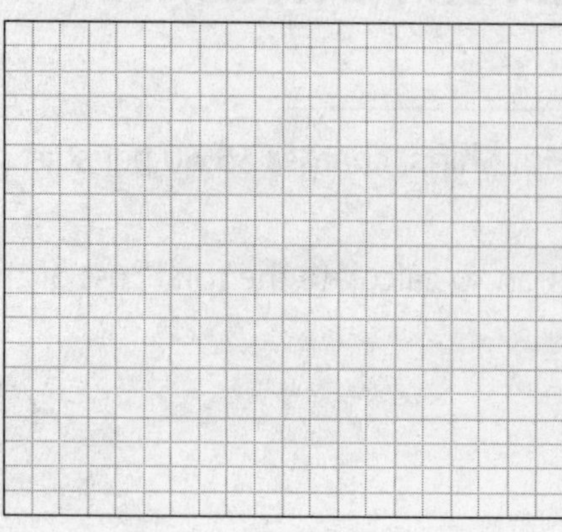

5. What is the average cost of a shirt if you order zero shirts? Use the equation to explain why this is the case

6. What is the lowest average cost you can get? Use your graph to explain why this is the case

7. How many shirts would need to be ordered for the average cost of a shirt from U.S. Shirts and Hot Shirts to be the same?

# Assignment for Simplifying Rational Expressions

**Name:** ______________________________

Factor the numerator and denominator of each rational expression and simplify completely.

1. $\dfrac{24x^3y^6z}{16x^2y^8z^3}$

2. $\dfrac{2x^3+10x^2}{3x^2+15x}$

3. $\dfrac{2x^2-8}{4x^2+4x-24}$

4. $\dfrac{8x^2-2x-3}{2x^2-7x-4}$

5. $\dfrac{4x^4-6x^3-18x^2}{6x^3-24x^2+18x}$

6. $\dfrac{2x^2y-6xy}{x^2y-2xy-3y}$

# Assignment for Multiplying Rational Expressions

**Name:** ____________________________

Compute each product and simplify completely.

1. $\frac{x^3+3x^2}{x-2} \bullet \frac{2x-4}{4x^2+12x}$

2. $\frac{b^2+2b-3}{2b+5} \bullet \frac{2b^2-b-15}{b^2-9}$

3. $\frac{2t-10}{t^2-16} \bullet \frac{t+4}{6t-30}$

4. $\frac{20p^2-50p-30}{2p^5-9p^4-5p^3} \bullet \frac{p^3-5p^2}{5p-15}$

5. $\frac{4x^2+x-3}{2x^2-3x-5} \bullet \frac{6x^2-11x-10}{12x^2-x-6}$

# Assignment for Dividing Rational Expressions

**Name:** ____________________________

Compute each quotient and simplify completely.

1. $\frac{4x-8}{x^2-3x+2} \div \frac{3x-6}{x-1}$

2. $\frac{f^2-9}{5f+10} \div \frac{f-3}{5f^2-20}$

3. $\frac{3a^2-6a}{a^2-6a+9} \div \frac{a^2-4}{a^2-a-6}$

4. $\frac{3w^2-8w+4}{9w^2-4} \div \frac{3w^2-5w-2}{9w^2-3w-2}$

5. $\dfrac{x+2y}{2x^2+3xy+y^2} \div \dfrac{2x^2+5xy+2y^2}{x+y}$

# Assignment for Adding and Subtracting Rational Expressions

**Name:** ____________________________

Compute each sum or difference and simplify completely.

1. $\frac{x^2-3}{2x^2-x-3}-\frac{-3x^2+4x}{2x^2-x-3}$

2. $\frac{y-1}{y+1}-\frac{y+1}{y-1}$

3. $\frac{10}{u^2-5u-14}+\frac{2}{u-7}$

4. $\frac{k-1}{k-2}-\frac{k-4}{k+1}$

5. $\dfrac{2x+1}{x^2+8x+16}-\dfrac{3}{x^2-16}$

# Assignment for Solving Rational Expressions

**Name:** ______________________________

Solve the following rational equations for the variable and check the solutions.

1. $\dfrac{16 - x}{5 + x} = 2$

2. $\dfrac{d}{d+1} + \dfrac{2}{d-3} = \dfrac{4}{d^2 - 2d - 3}$

3. $\dfrac{4r+1}{r+1} = \dfrac{12}{r^2-1} + 3$

4. $\dfrac{x}{x+1} + \dfrac{x}{x-2} = 2$

5. $\dfrac{1}{c} + \dfrac{c}{c+2} = 1$

# Assignment for Winning Percentage

**Name:** ______________________________

Your team's star basketball player is having some difficulty at the free throw line this year. So far this season, she has only made 6 shots out of 16 attempts.

1. What is her current free throw percentage?

2. What will be her free throw percentage if she makes her next 10 free throw attempts?

3. What will be her free throw percentage if she makes her next 20 free throw attempts?

4. Write an equation to her free throw percentage from the number of additional free throws made in a row.

5. Graph the function below.

6. How many additional free throws would she need to make in a row to have a free throw percentage of 80%? Solve algebraically.

7. How many additional free throws would she need to make in a row to have a free throw percentage of 90%? Solve algebraically.

8. How many additional free throws would she need to make in a row to have a free throw percentage of 100%? Solve algebraically.

## Assignment for The Five-Number Summary

**Name:** ______________________________

The Billboard top 20 albums on June 7, 2003 and June 4, 1994 are shown below.

| June 7, 2003 | | |
|---|---|---|
| **Rank** | **Weeks on Chart** | **Artist and Title** |
| 1 | 1 | Staind, 14 Shades of Grey |
| 2 | 1 | Deftones, Deftones |
| 3 | 6 | Kelly Clarkson, Thankful |
| 4 | 12 | Evanescence, Fallen |
| 5 | 16 | 50 Cent, Get Rich Or Die Tryin' |
| 6 | 3 | Soundtrack, The Matrix Reloaded |
| 7 | 65 | Norah Jones, Come Away With Me |
| 8 | 5 | Soundtrack, The Lizzie McGuire Movie |
| 9 | 1 | David Banner, Mississippi: The Album |
| 10 | 8 | Cher, The Very Best of Cher |
| 11 | 4 | Soundtrack, American Idol Season 2: All-Time Classic American Love Songs |
| 12 | 1 | Ricky Martin, Almas Del Silencio |
| 13 | 9 | Linkin Park, Meteora |
| 14 | 1 | Jo Dee Messina, Greatest Hits |
| 15 | 3 | Jack Johnson, On And On |
| 16 | 3 | The Isley Brothers Featuring Ronald Isley, Body Kiss |
| 17 | 1 | Weird Al Yankovic, Poodle Hat |
| 18 | 9 | Celine Dion, One Heart |
| 19 | 28 | Sean Paul, Dutty Rock |
| 20 | 9 | Various Artists, NOW 12 |

| June 4, 1994 | | |
|---|---|---|
| **Rank** | **Weeks on Chart** | **Artist and Title** |
| 1 | 44 | Soundtrack, The Crow |
| 2 | 115 | Tim McGraw, Not A Moment Too Soon |
| 3 | 102 | Ace Of Base, The Sign |
| 4 | 53 | Benedictine Monks of Santo Domingo De Silos, Chant |
| 5 | 36 | Soundtrack, Above The Rim |
| 6 | 93 | Counting Crows, August |
| 7 | 51 | Pink Floyd, The Division Bell |
| 8 | 65 | R Kelly, 12 Play |
| 9 | 83 | Reba McEntire, Read My Mind |
| 10 | 72 | All-4-One, All-4-One |
| 11 | 26 | Indigo Girls, Swamp Ophelia |
| 12 | 96 | Toni Braxton, Toni Braxton |
| 13 | 128 | Mariah Carey, Music Box |
| 14 | 63 | Enigma, The Cross Of Changes |
| 15 | 114 | Yanni, Live At The Acropolis |
| 16 | 149 | Celine Dion, The Colour Of My Love |
| 17 | 89 | The Smashing Pumpkins, Siamese Dream |
| 18 | 17 | Erasure, I Say I Say I Say |
| 19 | 75 | Soundgarden, Superunknown |
| 20 | 44 | Travis Tritt, Ten Feet Tall |

Source: http://www.billboard.com

1. Examine the number of weeks on the chart for the top 20 albums. Determine the minimum, maximum, median, 1st quartile, and 3rd quartile values for each data set.

2. Create box and whisker plots for each data set on the same graph.

3. What can you conclude about the data based on the graphs? Provide an explanation why the data appears as it does.

## Assignment for Mean and Standard Deviation

**Name:** ______________________________

In 2004, The US Fire Administration performed a census to determine the number of fire departments in the United States. The results of the census as of April 2004 are listed below.

| State | Number of Fire Departments | State | Number of Fire Departments | State | Number of Fire Departments |
|---|---|---|---|---|---|
| AL | 621 | KY | 564 | ND | 240 |
| AK | 105 | LA | 309 | OH | 1014 |
| AZ | 204 | ME | 292 | OK | 534 |
| AR | 592 | MD | 221 | OR | 255 |
| CA | 725 | MA | 331 | PA | 1532 |
| CO | 286 | MI | 846 | RI | 66 |
| CT | 238 | MN | 608 | SC | 317 |
| DE | 42 | MS | 300 | SD | 256 |
| DC | 2 | MO | 617 | TN | 478 |
| FL | 428 | MT | 224 | TX | 1115 |
| GA | 421 | NE | 301 | UT | 152 |
| HI | 10 | NV | 69 | VT | 170 |
| ID | 141 | NH | 189 | VA | 424 |
| IL | 949 | NJ | 597 | WA | 368 |
| IN | 615 | NM | 223 | WV | 334 |
| IA | 593 | NY | 1367 | WI | 661 |
| KS | 432 | NC | 771 | WY | 91 |

Source: http://www.usfa.fema.gov/

1. Determine the mean, median, and standard deviation for the data above and describe what each means in the problem.

1. Determine the mean and standard deviation for each problem.

   a. The top 20 points leaders in the 2004 Daytona 500 are shown below.

| Driver | Points |
|---|---|
| Dale Earnhardt Jr. | 185 |
| Tony Stewart | 180 |
| Scott Wimmer | 170 |
| Kevin Harvick | 165 |
| Jimmie Johnson | 160 |
| Joe Nemechek | 150 |
| Elliott Sadler | 146 |
| Jeff Gordon | 147 |
| Matt Kenseth | 143 |
| Dale Jarrett | 134 |

| Driver | Points |
|---|---|
| Bobby Labonte | 130 |
| Greg Biffle | 127 |
| John Andretti | 129 |
| Casey Mears | 121 |
| Dave Blaney | 118 |
| Kurt Busch | 115 |
| Ward Burton | 112 |
| Ricky Rudd | 109 |
| Brendan Gaughan | 106 |
| Terry Labonte | 108 |

Source: http://msn.espn.go.com/

   b. The Top Grossing Movies at the USA Box Office as of May 2004 are shown below.

| Movie | Total Box Office Gross |
|---|---|
| Titanic | $600,743,440 |
| Star Wars: A New Hope | $460,935,655 |
| E.T. The Extra-Terrestrial | $434,949,459 |
| Star Wars: The Phantom Menace | $431,065,444 |
| Spider-Man | $403,706,375 |
| The Return Of The King | $371,147,794 |
| Jurassic Park | $356,763,175 |
| The Two Towers | $340,478,898 |
| Finding Nemo | $339,714,367 |
| Forrest Gump | $329,452,287 |

| Movie | Total Box Office Gross |
|---|---|
| The Lion King | $328,423,001 |
| Harry Potter And The Sorcerer's Stone | $317,557,891 |
| The Fellowship Of The Ring | $313,837,577 |
| Star Wars: Attack Of The Clones | $310,675,583 |
| Star Wars: Return Of The Jedi | $309,064,373 |
| Independence Day | $306,200,000 |
| Pirates Of The Caribbean | $305,411,224 |
| The Sixth Sense | $293,501,675 |
| Star Wars: The Empire Strikes Back | $290,158,751 |
| Home Alone | $285,761,243 |

Source: http://www.imdb.com/

# Assignment for Binomial Probabilities

**Name:** ______________________________

1. During the 2002-2003 NHL season, Alexander Mogilny of the Toronto Maple Leafs had a shooting percentage of 20%. This means that he scored a goal 20% of the time he shot the puck on the net. During one game, he has five shots on the net.

   a. What is the probability that he will have zero goals in the game?

   b. What is the probability that he will have exactly one goal in the game?

   c. What is the probability that he will have exactly two goals in the game?

   d. What is the probability that he will have exactly three goals in the game?

   e. What is the probability that he will have exactly four goals in the game?

   f. What is the probability that he will have exactly five goals in the game?

   g. What is the probability that he will have at least three goals in the game?

2. You buy many of your CDs from a used CD store. Based on what you have observed, about 10% of the CDs in the store are scratched. If you pick out 7 CDs, what is the probability that

   a. none of the CDs are scratched?

   b. exactly one CD is scratched?

   c. exactly two CDs are scratched?

   d. exactly three CDs are scratched?

   e. exactly four CDs are scratched?

   f. exactly five CDs are scratched?

   g. exactly six CDs are scratched?

   h. exactly seven CDs are scratched?

 

# Assignment for Binomial Probability Distributions

**Name:** ______________________________

1. White Brite toothpaste claims that 4 out of 5 dentists recommend it.

   a. If 150 dentists are chosen at random, how many do you expect to recommend White Brite?

   b. Calculate the mean and standard deviation for a binomial probability with n = 150 and p = 0.80.

2. During the 2003-2004 NBA season, Shaquille O'Neal led the league with a 0.578 field goal percentage.

   a. If Shaq attempts 50 shots in a game, how many shots do you expect him to make?

   b. Calculate the mean and standard deviation for a binomial probability with n = 50 and p = 0.578.

3. As a promotion, the band Deck of Jack is randomly inserting autographs into their latest CD. Their web site states that one out of every 20 CDs will include an autograph.

   a. If Deck of Jack sells 750,000 CDs, how many will contain autographs?

   b. Calculate the mean and standard deviation for a binomial probability with n = 750,000 and p = 0.05.

4. Wheels bus service claims that 85% of the seats on all of their buses are booked for a specific date.

   a. If Wheels has 200 buses and each one seats 48 people, how many tickets do you expect Wheels to sell?

   b. Calculate the mean and standard deviation for a binomial probability with $n = 9600$ and $p = 0.85$.

5. From past experience, Calc-U-Late computers estimates that 35% of attendees at exhibits visit their booth.

   a. If the national computer exhibit has 75,000 attendees, how many do you expect to vist Calc-U-Late's booth?

   b. Calculate the mean and standard deviation for a binomial probability with $n = 75{,}000$ and $p = 0.35$.

6. Due to a manufacturing error, 1% of the Knights of the Round Table action figures are packaged without a sword.

   a. If 350,000 action figures are shipped, how many do you expect to be missing a sword?

   b. Calculate the mean and standard deviation for a binomial probability with $n = 350{,}000$ and $p = 0.01$.

# Assignment for The Normal Curve

**Name:** ______________________________

1. Estimate the mean and standard deviation of each of the normal curves below.

 a.

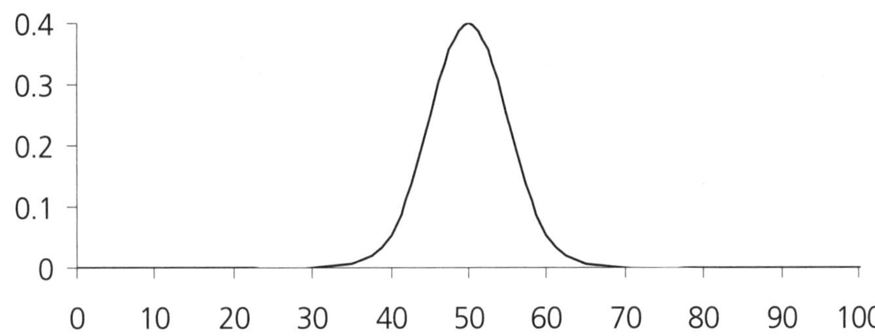


 b.

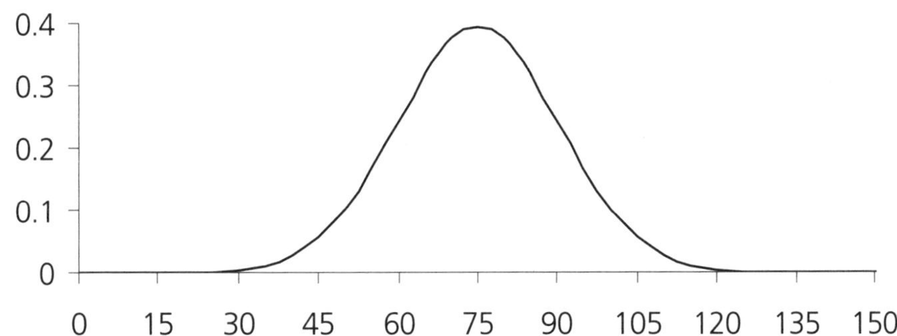


 c.

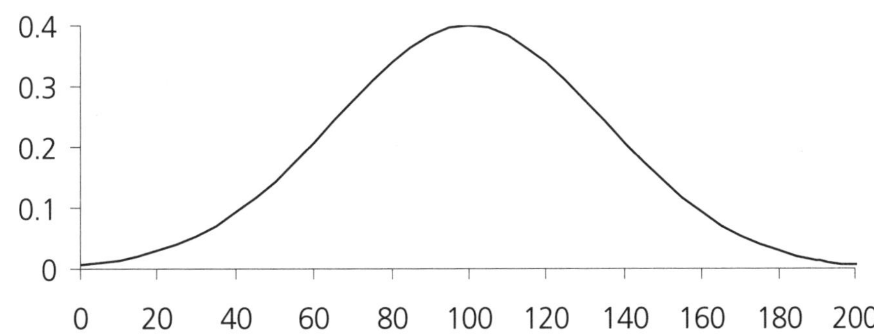

d.

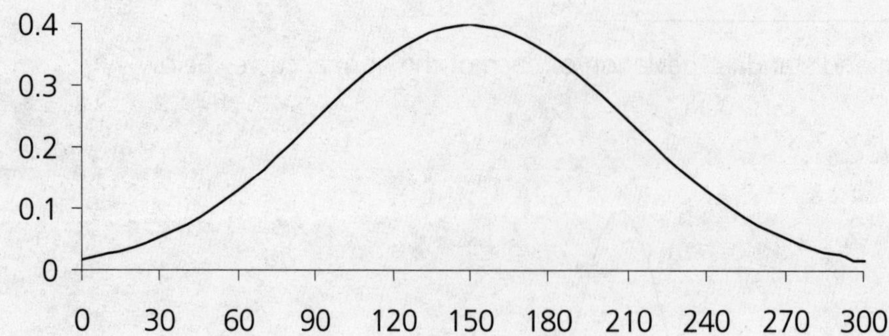
0.4
0.3
0.2
0.1
0
0 30 60 90 120 150 180 210 240 270 300

e.

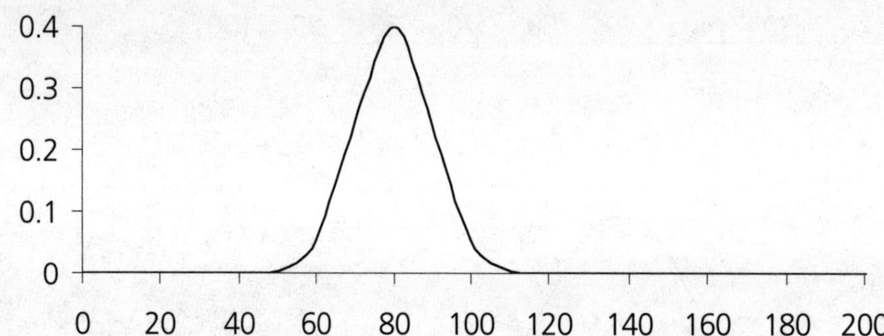
0.4
0.3
0.2
0.1
0
0 20 40 60 80 100 120 140 160 180 200

f.

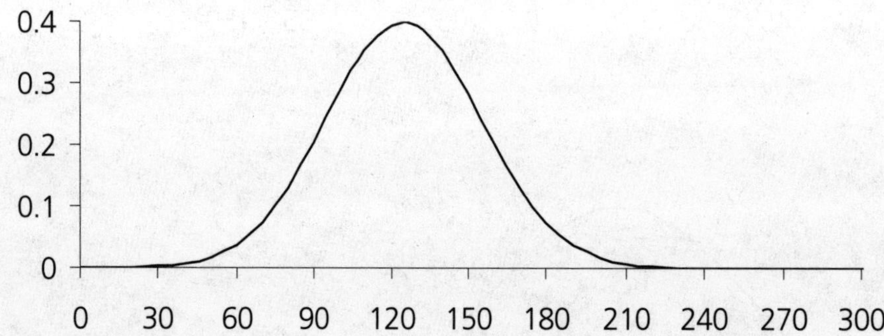
0.4
0.3
0.2
0.1
0
0 30 60 90 120 150 180 210 240 270 300

# Assignment for Standard Scores

**Name:** ______________________________

1. The Graduate Record Exam (GRE) is a test consisting of three sections (verbal, quantitative, and analytical) that is often required for admission to graduate school. A summary of the mean and standard deviation for each section appears below.

| | **Verbal** | | **Quantitative** | | **Analytical** | |
|---|---|---|---|---|---|---|
| **Year** | **Mean** | **Standard Deviation** | **Mean** | **Standard Deviation** | **Mean** | **Standard Deviation** |
| 1996 | 473 | 114 | 558 | 139 | 549 | 131 |
| 1997 | 472 | 113 | 562 | 139 | 548 | 129 |
| 1998 | 471 | 113 | 569 | 141 | 543 | 133 |
| 1999 | 468 | 114 | 565 | 143 | 542 | 133 |
| 2000 | 465 | 116 | 578 | 147 | 562 | 141 |

Calculate the standard score (z-score) for each situation below and explain what the standard score means.

a. Tony scored 500 on the Verbal section in 1997

b. Maria scored 520 on the Analytical section in 2000

c. Jerome scored 700 on the Quantitative section in 1999

2. In 2000, 68% of the Verbal section scores fell between what two scores?

3. In 2000, 95% of the Verbal section scores fell between what two scores?

4. In 2000, 99.7% of the Verbal section scores fell between what two scores?

5. Estimate the probability of each of the following using the table of probabilities:
   a. Scoring less than 700 on the Analytical section in 2000

   b. Scoring more than 700 on the Analytical section in 2000

   c. Scoring less than 400 on the Quantitative section in 1999

   d. Scoring more than 400 on the Quantitative section in 1999

   e. Scoring more than 650 on the Verbal section in 2000

# Assignment for Confidence Intervals

**Name:** ______________________________

1. You are writing an article for the school newspaper on electives. You decide to conduct a survey of some students in your school. You ask 128 students if they feel the school offers enough electives. Of the 128 students you asked, 76 said they felt the school did offer enough electives.

   a. What is the sample proportion for your survey?

   b. Find a 90% confidence interval.

   c. Find a 99% confidence interval.

2. In a survey of 2500 Americans, 841 responded that they donated to schools, colleges, or universities.

   a. What is the sample proportion for your survey?

   b. Find a 95% confidence interval.

   c. Find a 99.9% confidence interval.

3. A recent survey of parents of kids ages 2 to 11 found that when making a difficult parenting decision, 70% rely on their gut instinct as opposed to expert advice. Give a 99% confidence interval for the proportion of parents who rely on their gut instincts if the sample size is:

   a. 100

   b. 1000

   c. 100,000

3. Research has shown that 83% of employees use their work computers for personal use. Give a 95% confidence interval for the employee population if the sample size is:

   a. 50

   b. 500

   c. 5000

# Assignment for Hypothesis Testing

**Name:** ______________________________

You and your friend are watching TV at your house when you see a program about supernatural phenomena. Upon hearing this, your friend claims that she is actually psychic. Having known her for as long as you have, you don't believe her.

In order to test her claim, you decide to go through a deck of cards and see if she can guess the suit of each card. She says that she can use her psychic powers to guess the correct suit of the card more often than by random chance.

You decide to do a hypothesis test with a significance level of $\alpha = 0.05$

1. Formulate the hypothesis for the experiment.

   a. What is the null hypothesis, $H_0$?

   b. What is the alternate hypothesis, $H_\alpha$?

2. You go through a standard deck of cards (52 cards) a total of 4 times. You monitor the number of correct and incorrect guesses.

   a. Describe your test statistic using the values for *p* and *n*.

   b. Determine the mean and standard deviation for the binomial experiment.

3. After going through the deck 4 times, your friend guessed correctly 62 times.
   a. Determine the standard score for the sample statistic.

   b. Find the p-value, the probability of getting a result at least as extreme as the observed result.

4. Compare your calculated p-value to the predetermined significance level of $\alpha = 0.05$. What conclusion can you draw?

5. Your friend says you made a mistake counting and she actually guessed correctly 63 times. Does this change your conclusion?

# Assignment for Parts of a Circle

**Name:** ____________________________

1. Name at least one example of each term in the figure below. Some terms have more than one answer. (For example, one chord in the figure is line segment *GM*, and there are others.)

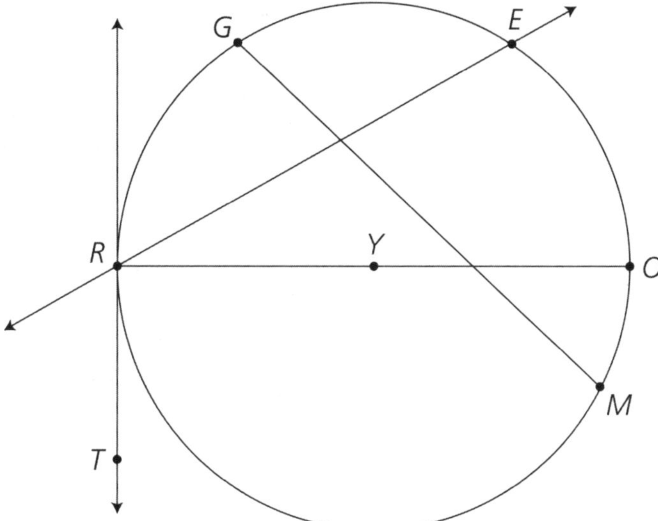


a. Chord:

b. Radius:

c. Diameter:

d. Tangent:

e. Arc:

f. Major arc:

g. Minor arc:

h. Semicircle:

i. Secant:

2. In this circle, draw:

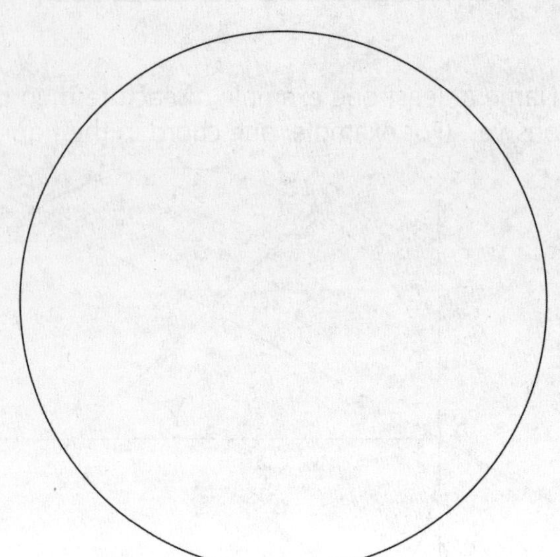

a. Tangent *JB*

b. Radius *RB*

c. Diameter *BH*

d. Chord *HI*

e. Secant *HJ*

3. Name *all* the examples of each term in the circle in problem 2.

a. Major arc:

b. Minor arc:

c. Semicircle:

# Assignment for Angles of a Circle

**Name:** ____________________________

1. The center of this circle is point $S$.

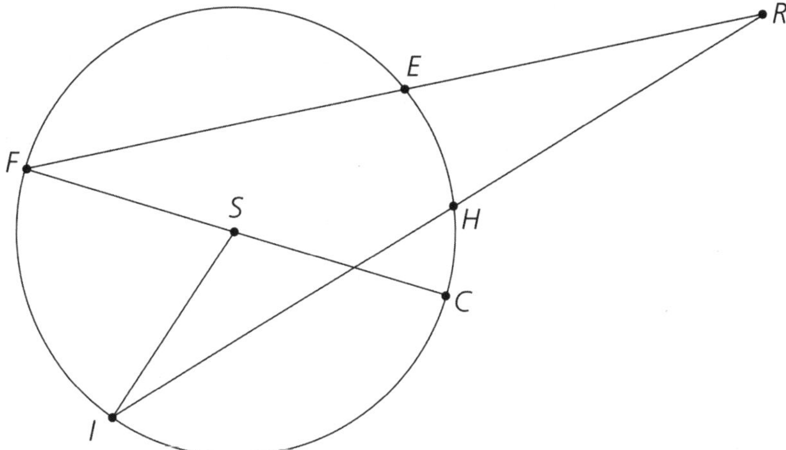


   a. If the measure of arc $CE$ is 59°, find the measure of arc $CFE$.

   b. If the measure of $\angle CSI$ is 124°, find the measure of arc $FI$.

   c. If the measure of arc $CE$ is 55°, find the measure of $\angle EFC$.

   d. If the measure of $\angle FSI$ is 71°, find the measure of arc $IC$.

   e. What is the measure of arc $FEC$?

2. If $m\angle ANG = 74°$, find $m\angle AEG$ and the measure of arc $ANG$.

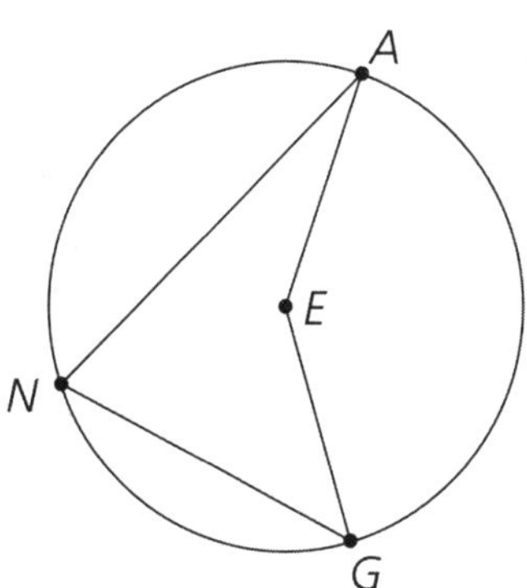

3. The measure of arc $CA$ is 105°, the measure of arc $EA$ is 47°, and the measure of arc $TE$ is 100°. Find $m\angle ETC$, $m\angle TCE$, $m\angle CAE$, and $m\angle TEA$.

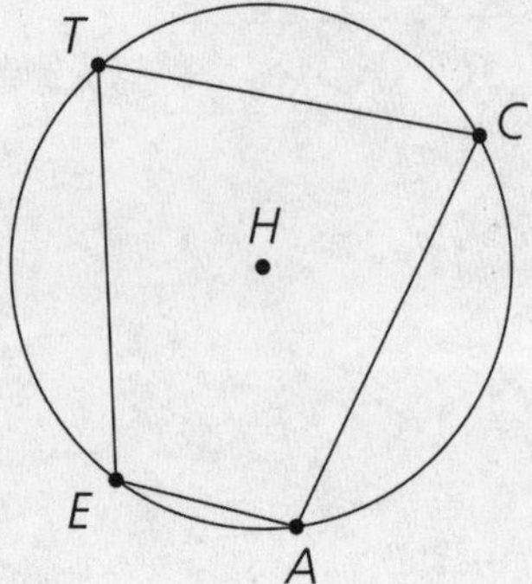

# Assignment for Interior Angles of a Circle

**Name:** ____________________________

1. Find the value of $x$.

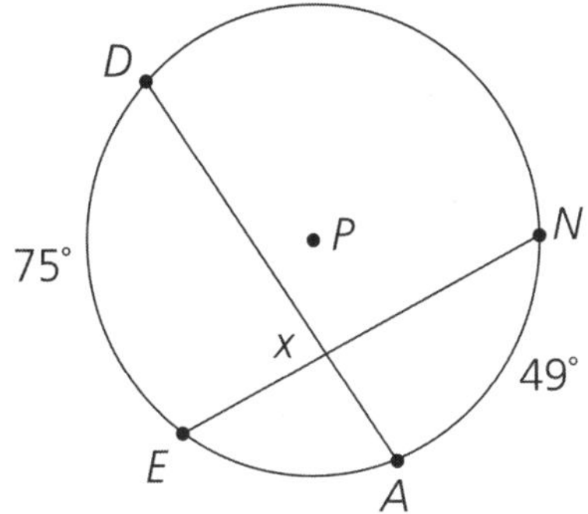


2. Find the measure of arc $DE$.

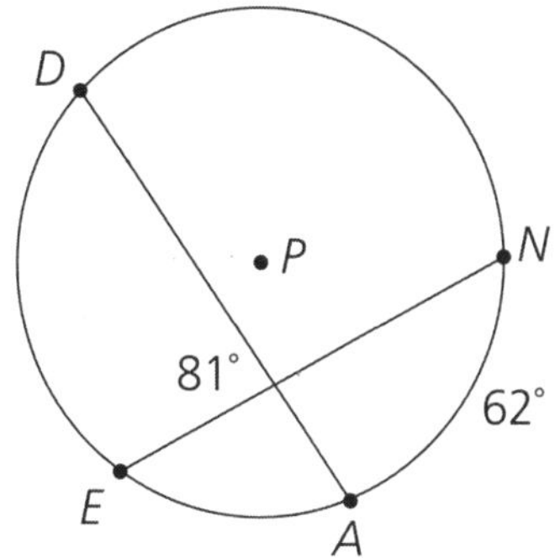


3. Find the measure of arc $EA$.

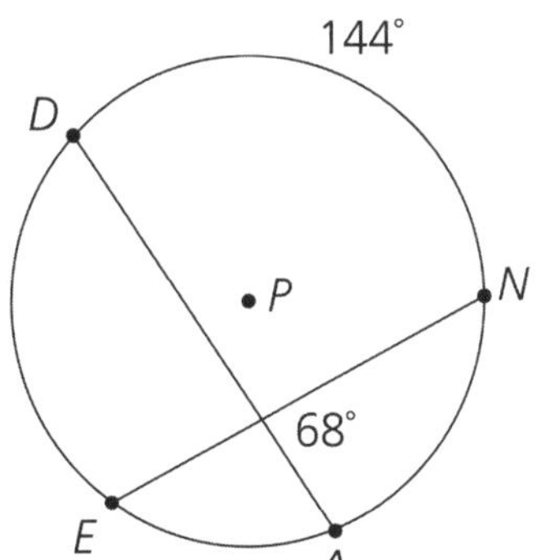

4. Find the value of $x$, in circle $O$.

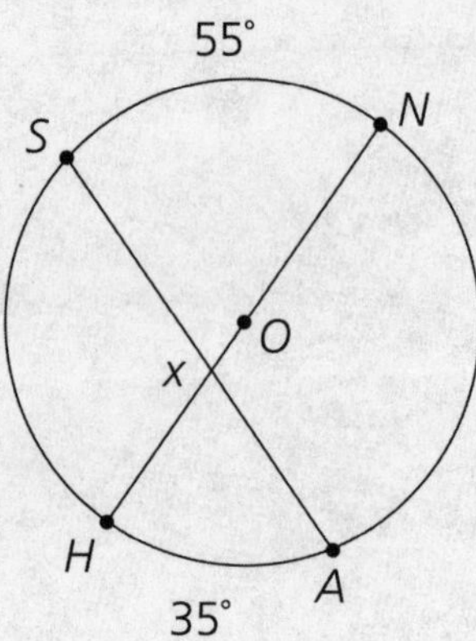


5. Find the value of $x$, in circle $O$.

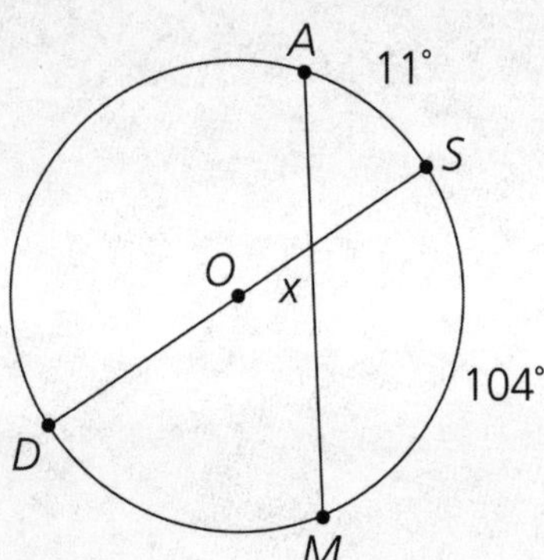


6. Find the measure of arc $NY$.

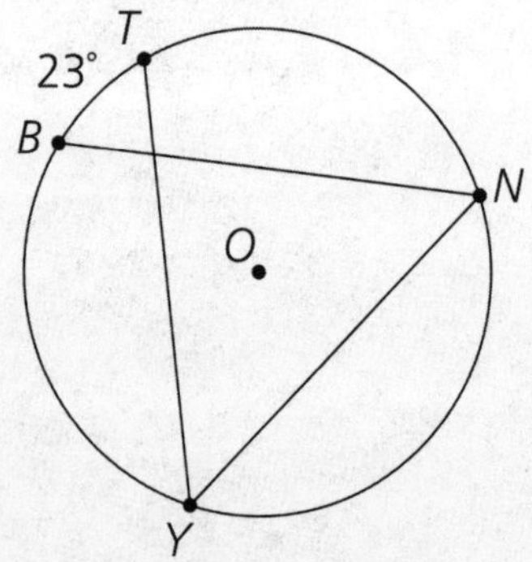


$m\angle TYN = 54°$
$m\angle BNY = 54°$

# Assignment for Exterior Angles of a Circle

**Name:** ______________________________

1. The measure of arc *ER* is 38° and the measure of arc *OT* is 121°. Find the measure of exterior angle *U*.

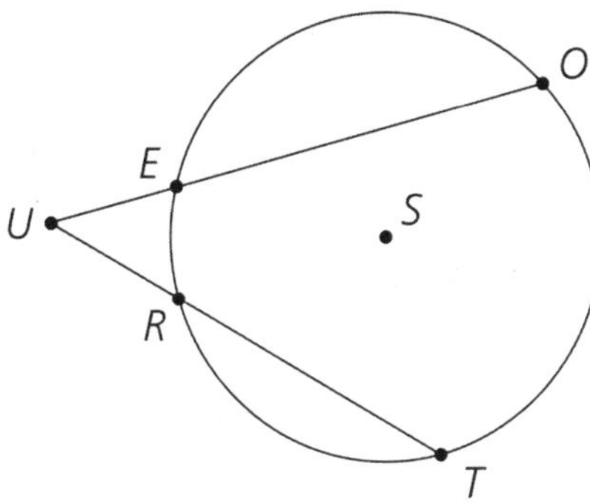


2. Segment *OE* is the diameter of circle *S* and the measure of arc *OT* is 132°. Find the measure of exterior angle *U*.

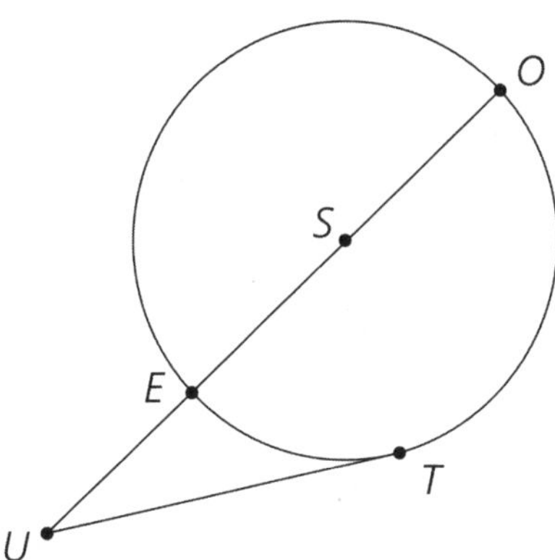


3. In this figure, *HO* = *ES*, the measure of arc *OH* is 41°, and the measure of arc *HE* is 171°. Find the measure of exterior angle *U*.

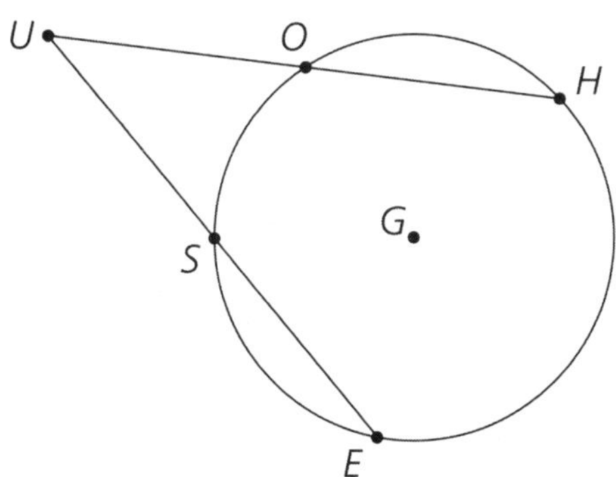

4. If the measure of arc *HE* is 99°, find the measure of exterior angle *U*.

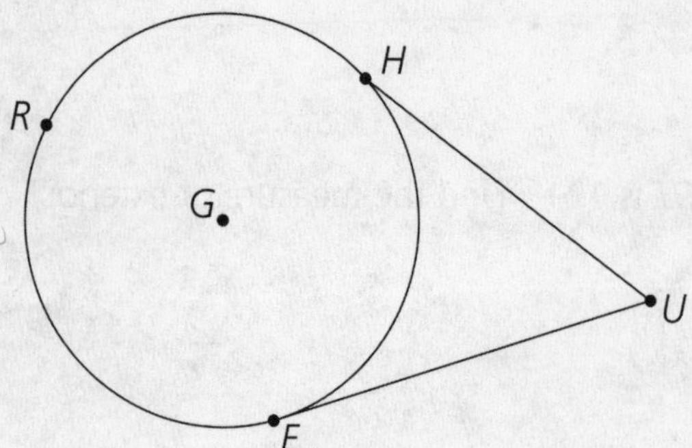


5. The measure of exterior angle *C* is 57° and the measure of arc *RE* is 141°. Find the measure of arc *IL*.

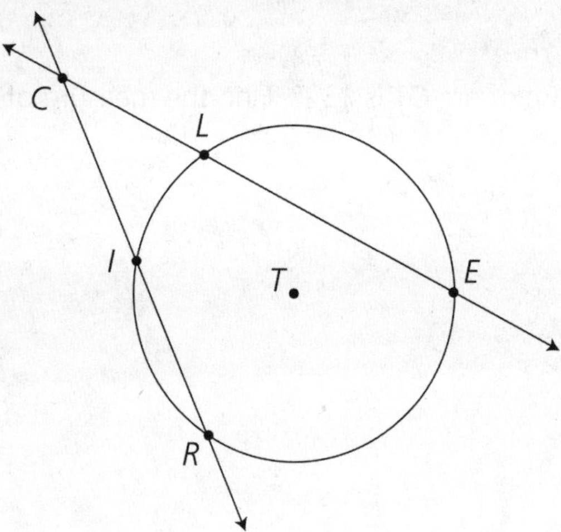


6. In this figure, *CB* || *DE*, the measure of arc *CB* is 54°, the measure of arc *DE* is 86°, and the measure of exterior angle *F* is 24°. Find the measure of arc *GI*.

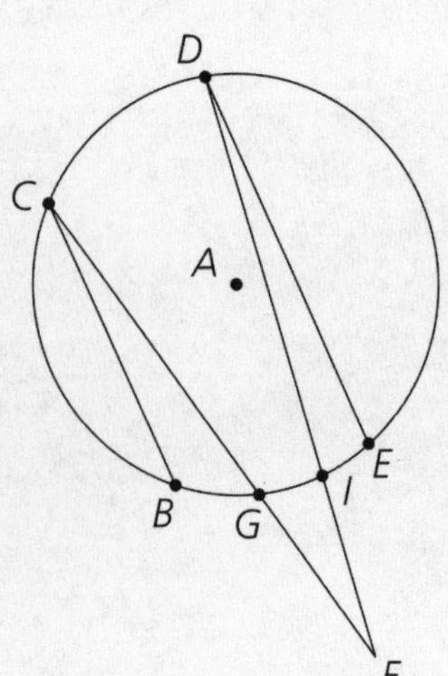

# Assignment for Arcs of a Circle

**Name:** ____________________________

1. Shade the arc determined by each of the chords shown in the following circle.

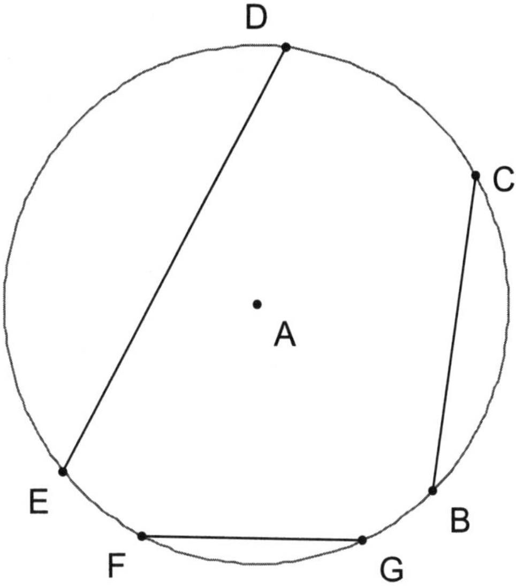


2. Using the circle pictured above, name at least two different minor arcs, making sure to use the correct naming convention.

3. Using the circle pictured above, name at least two major arcs, making sure to use the correct naming convention.

4. Is it possible for the measure of arc ED to be 180°? Why or why not?

5. Given that the measure of arc CB is 76°, find the measure of arc CEB.

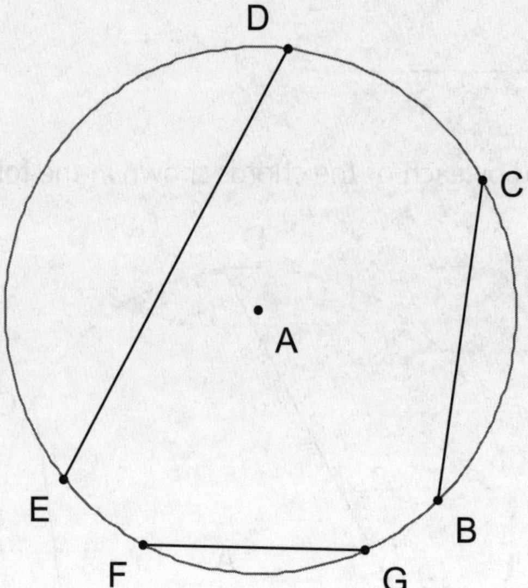


6. Given that the measure of arc DEB is 231°, find the measure of arc EB.

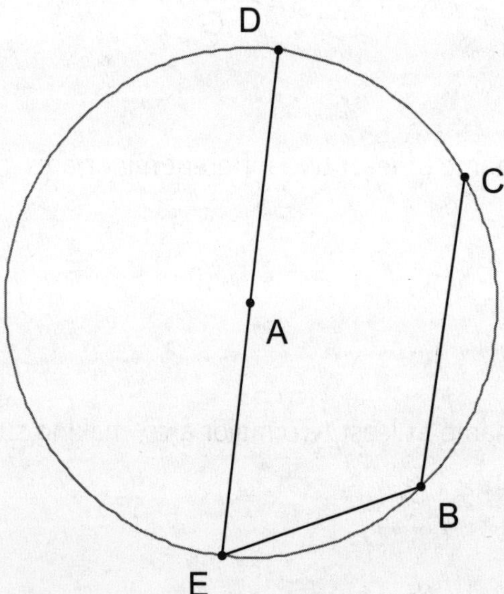


a. Find the measure of arc BD.

# Assignment for Arc Length

**Name:** ______________________________

1. Describe the difference between the *measure* of minor arc *BC* and the *length* of minor arc *BC* in circle *A*. Use any tools you choose to accomplish this.

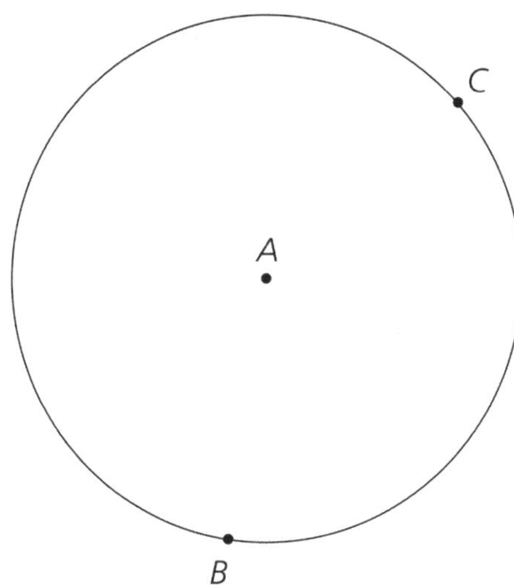


2. The radius of circle *E* is 16 cm and the measure of angle *JSB* is 40°. Find the *length* (not the *measure*) of arc *JB*.

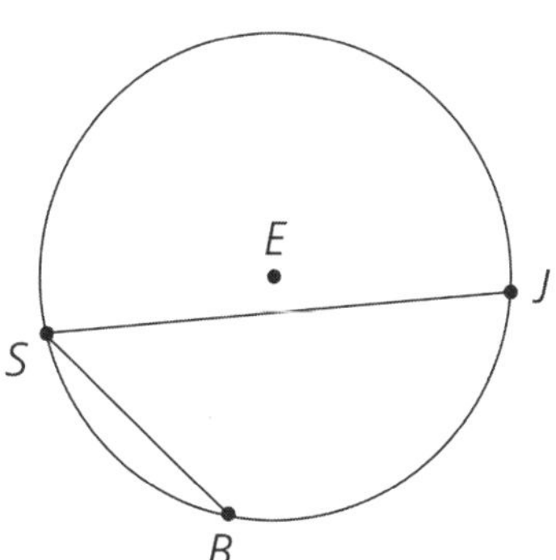

3. The radius of circle *I* is 6 mm and the measure of arc *HC* is 80°. Find the *length* (not the *measure*) of arc *SC*.

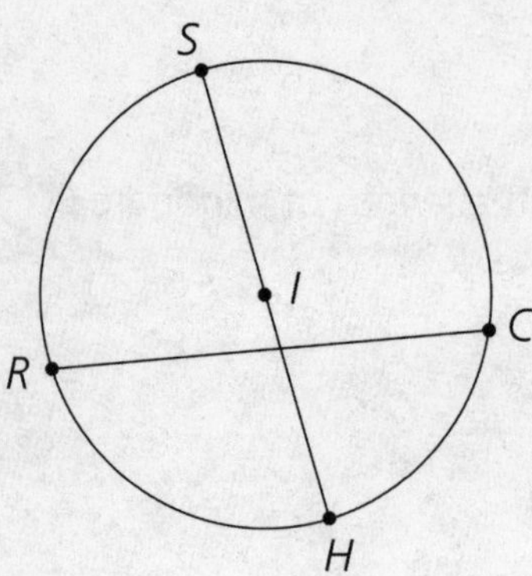


4. The length of arc *SJ* is 24π cm and $m\angle JOS = 80°$. Find the length of a diameter in circle *H*.

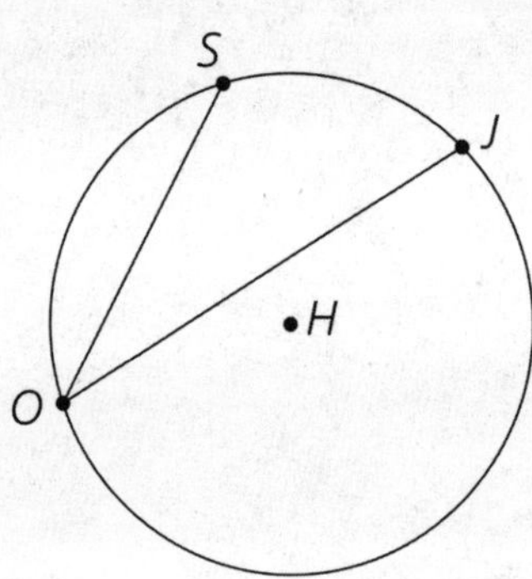


5. The length of arc *WH* is 10π mm, line *WT* is parallel to line *HI*, and the radius of circle *E* is 15 mm. Find the measure of arc *TI*.

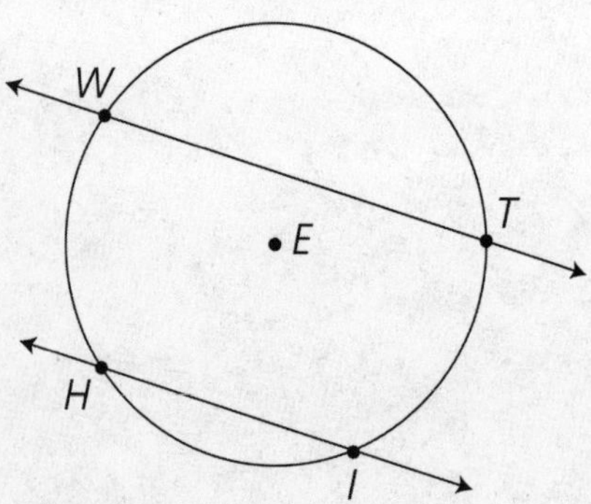

# Assignment for Sectors of a Circle and Segments of a Circle

**Name:** ______________________________

1. If the radius of circle *A* is 18 cm and the measure of central $\angle BAC$ is 60°, find the area of the sector of the circle determined by the radii AB and AC. (Hint: First solve for the height of triangle *ABC* by using what you know about the length of sides in special right triangles.)

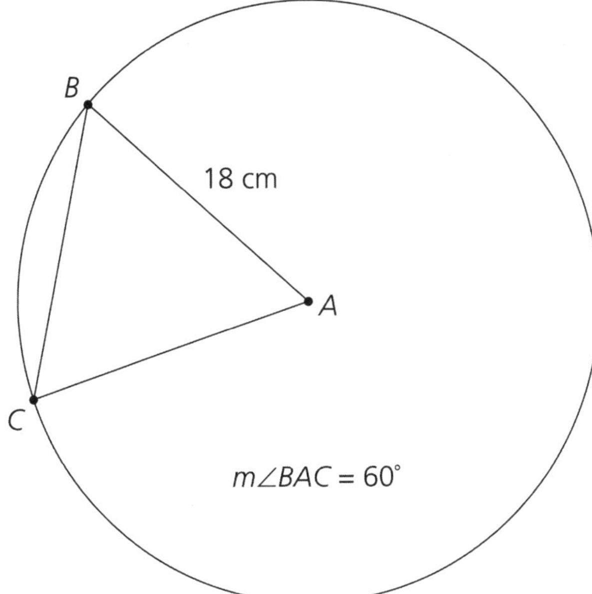


a. Next, determine the area of the segment of the circle determined by the chord *BC*.

2. If the radius of circle $A$ is 22 cm and the measure of central $\angle RAT$ is 90°, find the area of the sector of the circle determined by the two radii AB and AT. (Hint: First solve for the height of $\Delta RAT$ by using what you know about the length of sides in special right triangles.)

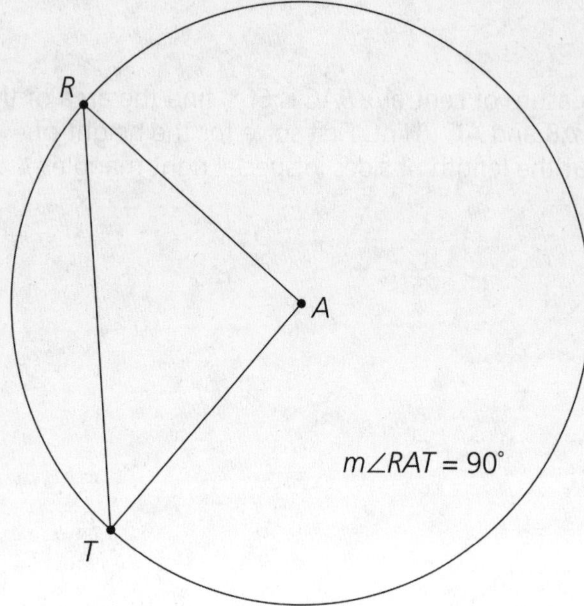


a. Now, find the area of the segment of the circle determined by chord $RT$.

# Assignment for Chords of a Circle

**Name:** ______________________________

1. The center of this circle is point $T$.

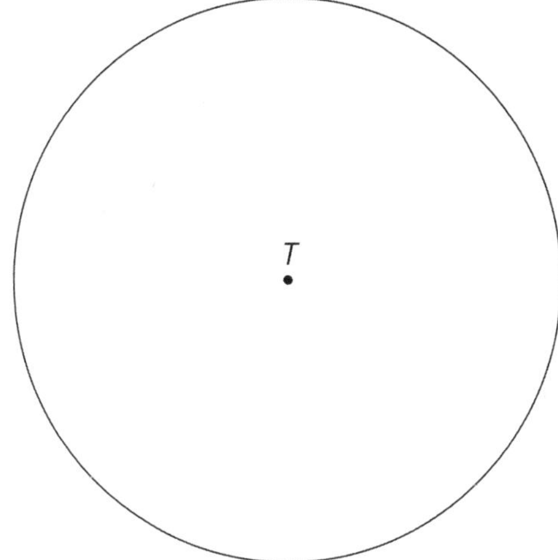


a. Draw an inscribed right angle in circle $T$. Label each point where the angle intersects the circle. What is the name of the right angle?

b. Draw the chord determined by the inscribed right angle. What is the name of the chord?

c. What else do you know about the chord determined by an inscribed right angle?

d. Draw a second inscribed right angle in circle $T$. Label each point where the angle intersects the circle. What is the name of the second right angle?

e. Draw the chord determined by the second inscribed right angle. What is the name of the chord?

f. What else do you know about the chord determined by the second inscribed right angle?

g. Do you think every inscribed right angle will determine the longest chord of the circle (that is, the diameter)? Why or why not?

2. Below is a section of a circle. Locate the center of the circle by construction. (Draw two chords and construct their perpendicular bisectors.)

3. Use only the information given in this circle to determine which chord is longer, *IH* or *JK*. Justify your conclusion.

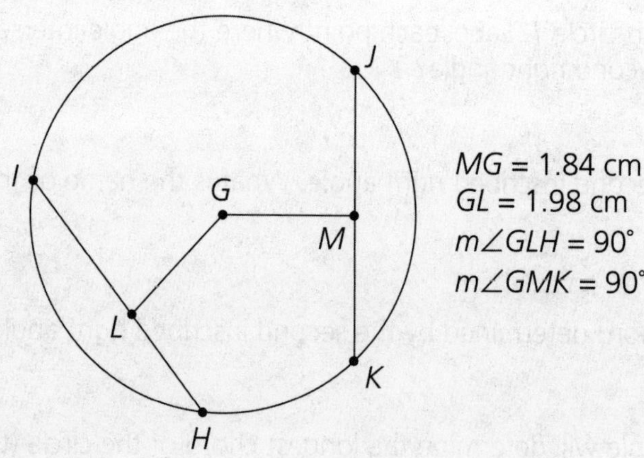

4. In this circle, draw two chords that are equidistant from the center of the circle. Are the chords the same length? Justify your conclusion.

5. In this circle, draw two chords that are the same length. Are the chords equidistant from the center of the circle? Justify your conclusion.

# Assignment for Tangents in a Circle

**Name:** ______________________________

1. In the space below, create a circle O with a tangent line drawn.

2. Label the point of tangency as point A.
3. Label another point on the tangent as point B.
4. Create the second tangent line to the circle that passes through point B.
5. Label this second point of tangency as point C.
6. Create the radii OA and OC.
7. What is the m∠OAB?

8. What is the m∠OCB?

9. Use a protractor to measure ∠AOC.
10. What is the m∠ABC? Explain how you found this measure.

# Assignment for Tower Drop

**Name:** ______________________

Sam is on the track team in college. His favorite event is the 400-meter hurdles. Today he is attempting to qualify for the regional track meet. If his time is under a minute, he will qualify, but he has never been so nervous about an event. When the gun sounds, Sam speeds from the start and about 5 seconds later realizes that he is going out too fast. He slows his pace a bit. There are ten hurdles to clear. Sam takes the first three hurdles with no trouble, but at the fourth one, his back leg just catches the hurdle. His steps are off going into the fifth hurdle, so he has to slow a bit more. On the sixth hurdle, he trips and falls. He quickly gets back up and finishes the race, but his time is 20 seconds off the qualifying time.

1. Sketch a graph showing the height of Sam's hips above the ground over the time he was running the 400-meter hurdles.

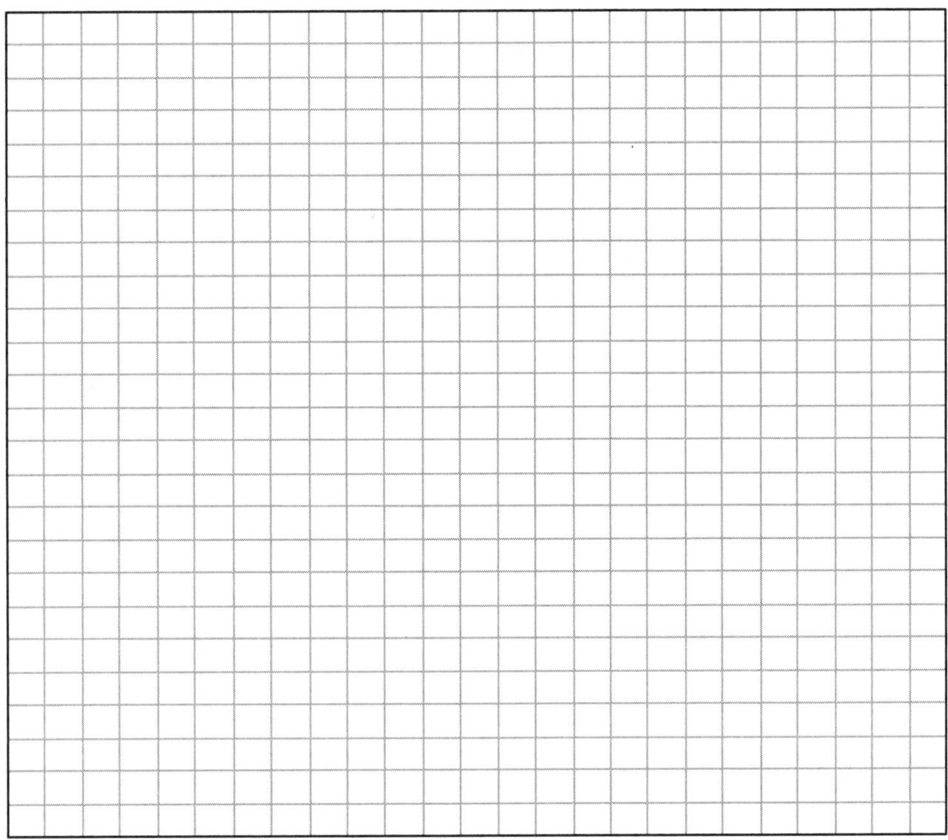

2. Does the graph you created model a periodic function? Why or why not?

3. What are the similarities and differences between the graph you created for the *Tower Drop* problem in the textbook and the hurdles problem in this assignment?

# Assignment 1 for Gum on a Bicycle Wheel

**Name:** ______________________________

In the *Gum on a Bicycle Wheel* problem, you analyzed the height of the gum as the bicycle wheel goes around based on the number of revolutions of the wheel. Now consider the inverse situation. Given the height of gum on the bicycle wheel, determine the amount of revolution of the bicycle wheel. Recall that the wheel has a diameter of 28 inches.

1. When the gum is at a height of 7 inches, what is the amount of revolution of the bicycle wheel?

   a. Give another possible answer to the question. If there is no other possible answer, explain why not.

   b. How many possible answers are there to the question?

2. The gum on the bicycle wheel is approximately 1.9 inches from the ground after a revolution of 30°.

   a. When is the next time the gum will be approximately 1.9 inches from the ground?

   b. When else will the gum be approximately 1.9 inches from the ground?

3. The gum on the bicycle wheel is at a certain height after a revolution of 100°. When else will it be at that height?

4. The gum on the bicycle wheel is at a certain height after a revolution of $x$°. When else will it be at that height?

# Assignment 2 for Gum on a Bicycle Wheel

**Name:** ___________________________

1. In the *Gum on a Bicycle Wheel* problem, you analyzed the height of the gum as the bicycle wheel went around based on the number of revolutions of the wheel. This time, consider the distance the bicycle has traveled as the wheel goes around.

   a. How far will the bicycle travel after one revolution of the wheel? Recall that the wheel has a 28-inch diameter.

   b. If you were to graph the distance traveled based on the number of revolutions, what would you expect the graph to look like?

2. Create a table of values for the distance the bicycle travels with respect to the number of revolutions of the wheel.

| Revolutions of the Bicycle Wheel | Distance Traveled |
|---|---|
| revolutions | inches |
| | |
| | |
| | |
| | |
| | |
| | |
| | |
| | |
| | |
| | |
| | |
| | |

3. Create a graph showing the distance the bicycle travels with respect to the number of revolutions of the wheel.

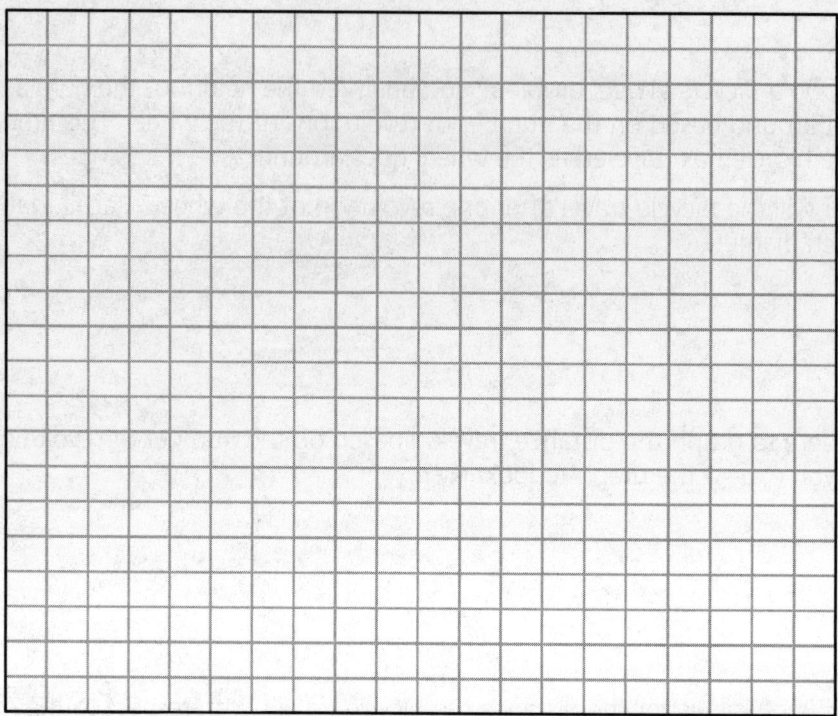

# Assignment for The Ferris Wheel

**Name:** ____________________________

You are at the amusement park baby-sitting your six-year old sister and her friend. As they are riding the Kiddyland Ferris wheel, you read a plaque about the Ferris wheel's size. The ride is similar to the large Ferris wheel in that they both complete one revolution over 2 minutes. However, at 25 feet in diameter, the Kiddyland Ferris wheel has only half the diameter of the larger ride.

1. Complete this table to show the distance your sister and her friend will be from the ground over the first three revolutions of their ride on the Kiddyland Ferris wheel.

| Time when the Ferris wheel reaches the point | | | Distance from the ground |
|---|---|---|---|
| 1st revolution | 2nd revolution | 3rd revolution | |
| seconds | seconds | seconds | feet |
| 0 | | | |
| 10 | | | |
| 20 | | | |
| 30 | | | |
| 40 | | | |
| 50 | | | |
| 60 | | | |
| 70 | | | |
| 80 | | | |
| 90 | | | |
| 100 | | | |
| 110 | | | |

2. How does this table differ from the table you created for *The Ferris Wheel* problem in the textbook?

3. Create a graph of the height of your sister and her friend above the ground with respect to the time since the ride started.

4. What are the similarities and difference between this graph and the graph you created for *The Ferris Wheel* problem in the textbook?

# Assignment for The Unit Circle

**Name:** ____________________________

1. Sketch a unit circle with a central angle measuring 60°.

   a. Calculate the length (not the measure) of the arc inscribed in the central angle.

   b. Calculate the area of the sector bound by the central angle and its intercepted arc.

2. Sketch a circle with a radius measuring 3 units with a central angle measuring 60°.

   a. Calculate the length (not the measure) of the arc inscribed in the central angle.

   b. Calculate the area of the sector bound by the central angle and its intercepted arc.

3. Sketch a circle with a radius measuring 4 units with a central angle measuring 150°.

   a. Calculate the length (not the measure) of the arc inscribed in the central angle.

   b. Calculate the area of the sector bound by the central angle and its intercepted arc.

4. Sketch a circle with a radius measuring 3 units with a central angle measuring 120°.

   a. Calculate the length (not the measure) of the arc inscribed in the central angle.

   b. Calculate the area of the sector bound by the central angle and its intercepted arc.

# Assignment for Radians

**Name:** ____________________________

1. Draw angles in standard position with the given measures.

a. $\frac{5\pi}{4}$ radians

b. $-\frac{\pi}{3}$ radians

c. $\frac{5\pi}{2}$ radians

d. $\frac{7\pi}{6}$ radians

e. $\frac{\pi}{2}$ radians

f. $-2\pi$ radians

2. Write each angle measure in radians.

a. 150°

b. -30°

c. 300°

d. -270°

e. 135°

f. 540°

3. Write each angle measure in degrees.

a. $\frac{3\pi}{4}$ radians

b. $-\frac{\pi}{6}$ radians

c. $\frac{9\pi}{2}$ radians

d. $\frac{2\pi}{3}$ radians

e. $-\frac{\pi}{8}$ radians

f. $\pi$ radians

# Assignment for The Sine and Cosine Functions

**Name:** ______________________________

1. Use your calculator (in radian mode) to evaluate each of the following.

   a. $\sin\frac{\pi}{6}$ b. $\sin\left(\frac{\pi}{6}+2\pi\right)=\sin\frac{13\pi}{6}$

   c. $\sin\frac{\pi}{4}$ d. $\sin\left(\frac{\pi}{4}+2\pi\right)=\sin\frac{9\pi}{4}$

   e. $\sin\frac{2\pi}{3}$ f. $\sin\left(\frac{2\pi}{3}+2\pi\right)=\sin\frac{8\pi}{3}$

   g. $\sin\frac{\pi}{2}$ h. $\sin\left(\frac{\pi}{2}+2\pi\right)=\sin\frac{5\pi}{2}$

2. What do you think is true about $\sin\theta$ and $\sin(\theta+2\pi)$?

3. Use your calculator (in radian mode) to evaluate each of the following.

   a. $\sin\frac{\pi}{6}$ b. $\sin\left(\frac{\pi}{6}+\pi\right)=\sin\frac{7\pi}{6}$

   c. $\sin\frac{\pi}{4}$ d. $\sin\left(\frac{\pi}{4}+\pi\right)=\sin\frac{5\pi}{4}$

   e. $\sin\frac{2\pi}{3}$ f. $\sin\left(\frac{2\pi}{3}+\pi\right)=\sin\frac{5\pi}{3}$

   g. $\sin\frac{\pi}{2}$ h. $\sin\left(\frac{\pi}{2}+\pi\right)=\sin\frac{3\pi}{2}$

4. What do you think is true about $\sin\theta$ and $\sin(\theta+\pi)$?

5. Use your calculator (in radian mode) to evaluate each of the following.

a. $\cos\frac{\pi}{3}$

b. $\cos\left(\frac{\pi}{3}+2\pi\right)=\cos\frac{7\pi}{3}$

c. $\cos\frac{3\pi}{4}$

d. $\cos\left(\frac{3\pi}{4}+2\pi\right)=\cos\frac{11\pi}{4}$

e. $\cos\frac{5\pi}{6}$

f. $\cos\left(\frac{5\pi}{6}+2\pi\right)=\cos\frac{17\pi}{6}$

g. $\cos\frac{\pi}{2}$

h. $\cos\left(\frac{\pi}{2}+2\pi\right)=\cos\frac{5\pi}{2}$

6. What do you think is true about $\cos\theta$ and $\cos(\theta+2\pi)$?

7. Use your calculator (in radian mode) to evaluate each of the following.

a. $\cos\frac{\pi}{3}$

b. $\cos\left(\frac{\pi}{3}+\pi\right)=\cos\frac{4\pi}{3}$

c. $\cos\frac{3\pi}{4}$

d. $\cos\left(\frac{3\pi}{4}+\pi\right)=\cos\frac{7\pi}{4}$

e. $\cos\frac{5\pi}{6}$

f. $\cos\left(\frac{5\pi}{6}+\pi\right)=\cos\frac{11\pi}{6}$

g. $\cos\frac{\pi}{2}$

h. $\cos\left(\frac{\pi}{2}+\pi\right)=\cos\frac{3\pi}{2}$

8. What do you think is true about $\cos\theta$ and $\cos(\theta+2\pi)$?

# Assignment for What's Your Sine?

**Name:** ______________________________

Sketch the graph of each sinusoidal function by transforming the parent sine function. Write a description of the transformations from the parent function, $y = \sin\theta$.

1. $y = 2\sin\theta - 1$

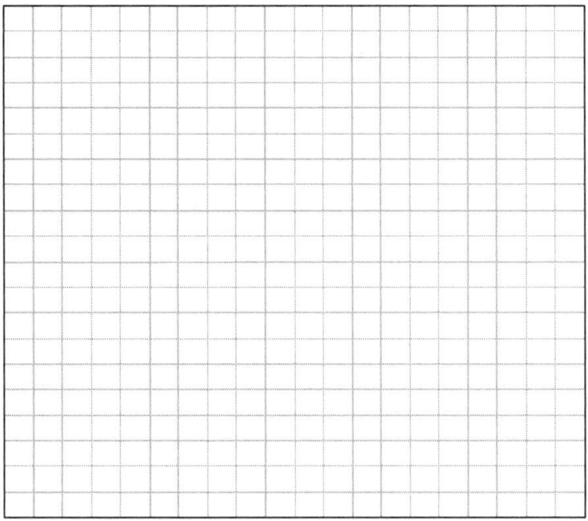

2. $y = 3\sin(\theta - \frac{\pi}{2})$

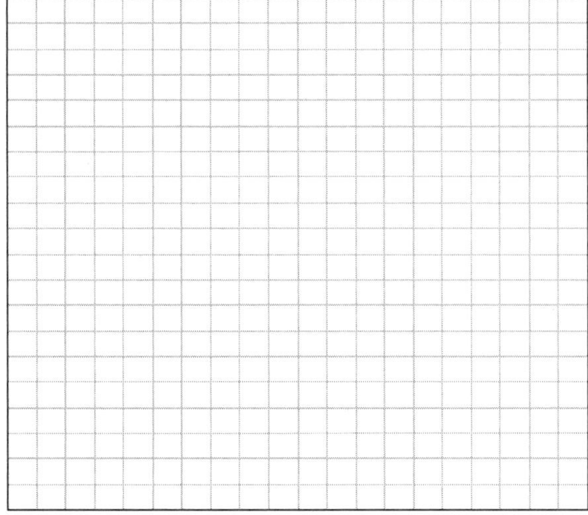

Write the algebraic function for each graph shown based on the transformations performed on the parent sine function. Also write a description of the transformation from the parent function, $y = \sin\theta$.

3.

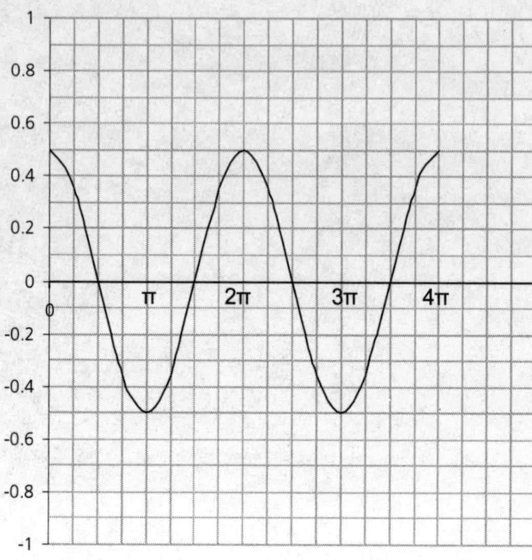

4.

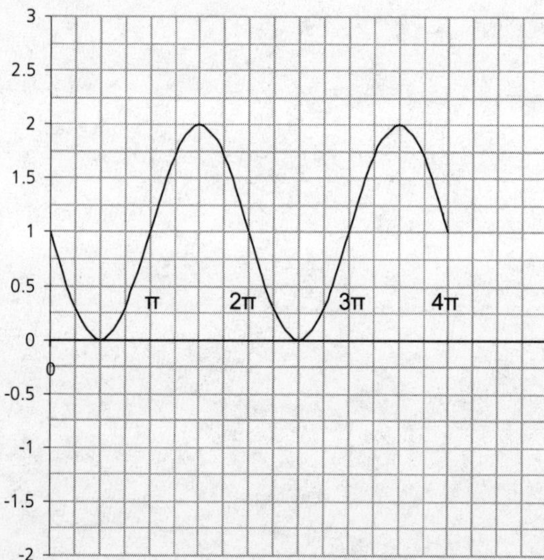

# Assignment for Rabbit Population

**Name:** ______________________________

Some people believe in the biorhythm theory. This theory states that a person's functioning is controlled by three factors: a physical cycle, an emotional cycle, and an intellectual cycle. Each factor can be modeled using sinusoidal functions. A value of 1 indicates a day for which a cycle is at a high point, and -1 indicates a low point. A value of 0 indicates a day of vulnerability or risk. The physical cycle has a period of 23 days, the emotional cycle has a period of 28 days, and the intellectual cycle has a period of 33 days.

1. The function modeling the physical cycle is $y_{physical} = \sin\left(\frac{2\pi}{23}x\right)$, where $x$ is the number of days since you were born. Explain why this mathematical function has a period of 23 days.

2. Write functions for the emotional and intellectual cycles.

3. Determine the number of days you have been alive. Remember to consider leap years.

4. Evaluate each of the three functions using your answer from question 3 as the input. What should this indicate about your physical, emotional, and intellectual states?

5. Make a graph showing the curves for each function for all the days in the current month.

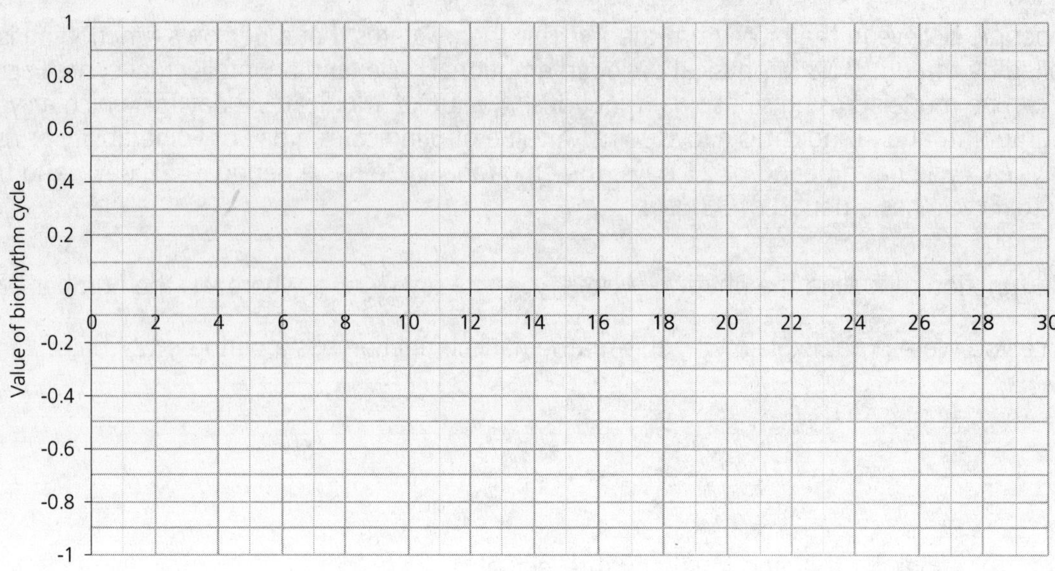


6. Based on your graphs, what would someone who believes in biorhythms think is your best day this month? What is the worst day? Why?

# Assignment for Seasonal Affective Disorder

**Name:** ____________________________

The average daily temperature for a given location can be modeled with a sinusoidal function. Below is data regarding the average temperature in Denver, Colorado throughout the year.

| Date | Day | Average Temperature |
|---|---|---|
| Dec. 31 | 0 | 29° |
| Jan. 10 | 10 | 29° |
| Jan. 20 | 20 | 29° |
| Jan. 30 | 30 | 30° |
| Feb. 9 | 40 | 32° |
| Feb. 19 | 50 | 34° |
| Mar. 1 | 60 | 36° |
| Mar. 11 | 70 | 38° |
| Mar. 21 | 80 | 41° |
| Mar. 31 | 90 | 43° |
| Apr. 10 | 100 | 46° |
| Apr. 20 | 110 | 49° |
| Apr. 30 | 120 | 52° |
| May 10 | 130 | 55° |
| May 20 | 140 | 59° |
| May 30 | 150 | 62° |
| June 9 | 160 | 66° |
| June 19 | 170 | 69° |
| June 29 | 180 | 71° |

| Date | Day | Average Temperature |
|---|---|---|
| July 9 | 190 | 73° |
| July 19 | 200 | 74° |
| July 29 | 210 | 74° |
| Aug. 8 | 220 | 73° |
| Aug. 18 | 230 | 72° |
| Aug. 28 | 240 | 69° |
| Sept. 7 | 250 | 65° |
| Sept. 17 | 260 | 62° |
| Sept. 27 | 270 | 58° |
| Oct. 7 | 280 | 55° |
| Oct. 17 | 290 | 51° |
| Oct. 27 | 300 | 46° |
| Nov. 6 | 310 | 41° |
| Nov. 16 | 320 | 37° |
| Nov. 26 | 330 | 34° |
| Dec. 6 | 340 | 32° |
| Dec. 16 | 350 | 30° |
| Dec. 26 | 360 | 29° |

1. Create a scatterplot of the data for the average daily temperature in Denver.

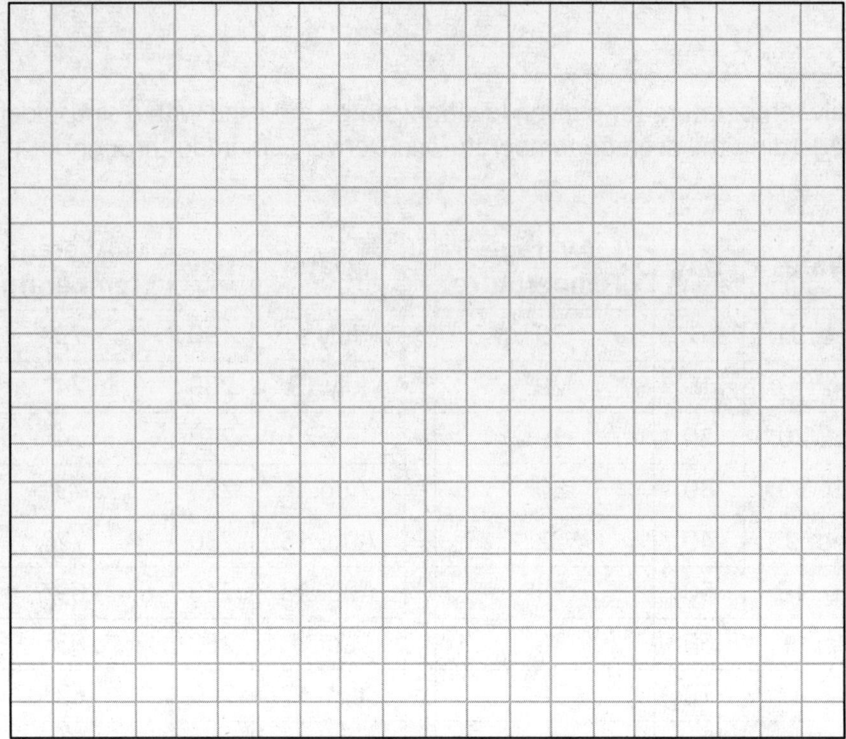

2. Write a sinusoidal equation in the form $f(x) = A \sin (Bx + C) + D$ that models the data.

3. What do each of the parameters of the equation represent in terms of the data?
   a. What does A represent?
   b. What does B represent?
   c. What does C represent?
   d. What does D represent?

# Assignment for The Tangent Function

**Name:** ______________________________

Sketch the graph of each tangent function by transforming the parent tangent function. Write a description of the transformations from the parent function, $y = \tan\ \theta$.

1. $y = \tan\ \theta + 1$

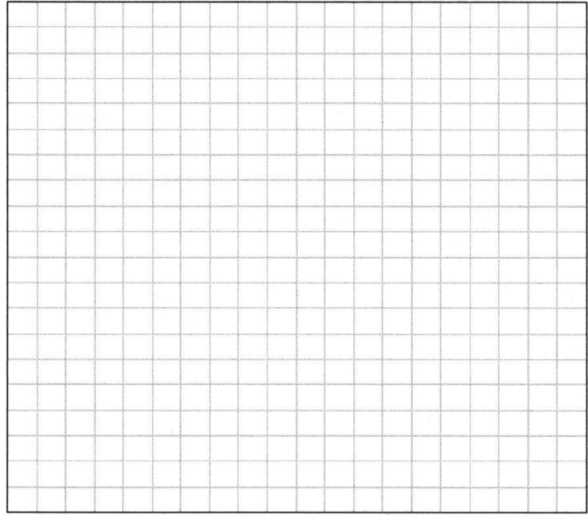

2. $y = \tan\ (\theta - \frac{\pi}{2})$

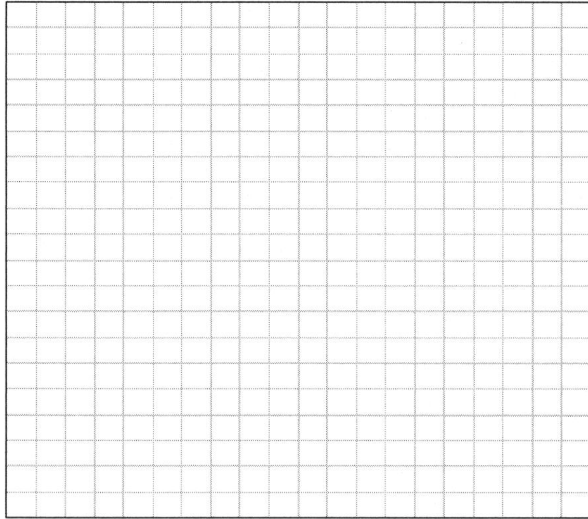

3. For what angle is the value of the tangent function $\sqrt{3}$?

   a. Give another possible answer to the question. If there is no other possible answer, explain why not.

   b. How many possible answers are there to the question?

4. Given $\tan\frac{\pi}{12} \approx 0.268$, for what other angles is this tangent function value approximately 0.268?

 

# Assignment for The Inverse Sine Function

**Name:** ______________________________

Use the inverse sine function to find the measure of $\theta$.

1.

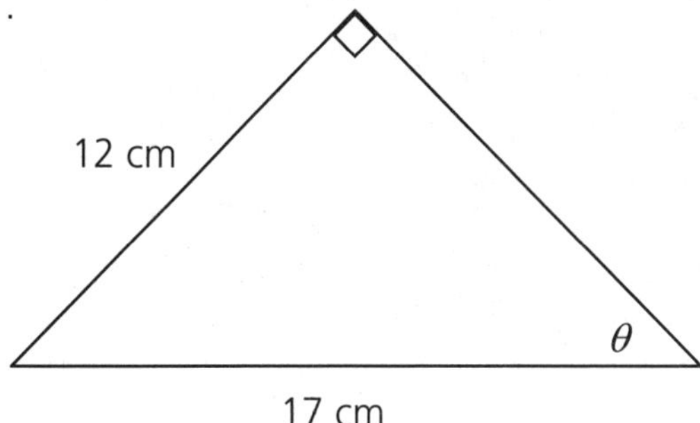


2.

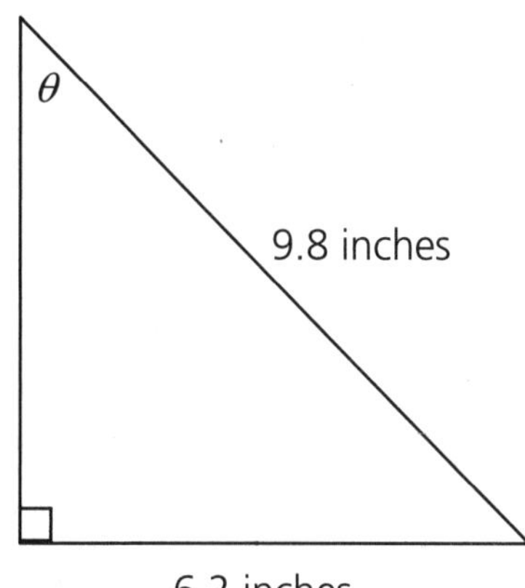


3.

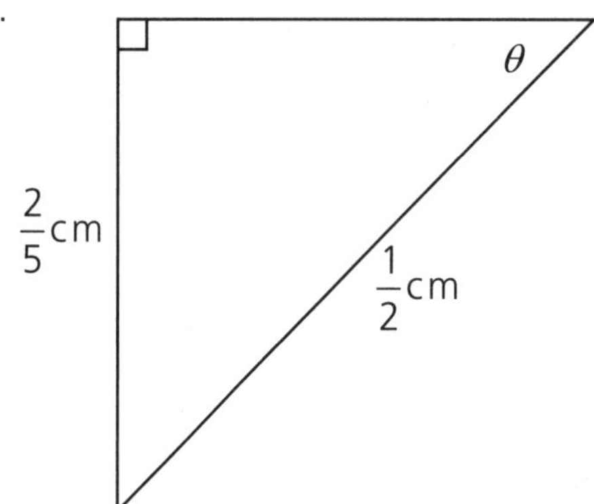

4. The law requires that wheelchair ramps have a ramp angle less than or equal to $8\frac{1}{3}°$. A ramp that is 7 feet in length is used for a 1 foot vertical rise. Is this ramp in compliance with the law? Hint: Drawing a diagram will help.

# Assignment for The Inverse Cosine Function

**Name:** ______________________________

Use the inverse cosine function to find the measure of $\theta$.

1.

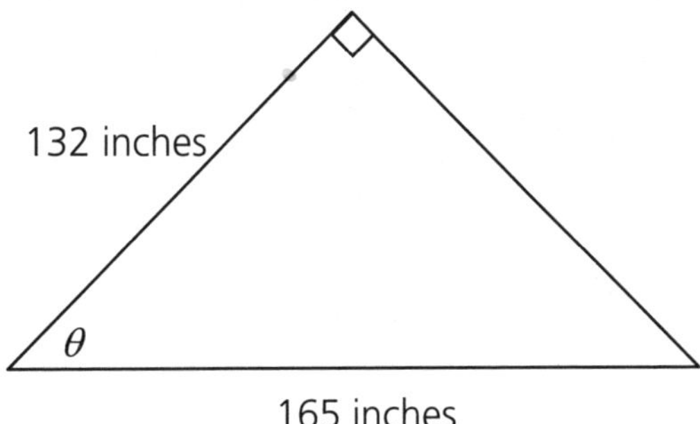


2.

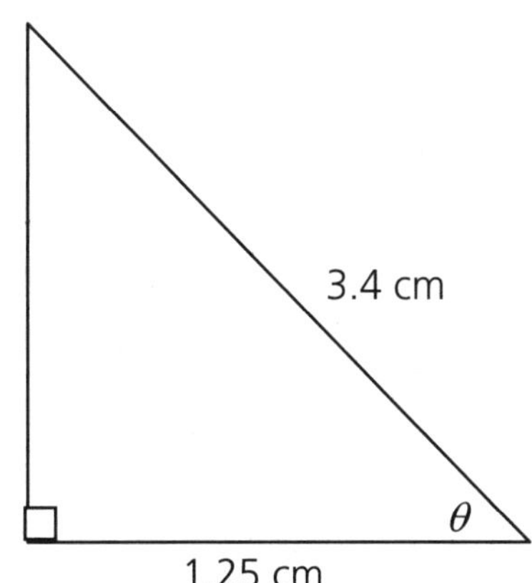


3.

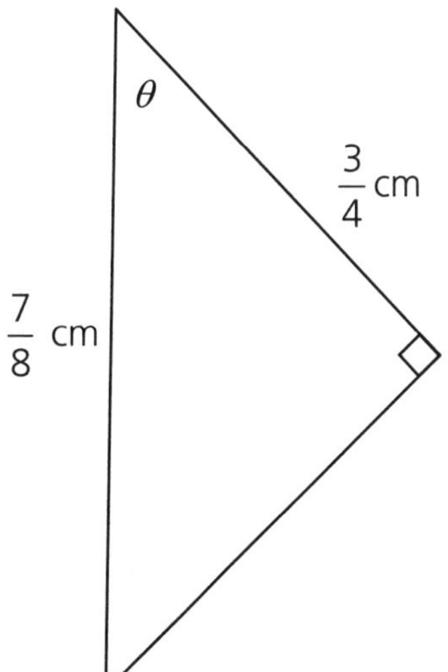

4. Paint Boys Painting Company has set a safety guideline for their painters about using ladders. They instruct the painters to place the ladders at an angle measuring between 55° and 75° from the level ground. If Diana places the foot of her 30-foot ladder 6 feet from the base of the wall, is the ladder placed safely?

5. The Leaning Tower of Pisa is approximately 55 meters in length and leans about 4.4 meters. At what angle does the tower lean off of vertical?

6. Before efforts to keep the Leaning Tower of Pisa from falling over, it leaned about 4.9 meters. At what angle did the tower lean off of vertical then?

# Assignment for The Inverse Tangent Function

**Name:** ____________________________

Use the inverse tangent function to find the measure of $\theta$.

1.

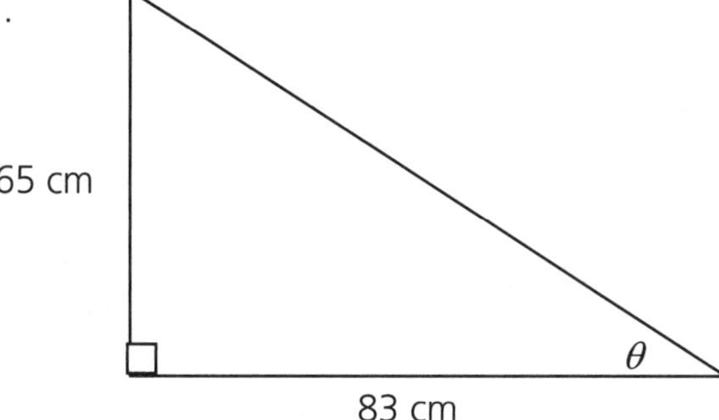


2.

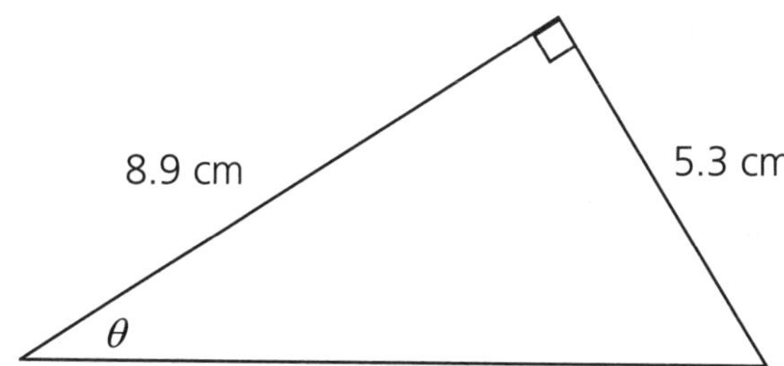


3.

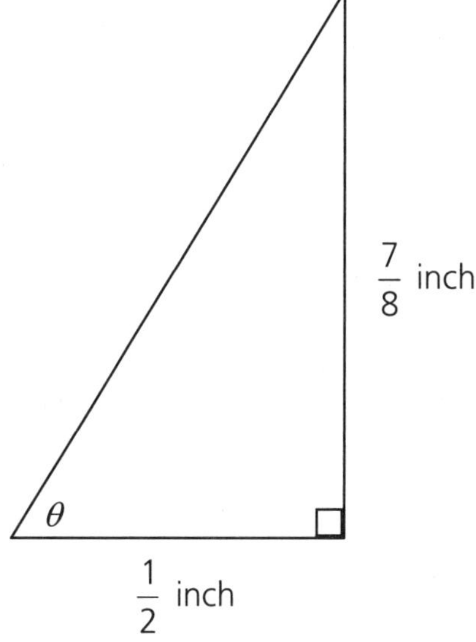

4. An airplane is making its approach for landing. The plane is currently at an altitude of 5000 feet. The horizontal distance from the plane to the runway is 32,000 feet. What is the angle of depression from the airplane to the runway?

 

# Assignment for The Law of Sines

**Name:** ______________________

1. Use the law of sines to determine the value of $x$.

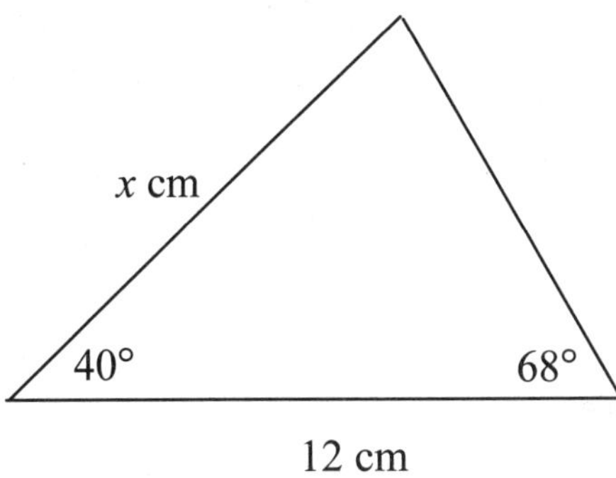


2. Use the law of sines to determine the value of $\theta$.

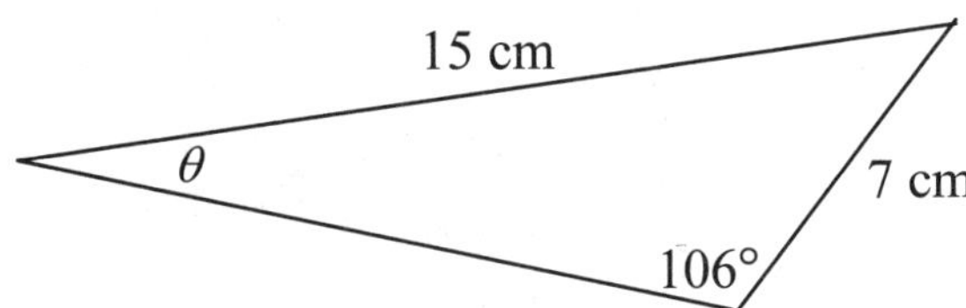


3. A disabled ship (at point S) is sighted from two different lighthouses that are 8 miles apart (at points $L_1$ and $L_2$). If $m\angle S\,L_1\,L_2 = 44°$ and $m\angle S\,L_2\,L_1 = 66°$, find the distance from the ship to the nearest lighthouse.

# Assignment for The Law of Cosines

**Name:** ______________________________

1. Use the law of cosines to determine the value of $\theta$.

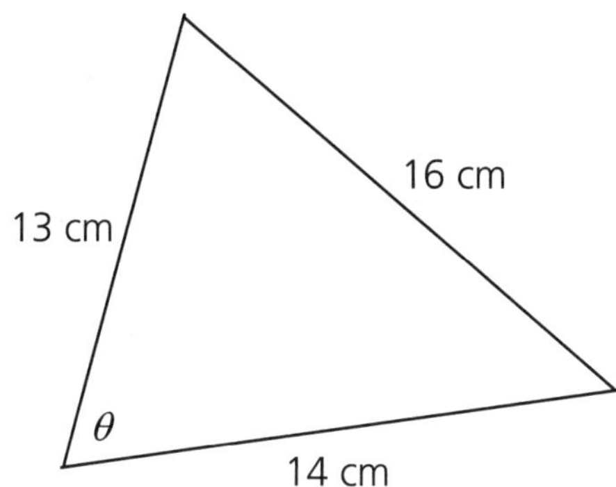


2. Use the law of cosines to determine the value of $x$.

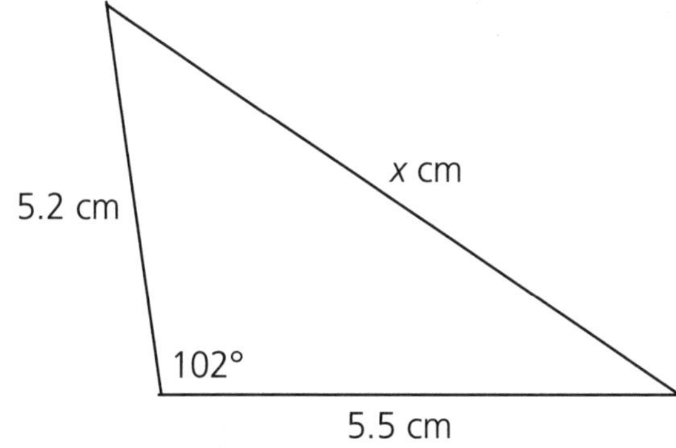


3. Lifeguard stands are set up on the beach as shown in the diagram with 88 yards between stands A and B and 80 yards between stands B and C. The stands make an angle of 150° at vertex B. A buoy line needs to be set up from lifeguard stand A to lifeguard stand C to mark a division between more shallow and deeper waters. About how long must the buoy line be?

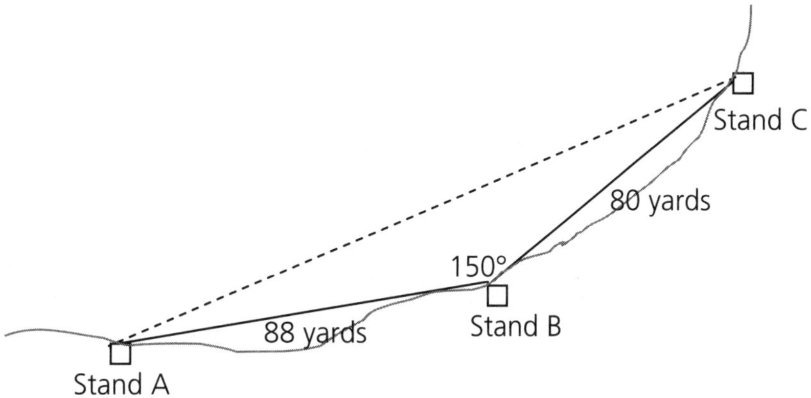

# Assignment for Solving Trigonometric Equations

**Name:** ____________________________

1. Solve $\cos\theta = \frac{1}{2}$ over the domain $0 \le \theta \le 2\pi$.

2. Solve $\sin\theta = 0.875$ over the domain $-\frac{\pi}{2} \le \theta \le \frac{\pi}{2}$.

3. Solve $\tan\theta = \frac{1}{\sqrt{3}}$ over the domain of all real numbers.

4. Solve $8\sin\theta - 4 = 3$ over the domain of all real numbers.

5. Solve $5\cos\theta + 1 = 3$ over the domain of $0 \leq \theta \leq \pi$.

# Assignment for Round and Round

**Name:** ______________________________

1. Identify the center and radius for each circle defined below.

   a. $(x - 7)^2 + (y + 2)^2 = 16$

   b. $(x + 4)^2 + (y + 1)^2 = 81$

   c. $(x - 1)^2 + (y - 10)^2 = 4$

   d. $(x + 4)^2 + (y - 9)^2 = 64$

2. Write the equation of a circle based on the given information.

   a. Center (0, -4); radius = 10

   b. Center (1, 5); radius = 15

   c. Center (-2, 1); radius = 6

   d. Center (-3, -5); radius = 11

3. Determine the domain, range, and x- and y-intercepts for each circle.

   a. $(x - 1)^2 + (y + 6)^2 = 25$

   b. $(x + 2)^2 + (y + 3)^2 = 144$

## Assignment for It's a Stretch

**Name:** ______________________________

1. Sketch the graph that results if a circle is dilated along the x-axis by a factor of 3 and along the y-axis by a factor of 2. Write the equation representing the graph.

2. Sketch the graph that results if a circle is dilated along the x-axis by a factor of 5 and along the y-axis by a factor of 1.5. Write the equation representing the graph.

3. Sketch the graph that results if a circle is dilated along the x-axis by a factor of 3.5 and along the y-axis by a factor of 6. Write the equation representing the graph.

4. Sketch the graph that results if a circle is dilated along the x-axis by a factor of 8 and along the y-axis by a factor of 2.5. Write the equation representing the graph.

# Assignment for Try to Focus

**Name:** ______________________________

Compute the coordinates of the foci for each ellipse

1. $\frac{x^2}{25}+\frac{y^2}{9}=1$

2. $\frac{x^2}{4}+\frac{y^2}{49}=1$

3. $\frac{x^2}{100}+\frac{y^2}{225}=1$

# Assignment for Whispering Galleries

**Name:** ______________________________

Write the equation of an ellipse from the given information and sketch the graph

1. Vertices at (-5, 0) and (5, 0). Co-vertices at (0, -2) and (0, 2).

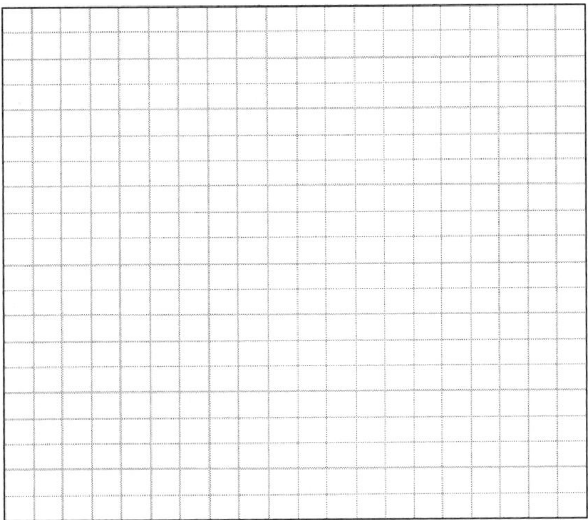

2. Vertices at (0, -7) and (0, 7). Foci at (0, -3) and (0, 3).

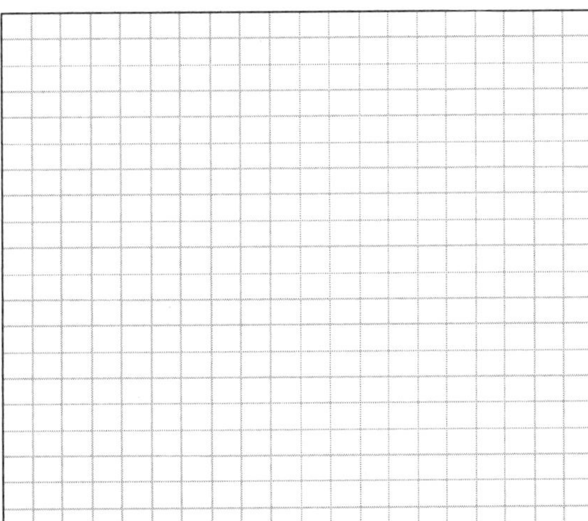

3. Foci at (-5, 0) and (5, 0). Co-vertices at (0, -4) and (0, 4).

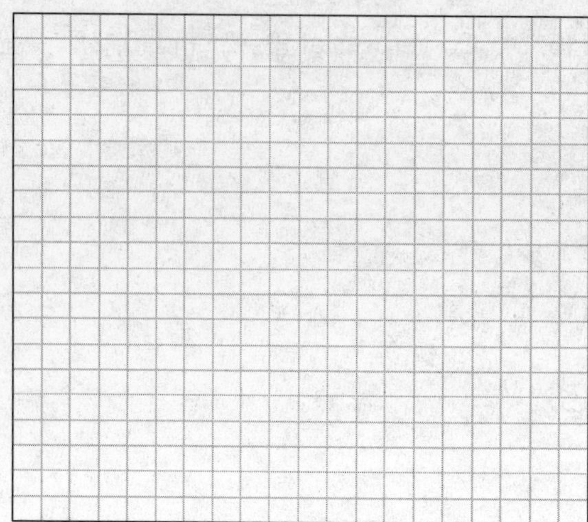

4. Vertices at (0, -8) and (0, 8). Co-vertices at (-3, 0) and (3, 0).

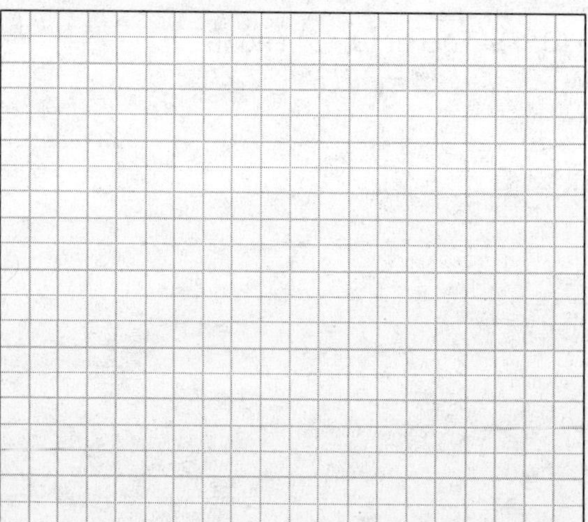

# Assignment for The Law of Orbits

**Name:** ______________________________

For the ellipse modeling the orbit of Pluto, the distance between the vertices is approximately 78.88 AU (Astronomical units) and the distance between the foci is approximately 19.642 AU. One Astronomical unit is approximately 93,000,000 miles. Assume the sun appears at the left focus.

1. For the ellipse modeling the orbit of Mercury, what are the coordinates of the vertices?

2. What are the coordinates of the foci?

3. What are the coordinates of the co-vertices?

5. Write the equation for an ellipse modeling the orbit of Mercury centered at the origin with the major axis along the x-axis and the sun to the left of the origin.

6. Graph the ellipse modeling the orbit of Mercury.

# Assignment for Equations of Hyperbolas

**Name:** ______________________________

Graph each hyperbola and calculate the coordinates of the vertices, co-vertices and foci and the equations for the asymptotes, transverse and conjugate axes.

1. $\frac{x^2}{16} - \frac{y^2}{36} = 1$

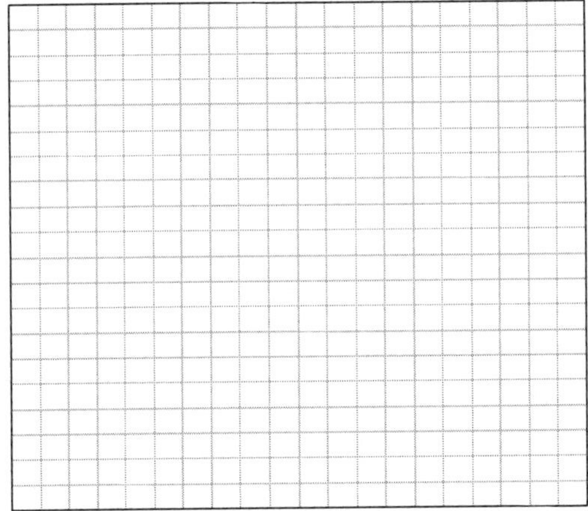

2. $\frac{x^2}{81} - \frac{y^2}{9} = 1$

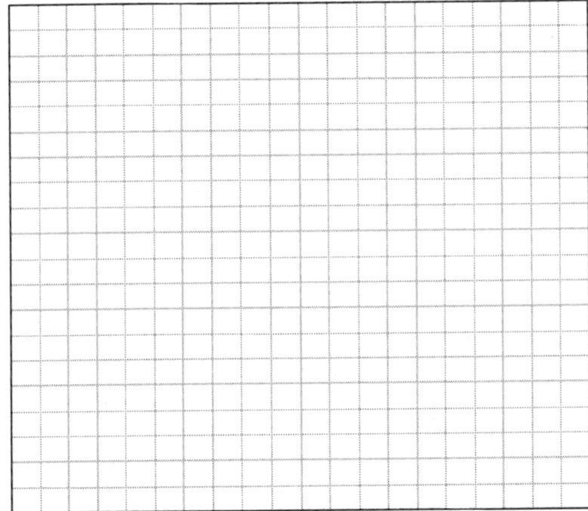

3. $\frac{x^2}{121} - \frac{y^2}{64} = 1$

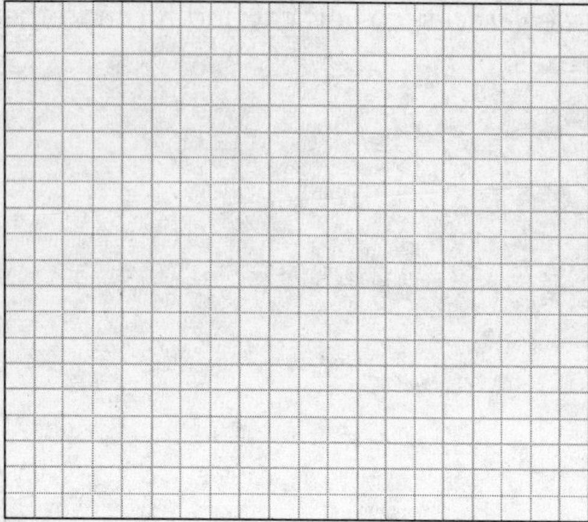

4. $\frac{x^2}{4} - \frac{y^2}{196} = 1$

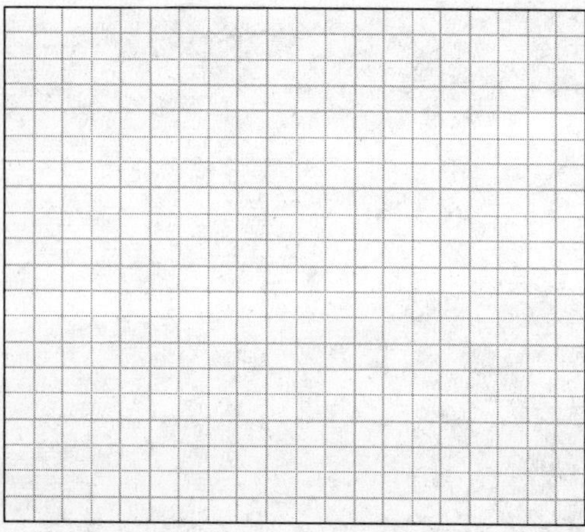

# Assignment for Change of Perspective - Hyperbolas

**Name:** ______________________________

1. Determine the coordinates of the vertices, co-vertices and foci and the equations of the asymptotes, transverse and conjugate axes for each hyperbola.

   a. $\frac{y^2}{25} - \frac{x^2}{9} = 1$

   b. $\frac{x^2}{49} - \frac{y^2}{100} = 1$

2. Determine the equation of the hyperbola defined by the given information
   a. Vertices at (0, -12) and (0, 12); Co-vertices at (-9, 0) and (9, 0)

   b. Vertices at (-8, 0) and (8, 0); Foci at (-13.60, 0) and (13.60, 0)

   c. Co-vertices at (-13, 0) and (13, 0); Foci at (0, -13.34) and (0, 13.34)

   d. Vertices at (-2, 0) and (2, 0); Co-vertices at (0, -15) and (0, 15)

# Assignment for Transforming Hyperbolas

**Name:** ______________________________

Graph each hyperbola and determine the coordinates of the center, vertices, co-vertices and foci and the equations for the asymptotes, transverse and conjugate axes for each hyperbola

1. $\frac{(x-1)^2}{25} - \frac{(y+2)^2}{9} = 1$

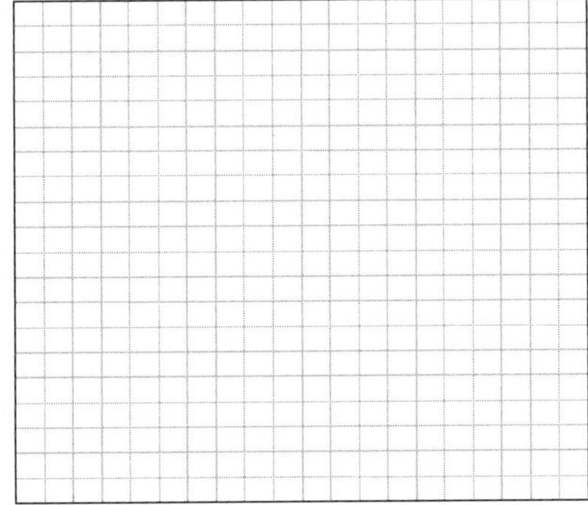

2. $\frac{(y+3)^2}{36} - \frac{(x+1)^2}{25} = 1$

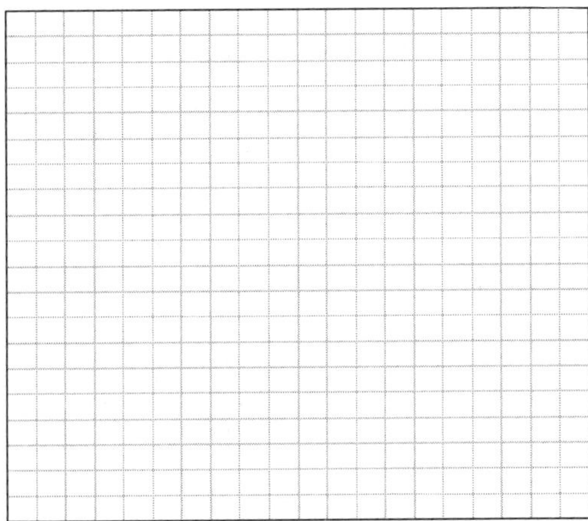

3. $\frac{(y-4)^2}{16} - \frac{(x+2)^2}{9} = 1$

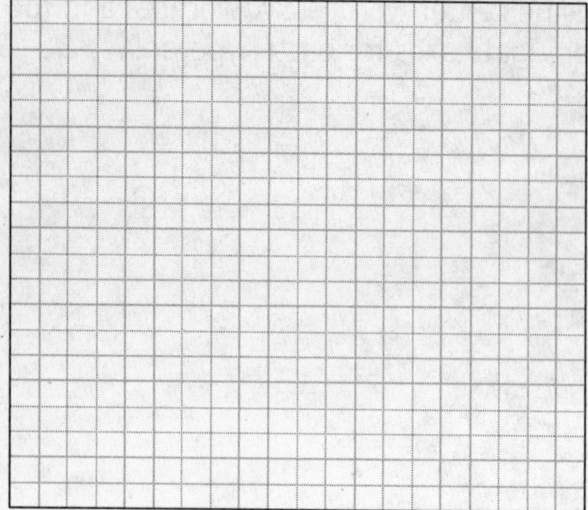

4. $\frac{(x-5)^2}{25} - \frac{(y-3)^2}{49} = 1$

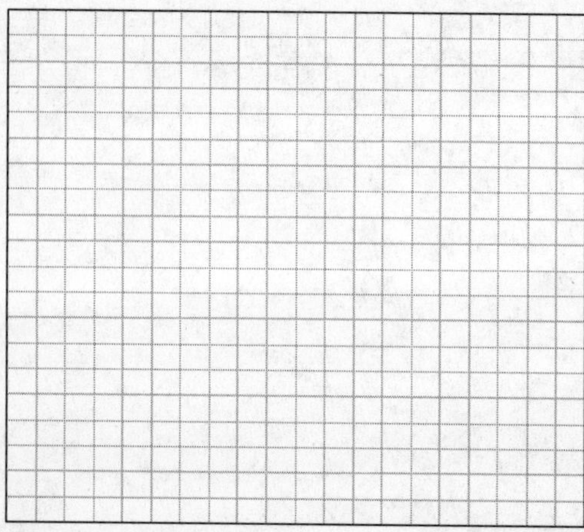

# Assignment for Flashlight

**Name:** ______________________________

Determine the coordinates of the focus and the equation for the directrix for each equation.

1. $y=\frac{1}{8}x^2$

2. $y=7x^2$

3. $y=\frac{1}{15}x^2$

4. $y=3x^2$

# Assignment for Radio Waves

**Name:** ______________________________

Determine the equation of the parabola defined by the given information.

1. Focus at (0, 6) and directrix at y = -6

2. Focus at (0, -3) and directrix at y = 3

3. Focus at $(0, \frac{1}{2})$ and directrix at $y = -\frac{1}{2}$

4. Focus at $(0, -\frac{1}{8})$ and directrix at $y = \frac{1}{8}$

5. Focus at (0, 7) and directrix at y = -7

# Assignment for Change of Perspective - Parabolas

**Name:** ______________________________

1. Determine the coordinates of the focus and the equation of the directrix for each parabola.

   a. $x = -\frac{1}{9}y^2$

   b. $y = \frac{1}{15}x^2$

   c. $x = \frac{1}{3}y^2$

2. Determine the equation of the parabola defined by the given focus or directrix with the vertex at the origin.

   a. Focus at (12, 0)

   b. Directrix at $x = 5$

   c. Focus at (0, 1.5)

# Assignment for Transforming Parabolas

**Name:** ______________________________

Graph each parabola and compute the coordinates of the vertex and focus and the equations for the axis of symmetry and directrix.

1. $x+2=\frac{1}{3}(y-5)^2$

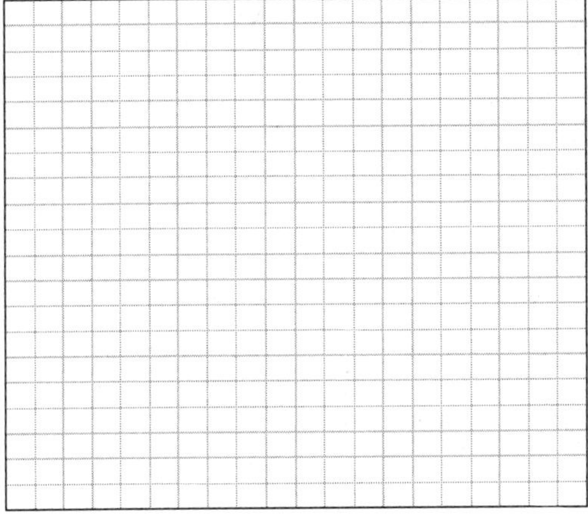

2. $y+1=\frac{1}{5}(x-3)^2$

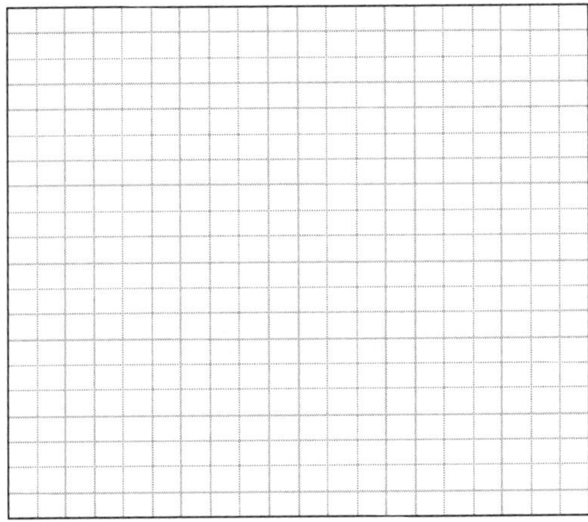

3. $y - 4 = \frac{1}{8}(x - 2)^2$

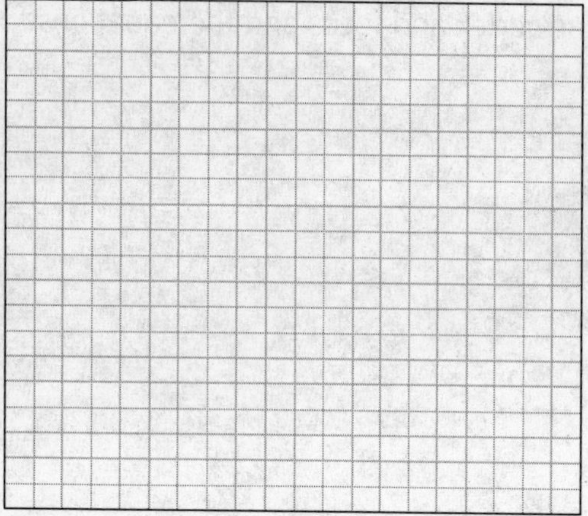

4. $x + 1 = \frac{1}{6}(y + 3)^2$

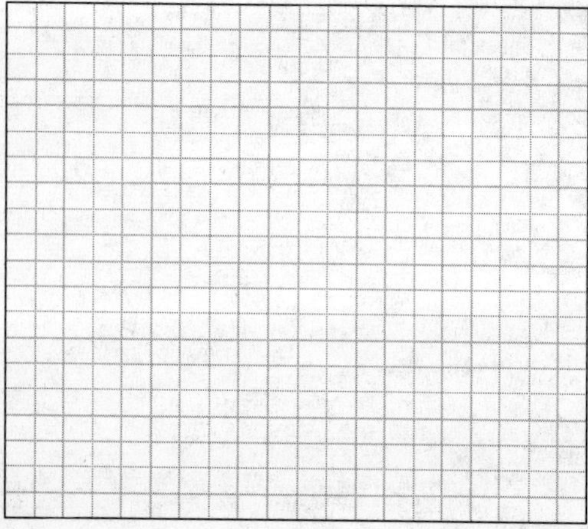

# Assignment for Get to the Hearth of the Matter

**Name** ______________________________

1. The client who built the Hearth on the patio decided to add a chimney to the hearth. How many measurements will you need to take in order to design and build the chimney? What are the measurements you will need?

2. You and the client decided to build a chimney with a square base 18″ on each side. The height of the chimney will be 4 feet. Find the volume of the chimney.

3. What should be the units on the calculation you performed above? Why?

4. A point is a zero dimensional object. As in the story of *Flatland*, what kind of trail will a point leave in its wake as it moves in a straight path?

5. A line is a one-dimensional object. What is a two-dimensional object analogous to a line? Explain your response.

6. A plane is a two-dimensional object. What is the three-dimensional equivalent to a plane?

7. Romero makes the claim that every two-dimensional shape has a three-dimensional counterpart. Do you agree or disagree with this claim? If you agree, provide at least three examples. If you disagree, explain why and provide a counterexample.

# Assignment for Moving from Two Dimensions to Three

**Name** ______________________________

1. Your school places on order for copy paper. Each ream of paper (500 sheets) measures 8.5″ by 11″ by 3″. What is the volume of each ream of paper?

2. A box of paper contains two stacks of 5 reams each (10 reams of paper). Calculate the internal measurements of the box based on the measures of one ream of paper.

3. What is the approximate volume of each box if the cardboard adds about 0.4 inches on each side of the box?

4. If your school orders 30 boxes of paper for the start of a new school year, will all the boxes fit in the storage closet near the copy room? The storage closet measures 3 feet by 3 feet by 6.5 feet.

5. A couple about to get married decide to have a sugar sculpture at their reception instead of an ice sculpture. To carve the sculpture, the artist begins with a rectangular block of sugar, shaves some sugar from the block, and creates a hexagonal prism with a regular base as the starting shape. Sketch such a prism in the space below.

   a. If each side of the base of the hexagonal prism in the last problem measures 2 decimeters, what is the area of the base?

   b. If the height of the prism is 7.5 decimeters, calculate the amount of sugar in the prism.

   c. How many vertices does this prism have?

   d. How many edges?

   e. How many lateral edges?

   f. How many bases does the prism have?

   g. How many lateral faces?

   h. How many total faces?

## Assignment for Nets Aren't Just for Fishing Anymore

**Name** ____________________________

1. Create the net for the package in the last problem from the section on nets in your textbook.

2. If each dimension of the package is doubled, what happens to the amount of wrapping paper needed to wrap the package?

3. If each dimension of the package is doubled, what happens to the amount of ribbon needed for the package?

4. What is the original volume of the package?

5. If each dimension of the package is doubled, what happens to the volume of the package?

6. Create the net for an oblique triangular prism?

7. Compare and contrast the net for the oblique triangular prism for that of the right triangular prism you created during the lesson.

# Assignment for Polyhedra

**Name** ______________________________

1. Create the net for a regular tetrahedron.

2. If each edge of the regular tetrahedron above is 2 feet, calculate the surface area of the tetrahedron.

3. Create the net for a right square pyramid in the space below.

4. Create the net for an oblique square pyramid in the space below.

5. Compare and contrast the nets for the right and the oblique square pyramids.

# Assignment for Volume of Prisms and Pyramids

**Name** ______________________________

1. Find the volume of each solid below.

   a. Triangular Prism

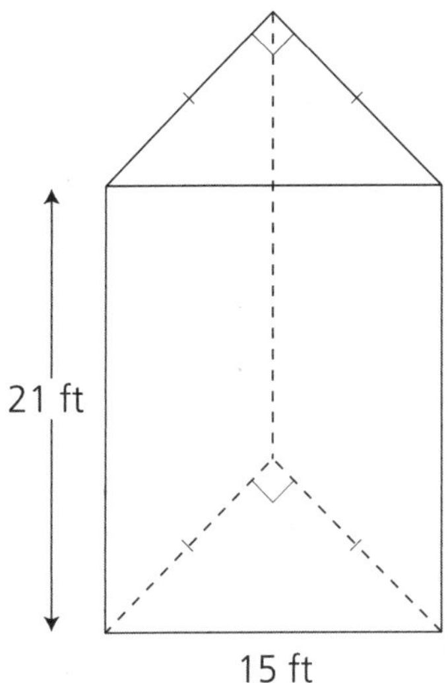


   b. Trapezoidal Prism

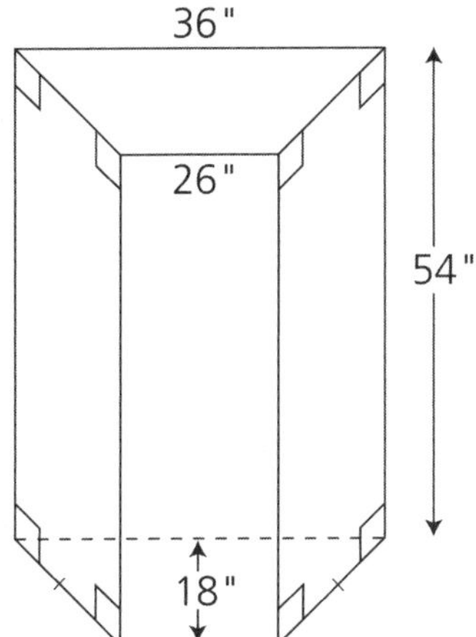

2. You are planning to build a shed in your back yard. At the local home supply store, you see two stacks of 8′ by 8′ plywood. One stack is neat and tidy with one sheet stacked directly atop the other. One stack is starting to slant so each sheet of plywood is nudged over a bit from the one below it. If the stacks are the same height, which one contains more plywood? Explain your response.

3. Find the volume of a cube with each edge measuring 3 inches.

4. Find the surface area of the cube above.

5. Your friend states that the surface area of a cube is twice as big as the volume. Do you agree or disagree? Explain.

# Assignment for More Three-Dimensional Figures

**Name** ______________________________

1. Calculate the lateral area and the surface area for each of the following cones or cylinders.

   a.

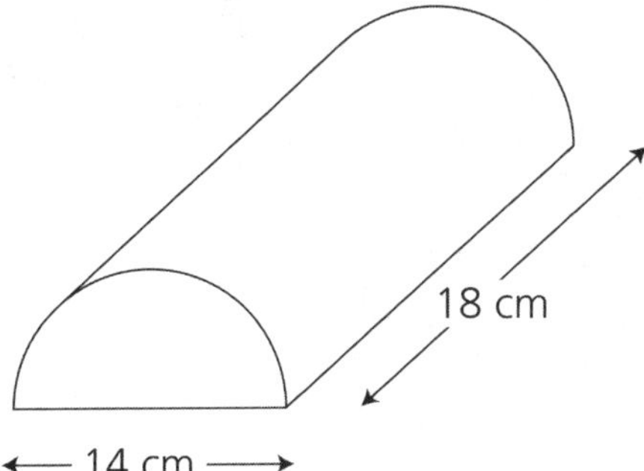


   b.

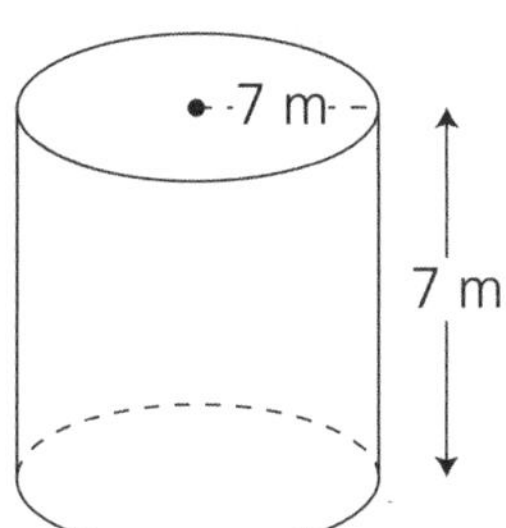

c.

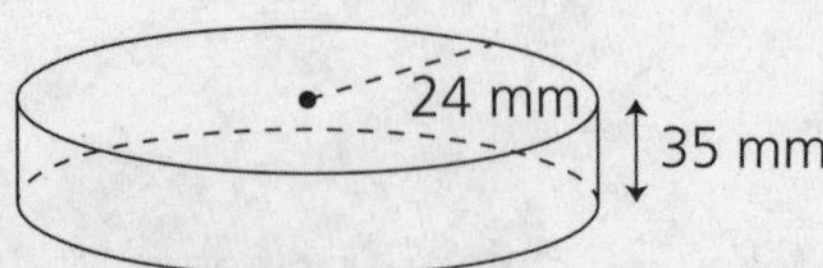


d. The diameter of the base of a cone and the height of the cone match the dimensions of the cylinder in problem b. Find the volume of the cone.

e. The circumference of the base of a cone is 20 cm. The height of the cone is 15 cm. Find the surface area of the cone.

# Assignment for Volume of Cylinders and Cones

**Name** ____________________

1. Calculate the volume of the following cylinders.

   a. The diameter of the beaker is 10 cm and the height is 10 cm.

   b. The tape is 1 inch thick and has a radius of 2 inches.

   c. The tire has a radius of 22 inches and a thickness of 5 inches.

2. Calculate the volume of the following cones.

   a. The orange traffic cone has a diameter of approximately 9 inches on conical portion (not including the base). The height of the cone is about 30 inches.

   b. A funnel used to change the oil in a car has a circumference of the base that is approximately 60 centimeters. The height of this funnel (excluding the tip) is about 25 centimeters.

   c. A mini-ice cream cone has a diameter of 3.5 cm and a height of 6 cm. Calculate its volume.

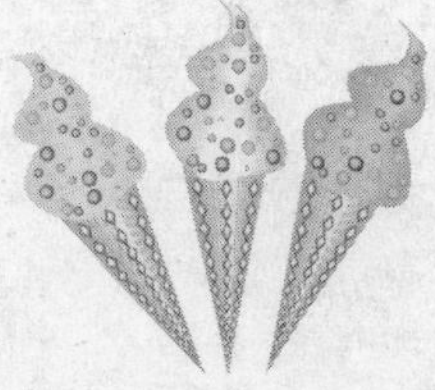

3. Take a regular sized sheet of paper and roll it along the longer edge to make a cylinder. Sketch the cylinder below.

   a. Take the same sheet of paper and roll it along the shorter edge to make a different cylinder than is sketched above. Sketch the new cylinder.

   b. If each of the cylinders above was filled with peanuts, would they hold the same amounts or would one of the two hold more than the other. Explain.

   c. Compare the surface areas of each.

d. Compare the volumes of each.

## Assignment for Spheres

**Name** ______________________________

1. A can of tennis balls fits 3 balls quite neatly. If the radius of each tennis ball is approximately 3 cm, find:
   a. The volume of each tennis ball

   b. The volume of the tennis balls in the can

   c. The height of the can

   d. The volume of the can

   e. The amount of wasted or "left-over" space in the can of tennis balls.

2. If the radius of each tennis ball is approximately *r* cm, find:
   a. The volume of each tennis ball

   b. The volume of the tennis balls in the can

c. The height of the can

d. The volume of the can

e. The amount of wasted or "left-over" space in the can of tennis balls.

3. If the radius of the tennis ball is changed, will the overall amount of wasted space change? Explain. Will the amount of wasted space change in relative to the volume of the cylinder change? Explain.

4. A new umbrella design was created in the shape of a hemisphere with a special plastic coating on the material to better repel water. The diameter of the umbrella is about 1 yard. Because the umbrella is still in its beginning stages, the manufacturer only produces 200 of them to be sold in select markets. How much of the specially coated material must be produced for the manufacture of these umbrellas?

# Assignment for Applications of Volume

**Name:** ____________________________

1. The bed of a pickup truck is 9 ft in length, 4 ft in width, and 2 ft in height. A single piece of firewood is pictured below. Find how many pieces of such firewood will fit in the bed of the truck. (Do not stack the wood above the side walls of the truck bed.) Show your work.

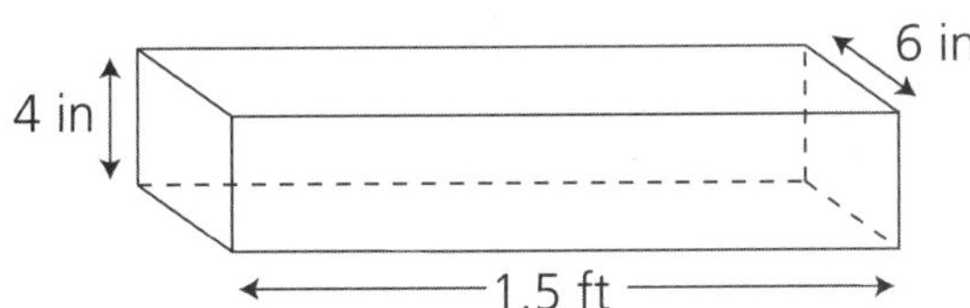


2. Imagine a brand new tube of toothpaste. How many times do you expect you'd be able to brush your teeth with this tube? Describe how you could arrive at your answer without ever taking off the lid. Make some estimates when you brush your teeth at home tonight, and estimate a solution to this problem.

3. Weather people use rain gauges to record the amount of rainfall. A rain gauge is a rather simple construction: a funnel located at the top of a hollow cylindrical tube. In the illustration below, the area of the mouth of the funnel is ten times the area of the top of the cylindrical hollow tube, and the diameter of the tube is one inch. If 2.5 inches of rain has fallen today, how high will the water be in the tube?

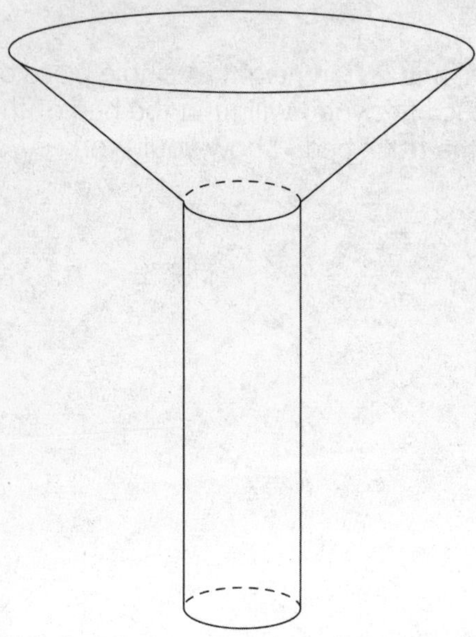

4. Your municipality is replacing the sewage pipes in the community. Which of the two plans under consider if most efficient and why?

   Plan 1: Install one large pipe with a radius of 50 cm

   Plan 2: Install two pipes; one with a radius of 30 cm and one with a radius of 40 cm

5. The local coffee barista decides to sell doughnuts with the coffee. They will sell mini-doughnuts (doughnut holes) and regular sized, regular shaped doughnuts. The regular doughnuts are 4 inches in diameter with a 1-inch hole in the center. The doughnut holes have a 3 cm diameter. Each doughnut hole is half the price of a regular doughnut. Who benefits from this pricing scheme – you or the shop owner? Explain.

# Assignment for Using Matrices with Transformations #1

**Name:** ______________________

*Part 1*

In class, you found the transformation matrix for reflecting over the **x-axis.** In tonight's homework assignment, you will find the transformation matrix for a reflection over the **y-axis**.

1. Where does (1,0) map when it is reflected over the y-axis?

2. Where does (0,1) map when it is reflected over the y-axis?

3. Create your 2x2 transformation matrix from the answers to questions 1 and 2.

4. Test that you have the correct matrix by multiplying this matrix by any chosen point. How does multiplication show you whether your transformation matrix is correct or not?

5. Confirm that you have the correct matrix by multiplying this matrix by the matrix for the triangle you created in class today. (If you do not have the graph of the triangle and forget the coordinates of its vertices, create a new triangle on a coordinate grid and use its matrix to complete this question.) How does multiplication confirm whether your transformation matrix is correct?

6. Now use the same transformation matrix to calculate the general result of reflecting over the y-axis with the most general point $\begin{bmatrix} x \\ y \end{bmatrix}$. Does this result agree with the result you obtained using the mapping notation (→ notation)? How do you know?

*Part 2*

In class, you found the transformation matrix for reflecting over the line **y = x.** In tonight's homework assignment, you will find the transformation matrix for a reflection over the line **y = -x**.

1. Where does (1,0) map when it is reflected over y = -x?

2. Where does (0,1) map when it is reflected over y = -x?

3. Create your 2x2 transformation matrix from the answers to questions 1 and 2.

4. Test that you have the correct matrix by multiplying this matrix by any chosen point.

5. Confirm that you have the correct matrix by multiplying this matrix by the matrix for a chosen polygon. (Create a polygon on a coordinate grid and use its matrix to complete this question.) How does multiplication confirm whether your transformation matrix is correct?

6. Now use the same transformation matrix to calculate the general result of reflecting over the line y = -x with the most general point $\begin{bmatrix} x \\ y \end{bmatrix}$. Does your result here agree with the result from using the mapping notation (→ notation)? How do you know?

# Assignment for Using Matrices with Transformations #2

**Name:** ______________________________

In class, you created the transformation matrices for rotations of 180° and 90 °. In tonight's assignment, you will create the transformation matrices for other rotations.

1. Where does (1,0) map under a three-quarter turn?

2. Where does (0,1) map under a three-quarter turn?

3. Show your transformation matrix for a 270° turn below.

4. Use any point of your choosing to test whether your transformation matrix is correct. Multiply the matrix for this point by the transformation matrix above. Keep in mind where the point should land if you rotate it by 270°. Does the matrix you get as your result represent the point that it should?

5. Use a triangle or any polygon you have graphed whose vertices are not on either axis to confirm that your result is correct. Do this by multiplying your transformation matrix by the polygon's matrix. Keep in mind where the vertices of the polygon should land if you rotate it by 270°. Does the matrix you get as your result represent this polygon?

[ ] * [ ] = [ ]

6. Use the same transformation matrix to calculate the general result of rotating by 270°, by multiplying by the matrix $\begin{bmatrix} x \\ y \end{bmatrix}$. Keep in mind where the point should land if you rotate it by 270°. Does the matrix you get as your result represent the point that it should? Why?

7. Does your result here agree with the result from using the mapping notation back in the section on rotations? Explain how you know?

8. Create the transformation matrix for a rotation of -90°. How do you know your matrix is correct? (Hint: You should be able to create this matrix from what you have previously done.)

9. What is the transformation matrix for a rotation of -180°? How do you know?

# Assignment for Using Matrices with Transformations #3

**Name:** ______________________________

In class, you created the transformations matrices for dilations by a scale factor of 2 and by a scale factor of ½ with respect to the origin. In tonight's assignment, you will also create transformation matrices for dilations. However, the dilations will use different scale factors.

1. As you did in class today, form the quadrilateral A(0,0); B(1,0); C(0,1); D(2,2). Dilate the original quadrilateral ABCD by a factor of 3 with respect to the origin. Create the dilated quadrilateral.

2. Provide the coordinates of each of the vertices.

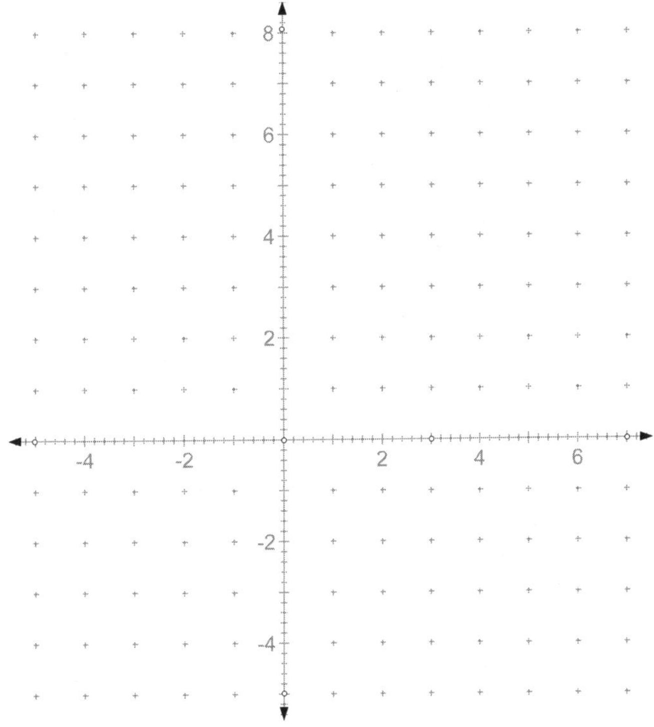


3. Under this dilation (1, 0) → (    ,    ).

4. And (0, 1) → (    ,    ).

5. So the transformation matrix for a dilation of scale factor 3 is $\begin{bmatrix} & \\ & \end{bmatrix}$.

6. Dilate quadrilateral ABC by a factor of ¼ with respect to the origin. What are the coordinates of the new vertices?

7. What is the transformation matrix for this dilation?

10. If you dilated by a scale factor "k", what would the transformation matrix become for this dilation? How do you know?

# Assignment for Glide Reflections

**Name:** ______________________________

Use reflections to attempt to complete the pool shots pictured below. If the shot cannot be made, explain why.

Assume the shooter puts no spin on the ball, and assume you hit the cue ball into the numbered balls "head on," so that the center of the cue ball would pass through the center of the numbered ball if that were possible.

1. Use a single reflection.

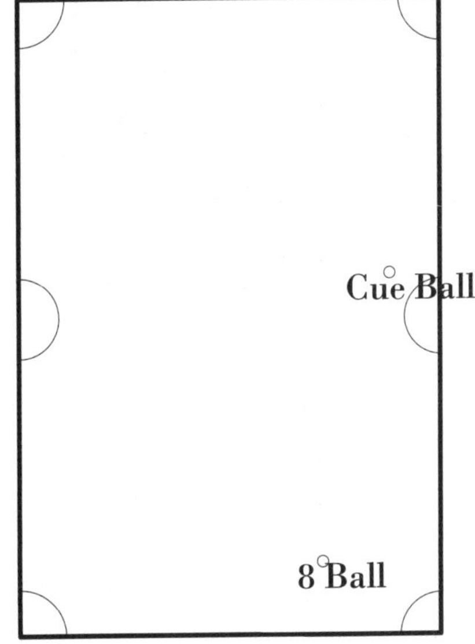


Shoot for this pocket.

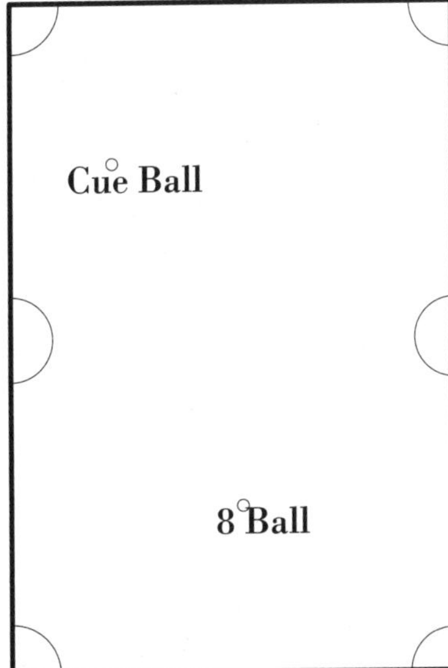


**Shoot for this pocket**

2. Use double reflections (translations or rotations).

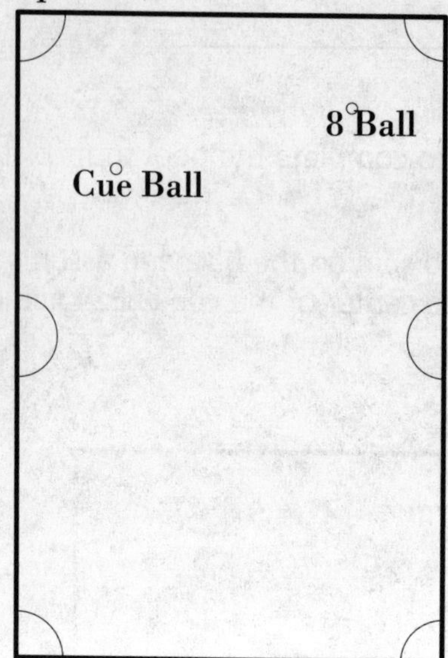


3. Use a glide reflection on this shot.

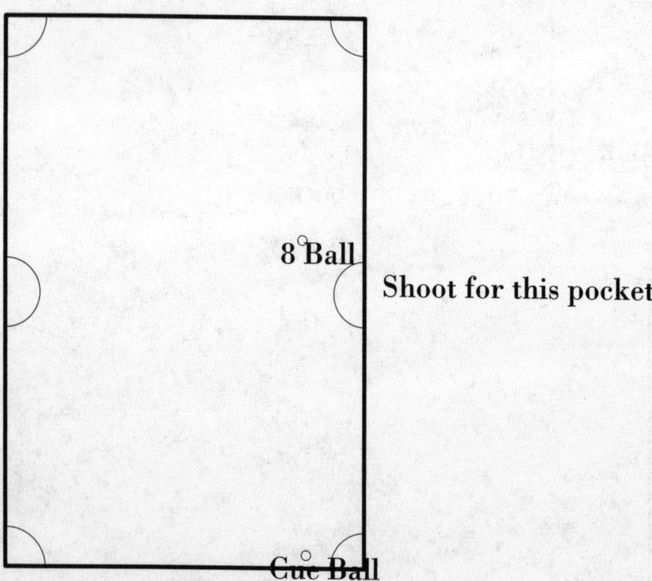

# Assignment for Tessellations

**Name:** ______________________________

Tessellations with regular polygons do not have as much aesthetic value as an M.C. Escher print. In this activity, your goal will be to use your creativity to make an interesting design that will tessellate.

1. Start with a **regular** polygon that will tessellate the plane. Which polygons can you start with?

2. Create a copy of the regular polygon you chose on an index card or tagboard paper.

3. Use your scissors to cut an irregular shape from one side of your polygon and translate it to the side directly across from it. (Note that in the equilateral triangle, this is not possible.) Tape the shape down. Here is an example:

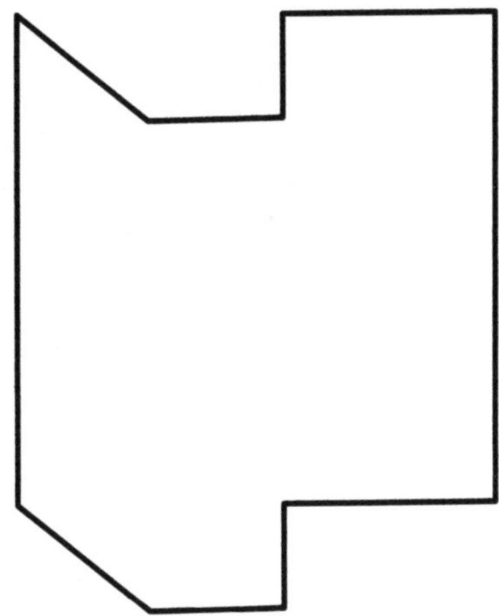

4. Alternatively, cut a different shape from another side of your polygon. Rotate this shape around a vertex or around the midpoint of one side and place it adjacent to a new side of your polygon. Here are examples:

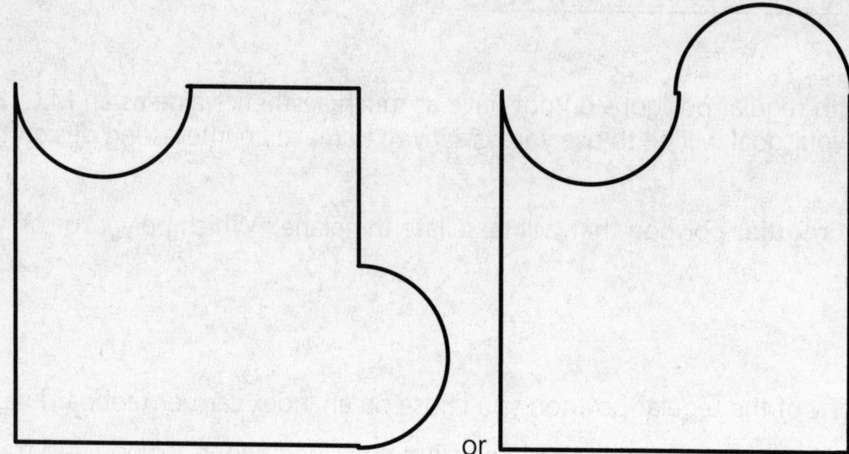


5. Use some creativity to create an animal or object from your new shape. Tessellate the plane with your new shape. Trace a part of that tessellation below.

6. Explain why your shape will tessellate the plane even though it differs from the original polygon you chose.

# Assignment for Putting It All Together

**Name:** ______________________________

In class, you examined the following tessellation of trapezoids:

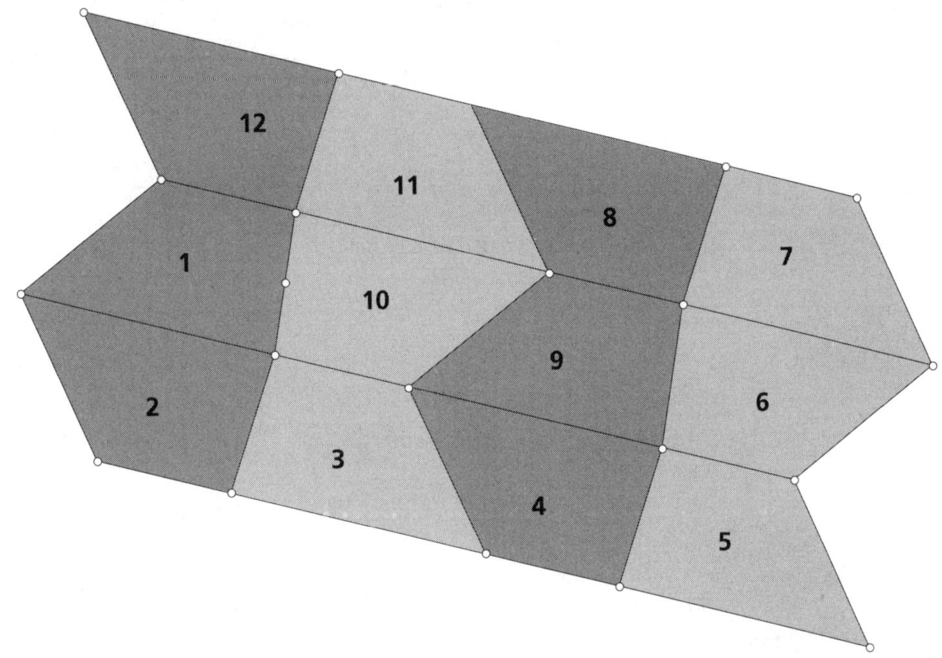


You were asked a series of questions about mapping one trapezoid to another. In this assignment, you will respond to similar questions. However, you will work with a different set of trapezoids.

1. Transform figure 2 to figure 6 using any transformation(s).

   a. How many transformations did you need to perform to map figure 2 onto figure 6?

b. Could you have performed this mapping using fewer transformations? If so, show this below. If not, explain why not?

c. Is this transformation an isometry? Why or why not?

2. Transform figure 2 to figure 6 using ONLY reflections.

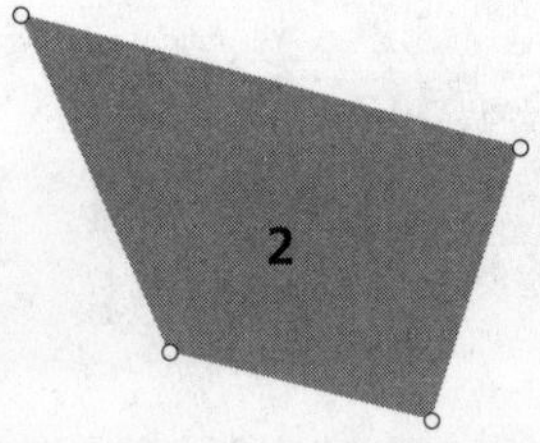


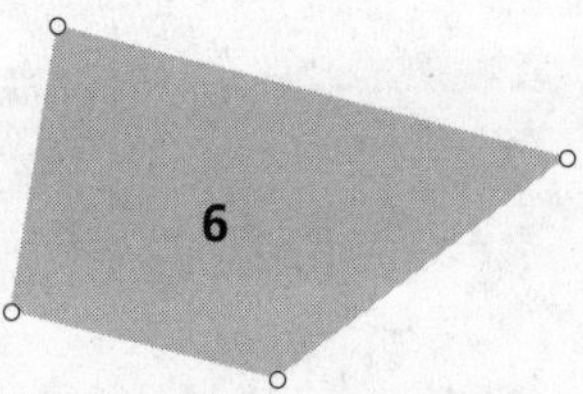


a. How many reflections did you need to transform figure 2 to figure 6?

b. Could you have used fewer reflections to transform figure 2 to figure 6? Why or why not?